International Conference on Frontiers of Sensors Technologies 2016 (ICFST 2016)

MATEC Web of Conferences Volume 59 (2016)

Hong Kong, China
12 - 14 March 2016

Editors:

Haider F. Abdul Amir
Alexander M. Korsunsky
Zhihua Guo

ISBN: 978-1-5108-2465-2

Printed from e-media with permission by:

Curran Associates, Inc.
57 Morehouse Lane
Red Hook, NY 12571

Some format issues inherent in the e-media version may also appear in this print version.

Printed by Curran Associates, Inc. (2016)

For additional information, please contact EDP Sciences – Web of Conferences at the address below.

EDP Sciences – Web of Conferences
17, Avenue du Hoggar
Parc d'Activité de Courtabœuf
BP 112
F-91944 Les Ulis Cedex A
France

Phone: +33 (0) 1 69 18 75 75
Fax: +33 (0) 1 69 28 84 91

contact-edps@webofconferences.org

Additional copies of this publication are available from:

Curran Associates, Inc.
57 Morehouse Lane
Red Hook, NY 12571 USA
Phone: 845-758-0400
Fax: 845-758-2633
Email: curran@proceedings.com
Web: www.proceedings.com

TABLE OF CONTENTS

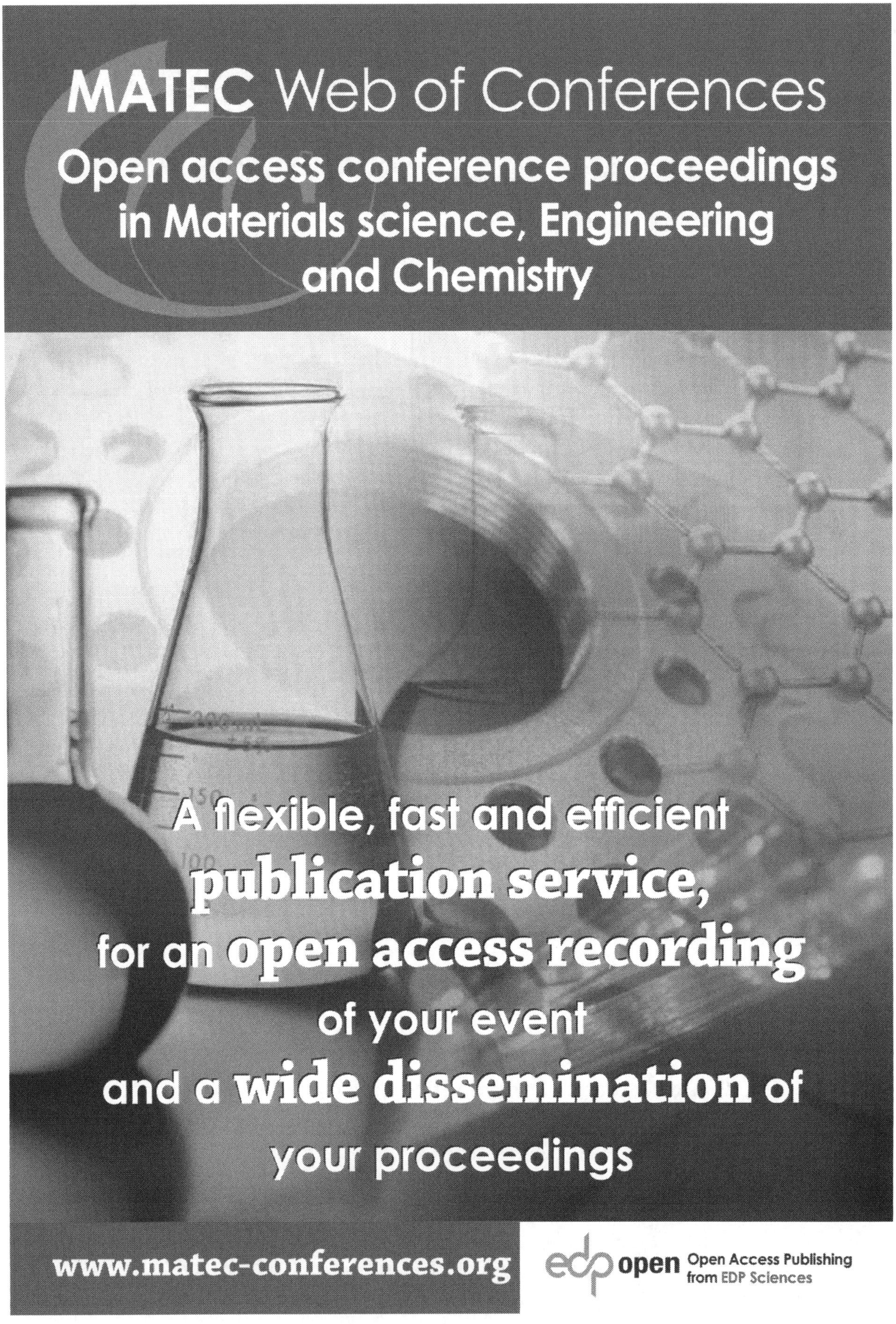
MATEC Web of Conferences
Open access conference proceedings in Materials science, Engineering and Chemistry
A flexible, fast and efficient publication service,
for an open access recording of your event
and a wide dissemination of your proceedings
www.matec-conferences.org
edp open Open Access Publishing from EDP Sciences

ISSN (electronic edition): 2261-236X

Committees

Conference Chairs

Prof. Haider F. Abdul Amir, University Teknologi Mara in Shah Alam, Malaysia

Program Committee Chairs

Prof. Alexander M. Korsunsky MA DPhil CPhys Minst, Trinity College, Oxford, UK

Prof. Zhihua Guo, Southwest Jiaotong University, China

International Technical Committees and Reviewers

Prof. Tjokorda Gde Tirta Nindhia, Udayana University, Indonesia
Dr. C. Makabe, University of the Ryukyus, Japan
Dr. Amir H. Alavi, Michigan State University, USA
Assoc. Prof. Burçak Karagüzel Kayaoğlu, Istanbul Technical University, Turkey
Asst. Prof. Dr. Chakchai So-in, Khon Kaen University, Thailand
Asst. Prof. Kaykhosrow Khojier, Islamic Azad University, Iran
Assoc. Prof. Christopher Shearwood, Nanyang Technological University, Singapore
Prof.Tatsujiro Miyazaki,University of the Ryukyus, Japan
Prof. Aleksandra Lobnik, Institute for Textile Chemistry, Slovenia
Prof. Chien-Kuo Hsieh, Ming Chi University of Technology, Taiwan
Cheng-Shane Chu, Ming Chi University of Technology, Taiwan
Prof. Masayuki Hirayama, Nihon University, Japan
Dr. Hadi Salehi, Michigan State University, USA

Preface

2016 International Conference on Frontiers of Sensors Technologies (ICFST 2016) was held during March 12-14, 2016 in Hong Kong.

The conference provides a platform for researchers, scientists, developers, and practitioner to discuss the topics on frontiers of sensors technologies and share state-of-art research results, for exploring new areas of research and development, and to discuss emerging issues in the fields of sensors technologies with participants from all over the world, both from academia and from industry. And their research success is reflected in every paper received, with participants coming from several countries, allowing a real multinational and multicultural exchange of experiences and ideas.

The proceedings of ICFST consist of 48 selected papers which were submitted to the conference from universities, research institutes and industries. All of the papers were subjected to peer-review by conference committee members and international reviewers. The papers selected for the proceedings depended on their quality and their relevancy to the conference. The proceedings tend to present to the readers exactly the frontiers of sensors technologies, which cover the following areas: electronic engineering and sensing technology, power system, power system, power system, design and implementation of robot, mechanical design and manufacturing engineering, wireless communication and satellite engineering, image processing.

We'd like to express our sincere gratitude to everyone who has contributed to ICFST 2016, and it could only succeed as a team effort. And we want to thank the conference chairs, the program chairs especially, for putting everything of the conference together, as well as all the technical committee members and reviewers for their excellent work in reviewing the papers and their efforts.

Prof.Haider F.Abdul Amir
University Mara in Shah Alam, Malaysia

Prof Zhihua Guo
Southwest Jiaotong University, China

March 12, 2016.

Statement of Peer review

In submitting conference proceedings to MATEC Web of Conferences, I certify to the Publisher that I adhere to the **Policy on Publishing Integrity** of the journal in order to safeguard good scientific practice in publishing.

1. All articles have been subjected to peer review administered by the proceedings editors.
2. Reviews have been conducted by expert referees, who have been requested to provide unbiased and constructive comments aimed, whenever possible, at improving the work.
3. Proceedings editors have taken all reasonable steps to ensure the quality of the materials they publish and their decision to accept or reject a paper for publication has been based only on the merits of the work and the relevance to the journal.

Title, date and place of the conference

2016 International Conference on Frontiers of Sensors Technologies

Hong Kong March 12-14, 2016

Proceedings editor(s):

Prof.Haider F.Abdul Amir
University Mara in Shah Alam, Malaysia
Email:haider@salam.uitm.edu.my

Prof.Alexander M.Korsunsky
Trinity College, Oxford, UK
Email:alexander.korsunsky@eng.ox.ac.uk

Prof.Zhihua Guo
Southwest Jiaotong University,China
Email:gzh@home.swjtu.edu.cn

Date and editor's signature 31 March 2016

Hydrogen peroxide sensing based on carbon quantum dots

Cheng-Shane Chu , Meng-Wei Hsieh and Zhi-Ren Su

Ming Chi University of Technology, 24301 New Taipei City, Taiwan

Abstract. In this work, carbon quantum dots (CQDs) have been synthesized by one step hydrothermal method to detect hydrogen peroxide (H_2O_2). The optical H_2O_2 sensors were based on the fluorescent quenching of the carbon quantum dots. The optical H_2O_2 sensors demonstrate capable of detecting H_2O_2 over the range of 0-88.2 mM and 0-60 mM. The H_2O_2 sensitivities of wavelength shift to changes in the H_2O_2 concentration were found to be 0.18 nm/mM and 0.26 nm/mM. These results of the optical sensors showed that the method can be used in practice detection of H_2O_2 and could offer a new approach for developing a new optical biosensor. The CQDs exhibit good emission property and high stability, as well as excitation-independent emission behavior. Moreover, it is attractive that CQDs can be used as an effective fluorescent probe for the detection of H_2O_2 with linear Stern-Volmer plot and wavelength shift in an aqueous solution.

1 Introduction

Hydrogen peroxide (H_2O_2) plays an essential role as an oxidizing, bleaching and sterilizing agent in biochemical and chemical industry [1]. High concentrations of H_2O_2 would cause irritation to dyes and shin, it affects human health [2]. On the other hand, the determination of H_2O_2 concentration is important in food [3], clinical [4] and environmental applications [5]. There are many methods which can be used for the quantitative analysis of H_2O_2 in solution, such as electrochemical method [6] and spectroscopy [7-12]. The electrochemical methods are well suited determining lower concentrations of H_2O_2, however, they can suffer interference from other reactive oxygen species. Furthermore, electrochemical detection methods are all susceptible in the environment with significant electromagnetic interference. On the other hand, the field of spectroscopic detection method can be subdivided into chemiluminescence [7-10], and spectrofluorometry [2, 11]. The spectroscopic techniques can also be used for detection of H_2O_2 at low concentrations. The spectroscopic techniques have been integrated with luminescent quantum dots (QDs), making possible the development of optical sensors with high sensitivity and many of advantages of luminescent QDs, such as small size, high selectivity and immunity to electromagnetic interference.

In the last decade, there has been a focus on the luminescent carbon quantum dots (CQDs) to further improve the optical sensing technique due to their unique optical properties caused by the quantum confinement effect. The luminescent carbon quantum dots (CQDs) were first produced in 2006 [12]. Compare to semiconductor quantum dots, luminescent CQDs were considered as a brand new class of luminescent due to their small size, high quantum yield, low cytotoxicity, possess fine biocompatibility, high photostability and environment friendliness, which make them suitable chemical and biological analyses. There are several methods to prepare CQDs have been reported [13-17]. However, these methods usually require not only complicated equipment and expensive materials but also complicated post treatment to enhance the quantum yield and water solubility of CQDs. Meanwhile, the isolation, purification and functionalization processes based on these methods are all complicated and time consuming. Therefore, the hydrothermal method has been widely used to prepare various luminescent QDs because of the high reactivity of the reactants, easy control of the solution, low level of air pollution and low energy consumption under hydrothermal conditions [18].

In this work, we have presented a new and simple H_2O_2 detecting method based on the fluorescence quenching and wavelength shift of CQDs. Fluorescence quenching refers to any process which decreases the fluorescence intensity of a certain fluorophore due to various molecular interactions. The optical sensor was composed of CQDs without any complex processes of functionalization or conjugation. The obtained optical biosensor showed the linear detection range and wavelength shift. The sensitivity enables the use of the QDs as H_2O_2 sensors and provides a versatile fluorescent reporter for the activities of oxidases and for the detection of their substrate.

2 Experiments

2.1 Synthesis of CQDs

All reagents used were of analytical grade without further purification. The CQDs were synthesized from sodium citrate through a simple, convenient and one-step hydrothermal method. Briefly, sodium citrate (0.3 g), NH_4HCO_3 (3.1 g) and water (20 mL) were sealed into a Teflon equipped stainless steel autoclave, which was then placed in a drying oven followed by hydrothermal treatment at varying temperatures for 4h. After the reaction, the autoclave was cooled to room temperature. The purification of the CQDs was conducted through a dialysis tube for about 24 h in dark.

Figure 1 (a) shows a TEM image of the CQDs. From inspection, the average diameter of the QDs is 5 nm. Figure 1(b) presents the energy-dispersive X-ray spectroscopy (EDX) results for the composition of the CQDs. The nanoparticles are composed principally of O, C, and Cu elements, where the x axis represents energy (keV) and the y axis represents the counts per second per electron (basically X-ray intensity). The Cu content originates from the copper grid, the C content originates from the CQDs.

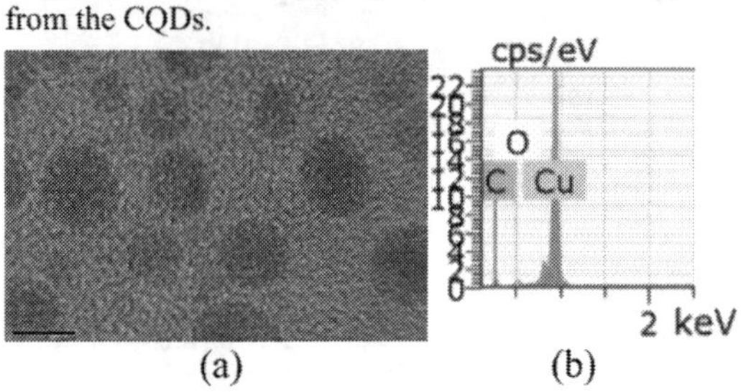

Fig. 1 (a)TEM image showing CQDs at a resolution of 5 nm (b) EDX analysis results for CQDs.

2.2 Instrumentation

Figure 2 shows a schematic illustration of the experimental arrangement used to characterize the performance of the optical H_2O_2 sensors. In the sensing experiments, the fluorescence excitation was provided by an LED (NSHU591B, NICHIA, COPR.) with a central wavelength of 365 nm driven by an arbitrary waveform generator (TGA1240, Thurlby Thandar Instruments (TTI) Ltd.) at 10 kHz in pulse signal mode. The emission measurements were acquired using a USB 4000 spectrometer (Ocean Optics).

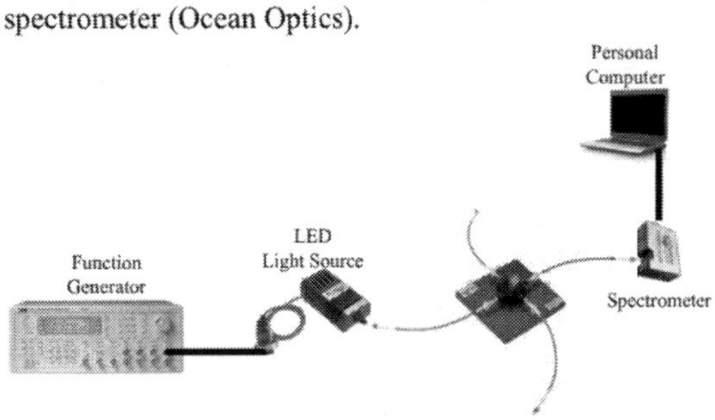

Fig. 2 Schematic diagram showing experimental arrangement used for characterization.

3 Results and Disscussion

3.1 Optical H_2O_2 sensing properties

Figures 3 (a) and (b) present the room-temperature fluorescence spectra of the optical H_2O_2 sensors under H_2O_2 concentrations ranging from 0-88 and 0-60 mM, respectively. Note that in acquiring the relative fluorescence intensity data, the integration time of the CCD spectrometer was set at 10 ms, and as a result, the optical H_2O_2 sensors exhibit strong fluorescent emissions at 604 nm and 570 nm. Figure 3 shows that the relative fluorescence intensity of the optical H_2O_2 sensors decreases significantly as the H_2O_2 concentration increases.

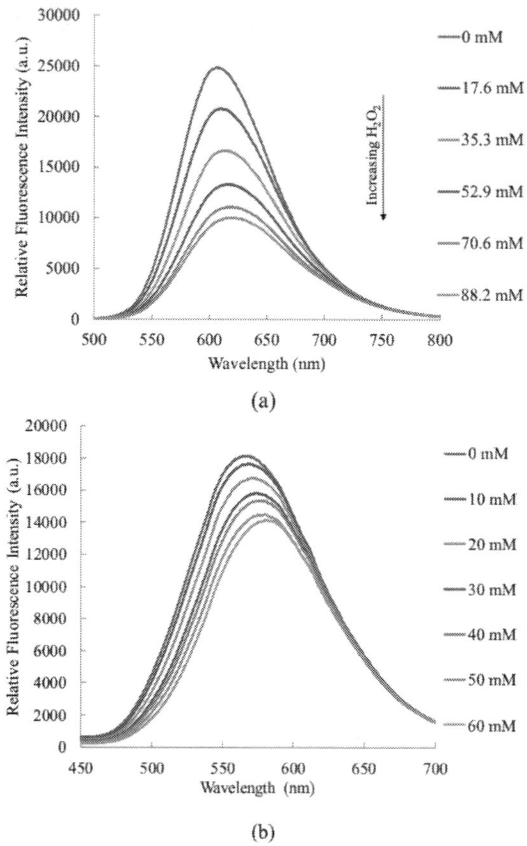

Fig. 3 Emission spectra of optical sensors under different H2O2 concentrations: (a) red emission, (b) yellow emission.

The calibration curves of optical H_2O_2 sensors were shown in Fig. 4. It is seen that the fluorescence intensity reduces exponentially as the H_2O_2 concentration increased. In other words, the fluorescence intensity is reduced as a result H_2O_2 quenching. The quenching effect can be described by the Stern-Volmer equation [19]:

$$I_0/I = 1 + K_{SV}[Q] \quad (1)$$

where I_0 and I represent the steady-state fluorescence intensities in the absence and presence of H_2O_2, respectively; K_{SV} is the Stern-Volmer quenching constant; $[Q]$ represents the concentration of quencher (H_2O_2). H_2O_2 is a good quencher for the CQDs and from the Stern-Volmer plot, the concentration of H_2O_2 could be estimated.

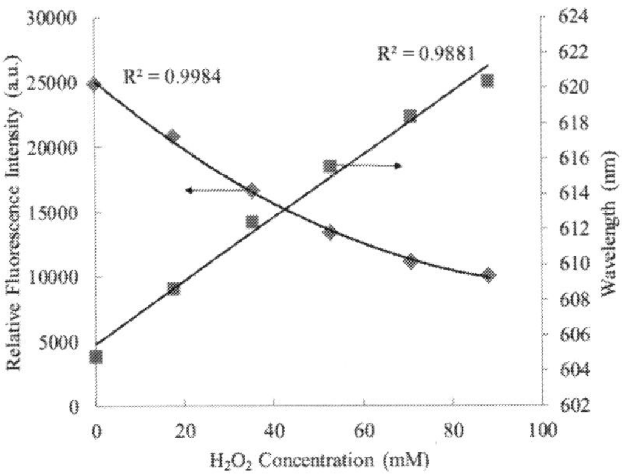

(a)

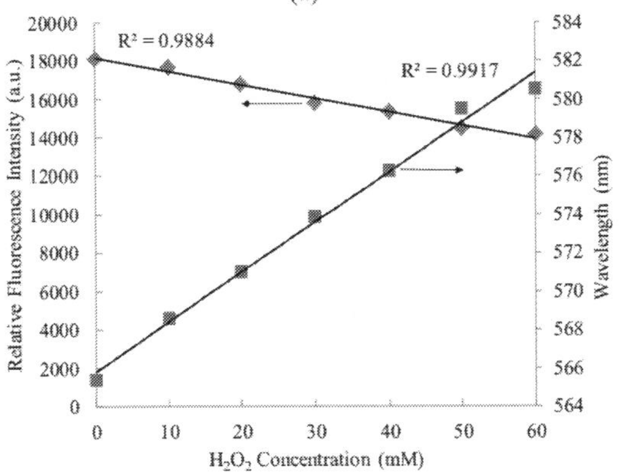

(b)

Fig. 4 Calibration curves of optical H2O2 sensors: (a) red emission, (b) yellow emission.

Figures 5 have been shown that the peak emission wavelength of CQDs, have a linear dependence on H2O2 concentration in the environmental range. It can be seen that the peak emission wavelength increases linearly as the H2O2 concentration increase. The peak wavelength, on the other hand, is shifted towards longer wavelengths. Either effect can be used to obtain H2O2 concentration information. The optical H2O2 sensors demonstrate capable of detecting H2O2 over the range of 0-88.2 mM and 0-60 mM, the wavelength shift to changes in the H2O2 concentration were found to be 0.18 nm/mM and 0.26 nm/mM, respectively. However, simple fluorescence intensity measurements are prone to error due to optical power fluctuations. Due to the presence of a wavelength shift is proportional to H2O2 concentration and independent of system optical power level.

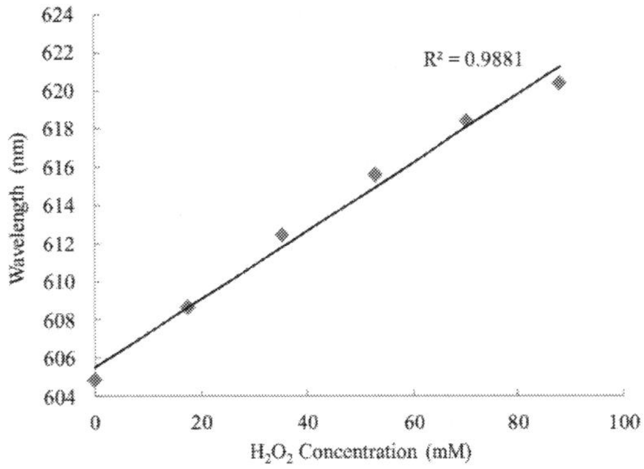

(a)

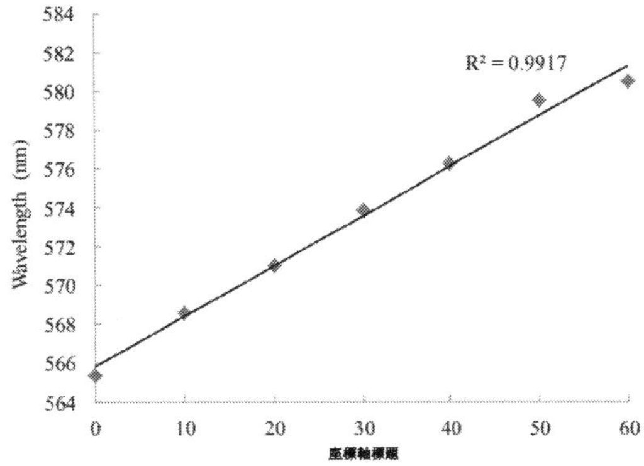

(b)

Fig. 4 Variation of peak emission wavelength with H2O2 concentration: (a) red emission, (b) yellow emission.

4 Conclusions

In this study, we synthesized water-soluble luminescent CQDs by one step hydrothermal method using a common carbon resource. The luminescent CQDs have been successfully applied to develop new fluorescent biosensors for sensitive detection of H_2O_2. The resulting sensors exhibited the linear Stern-Volmer plots for the determination of H_2O_2. The H_2O_2 sensitivities of wavelength shift to changes in the H_2O_2 concentration were found to be 0.18 nm/mM and 0.26 nm/mM. The optical sensors for H_2O_2 exhibit good analytical performance with low cost, convenient and sensitive detection. The CQDs may also be a promising candidate for biochemistry and biological applications.

Acknowledgement

The authors gratefully acknowledge the financial support provided to this study by the Ministry of Science and Technology of Taiwan under Grant No. MOST 104-2221-E-131-029.

References

1. K. C. Lin, T. H. Tasi, S. M. Chen, Bioesns. Bioelectron., **26** (2010).

2. A. Lobnik, M. Cajilakovic, Sens. Actuators B Chem., **74** (2001).
3. K. Kriz, M. Anderlung, D. Kriz, Bioesns. Bioelectron., **16** (2001).
4. K. F. Zhou, Y. H. Zhu, X. L. Yang, J. Luo, C. Z. Li, S. R. Luan, Electrochim. Acta, **55** (2010).
5. Z. H. Wang, F. Liu, X. Teng, C. X. Zhao, C. Lu, Analyst, **136** (2011).
6. S. A. G. Evans, J. M. Elliott, L. M. Andrews, P. N. Bartlett, P. J. Doyle, G. Denuault, Anal. Chem., **74** (2002).
7. K. M. Wang, J. Li, X. H. Yang, F. L. Shen, X. Wang, Sens. Actuators B Chem., **65** (2000).
8. S. Hanalka, J. M. Lin, M. Yamada, Anal. Chim. Acta., **426** (2001).
9. J. Yuan, W. Guo, E. Wang, Anal. Chem., **80** (2008).
10. A. N. Diaz, M. C. R. Peinado, M. C. T. Minguez, Anal. Chim. Acta., **363** (1998).
11. M. J. Zhou, Z. J. Diwu, N. PanchukVoloshina, R. P. Haugland, Anal. Biochem., **253** (1997).
12. Y. P. Sun, B. Zhou, Y. Lin, W. Wang, K. A. S. Fernando, P. Pathak, M. J. Meziani, B. A. Harruff, X. Wang, H. F. Wang, P. J. G. Luo, H. Yang, M. E. Kose, B. L. Chen, L. M. Veca, S. Y. Xie, J. Am. Chem. Soc., **128** (2006).
13. S. L. Hu, K. Y. Niu, J. Sun, J. Yang, N. Q. Zhao, X. W. Du, J. Mater. Chem., **19** (2009).
14. Q. L. Zhao, Z. L. Zhang, B. H. Huang, J. Peng, M. Zhang, D. W. Pang, Chem. Commun., **41** (2008).
15. L. Y. Zheng, Y. W. Chi, Y. Q. Dong, J. P. Lin, B. B. Wang, J. Am. Chem. Soc., **131** (2009).
16. J. G. Zhou, C. Booker, R. Y. Li, X. T. Zhou, T. K. Sham, X. L. Sun, Z. F. Ding, J. Am. Chem. Soc., **129** (2007).
17. L. Tian, D. Ghosh, W. Chen, S. Pradhan, X. J. Chang, S. W. Chen, Chem. Mat., 21 (2009).
18. S. H. Feng, R. R. Xu, Accounts Chem. Res., **34** (2001).
19. J. R. Lakowicz, *Principles of Fluorescence Spectroscopy*, 2nd ed; Kluwer Academic/Plenum Press, New York, 1999, Chapter 8 and 9.

Sensor Selection method for IoT systems – focusing on embedded system requirements

Masayuki HIRAYAMA
College of Science *and* Technology, *NIHON* University
Narashinodai *7-24-1*, Funabashi CHIBA *274-8501*, JAPAN

Abstract. Recently, various types of sensors have been developed. Using these sensors, IoT systems have become hot topics in embedded system domain. However, sensor selections for embedded systems are not well discussed up to now. This paper focuses on embedded system's features and architecture, and proposes a sensor selection method which is composed seven steps. In addition, we applied the proposed method to a simple example – a sensor selection for computer scored answer sheet reader unit. From this case study, an idea to use FTA in sensor selection is also discussed.

1 Introduction

Recently, various types of sensors have been developed and widely used. With these technical breakthroughs, IoT (Internet of Things) systems, which can realize and create various types of additional value, have been proposed and developed using these sensors [1][2]. In most IoT systems, embedded systems are used to collect or detect various data. An embedded system consists of a microcomputer and a wide variety of sensors. It detects various types of data through the sensors and processes data and information. However, if engineers do not pay attention to the sensors applied in the systems, the expected system functions may not be realized. We have developed embedded systems and discussed their problems and their solutions in terms of their architecture. This paper focuses on sensor selection for embedded systems and proposes a sensor-selection method.

2 Embedded-system architecture as a sensor-driving facility

2.1 System architecture

In an IoT system, the embedded system is responsible for driving the sensor mechanism, which provides the following functions.

1. A microcomputer unit connects the sensor-driving circuits and activates the sensors by sending a start signal.
2. The microcomputer unit receives data collected by sensors via amplifier circuits.
3. Data or signals are recorded in the microcomputer's memory area, and then processed by the embedded software.
4. The processed data are transferred to IT systems via data communication networks.

These embedded-system functions and procedures are initially based on the functional and safety requirements of the target system. In addition, system sensors vary according to the expected functions and safety requirements of the system. Therefore, when selecting system sensors, the sensor's performance specification data should not only be considered, but also the system requirements and use cases.

Moreover, the sensor selection has a significant influence on the hardware and software architecture of the embedded system. From a hardware point of view, if the sensed data level is relatively low, amplifier circuits are necessary. Moreover, if the sensed data contains noise, then noise-reduction circuits are also required. On the other hand, if a multi-sensing mechanism is applied to a system to reinforce the system safety properties, one should pay careful attention to the pin assignments of the microcomputer.

From the software point of view, sensor selection has a significant influence on the data-sensing timing control and the data-collection process. These procedures are mainly accomplished by interrupt routines and a task-priority mechanism in the embedded software framework. Thus, to realize suitable data sensing, the software architecture should also be considered. Furthermore, when removing noise from the sensed data with a software procedure, the software architecture and algorithm are also affected. Thus, sensor selection has a large influence, not only on the system functions but also

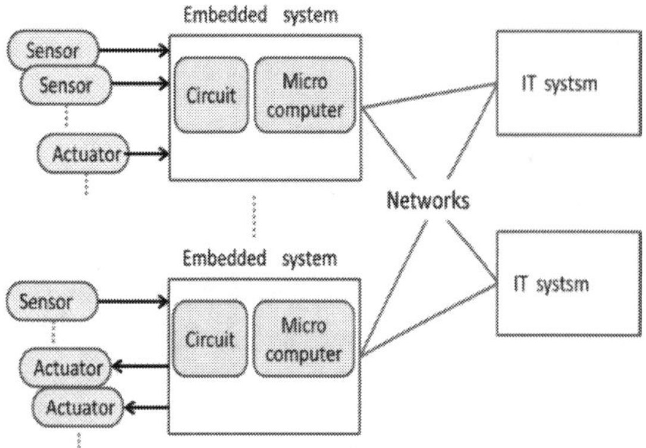

Figure.1 Architecture of embedded systems on the embedded system of an IoT system that mainly drives sensing units.

2.2 Problems

As mentioned above, the sensor selection has a large influence on the embedded system architecture, as well as on the IoT system functions. However, there has been little discussion on the basic process of sensor selection from an embedded-system viewpoint. As a result, engineers tend to select sensors for their systems without sufficient consideration; they mainly refer to their own and other engineers' experience. Thus, in the worst case, a sensor may be changed during the development phase, possibly resulting in a software or hardware architecture change.

The above sensor-selection process poses the following problems.

1. The sensor-selection process and its steps are not clearly defined.
2. The evaluation viewpoint of the sensor selection is not discussed sufficiently
3. The relation between the selected sensor and the embedded-system architecture as a sensor-driving unit is not discussed well.

Visual sensors, sound sensors, and other types of sensors are applied in terms of system development. This variety of sensors and basic physics makes sensor selection difficult, and as a result, the sensor-selection method cannot be clearly defined. Thus, it is important to define the general process and steps for sensor selection, which are commonly applied to various sensor types and physics.

3 Sensor-selection method

3.1 Basic sensor-selection steps and characteristics architecture

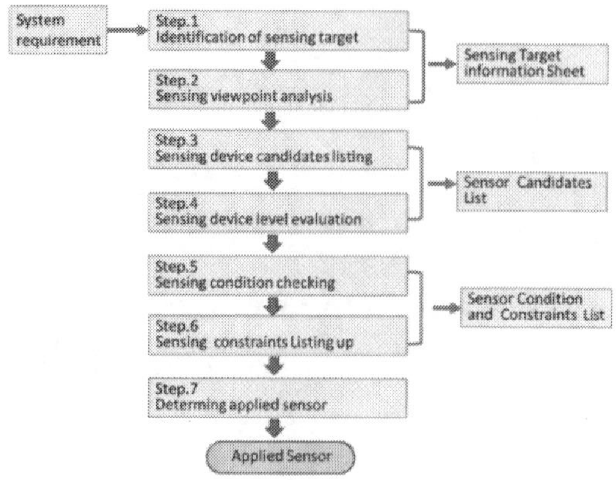

Figure.2 Sensor selection process flow

a. Basic steps

As Figure 2 shows, the proposed sensor-selection procedure is composed of the following seven steps.

Step 1. Identification of the target to be sensed
By considering the IoT system specifications and requirements, the type of sensing data, information, and situations can be found.
Step 2. Sensing viewpoint analysis
Which of the five human senses—eyesight, hearing, touch, smell, or taste—is suitable for the sensing target is determined.
Step 3. Sensing-device candidate list
Sensor types and physics are examined while considering the sensing viewpoints and target. Then, the sensor-device candidates are listed.
Step 4. Sensing-device level evaluation
There are three types of sensors. The first is a presence-judging sensor that can detect the presence of a phenomenon or a target object. The second is a scale-measure sensor that can measure the feature-value level of a target phenomenon or object. The third is a physical-degree sensor than can determine the degree of a physical phenomenon. Considering the system specification and listed sensor's physics, the required sensing-device level should be evaluated.
Step 5. Sensing condition checking
The detailed conditions for sensor selection are compared with the system specifications. In this step, use-case analysis is also helpful.
Step 6. List sensor constraints
After examining the actual system's use cases and safety viewpoint, the sensing constraints are extracted.
Step 7. Define the applied sensor
Considering the above six steps, the sensors for the system are defined.

b. Characteristics of the proposed steps

Many sensors types and physics are available. Thus, it is important to select sensors with a unique procedure without any restriction of sensor types. In the above procedure, sensor-selection sheets are also proposed. Engineers describe and summarize their sensor selection examination or discussion results in these sheets. This

DOI: 10.1051/matecconf/20165901002

makes sensor-selection easier for a less-experienced engineer. These sensor-selection sheets not only contain information from the sensor-data sheets, but also requirements and constraints from the IoT system users' viewpoint. Referring to these sheets, engineers can confirm the reasons for the sensor selection. The detailed notation for the sensor-selection sheets is depicted in the next section.

3.2 Case study and Sensor-selection notation

In this subsection, a simple example of the proposed method using a computer-scored answer-sheet reader unit is presented. The example application system automatically reads four figures printed in a bubble pattern on each answer sheet. These four figures show a matriculation number. Each bubble pattern on a sheet is two or three square millimeters, and the four laterally aligned bubble patterns are represented by four figures. The reading unit detects each bubble pattern using a sensor and identifies the colors of the bubble pattern—white or black. Figure 3 shows the example sensor-selection sheets for this application.

(1) Steps 1 and 2

In Step 1 and Step 2, as Table.1 shows, we can determine that the sensing target for this system is a bubble pattern printed on answer sheets. We can observe from the sensor-selection sheets that the bubble-pattern size is two or three square millimetres, and its color is white or black. In order to detect the bubble pattern, a visual sensor is the preferred alternative.

(2) Steps 3 and 4

As shown in Table.2, Step 3 includes various types of visual sensor—e.g., a CCD (charge-coupled device) camera, infrared array, color-metric sensor, etc. In the candidate sensor sheets, the comparison and evaluation of each sensor are described from the viewpoints of accuracy, limitation, and stability. For the sensing-device level evaluation in Step 4, the CCD belongs to a physical-degree sensor and the infrared array is a presence-judgment sensor. The color-metric sensor is a scale-measure sensor.

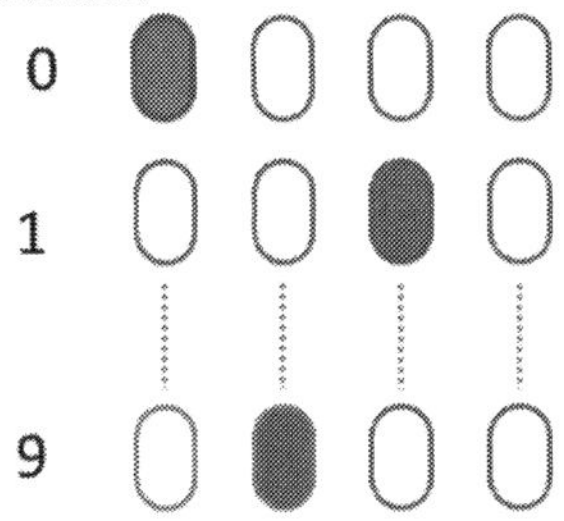

Figure.3 Bubble patterns on answer sheet

(3) Step 5

In Step 5, as shown in Table.3, the sensing condition—e.g., sensing area, sensing time constraints, detection accuracy, noise—and the sensing constraints—e.g., the privacy, safety, and cost of the sensing unit—are described. Confirming from Figure 3.1 to Figure 3.3, the infrared array sensor is the most suitable sensor for reading the bubble pattern in this system.

Table.1 Sensing Target information sheet

Sensing Target information Sheet(1)

Target Objects or Target Data

1	Bubble pattern printed on answer sheets	Visual sensor

Characteristics / Features of the Target

1	Bubble pattern 2.0 – 3.0 m^2	
2	Colour white or black	
3	Four laterally aligned bubble patterns are represented	
4	Four figures show a matriculation number	

Type of senses
TU: Touch
HS: Hearing and sound
SM: Smell
TA: Taste
EY: Eyesight

4 Discussion

In this research, referring to the above sensor-selection results, a computer-scored answer-sheet reader unit with an infrared array sensor was actually developed. Reviewing the experimental development, the problems of the proposed method are discussed here. In the system-evaluation phase, the following system problems were confirmed.

- There are differences in the bubble-pattern detection rate depending on the environment of the system installation site.
- The bubble patterns cannot be properly identified when the answer sheets are slightly out of alignment in the correct position.

In the proposed method, a sensor was selected by analysing the system requirement specifications. However, the system environment situation was not considered, because it was not described in the system specification. As a result, a deficient system situation that might not be well handled by the selected sensors could occur. In order to solve these problems, the system's

DOI: 10.1051/matecconf/20165901002

Table.2 Sensor Candidates list

Candidate No.	Device Category	Device Type	Accuracy	Sensing Limit	Measurement Stability
1	CCD Camera	Physical degree sensor	High	<80 cm	Depend on Graphics processing unit
2	Infracted array	Presence judgement sensor	High	<50 cm	High
3	Colour metric	Scale measurement sensor	Medium	<30cm	Depend on Graphics processing unit

Table.3 Sensing condition and constraints list

Detail condition and constraints for sensor Condition

Type of sensing data	Digital signal - black =1, white =0
Sensing area	Under day light Brightness
Time constraints	5msec / 1 bubble pattern detection
Accuracy	0.95 (detection error ratio is under 5 %)
Disturbance and noise	Ambient light effects, miss-filling of bubble pattern
Basic principal of sensing	Detection of white bubble or black bubble

External Constraints

Privacy	Don't care
Safety	Don't care
Cost	Electric and machine cost in total < $100 Sensor cost <$10

invalid behaviour as well as its valid behavior described in the specification should be considered in Step 1 of the proposed sensor-selection steps.

At this point particularly, as Figure 4 shows, a fault-tree analysis of the target system is an effective tool. The sensor candidates that can respond to the system's invalid behavior are extracted. In the case of the computer-scored answer-sheet reader unit, the most undesirable situation was "Cannot read bubble pattern." This situation is caused by the following three conditions:

1. Incorrect marking,
2. Incorrect answer-sheet setting, and
3. Incorrect environment for the system, or invalid system setting.

Of those, condition 1 is difficult to control with the developed system. However, misclassification of the bubble patterns can be reduced by installing a pattern-detection mechanism that uses a sensor to check the sheet's position, or a sensor that evaluates the environmental brightness.

5 Related work

As for sensor selection, several researches and ideas have been reported. Saini proposed a context based multimedia sensor selection method[3]. In the method, sensor for multimedia system is selected by two stages. In the first stage, in order to provide meaningful information for query evaluation, context and availability map are used. In the second stage, sensors are divided and their priorities are evaluated. The proposed method is quite useful for multimedia system's sensor selection, however, it is insufficient for other types of systems. On the other hand, Cao proposed a multiobjective optimization technique for sensor selection[4]. The proposed technique was mainly focusing on Wireless Sensor Network's sensor selection. It provided a simulation model and simulation algorism for sensor selection. This research is also insufficient for applying other system domains, because the proposed simulation model was deeply depend on Wireless Sensor Network..

Elahi proposed a sensor ranking method[5]. In the method, single-period-model and multi-period-model for prediction model are discussed. Using their method, suitable sensors are searched with matching a content-based sensor search.

6 Conclusion

This paper focused on IoT systems, which have been widely developed and used in recent years. A sensor-selection method and a simple application example were introduced to select a suitable system design. The proposed method was a basic idea for selecting a suitable sensor for the system requirements. Sensor selection considerably influenced the system hardware and

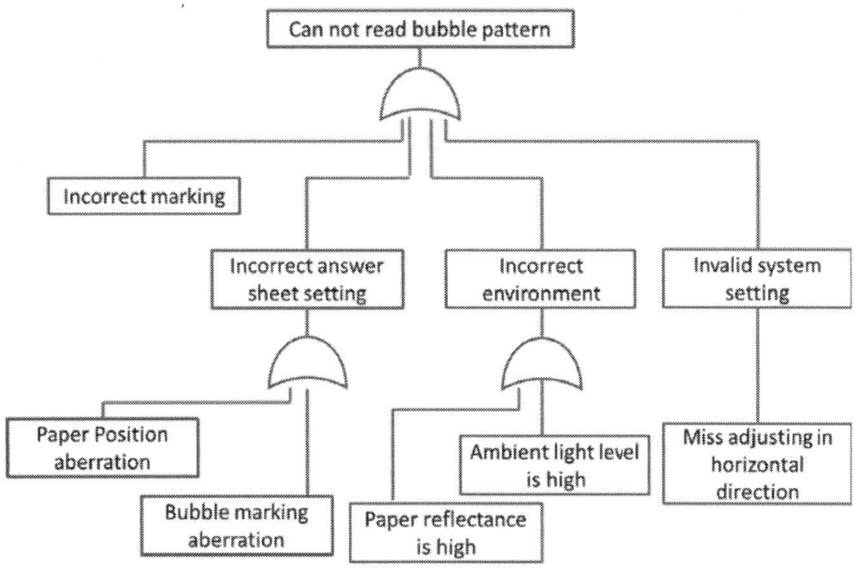

Figure.4 Fault Tree Analysis for bubble pattern sensing unit

software, which provided essential mechanisms for the embedded system. Thus, while considering the sensor-selection condition from the system specifications, the sensor's effects on the embedded system's hardware and software should be analysed.

Moreover, skilled engineers are not always in charge of sensor selection. Sometimes, a less-skilled engineer might choose the sensors. Thus, it is important to be able to select appropriate system sensors in spite of the engineer's experience. Therefore, to more intelligently select a sensor, the following information is quite important.

- A sensor specification notation which describes the sensor's detailed features;
- A formal and visualized representation of the sensor information.

If these techniques become a practical reality, an automatic sensor-selection system may be realized.

References

1. M.Khatu,N.Kaimal,P.Jadhav and S.Rizvi, "Implementation of Internet of Things or Home Automation",International Journal of Emerging Engineering Research and Technology, Vol.3, Fer. 2015,pp7-11
2. D.Giusto,A.Iera,G.Morabito,L.Atzori,"The Internet of Things"< springer,2010. ISBN:978-1-4419-1673-0
3. Mukesh Saini, Mohan Kankanhalli,"Context-Based Multimedia Sensor Selection Method",Proc. of AVSS09, pp. 262-267, (2009)
4. Nianxia Cao, Engin Masazade,Pramod K. Varshney,"A Mutiobjective Optimization based Sensor Selection Method for Target Tracking in Wireless Sensor Networks", Proc. of 16th International conference on Information Fusion, pp.974-980,(2013)
5. B.M.Elahi,K.Romer,B.Ostermajer,M.Fahrmair and W.Kellerer,"Sensor Ranking: A Primitive for Efficient Context-based Sensor Search",in Proc. Of the 2009 International Conference on Information Processing in sensor Networks, pp.217-228

MATEC Web of Conferences **59**, 01003 (2016) DOI: 10.1051/matecconf/20165901003
ICFST 2016

An Improved Fast Compressive Tracking Algorithm Based on Online Random Forest Classifier

Jintao Xiong, Pan Jiang, Jianyu Yang, Zhibin Zhong, Ran Zou and Baozhong Zhu

University of Electronic Science and Technology of China, Chengdu 611731, Sichuan, P. R. China

Abstract. The fast compressive tracking (FCT) algorithm is a simple and efficient algorithm, which is proposed in recent years. But, it is difficult to deal with the factors such as occlusion, appearance changes, pose variation, etc in processing. The reasons are that, Firstly, even if the naive Bayes classifier is fast in training, it is not robust concerning the noise. Secondly, the parameters are required to vary with the unique environment for accurate tracking. In this paper, we propose an improved fast compressive tracking algorithm based on online random forest (FCT-ORF) for robust visual tracking. Firstly, we combine ideas with the adaptive compressive sensing theory regarding the weighted random projection to exploit both local and discriminative information of the object. The second reason is the online random forest classifier for online tracking which is demonstrated with more robust to the noise adaptively and high computational efficiency. The experimental results show that the algorithm we have proposed has a better performance in the field of occlusion, appearance changes, and pose variation than the fast compressive tracking algorithm's contribution.

Index Terms—fast compressive tracking, naive Byes classifier, online, random forest.

1 Introduction

In the current target tracking, the visual tracking algorithms are roughly divided into two categories, i.e. generative or discriminative. Most generative algorithms only take the holistic representation into consideration and do not make full use of the discriminative information between the target and the background. So these algorithms are difficult to represent the target correctly under occlusion and pose variation.

Besides, discriminative model regards the tracking problem as a binary classification problem by obtaining the decision boundary of target and background to distinguish them i.e. such as online boosting, MIL tracker, online Semi-Boost. In recent years, Kai Hua Zhang raised the ideas on generative and discriminative model and proposed the fast compressive tracking algorithm [1], which is also demonstrated not efficient enough to deal with noise, thus suffering from drifting problem.

2 Compressive tracking

The compressive sensing theory states that the high-dimension of the feature space can be projected to low-dimension space which contains enough information to reconstruct the original high-dimension through random projection matrix.

$$v = Rx . \qquad (1)$$

Where $R \in r^{n*m}$ is a random projection matrix, and $n << m$. According to Johnson-Lindenstrauss [2] lemma, for the two K-sparse vectors, the following property can be obtained to ensure the integrity of important information.

$$(1-\varepsilon)\|x_1 - x_2\|_{l_2}^2 \le \|Rx_1 - Rx_2\|_{l_2}^2 \le (1+\varepsilon)\|x_1 - x_2\|_{l_2}^2 \quad (2)$$

$$r_{ij} = \sqrt{\rho} \times \begin{cases} 1, \ with\ Probability \dfrac{1}{2\rho} \\ 0, with\ Probability 1-\dfrac{1}{\rho} \\ -1, with\ Probability \dfrac{1}{2\rho} \end{cases} \quad (3)$$

To deal with the scale change problems, series of rectangle filters are used to convolve the original image, the rectangle filters are defined as:

$$h_{i,j}(x,y) = \begin{cases} 1, 1 \le x \le i, 1 \le y \le j \\ 0, otherwise \end{cases} \quad (4)$$

Diaconis and Freedman [3] show that the low-dimension vectors which are projected from the high-dimension through random measure matrix are almost Gaussian.

$$p(v_i|y=1) \sim N(\mu_i^1, \sigma_i^1) \quad (5)$$

$$p(v_i|y=0) \sim N(\mu_i^0, \sigma_i^0) \quad (6)$$

Where $\mu_i^1(\mu_i^0)$ and $\sigma_i^1(\sigma_i^0)$ are mean and standard deviation of the positive (negative) class. Naive Bayes classifier is a generative model. We assume all the elements are independently distributed. We can estimate the unknown parameters by the maximum likelihood through:

$$L(\mu_i^1, \sigma_i^1, \mu_i^0, \sigma_i^0) = \prod p(v_i|y)p(y) \quad (7)$$

$$\mu^1 = \frac{1}{n}\sum_k^{n-1} v_i(k) \quad (8)$$

$$\sigma^1 = \sqrt{\frac{1}{n}\sum_k^{n-1}(v_i(k) - \mu^1)^2} \quad (9)$$

Taking the introduction of new samples into account in each new frame , the scale parameters are incrementally updated by:

$$\mu_i^1 = \lambda\mu_i^1 + (1-\lambda)\mu^1 \quad (10)$$

$$\sigma_i^1 = \sqrt{\lambda(\sigma_i^1)^2 + (1-\lambda)(\sigma^1)^2 + \lambda(1-\lambda)(\mu_i^1 - \mu^1)^2} \quad (11)$$

In the end, we can model with a naive Bayes classifier to predict the position from the candidate which is most likely to be the target sample. The main components of the proposed compressive tracking algorithm are shown by Figure. 1.

$$H(v) = \sum_{i=1}^{n} \log\left(\frac{p(v_i|y=1)}{p(v_i|y=0)}\right) \quad (12)$$

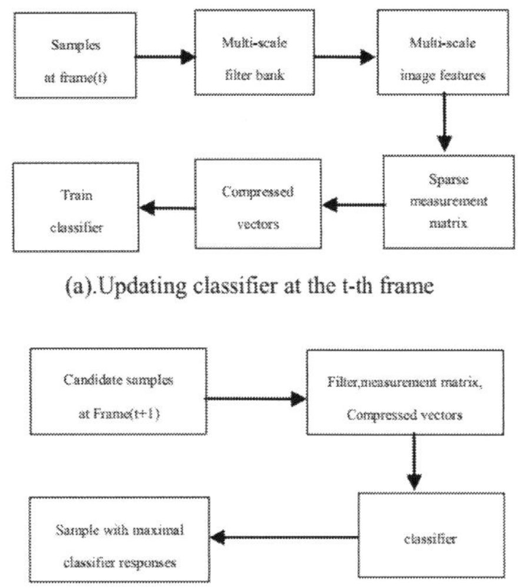

(a).Updating classifier at the t-th frame

(b).Tracking at the (t+1)-th frame

Figure. 1. Main components of the proposed compressive tracking algorithm

3 Online random forest

Integrated learning is a robust machine learning algorithms, especially random forest, which has been proved strong anti-interference ability, therefore, when the wrong obstruction sample is updated, it will not cause great impact. Online random forest is mainly with the idea of online learning and integrated learning.

3.1 Random forest

Random forest is composed by a lot of classification and regression trees.

The randomness of the random forest is not only reflected in the choice of the samples, but also in the choice of the random tests. Random forest selects a series of tests randomly from the feature set, and then picking the best test among them according to the maximum information gain. It is the reason why the random forest is not easy to over-fit.

Given the training set X and labels Y, and with an ensemble of trees, $t_1(x)...t_k(x)$ we define the average number of votes at a sample (x, y):

$$N_j = \frac{1}{K}\sum_{i=1}^{K} I\left(t_i(x) = j\right) \quad (13)$$

Where I is the indicator function and K is the number of the trees, we believe that the larger distance between the average number of the right class and any other class, the more confidence the classifier is. So we define the generalization error.

$$\mathrm{E} = P_{X,Y}\left((N_Y - max_{j \neq Y} N_j) < 0\right) \quad (14)$$

It also has been shown by Breiman [4] that as the number of trees increases, for almost surely all sequences θ, E converges to

$$E_1 = P_\theta\left(h(X,\theta) = Y\right) \quad (15)$$

$$E_2 = max_{j \neq Y} P_\theta\left(h(X,\theta) = j\right) \quad (16)$$

$$P_{X,Y}((E_1 - E_2) < 0) \quad (17)$$

This result explains that the random forest do not over-fit as more trees as added. It has also been shown by Breiman [5] that this error has an upper bound

$$E \le \rho\frac{1-S^2}{S^2} \quad (18)$$

Where ρ is the mean correlation between pairs of trees in the forest and s is the strength of the ensemble.

3.2 Online learning

In the tracking process, the samples are updated with the image sequences continuously. We cannot get the entire training set in advance. What we need to do is to train the classifier with the samples which arrive sequentially, so that it would converge to the off-line model. Online bagging provides a good idea for the online updated.

Given a training set T of size N, the probability of every sample which is selected randomly k times is

$$P(\mathrm{K} = \mathrm{k}) = C_N^k\left(\frac{1}{N}\right)^k\left(\frac{1}{N}\right)^{N-k} \quad (19)$$

When $N \to \infty$, the probability distribution is a Poisson distribution, so each classifier is updated by each sample k times where k is generated randomly from the Poisson distribution.

In the online random forest, there exist another important problem that we should tell the node when it is appropriate to perform a split. Firstly, the samples are gathered over times, so we should know whether there are enough samples for a node to perform a split. Secondly, the splits are whether good enough for classification purpose. It has been shown by Amir Saffari [6] that if a node splits when $|R_j| > \alpha$ and $\exists s \in S : \Delta L(R_j, S) > \beta$, it will converge to the off-line model. Where α is the minimum number of samples. The β is the minimum information gain and $\Delta L(R_j, S)$ is representation of the information gain.

4 Improved method

The parameters of naive Bayes classifier will continue to be updated in the tracking process, However, when the target is in the event of partial occlusion or occlusion of all time, the noisy samples will be introduced into the model, and previous samples will be forgotten .The general solution is to judge when the target is under occlusion, so to design different update rates. But the finding of shelter materials will not only cause a large degree of complex algorithms, moreover, it does not guarantee the accuracy. However, it has been shown by Kai Hua Zhang [7] that the compressive tracking provides a good image representation way for the tracking problem. So we use it to make a discriminative representation of the samples in our algorithm.

The online random forest selects the partial samples and features from the total training set for the training of each trees, therefore, most of the trees will not be polluted by the noise. This is the most important reason why the online random forest is not straightforward to over-fit. Meanwhile, it has been proved theoretically in

(18) that the generalization error of random forest classifier has an upper bound.

Moreover, we can estimate the Out-Of-Bag-Error of the trees, through the Out-Of-Bag-Error, we propose to discard the tree which is polluted by the noisy samples, so it can reduce the influence of the factors such as occlusion, appearance changes, illumination change.

Eventually, when the target is completely blocked even severely blocked, there are no candidate positive samples in current frame. Therefore, it exists a situation that the output of each tree will be zero .the result can help us make a decision whether the target is completely blocked. What we need to do next is two things. Firstly, we should abandon the noisy samples we collect from the current frame. The second thing is to use the motion information of the target in the last several frames, including the target velocity, acceleration and other information to predict the target position.

The entire proposed algorithm is depicted in Algorithm.

Algorithm: improved fast compressive tracking based on online random forest

Require: the t-th and (t+1)-th image frame.
Require: the number of positive samples and negative samples in every frame N.
Require: the number of the trees in the forest M.
Require: the minimum number of the samples α .
Require: the minimum gain β .

(1). According to the position of the target in the t-th frame, we collect positive and negative samples in $P^c=\{p\|I(p)-I(t)\|<\gamma^c\}$, $N^{d,f}=\{p|\gamma^d<\|I(p)-I(t)\|<\gamma^e\}$ with $d<c<f$ and use a series of rectangle filters defined in (4) to convolve the original image and extract the compressed features trough the random projection matrix.

(2). Input the training samples into the random forest classifier.

For i=1: N
  For j=1: M
    K$\leftarrow$ $possion(1)$
    For k=1: K
      If (M>0)
        Find the node which the sample belongs to, and then update the node l=find leaf (i);
        If($|R_j|>\alpha$ and $\exists s\in S:\Delta L(R_j,S)>\beta$)
          Find the best test to create the left child and right child;
        End
        Else
          Estimate the Out-Of-Bag-Error (OOB);
        End
      End
    End
End

(3). Choose the candidate samples from the (t+1) frame in $F^f=\{p\|I(p)-I(t+1)\|<\gamma^f\}$ by shifting number of pixels Δ_f and input them into the trained classifier. For the candidate samples in the current frame, if all the responses of the trees are zero which means all the candidate samples are considered negative samples.

If (output [1…M] ==0)
  Predict the position of the target and discard the samples collected from the current frame through.

$$I_{t+1}=2*I_t-I_{t-1}$$

Else
  Find the tracking position $I(t+1)^*$ in the current frame with the most votes. Choose the candidate samples from the (t+1) frame in $F^g=\{p\|I(p)-I(t+1)^*\|<\gamma^g\}$ by shifting number of pixels Δ_g and input into the trained classifier. And find the tracking position $I(t+1)^*$ in the current frame with the most votes.
End

(4). select the trees from the forest randomly F
If (F.OOB>rand ())
  Discard (F);
  New (F);
End

5 Experimental result

In the experiment, the dimension of the compressed feature is set to be 50. The number of the trees is 100. The maximum depth of the tree is 5. And choose 10 tests from the test set to split the node. The minimum number of the samples is 100.

The dataset consists of two sequences. It is a surveillance video from the CMBC, public sequence

'Rotating Girl' respectively, which are presenting pose, scale variation and occlusion. As we can see from the result, our method tracks the target more accurately than FCT algorithm, when the target is under occlusion and pose variation. The results of the two sequences are showed in Figure 2 and Figure 3 respectively.

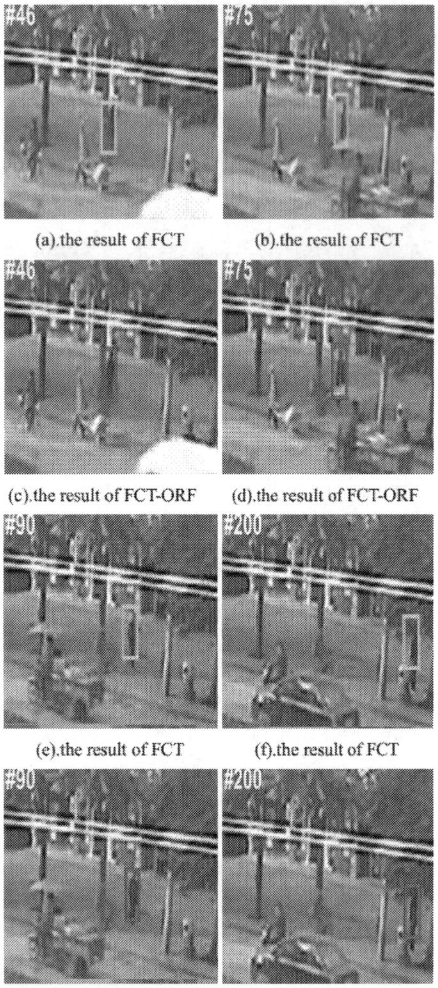

(a).the result of FCT (b).the result of FCT

(c).the result of FCT-ORF (d).the result of FCT-ORF

(e).the result of FCT (f).the result of FCT

(g).the result of FCT-ORF (h).the result of FCT-ORF

Figure.2.The results of the surveillance video from the CMBC.

In the sequence from the CMBC bank, the people we track is under the occlusion of the trees, the motorbike and the trash which means it will brow the noisy samples. If we do not discard the noisy samples and update the naïve Byes classifier with them. The result would be like this in the 46th 75th and 200th frame. However, Our method can handle with this problem well. Because even parts of the trees and the splitting of the nodes in the forest choose quite a lot of noise samples and the polluted features for the training, leading to failure of prediction. Other trees updated by the clean samples and features will still maintain a good classification performance. Finally, their common voting will also be able to predict the correct position of the target.

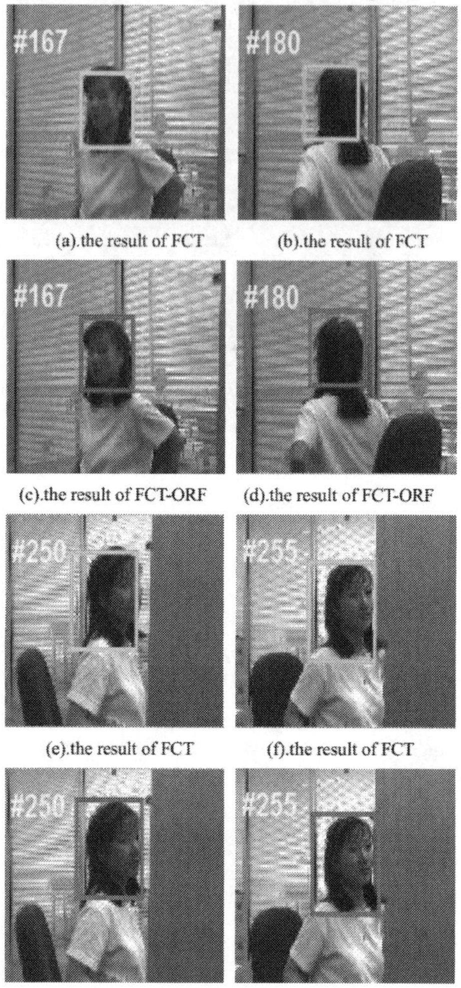

(a).the result of FCT (b).the result of FCT

(c).the result of FCT-ORF (d).the result of FCT-ORF

(e).the result of FCT (f).the result of FCT

(g).the result of FCT-ORF (h).the result of FCT-ORF

Figure.3.The results of the rotating girl.

In the compressive tracking algorithm, when the girl turns the face quickly, we need a faster update rate to adapt to the changes in the appearance. However, with the high update rate, it is easy to introduce the noisy. Thus, leading to the drift in the tracking. As you can see the result in the 180th frame. Meanwhile, we use a smaller update rate to do the experiment, the result is showed in the 250th and 255th frames, the drift occurs for the sake of the update rate, which cannot adapt to the changes in the appearance. However, our method abandons the trees, which is updated by original samples through estimating the Out-Of-Bag-Error adaptively. And the new trees are created with the current samples to accommodate with the changes in the appearance.

We evaluate the result with the center location error which is defined as the Euclidean distance between the central locations of the tracked objects and the manually labeled ground truth. The result of the sequences from the CMBC and Rotating girl are showed in Figure 4 and Figure 5 respectively.

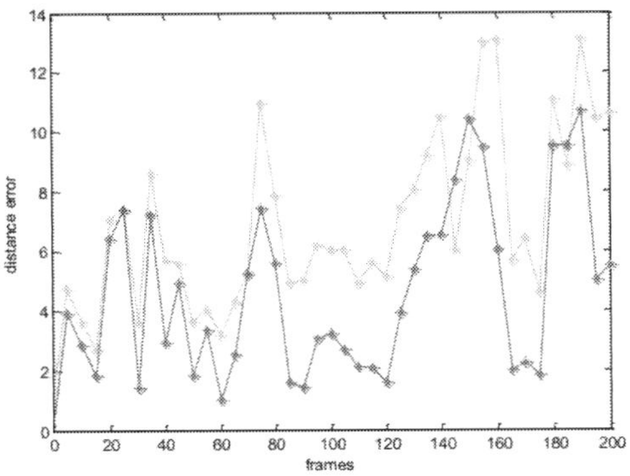

Figure.4. the sequences from the ICBC—FCT —FCT-ORF

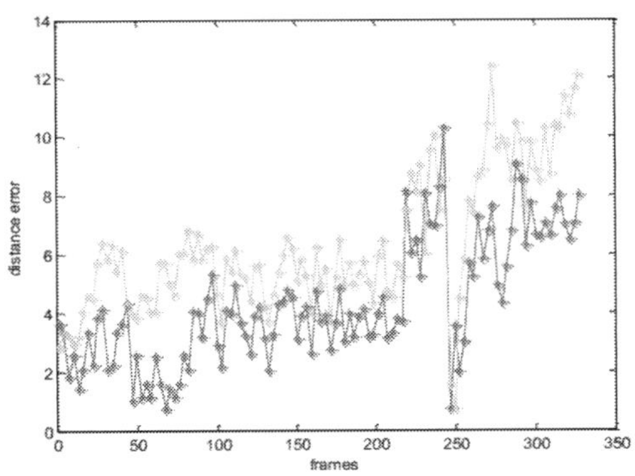

Figure.5. the rotating girl —FCT —FCT-ORF

6 Conclusion

In this paper, we proposed an improved compressive tracking algorithm based on online random forest. The simulation results show that our algorithm performs better than the original algorithms when the target is under occlusion, pose variation and appearance changes. The reasons are that we adopt the adaptive compressive sensing theory to make a discriminative representation of the target, moreover, the random forest classifier is robust to the noise for selecting the features and samples to train the decision trees randomly. Finally, The Out-Of-Bag-Error can tell us whether a decision tree is polluted completely and then discard it.

References

1. K. Zhang, L. Zhang, and M.-H. Yang, "Real-time compressive tracking," in Proc. Eur. Conf. Comput. Vis., 2012, pp. 864–877.
2. K. Zhang, L. Zhang, and M.-H. Yang, "Fast Compressive Tracking," in IEEE Transactions on Pattern Analysis and Machine Intelligence 36 (2014)1631-1643.
3. P. Diaconis and D. Freedman, "Asymptotics of graphical projection pursuit," Ann. Statist., vol. 12, pp. 793–815, 1984.
4. L. Breiman. Out-of-bag estimates. Technical report, 1996.
5. L. Breiman. Random forests. Machine Learning, 45(1):5–32, October 2001.
6. A. Saffari, C. Leistner, J. Santner, M. Godec, H. Bischof, "On-line Random Forest," in International Conference Computer Vision, 2009, pp,1393-1400.
7. H. Grabner, M.Grabner, and H. Bischof, "Real-time tracking via on-line boosting" in Proc. British Mach. Vis. Conf, 2006, pp. 47–56.

MATEC Web of Conferences **59**, 01004 (2016) DOI: 10.1051/matecconf/20165901004
ICFST 2016

Calibration and Distortion Field Compensation of Gradiometer and the Improvement in Object Remote Sensing

Xiangchao Hu[1,3], Hongfeng Pang[1,2,3], Liangrui Fu[1], Dixiang Chen[3] and Mengchun Pan[3]

[1] *Northwest Institute of Nuclear Technology, Xi'an,China*
[2] *Academy of Equipment, Beijing, China*
[3] *National University of Defense Technology, Changsha, China*

Abstract. Magnetometer, misalignment error and distortion field can reduce the accuracy of gradiometers. So, it is important to calibrate and compensate gradiometers error. Scale factor, bias, nonorthogonality, misalignment and distortion field should be considered. A gradiometer is connected by an aluminium frame, which contains two fluxgate magnetometers. A nonmagnetic rotation equipment is used to change gradiometer attitude, and the compensation parameters are estimated. Experiment results show that, after calibration and compensation, error of each axis is reduced from 888.4 nT, 1292.6 nT and 168.9 nT to 15.3 nT, 22.1 nT and 9.9 nT, respectively. It shows that the proposed method can calibrate gradiometer and compensate distortion field. After calibration and compensation, the object remote sensing performance is improved.

1 Introduction

Gradiometers are widely used for unexploded ordinance (UXO) detection, underwater exploration, and geologic research [1,2]. Magnetometer error, misalignment and distortion field influence the accuracy of gradiometers. It is important to calibrate gradiometer error and compensate distortion field. Scale factor, bias, nonorthogonality, and misalignment are hard to solve by mechanism [3,4]. In geomagnetic field circumstance, 0.1°misalignment error will lead to about 100 nT vector measurement error. Also, distortion field is hard to be avoided when it is deployed in vehicles [5,6]. Recently, magnetometer calibration has been developed using estimation methods [7-11]. However, there is little work in gradiometer calibration and compensation.

In this paper, a fluxgate gradiometer is designed, and an integrated calibration and compensation method is proposed. The true value of geomagnetic field intensity is obtained by a proton magnetometer. The nonmagnetic rotation equipment is used to change gradiometer attitude and the compensation parameters are estimated accurately. After integrated compensation, the measurement error of gradiometer is reduced obviously. In addition, the compensated gradiometer can be more effective for magnetic abnormity measurement.

2 Calibration and compensation theory

2.1 Magnetometer calibration and distortion field compensation

Considering of different scale factors, bias errors and nonorthogonality, the output of a three axis magnetometer can be expressed as [8]:

$$B_k = D(H_k + b') + \varepsilon_k \tag{1}$$

Where H_k is the measurement of an ideal three axis magnetometer. B_k is the measurement of nonorthogonal three axis magnetometer. $b' = [b_x^{'}\ b_y^{'}\ b_z^{'}]$ is bias vector. $D = [K_x\ D_{xy}\ D_{xz}; D_{xy}\ K_y\ D_{yz}; D_{xz}\ D_{yz}\ K_z]$ represent scale factor and nonorthogonality. ε_k are measurement noise; k is measurement point. Thus, the calibration model can be expressed as [12]: Considering of distortion field, the error model can be expressed as:

$$B_k = PD(H_k + b') + b_H + \varepsilon_k \tag{2}$$

Where P is soft material distortion field matrix, and b_H is hard material distortion field matrix. (2) can be expressed as:

$$\begin{bmatrix} B_k^x \\ B_k^y \\ B_k^z \end{bmatrix} = \begin{bmatrix} p_{11} & p_{12} & p_{13} \\ p_{21} & p_{22} & p_{23} \\ p_{31} & p_{32} & p_{33} \end{bmatrix} \cdot \begin{bmatrix} D_{11} & D_{12} & D_{13} \\ D_{21} & D_{22} & D_{23} \\ D_{31} & D_{32} & D_{33} \end{bmatrix} \cdot \begin{bmatrix} H_k^x + b_x^{'} \\ H_k^y + b_y^{'} \\ H_k^z + b_z^{'} \end{bmatrix} + \begin{bmatrix} b_H^x \\ b_H^y \\ b_H^z \end{bmatrix} + \begin{bmatrix} \varepsilon_k^x \\ \varepsilon_k^y \\ \varepsilon_k^z \end{bmatrix} \tag{3}$$

(3) can be expressed as:

$$\begin{bmatrix} B_k^x \\ B_k^y \\ B_k^z \end{bmatrix} = \begin{bmatrix} q_{11} & q_{12} & q_{13} \\ q_{21} & q_{22} & q_{23} \\ q_{31} & q_{32} & q_{33} \end{bmatrix} \begin{bmatrix} H_k^x + b_x^{'} \\ H_k^y + b_y^{'} \\ H_k^z + b_z^{'} \end{bmatrix} + \begin{bmatrix} b_H^x \\ b_H^y \\ b_H^z \end{bmatrix} + \begin{bmatrix} \varepsilon_k^x \\ \varepsilon_k^y \\ \varepsilon_k^z \end{bmatrix} \tag{4}$$

(4) can be expressed as:

$$B_k = qH_k + b + \varepsilon_k \tag{5}$$

(5) can be transformed as:

$$H_k = Q(B_k - b) \tag{6}$$

A further relationship can be obtained by squaring both sides of (6):

$$H_k^2 = [Q(B_k - b)]^T [Q(B_k - b)] \tag{7}$$

Q can be decomposed into an unitary matrix Q_A and upper triangular matrix Q_R. (7) can be expressed as:

$$H_k^2 = [Q_R(B_k - b)]^T [Q_R(B_k - b)] \tag{8}$$

Where H_k is the true value of geomagnetic field intensity. Equation (8) can be transformed as:

$$\begin{aligned} H_k^2 &= (B_k^x)^2 M_1 + (B_k^y)^2 M_2 + (B_k^z)^2 M_3 \\ &+ (B_k^x)(B_k^y)M_4 + (B_k^y)(B_k^z)M_5 + (B_k^z)(B_k^x)M_6 \\ &+ (B_k^x)M_7 + (B_k^y)M_8 + (B_k^z)M_9 + M_{10} \end{aligned} \tag{9}$$

Where (9) is a nonlinear equation. $M_1, \cdots M_{10}$ are expressed in [12]. There are nine unknown parameters to be estimate. There are N nonlinear equations when measurement point is N. Nine unknown parameters can be calculated by solving nonlinear equations. Also, the parameters can be estimated by nonlinear least square [13].

2.2 Misalignment calibration between magnetometers

After magnetometer calibration and compensation, the two magnetometers can be considered as orthogonal. Misalignment error is the coordinate error between magnetometers. Magnetic field projection on the first and second magnetometer coordinate are $B_1 = [B_{x1}, B_{y1}, B_{z1}]^T$ and $B_2 = [B_{x2}, B_{y2}, B_{z2}]^T$ respectively. As shown in Figure1, the relationship of projection can be expressed as [14]:

$$B_1 = T_1 T_2 T_3 B_2 \tag{10}$$

Where

$$T_1 = \begin{bmatrix} \cos\psi & 0 & -\sin\psi \\ 0 & 1 & 0 \\ \sin\psi & 1 & \cos\psi \end{bmatrix}$$

$$T_2 = \begin{bmatrix} 1 & 0 & 0 \\ 0 & \cos\phi & \sin\phi \\ 0 & -\sin\phi & \cos\phi \end{bmatrix}$$

$$T_3 = \begin{bmatrix} \cos\theta & \sin\theta & 0 \\ -\sin\theta & \cos\theta & 0 \\ 0 & 0 & 1 \end{bmatrix}$$

Where θ, ψ, ϕ are misalignment angles. The misalignment angles can be calculated by solving nonlinear equations [12].

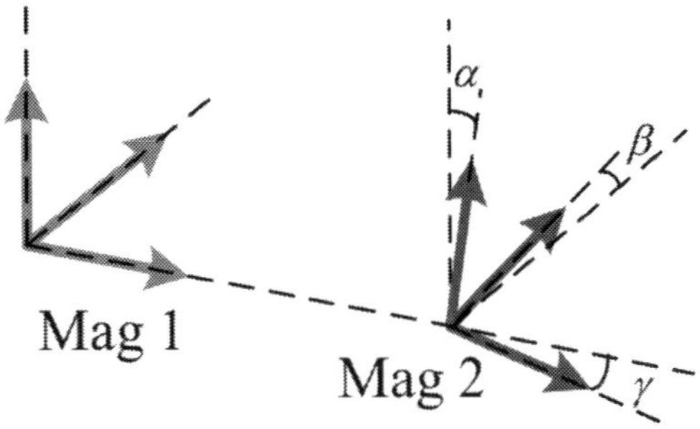

Figure 1. Misalignment error between magnetometers.

3 Experiment

3.1. Experimental system

The experimental system mainly contains a gradiometer, proton magnetometer, nonmagnetic rotation equipment and portable computer. The gradiometer consists of two fluxgate magnetometers, which is connected by an aluminium frame. There are installation interstice positions in the aluminium frame for magnetometers. Magnetometers are fixed by aluminium bolts. In order to guarantee little misalignment error between magnetometers, the installation interstices are parallel. The distance between magnetometers is 20 cm (Figure 2). Nonmagnetic rotation equipment is used to rotate the gradiometer. Soft material is a steel tube, and hard material is an annular magnet. The collection and process software is STL GradMag and Matlab respectively. Proton magnetometer is used to measure the true value of the magnetic field scalar. A site with stable magnetic field is chose as the calibration position. The gradiometer fixed on nonmagnetic rotation equipment is rotated symmetrically to get a batch of measurements.

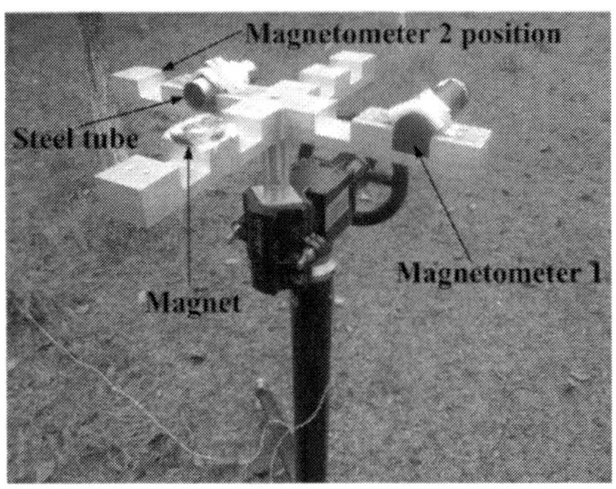

Figure 2. Gradiometers and experimental system.

3.2. Compensation results of magnetometers

Two magnetometers obtain a batch of measurements after the rotation of gradiometer. Firstly, magnetometers should be compensated using the scalar compensation method. The true intensity of geomagnetic field measured by proton magnetometer is about 48193 nT, and the magnetometers output intensity should be the same value no matter how to change the attitude. Thus, the integrated compensation performance can be evaluated.

The two magnetometers are compensated by solving nonlinear equations after the rotation. After compensation, RMS error of the first magnetometer is reduced from 1148.9 nT to 2.6 nT (Figure 3). The maximum error is reduced from 2141.1 nT to 5.1 nT. The calibration result of the second magnetometer is similar (Figure 4). After calibration, RMS error of the second magnetometer is reduced from 1355.5 nT to 2.6 nT. The maximum error is reduced from 2363.1 nT to 6.9 nT.

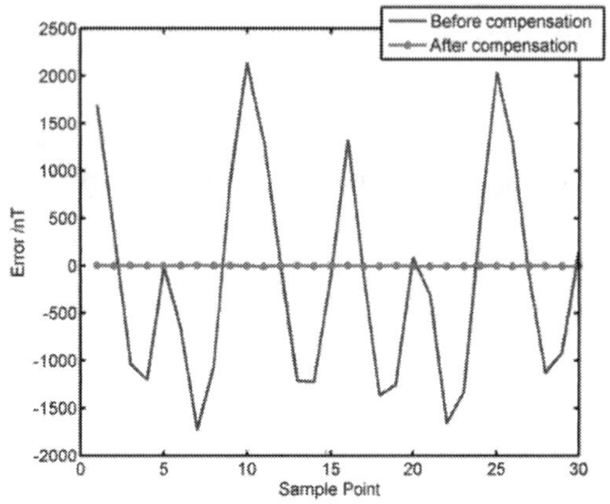

Figure 3. Compensation result of magnetometer 1.

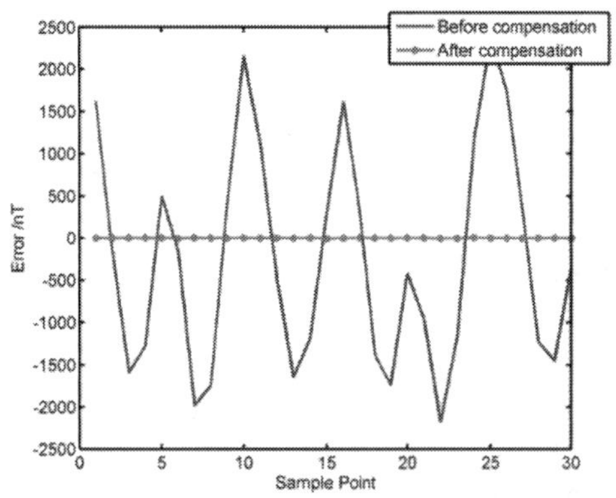

Figure 4. Compensation result of magnetometer 2.

3.3 Calibration results of misalignment

After the compensation of magnetometers, misalignment should be calibrated. There are two situations to analyze the calibration results of misalignment. (1) Misalignment calibration without distortion field compensation. (2) Misalignment calibration after distortion field compensation.

Figure 5 shows misalignment calibration result without distortion field compensation. Figure 5 (a), (c) and (e) show misalignment error before calibration, which reach 1300 nT. Figure 5 (b), (d) and (f) show misalignment error after calibration, which are reduced to about 750 nT. The details are shown in Table 1. It shows that the calibration accuracy is influenced without distortion field compensation, which should be improved. Figure 6 shows misalignment calibration result after distortion field compensation. Figure 6 (b), (d) and (f) show misalignment error after calibration, which are reduced to about 22 nT. Also, the details are shown in Table Ⅰ. After calibration, it is obvious that the accuracy is improved by one order. The estimated misalignment angles are $\alpha = -0.7364^{\circ}, \beta = -0.0046^{\circ}, \gamma = 0.0888^{\circ}$.

Table 1. Calibration result of the gradiometer in two situations.

	X axis (nT)	Y axis (nT)	Z axis (nT)
Before Calibration	888.4	1292.6	168.9
Situation 1	421.4	741.3	124.7
Situation 2	15.3	22.1	9.9

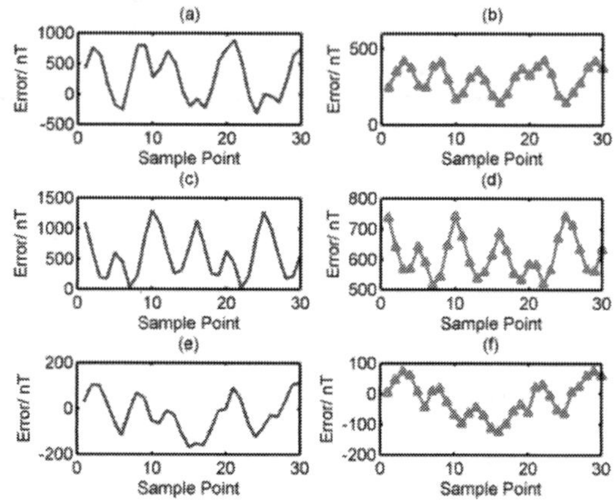

Figure 5. Calibration result of misalignment error without distortion field compensation.

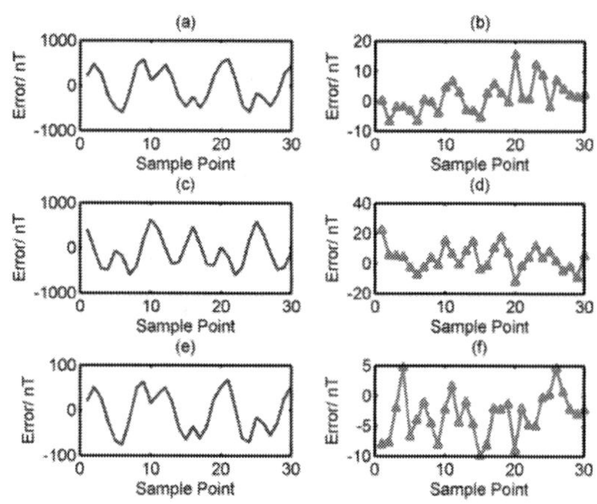

Figure 6. Calibration result of misalignment error after distortion field compensation.

 DOI: 10.1051/matecconf/20165901004

3.4 Calibration and compensation in other situations

In order to show the effectiveness of the proposed method, another situation is calibrated and compensated with different distortion field. In this situation, bigger soft material and hard material are deployed around the gradiometer (soft-iron 2 and hard-iron 2), which is shown in Figure 7. Similarly, the gradiometer is rotated symmetrically in 3D. Compensation performance of the first magnetometer is shown in Fig 8. After compensation, distortion field RMS error is reduced from 1891.4 nT to 4.1 nT. The RMS error of the second magnetometer is reduced from 1516.4 nT to 5.4 nT. As shown in Fig 9, after calibration, misalignment error is reduced to about 25 nT, thus proving the effectiveness of the proposed method.

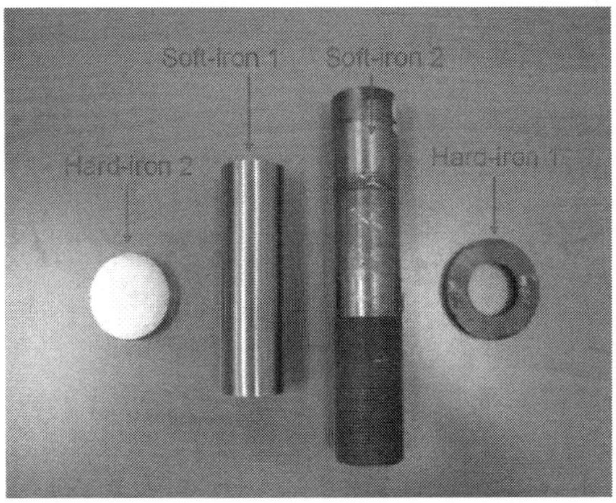

Figure 7. Different soft-material and hard-material.

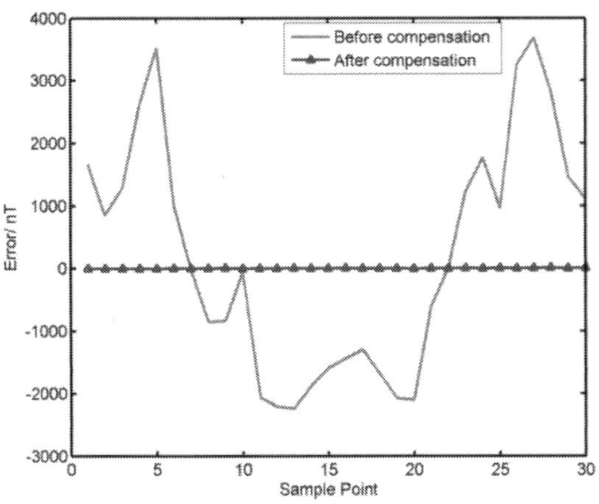

Figure 8. Compensation result of magnetometer 1 when bigger distortion fields are deployed.

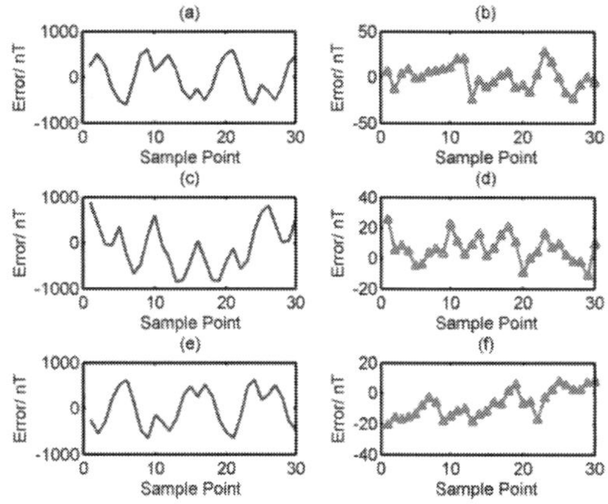

Figure 9. Calibration result of misalignment error when bigger distortion fields are deployed.

3.4 The improvement in magnetic object remote sensing

A magnetic abnormity object is measured by the gradiometer. There are 15 measurement positions, and the maximal distance is about 7 meters. Figure 10 shows the measured magnetic abnormity grads. Before calibration and compensation, system error reaches five hundreds of nT. After integrated compensation, there is little magnetic signal when object is removed far away. Magnetic abnormity grads become stronger when the object moves closer to the gradiometer. Thus, the magnetic abnormity grads are measured accurately after calibration and compensation.

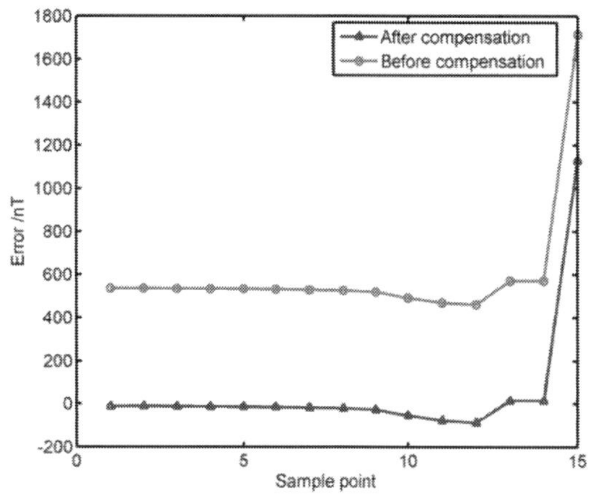

Figure 10. Magnetic abnormity grads measurement results.

4 Conclusions

Different scale factors, bias errors, nonorthogonality, misalignment and distortion field influence the accuracy of gradiometer. The error are calibrated and compensated by solving nonlinear equations. The accuracy is improved by at least one order. The calibration and compensation are still good when bigger distortion field are deployed. In addition, magnetic abnormity grads measurement results are compared. Thus, experimental results show that the proposed method can improve the accuracy of gradiometer, which is useful for magnetic object remote sensing.

Acknowledgment

This research is sponsored by National Natural Science Fund of China (Grant No. 51175507) and Young Scientist Fund of NINT (Grant No. KJYF3S01).

References

1. Y. Huang, L. H. Wu, F. Sun, IEEE Geosci. Remote. S. **11** (2013).
2. J. M. G. Merayo, P. Brauer, F. Primdahl, Sensors and Actuators A **120** (2005).
3. H. F. Pang, M. C. Pan, C. B. Wan, J. F. Chen, X. J. Zhu, F.L. Luo, IEEE Trans. Geosci. Remote Sens. 52 (2014).
4. H. F. Pang, D. X. Chen, M. C. Pan, Q. Zhang, S. T. Luo, J. Li, F. L. Luo, Sensor letters, **18** (2013).
5. D. Gebre-Egziabher, G. H. Elkaim, J. D. Powell, B. W. Parkinson, J. Aerospace. Eng. **19** (2006).
6. C. C. Foster, G. H. Elkaim, IEEE. T. Aero. Elec. Sys. **44** (2008).
7. D. G. Egziabh, J. Aircraft. **44** (2007).
8. J. L. Crassidis, K. L. Lai, R. R. Harman, J. Guidance Control Dynam. **28** (2005).
9. J. M. G. Merayo, P. Baruer, F. Primdahl, et al., Meas. Sci. Technol. **11** (2000).
10. J. M. G. Merayo, F. Primdahl, P. Brauer, et al., Sensors and Actuators A **89** (2001).
11. T. Pylvaninen, Appl. Math. Model. **32** (2008).
12. H. F. Pang, S. T. Luo, Q. Zhang, J. Li, D. X. Chen, M. C. Pan, F. L. Luo, Meas. Sci. Technol. **24** (2013).
13. H. F. Pang, J. Li, D. X Chen, M. C. Pan, S. T. Luo, Q. Zhang, F. L. Luo, Measurement **46** (2013).
14. J. Včelák, J. Appl. Phys. **99** (2006).

MATEC Web of Conferences **59**, 01005 (2016) DOI: 10.1051/matecconf/20165901005
ICFST 2016

Influence of the coupling between an atmospheric pressure ion mobility spectrometer and the low pressure ion inlet of a mass spectrometer on the mobility measurement

Frank Gunzer

Physics Department, German University in Cairo, Cairo, Egypt

Abstract. Ion mobility spectrometers (IMS) are versatile gas analyzers. Due to their small size and robustness, combined with a very high sensitivity, they are often used in gas sensing applications such as environmental monitoring. In order to improve the selectivity, they are typically combined with a mass spectrometer (MS). Since IMS works at atmospheric pressure, and MS works at vacuum, a special interface reducing the pressure over normally two stages has to be used. In this paper the influence of this coupling of different pressure areas on the IMS signal will be analyzed with help of finite elements method simulations.

1 Introduction

Ion mobility spectrometry is a well known technique in order to analyze gases. The corresponding spectrometers are small, robust, sensitive and fast. Their selectivity, i.e. the capability of giving clearly different responses for different substances, is relatively limited [1]. Thus, they are often combined with mass spectrometers. Then not only the mobility of a substance can be determined, but also the molecular mass of the analytes. The amount of information that can be used for identification is increased [2].

IMS work by employing a time-of-flight principle. The analytes are ionized, and these ions are accelerated with help of electric fields. Then they are, again with help of electric fields, driven towards the detector, which is located at a certain distance from the ionization location. By measuring the time the ions need to travel that distance, their mobility can be determined. Since IMS devices do not use vacuum (one factor that makes their size small), the ions constantly collide with ambient air molecules on the way to the detector. Thus it is their mobility that decides how fast they can reach the detector while drifting towards it [1].

MS can work with many principles [3], but all these different principles have in common that analyte ions are accelerated with help of electric fields in vacuum. One version is e.g. the time-of-flight mass spectrometry, where the ions, similar to IMS, travel a certain distance to a detector. Since they travel in vacuum, it is not the mobility and the collisions with air molecules that determine their velocity and thus their travel time, but their kinetic energy provided by the electric acceleration field and thus their mass. The flight time allows for a direct calculation of the molecular mass of the analytes.

Combining IMS and MS is relatively straight forward, since they both need analyte ions to obtain the corresponding information. One major obstacle is, however, the different pressure regimes. The ions have to be brought from an area of atmospheric pressure (1000 mbar) to the vacuum of the MS (ca. 5 x 10^{-7} mbar). This if often done via two stages, where an intermediate stage has a pressure of a few mbar.

The problem is that the opening at the interface of the IMS to the first pressure reduction stage changes the air flow in the IMS. Often, the IMS itself is then operated at reduced pressure. But then the ion chemistry is changed. This might be no problem if just the identification of analytes is to be performed, but might be problematic if further information is to be obtained that relies on the ion chemistry inside the IMS (e.g. clustering or dimer formation). Thus, atmospheric pressure IMS/MS combinations are very interesting and important, with the before-mentioned problems regarding the changed air flow in the IMS.

In this paper we have analyzed with help of finite elements method (FEM) simulations the change of the ion motion through the IMS when directly combined with an intermediate vacuum stage of a few mbar pressure. After a short description of the set up, the paper will show how the IMS response is directly affected by the pressure change caused by the interface to the reduced pressure stage.

2 Experimental

Figure 1 shows the typical set up of commercial IMS devices used in our laboratory [4-8]. On the left is region where the ions are formed.

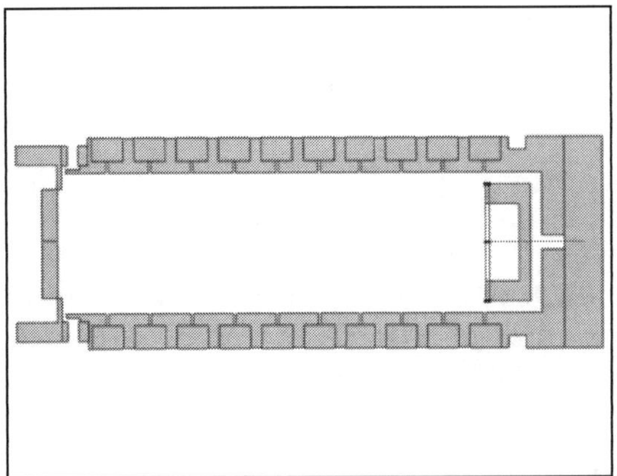

Figure 1. IMS set up. The center space is the drift region, enclosed on top and bottom by T-shaped electrodes. The detector is located on the right at the end of the drift region (here shown by two parallel, vertical lines at very small distance). In the middle, starting at the detector is the small tunnel leading into the low pressure area which is the last rectangle on the far right.

The largest part is the drift tube, where a constant electric field lets the ions drift towards the detector. In this drift region, the ions are separated according to their mobility, which allows them to reach different velocities when colliding with the ambient air molecules and thus to reach the detector after substance-specific drift times.

The connection into the low pressure stage can be realized by different means. One successful approach is to use a small diameter tunnel, where the diameter is ca. 200 µm (length 2cm) so that the high pressure can be maintained in the drift tube. The whole length is ca. 5.5 cm from left to right; the drift region is 5 cm wide (at a diameter of 2 cm). Inside the drift region, a low air current against the direction of the ions is supporting the separation process. This air current start on the right above and below the detector, where openings of ca. 1 mm diameter allow a flow of 200 ml/min to enter the drift region; the outlet is located on the left above and below the space where the ions are created.

The electrodes are supplied with a voltage difference of 2000 V between the entrance into the drift region and the detector, where a corresponding voltage divider leads to a uniform voltage drop from electrode to electrode when going from left to right. Thus a uniform electric field can be obtained at the centre of the drift region.

The ions are created in a small rectangular space in the middle of the region on the left of the drift region (diameter ca. 0.5 mm). This is the typical size when using an electron gun as we normally do in our experiments. The concentration is quite low with 10^{-13} mol/m^3 in order to be able to ignore Coulomb repulsion.

The simulation has been carried out by calculating the electric field with help of Maxwell's Equations. Ion-ion-repulsion in form of Coulomb Repulsion has, as said before, been ignored. The air currents have been simulated by solving typical computational fluid dynamics (CFD) equations for the k-ε-turbulence model [9]. The pressure at the outlets on the left were 10^5 Pa, the low pressure area was set to different pressures between 7.5 x 10^4 Pa and 10^2 Pa. The ions' motion has been calculated by solving the Navier-Stokes-Equations under the influence of the electric field and the convection flow field (i.e., the fields calculated in a previous step with help of Maxwell's Equations and CFD). The mesh size has been set to max element sizes of 0.05 mm, the mesh itself was created by the software (COMSOL Multiphysics, v. 4.3b, www.comsol.com).

3 Results and Discussion

The initial simulation used pressure values as they are typical in such a set up. On the outlet of the IMS 10^5 Pa pressure (outlet leads into ambient atmosphere), the low pressure side 400 Pa (typical pressure in reduced pressure stage), inlet into IMS provides 200 ml/min air. Figure 2 shows the air current calculated for that situation near the spot where the ions are created.

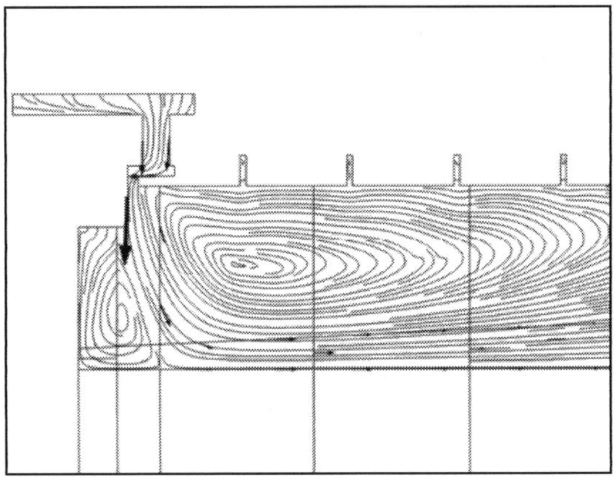

Figure 2. Calculated air current at the beginning of the drift region, only upper half of the device shown. The outlet at the top left of the drawing actually turns into an inlet (see arrow direction downward), and at the center of the device a current towards the right (i.e. the detector and the tunnel into the low pressure stage at 400 Pa pressure) forms.

As can be seen, the pressure difference is strong, so that at the outlet into ambient atmosphere actually the air is pushed into the device and not leaving it. At the center of the device, a current towards the right and thus towards the detector and the tunnel into the low pressure stage is created. Since the ions are created in our experiments near the center axis of the devices, the large turbulence in the drift region can actually be ignored; for standard set ups creating ion distributions over the whole width of the device this turbulence would be, however, problematic. The current in the middle is quite fast (much faster than the drift velocity of the ions, which is here typically 700 cm/s), and thus leads quickly to a deformation of the cloud. Figure 3 shows this deformation.

It becomes clear that the air current deforms the cloud and changes the effective velocity in comparison with an undisturbed cloud. The top picture in that figure shows a cloud a few hundred µs after leaving the region of its creation (vertical black line as spatial reference). The rectangular structure is still dominant, which resembles the shape of the ionization region. The deformation by diffusion is at this moment not very strong. The bottom picture in that figure shows the deformation together with arrows indicating the direction of the air current. The motion of the cloud is slower (the cloud is after the same time only on the black reference line), and the shape is clearly not rectangular.

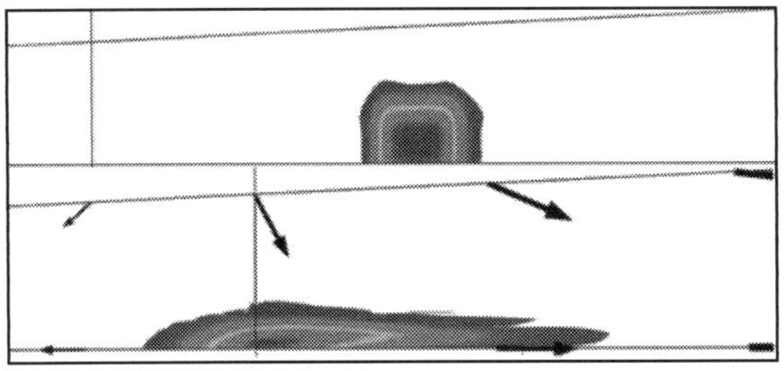

Figure 3. Top: undisturbed ion cloud. Bottom: Same cloud after the same travel time but this time disturbed by the air current. The cloud is closer to the reference line and does not show the same rectangular shape. Arrows show the direction of the air current. Due to the air flow against the motion of the ions induced by the electric field, the propagation into the direction of the detector is initially slower, but then faster.

The flow in the tunnel reaches a velocity of ca. 350 m/s due to the pressure difference. Up to a pressure of 10^3 Pa this situation does not change since the pressure difference is basically the pressure in the IMS; the contribution of the low pressure area is just too small (1 %). Thus, at these pressure differences the flow through the IMS is the same. Figure 4 shows the flow though the whole IMS in form of a velocity profile.

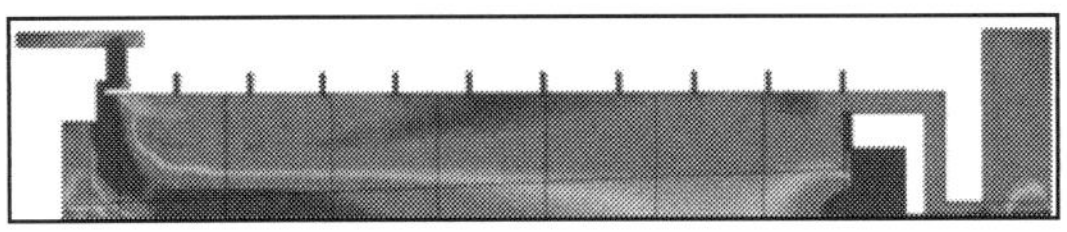

Figure 4. Velocity profile of the flow. Dark areas mean 10 m/s and more. In the tunnel, the velocity reaches 350 m/s.

Due to the small dimensions of the cloud, there is still a good time signal to be obtained after the tunnel, i.e. the influence of the disturbance on the final signal form is tolerable. The problem is that due to the fast travel time, the function of the IMS is reduced even if the obtained signal is still of good quality. The travel time is only a fraction of the undisturbed case, and thus different analytes would have less time to separate. The resolving power of the device would thus be drastically reduced. In order to see how much the effect is, we have tested the deformation of the obtained of the obtained signal at different pressure differences.

As indicated before, the situation only changes if the pressure difference between the two areas is not dominated by the high pressure area alone. Thus, if the pressure in the low-pressure stage becomes comparable to the one in the IMS, the flow through the tunnel becomes less fast and thus the disturbance created in the IMS should be lower.

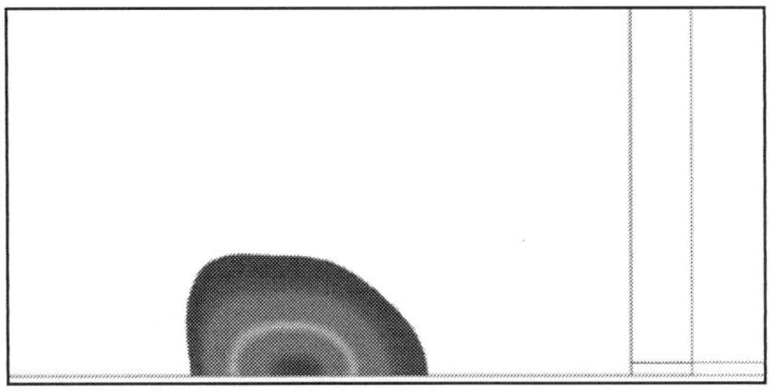

Figure 5. Ion cloud at the end of the drift region, pressure in low pressure stage is 7.5 x 10^4 Pa.

Interestingly, only at a pressure of 7.5 x 10^4 Pa is the low pressure area, the distortion becomes negligible. The influence on the flight time is, however, not. Figure 5 shows the ion cloud shortly before the detector in this case. The cloud has a nearly symmetric shape. Actually, the cloud remains largely undeformed in hat situation, and only in the last 10 mm of the drift region the deformation sets in. The tunnel is then near enough so that even this relatively low pressure difference shows an effect due to the increase flow velocity in direction of the detector. The remaining area is almost unaffected, the outflow regions near the region where the ions enter the drift region is really an outlet and not an inlet.

Despite the small influence on the shape of the ion cloud, there is still a relevant influence on the flight time. Figure 6 shows the signal that the detector would measure. It has been obtained by integrating the concentration over the detector line.

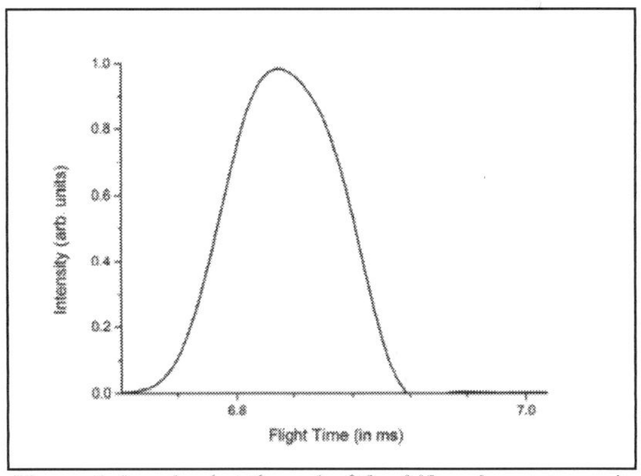

Figure 6. Ion cloud at the end of the drift region, pressure in low pressure stage is 7.5 x 10^4 Pa.

The signal looks basically normal, but the high intensity areas are slightly distorted. Here it shows already that the velocity field is non-uniform and thus this asymmetry appears in the signal. This is again a comparably small distortion, much larger becomes the distortion of the total flight time which is demonstrated in figure 7.

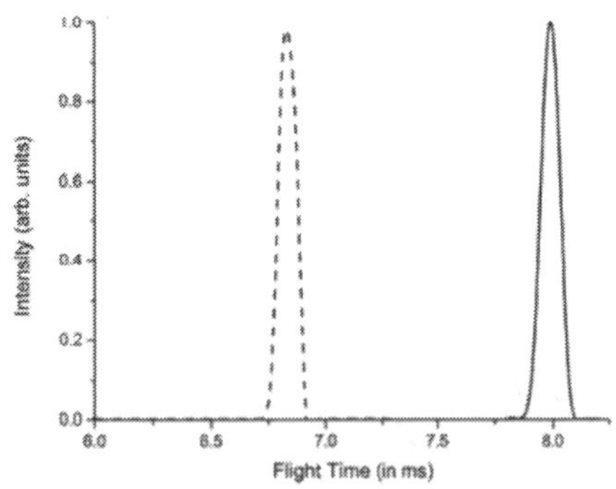

Figure 7. Ion cloud at the end of the drift region, pressure in low pressure stage is 7.5 x 10^4 Pa.

The flight time is reduced by 15 % from ca. 7.9 ms to ca. 6.8 ms. The width (FWHM), however, is changing similarly from 97 µs to 87 µs so that the resolving power (which is in IMS the ratio of the signal width to the flight time) is nearly unchanged (based on a Gauss-Peak fit, which is in the case of the disturbed signal slightly less accurate compared with the undisturbed peak, where it is very accurate). The change in resolving power will be stronger with shorter flight times when using the lower pressures on the other side of the tunnel, and similarly the distortions. In facts, when the pressure difference is very large, the ion cloud concentrates on the center axis of the device so that an ion mobility measurement might even become impossible since no ions collide with the detector but enter the tunnel instead. In the case shown before, with only a relatively small pressure drop of 250 mbar, the number of ions hitting the detector and not entering the tunnel is only a third of the total ion number, i.e. two thirds enter the tunnel.

4 Conclusions

In this paper we have shown the influence of a low pressure area connected to an atmospheric pressure IMS by a tunnel with 100 µm radius on the ion motion. Even at moderate pressure drops the ion cloud shows deformations with initially small changes of the resolving power. At a pressure reduction to 10^4 Pa the effects are already strong, and the motion of the ions is largely dominated by the air current in the IMS. The velocity of this current exceeds then the velocity caused by the electric field and drift through the IMS by far. Thus there are negative effects on the resolving power which is proportional to the drift time. Furthermore, the number of ions hitting the detector at the end of the drift tube would be very low compared to the number of ions entering the tunnel. Thus a combined IMS/MS measurement as targeted by such a set up would not work. The results have been obtained for small ion clouds as they are typically created by an electron used in out experimental set up. Using other ionization principles yielding larger ion clouds might then be beneficial. However, the distortions would be quite large, too, since the air current formed by larger pressure drop in the drift region is quite irregular, as has been shown in this publication. It becomes obvious that the use of ion optics into the low pressure region is extremely helpful, possibly absolutely necessary if the low pressure stage is to be operated at the typical 400 Pa pressure.

In order to verify the validity of these simulations, experiments are planned also with different set ups where corresponding simulations have been carried out in the past (e.g. using atmospheric ion funnels [10]). Should the experimental results confirm the simulations, it becomes evident that atmospheric IMS coupled via an intermediate stage to vacuum IMS is, although offering striking simplicity in its set up, coming at the price of IMS resolving power. In order to get the ions into the MS, the IMS itself or its simplicity has to be sacrificed. Finally, this paper has shown how helpful such simulations can be when combining multi-physics approaches, such as electric fields, air currents calculated by CFD as well as ion motion under the influence of these fields and subject to diffusion effects for the effective development of e.g. gas monitors for environmental relevant applications.

References

1. G.A. Eiceman, Z. Karpas, *Ion Mobility Spectrometry*, CRC Press, Boca Raton, USA, (2005)
2. A.B. Kanu, P. Dwivedi, M. Tam, L. Matz, H.H. Hill, J. Mass Spectrom. 43, 1–22 (2008)
3. E. Hoffman, V. Stroobant, *Mass Spectrometry - Principles and Applications,* John Wiley & sons, Chichester, England (2007)
4. F. Gunzer, S. Zimmermann, W. Baether, Anal. Chem. **82**, 3756 (2010)
5. W. Basther, S. Zimmermann, F. Gunzer, Int. J. Mobil. Spectrom. 13, 95 (2010)
6. W. Baether, S. Zimmermann, F. Gunzer, Sens. Act. B **171**, 1238 (2012)
7. F. Gunzer, Int. J. Ion Mobil. Spectrom. **18**, 41 (2015)
8. F. Gunzer, Analyst **140**, 6379 (2015)
9. J.F. Wendt, *Computational Fluid Dynamics - An Introduction*, Springer Berlin Heidelberg, Germany (2009)
10. Q. Yu, Z. Diao, K. Ni, X. Qian, F. Tang, X. Wang, Rapid Commun. Mass Spectrom. **29**, 1055 (2015)

Relaxation Time Coupling Method for Ultrasonic Sensor Design

Hu Yi, Wang Shu, Zhu Ming

School of Electronic Information and Communications, Huazhong University of Science and Technology, Wuhan 430074, China

Abstract. In our previous work [Sens. Actuator B: *Chem.* **203** 1-8 (2014)], a ultrasonic sensor has been proposed to detect gas compositions by the acoustic spectral peak location. However, the effective relaxation area constructed by existing theoretical model cannot match the detection method when there are more than one strong relaxational components in gas mixture. A method that calculates the relaxation time of a multi-component relaxation process is presented. Based on acquirement of decoupled relaxation times and their corresponding effective specific heat values, it calculates the whole relaxation time of a multi-component relaxation process. The method is based on the fact that for a multi-component gas mixture, such as occurs in many polyatomic gas mixtures, the relaxation process should be considered as a whole rather than being decoupled as some single relaxation processes. Acquiring relaxation time provides the approach to obtain the acoustic absorption spectral peak value directly.

1 Introduction

Intelligent ultrasonic sensor is applied widely to detect gas compositions [1]. Measuring acoustic relaxation absorption spectrum is a powerful method to characterize molecular relaxation process in gas mixture [1, 2, 3]. For the complexity of obtaining the whole spectrum, we proposed a method to employ the spectral peak location to detect gas composition. Based on this method, a gas detection sensor with two pairs of ultrasonic transducers are suggested to practical applications [4]. In a simple relaxation process, the acoustic relaxation absorption spectral peak location is determined by the relaxation time [5, 6]. Therefore, calculating the relaxation time in excitable gases is a key issue to apply the spectral peak location based method in gas detection.

Over the years, some relaxation time theories have been proposed [7-10]. However, these theories did not provide an appropriate physical model to demonstrate the theoretical derivation. Recently, based on the decoupling model in multi-component excitable gases [11], Zhang calculated the vibrational relaxation times of the specific decoupled relaxation processes [6]. However, a multi-component relaxation process should be considered as a whole physical process. Zhang's model can only calculate the decoupled relaxation times of every single relaxation processes. In practical detection, calculating the acoustic absorption spectral peak focuses on the whole relaxation time of a multi-component relaxation process.

Based on the relaxation time theories, we develop a model in this paper to calculate the whole relaxation time. The model can be applied to measure the spectral peak value, i.e. the effective relaxation frequency and the maximum acoustic absorption coefficient. In the paper, we propose a relaxation time coupling based method to obtain the whole relaxation time. Then we compare the predicted results based on the proposed method with the experimental data, and provide the experimental results in strong-strong relaxational gas mixtures.

2 Coupling relaxation times method

The acoustic relaxation absorption decoupling model [11] suggests that a multi-component vibrational relaxation process in excitable gases can be decoupled to some single relaxation processes. Effective isochoric specific heat C_V^{eff} of multi-component relaxation process can be characterized by the relaxation times as [6, 11]:

$$C_V^{\text{eff}} = C_V^{\infty} + \sum_{n=1}^{N} \frac{C_n^*}{1+i\omega\tau_n} \tag{1}$$

where C_V^{eff} represents the macroscopic 'footprint' of the inability of the internal degrees of freedom to follow the external temperature fluctuations [12]. C_V^{∞} is the high-frequency instantaneous limit value, $C_V^{\infty} \equiv \lim_{\omega\to\infty} C_V^{\text{eff}}(\omega)$. The C_V^{∞} is only dependent on the molecular symmetry. For a linear molecule, $C_V^{\infty} = 5\text{R}/2$, and for a nonlinear molecule, $C_V^{\infty} = 3\text{R}$, R is universal gas constant. There is another low-frequency limit $C_V^0 \equiv \lim_{\omega\to 0} C_V^{\text{eff}}(\omega)$. Therefore, the difference between the two limiting values represents the contribution of the relaxing internal degrees of freedom of the molecules, $C^{\text{int}} = C_V^0 - C_V^\infty = C_V^{\text{vib}}$ [5]. C_n^* and τ_n are the isochoric molar specific heat and the relaxation time for one of the decoupled single-relaxation processes, respectively.

Viewed from the gas vibration modes, we can write the C_V^{eff} as [3]:

$$C_V^{\text{eff}} = C_V^\infty + \sum_{j=1}^{N} a_j C_j^{\text{vib}} y_j \text{ , } y_j = \mathrm{d}T_j^{\text{vib}} / \mathrm{d}T \quad (2)$$

where y_j is the ratio between the temperature fluctuation of vibratioanl mode j ($\mathrm{d}T_j^{\text{vib}}$) and that of the external DOF ($\mathrm{d}T$). For a gas mixture with N kinds of vibrational modes, $C_V^{\text{vib}} = \sum_{j=1}^{N} a_j C_j^{\text{vib}}$, a_j is the mole fraction of mode j, C_j^{vib} is the vibrational isochoric molar specific heat of mode j. The C_j^{vib} is given by the Planck-Einstein function for a harmonic oscillator [2].

Based on Eq. (2) and the general relaxation equation provided in [11], we can obtain the expression of the algebraic relaxation equation in a matrix form [6]:

$$(\mathbf{R} + \mathrm{i}\omega\mathbf{I})\mathbf{y} = \mathbf{H} \quad (3)$$

where $\mathbf{y}_j = y_j$, $\mathbf{R}_{jj} = k_{jj}$, $\mathbf{R}_{jk} = -k_{jk}$, $\mathbf{H}_j = k_{jj} - \sum_{\substack{k=1 \\ k \neq j}}^{N} k_{jk}$, $\mathbf{I}$ is the identity matrix. Ref. [11] defined the matrix $\mathbf{R}$ as energy transition rate matrix. We can construct the matrix $\mathbf{R}$ by calculating the k_{jj} and k_{jk} as [6]:

$$k_{jj} = k_{10}(j) + \sum_{\substack{k=1 \\ k \neq j}}^{N} k_{01}^{10}(j,k) \frac{1 - \exp(-h\nu_j / k_B T_0)}{1 - \exp(-h\nu_k / k_B T_0)},$$

$$k_{jk} = k_{01}^{10}(j,k) \frac{\nu_k}{\nu_j} \frac{1 - \exp(-h\nu_j / k_B T_0)}{1 - \exp(-h\nu_k / k_B T_0)}, \quad (4)$$

where ν_j and ν_k are the vibrational frequency of mode j and k, respectively; h is Planck's constant; k_B is Boltzmann's constant; T_0 is the equilibrium temperature. $k_{10}(j)$ and $k_{01}^{10}(j,k)$ are the V-T transition rate and the V-V transition rate respectively. They can be calculated by the Tanczos equation [1, 3]. By using Eq. (1) and Eq. (3), we can obtain the relaxation times τ_n and the isochoric molar specific heat C_n^* of the decoupled single-relaxation processes. The reciprocals of the eigenvalues of the energy transition rate matrix $\mathbf{R}$ are exactly the relaxation times τ_n [6]. The detailed derivation is developed in [6].

For a multi-component relaxation process, refs. [6] and [11] decoupled the whole relaxation process to some single relaxation processes, and allocated the relaxation times to every specific relaxation process as Eq. (1). Based on the decoupling model, Zhang *et al.* suggested that there will be as many relaxation times as there are vibrational modes available to a gas, and the relaxation time of the primary relaxation process determines the spectral peak location in most cases [6]. However, we think that the "primary relaxation time" in [6] is actually the first relaxation time calculated based on the decoupling model. If there are two strong relaxational gases in mixture, C_1^* and C_2^* are in the same order of magnitude, meanwhile the first relaxation time τ_1 of C_1^* and the second relaxation time τ_2 of C_2^* in Eq. (6) are approximate. In this case, the whole relaxation time can not obtained by the decoupling model. Consequently, we proposes a coupling relaxation times method to obtain the whole relaxation time of a multimode relaxation process.

Actually, we consider the effective isochoric specific heat C_V^{eff} of a multi-component relaxation process can be given as [5]:

$$C_V^{\text{eff}} = C_V^\infty + \frac{C^{\text{int}}}{1 + i\omega\tau_{\text{m}}} \quad (5)$$

where τ_{m} is C^{int} is the difference between the C_V^0 and C_V^∞, represents the contribution of the relaxing internal degree of freedom of the molecules, τ_{m} is the whole relaxation time of the whole relaxation process [5].

Since there are no more than two significant relaxation processes appearing together in the available literatures, we focus on the case that C_1^* and C_2^* are both significant. Therefore we can rewrite the Eq. (1) as:

$$C_V^{\text{eff}} = C_V^\infty + \frac{C_1^*}{1 + i\omega\tau_1} + \frac{C_2^*}{1 + i\omega\tau_2} \quad (6)$$

where C_1^* and τ_1 are the isochoric molar specific heat and the relaxation time of the first significant relaxation process; C_2^* and τ_2 are the isochoric molar specific heat and the relaxation time of the second significant relaxation process. Combining Eq. (5) and (6), the equation can be obtained:

$$\frac{C^{\text{int}} - i\omega\tau_{\text{m}} C^{\text{int}}}{1 + \omega^2 \tau_{\text{m}}{}^2} = \frac{C_1^* - i\omega\tau_1 C_1^*}{1 + \omega^2 \tau_1{}^2} + \frac{C_2^* - i\omega\tau_2 C_2^*}{1 + \omega^2 \tau_2{}^2} \quad (7)$$

 DOI: 10.1051/matecconf/20165901006

To the two strong relaxation components gas mixtures, $\tau_1 \approx \tau_2$, thus the right hand side of the equation is $\dfrac{C_1^* + C_2^* - i\omega(\tau_1 C_1^* + \tau_2 C_2^*)}{1+\omega\tau_1\tau_2}$. In Eq. (7), since $\tau_1 \approx \tau_2$, C_1^* and C_2^* are much large than the other C_n^*, we have $C^{\mathrm{int}} \approx C_1^* + C_2^*$. Based on this approximation, we can conclude that $1+\omega^2\tau_{\mathrm{m}}{}^2 = 1+\omega\tau_1\tau_2$. Then from Eq. (7), τ_{m} is given by:

$$\tau_{\mathrm{m}} = \frac{C_1^*}{C_1^*+C_2^*}\tau_1 + \frac{C_2^*}{C_1^*+C_2^*}\tau_2 \qquad (8)$$

$C_n^* / \sum C_n^*$ can be considered as its contribution to the total vibrational heat capacity, the relaxation time τ_{m} is a linear combination of the decoupled relaxation times with the weighted factor $C_n^* / \sum C_n^*$:

$$\tau_{\mathrm{m}} = \sum_n^N \frac{C_n^*}{\sum C_n^*}\tau_n \qquad (9)$$

With the relaxation time τ_{m}, acoustic relaxation absorption spectra $\mu_{\mathrm{r}}(\omega)$ can be expressed as [13, 14]:

$$\mu_{\mathrm{r}}(\omega) = \pi\tau_{\mathrm{m}}\omega \mathrm{R} \frac{C^{\mathrm{int}}}{(\mathrm{R}+C_V^0)C_V^0 + \omega^2\tau_{\mathrm{m}}^2(\mathrm{R}+C_V^\infty)C_V^\infty} \qquad (10)$$

where $\omega = 2\pi f$ is the acoustic angular frequency, $\mu_{\mathrm{r}}(\omega)$ is the frequency dependence dimensionless acoustic relaxation absorption spectrum, R is universal gas constant. Based on Eq. (10), we can obtain the relaxation frequency and the maximal acoustic relaxation absorption coefficient, i.e. the acoustic relaxation absorption spectral peak value are:

$$f_{\mathrm{relax}} = \frac{1}{2\pi\tau}\left[(\mathrm{R}+C_V^0)C_V^0 / (\mathrm{R}+C_V^\infty)C_V^\infty\right]^{1/2},$$

$$\mu_{\mathrm{m}} = \mu_{\mathrm{r}}(2\pi f_{\mathrm{relax}}) \qquad (11)$$

3 Simulation results and discussion

Based on Eq. (11), we will validate the proposed method by comparing the spectral peak predicted by our method with the experimental data. We first consider the simulation in the mixture CO_2-N_2 and CH_4-N_2. They both have only one single strong relaxation process. The results are shown in Fig. 1.

As 60%CO_2-40%N_2 shown in Fig. 1(a), the curve represents the acoustic absorption spectrum predicted by the model given in [6], the diamond is its predicted peak. Then we select five pairs of the experimental data given by Ejakov *et al.* [15], they are given as the Plused-Circle in Fig. 1, and the other circles are also experimental data. With these five pairs of data, we synthesize them to a measured spectral peak by the algorithm proposed in [5]. The synthesized results are shown as Plus in Fig. 1(a). Finally, we calculate the peak value by our proposed method, the result is shown as Bold Cross in Fig. 1(a).

In Fig. 1(a), the spectral peak predicted by our proposed method (Bold Cross) is closer to the synthesized results (Plus) than the prediction in [6] (Diamond). Similarly, for the 60%CH_4-40%N_2 mixture, the spectral peak predicted by our proposed method is also closer to the synthesized results. The numerical values and relative errors are given in Table 1.

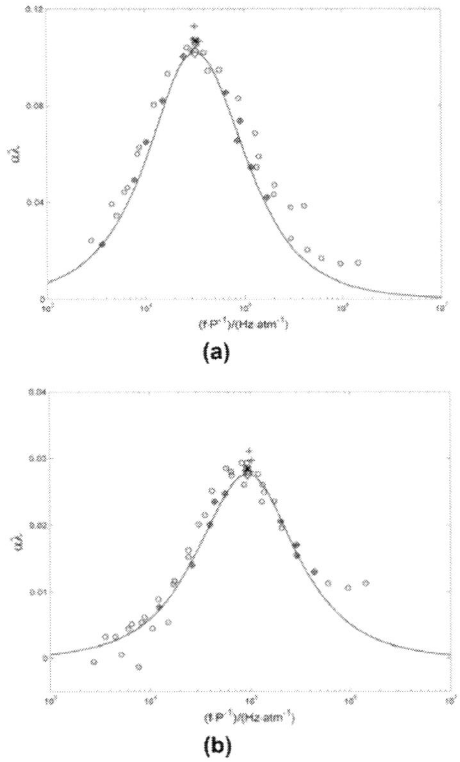

Figure 1. Comparisons of the synthesized results with the spectral peaks predicted by our method and the analytical model in [6]. **a:** mixture of 60%CO_2-40%N_2 at T=293.5K; **b:** mixture of 60%CH_4-40%N_2 at T=293.4K.

According to Table 1, for the 60%CO_2-40%N_2 mixture, predicting by our proposed method, the maximum relative error is 5.2732%, and the average relative error is 1.9367%. Yet for the spectral peak predicted by the analytical model in [6], the maximum relative error is 8.2840%, and the average relative error is 3.2347%. Similarly, for the CH_4-N_2 mixture, the maximum relative error and average relative error between the prediction results by our proposed method and the measurements are much smaller than the analytical model in [6]. Consequently, it is obvious that our proposed method can obtain the relaxation time more correctly than the model in [6].

Table 1. Calculation results of the spectral peaks in Fig. 1 for the mixtures of 60%CO_2-40%N_2 at T =293.5K and 60%CH_4-40%N_2 at T =293.4K. (3.232^4, 0.1098) means the values of x-axis acoustic frequency (Hz) with 1atm and y-axis dimensionless acoustic absorption coefficients $\mu_r(\omega)$ of the spectral peak, and the notation x^y implies $x \times 10^y$.

Mixtures	Synthesized result (Plus)	Our predictions (Bold cross)	Relative error	Model in [6] predictions (Diamond)	Relative error
	(3.232^4, 0.1098)		5.2732 %		8.2840 %
	(3.188^4, 0.1048)		0.4794 %		3.3531 %
60%CO_2 -40%N_2	(3.095^4, 0.1042)	(3.265^4, 0.1043)	0.0959 %	(3.285^4, 0.1014)	2.7613 %
	(2.752^4, 0.1031)		1.1505 %		1.6765 %
	(2.566^4, 0.1015)		2.6846 %		0.0986 %
	(9.697^4, 0.0309)		9.5745 %		13.187 %
	(1.022^5, 0.0295)		4.6010 %		8.0586 %
60%CH_4 -40%N_2	(8.931^4, 0.0279)	(9.358^4, 0.0282)	1.0638 %	(9.332^4, 0.0273)	2.1978 %
	(9.998^4, 0.0278)		1.4184 %		1.8315 %
	(9.038^4, 0.0274)		2.8369 %		0.3663 %

To validate our proposed method for the mixture with two strong relaxation gases in the mixture, five gas mixtures are considered: 76%CH_4-24%CO_2, 81%CH_4-19%CO_2, 86%CH_4-14%CO_2, 91%CH_4-9%CO_2, and 96%CH_4-4%CO_2. We calculate the single relaxation times by the model in [6] and the whole relaxation time by our proposed method, respectively. Then with the calculation results, the spectral peaks are predicted. Finally, we compare the predicted spectral peak value with the measurements data by the synthesizing algorithm [5]. The prediction results and relative errors are given in Table. 2.

As shown in Table. 2, when the concentration of CO_2 is 4%, the relative errors are 1.10% and 1.99%, respectively. This is because that the major component of the gas mixture is CH_4, the mixture can be thought as pure CH_4 actually. However when the concentration of CO_2 is 24%, the relative error between the measured result and the predicted by our proposed method is 5.14%, while the relative error is 17.37% for the model in [6]. The spectral peak value predicted by our proposed method are more accurate than that predicted by the model in [6].

4 Conclusion

In this paper, we improve the relaxation time model [6] to calculate the whole relaxation time of a multi-component relaxation process. Based on our proposed method, the spectral peak value can be obtained more correctly than the model in [6]. The relative error between the measured results and the predictions can be effectively decreased by the proposed method. Moreover, for most excitable gases, the relaxation time of the multi relaxation process should be considered as a whole rather than the decoupled first relaxation time that proposed in [6]. Consequently, the acoustic absorption spectral peak predicted by our proposed method can be more correctly than that predicted by the previous theory. The proposed coupling relaxation time method provides idea for intelligent ultrasonic sensor design in practical applications. This work is supported by National Natural Science Foundation of China (Grant Nos. 61371139, 61571201 and 61461008).

Table 2. Calculation results of the spectral peaks in Fig. 3 for the mixtures of CH_4-CO_2. The notation x^y implies $x \times 10^y$.

Mixtures	Synthesized results	Our method	Model in [6]	Relative error
76%CH_4-24%CO_2	(1.463^5, 0.05459)	(1.138^5, 0.05192)	(1.464^5, 0.04651)	5.14% and 17.37%
81%CH_4-19%CO_2	(1.390^5, 0.04970)	(1.118^5, 0.04753)	(1.377^5, 0.04363)	4.57% and 13.91%
86%CH_4-14%CO_2	(1.396^5, 0.04451)	(1.131^5, 0.04318)	(1.309^5, 0.04058)	3.08% and 9.68%
91%CH_4-9%CO_2	(1.298^5, 0.03931)	(1.165^5, 0.03889)	(1.271^5, 0.03726)	1.08% and 5.50%
96%CH_4-4%CO_2	(1.241^5, 0.03426)	(1.209^5, 0.03464)	(1.235^5, 0.03359)	1.10% and 1.99%

References

1. Phillips S, Dain Y, Lueptow R M *Meas. Sci. Technol.* **14** 70 (2003)
2. Dain Y, Lueptow R M *J. Acoust. Soc. Am.* **109** 1955 (2001)
3. Zhang K S, Wang S, Zhu M, Hu Y, Jia Y Q *Acta Phys. Sin.* **61** 174301 (in Chinese) (2012)
4. Hu Y, Wang S, Zhu M, Zhang K S, Liu T T, Xu D Y *Sens. Actuator B-Chem.* **203** 1 (2014)
5. Petculescu A G, Lueptow R M *Phys. Rev. Lett.* **94** 238301 (2005)
6. Zhang K S, Ou W H, Jiang X Q *J. Korean Phys. Soc.* **65** 1028 (2014)
7. Schwartz R N, Slawsky Z I, Herzfeld K F *J. Chem. Phys.* **20** 1591 (1952)

8. Tanczos F I *J. Chem. Phys.* **25** 439 (1956)
9. Dain Y, Lueptow R M *J. Acoust. Soc. Am.* **110** 2974 (2001)
10. Petculescu A G, Lueptow R M *J. Acoust. Soc. Am.* **117** 175 (2005)
11. Zhang K S, Wang S, Zhu M, Ding Y, Hu Y *Chin. Phys. B* **22** 014305 (2013)
12. Zhang K S, Wang S, Zhu M, Ding Y *Meas. Sci. Technol.* **24** 055002 (2013)
13. Herzfeld K F, Litovitz T A *Absorption and Dispersion of Ultrasonic Waves* (New York: Academic) p55-216 (1959)
14. Lambert J D *Vibrational and Rotational Relaxation in Gases* (Oxford: Clarendon) p1-114 (1977)
15. Ejakov S G, Phillips S, Dain Y, Lueptow R M, Visser J H *J. Acoust. Soc. Am.* **113** 1871 (2003)

MATEC Web of Conferences **59**, 01007 (2016) DOI: 10.1051/matecconf/20165901007
ICFST 2016

Hall Sensor Output Signal Fault-Detection & Safety Implementation Logic

SangHun, Lee[1], HongSeuk, Oh[2]

[1] *Intelligent automotive team, Daegu Mechatronics & Materials Institute, S.Korea*
[2] *2STARGROUPIND. LTD, S.Korea*

Abstract. Recently BLDC motors have been popular in various industrial applications and electric mobility. Recently BLDC motors have been popular in various industrial applications and electric mobility. In most brushless direct current (BLDC) motor drives, there are three hall sensors as a position reference. Low resolution hall effect sensor is popularly used to estimate the rotor position because of its good comprehensive performance such as low cost, high reliability and sufficient precision. Various possible faults may happen in a hall effect sensor. This paper presents a fault-tolerant operation method that allows the control of a BLDC motor with one faulty hall sensor and presents the hall sensor output fault-tolerant control strategy. The situations considered are when the output from a hall sensor stays continuously at low or high levels, or a short-time pulse appears on a hall sensor signal. For fault detection, identification of a faulty signal and generating a substitute signal, this method only needs the information from the hall sensors. There are a few research work on hall effect sensor failure of BLDC motor. The conventional fault diagnosis methods are signal analysis, model based analysis and knowledge based analysis. The proposed method is signal based analysis using a compensation signal for reconfiguration and therefore fault diagnosis can be fast. The proposed method is validated to execute the simulation using PSIM.

1 Introduction

Recently BLDC motors have been popular in various industrial applications and electric mobility. The BLDC motor is a DC motor type that has no brushes; therefore commutation is done electronically according to the permanent magnet rotor position based on the hall sensor signal.

Correct commutation and control of BLDC motors depend on the exact position detection of rotor. Sensors such as optical encoders, for high resolution applications, and hall effect sensors, for low resolution applications, are normally used to detect position of the permanent magnet rotor.

Rotor position detection techniques in sensorless control drives of the BLDC motor are mainly based on back-EMF sensing, back-EMF integration, free-wheeling diode conduction of the unexcited phase, flux linkage of motor, and third-harmonic analysis of back-EMF.

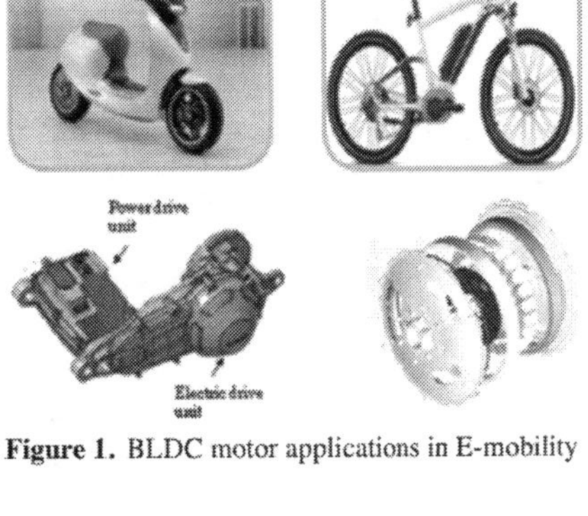

Figure 1. BLDC motor applications in E-mobility

Reduction of the BLDC motor manufacturing cost, the motor maintenance and possibility of the motor malfunctions due to failure or unbalanced positioning of sensors are advantages of sensorless control techniques; however difficulties of back-EMF sensing at low speeds and transient time, complexity of rotor detection algorithms and discontinuous response due to high commutation rates are the main drawbacks of sensorless techniques.

Low resolution hall effect sensor is popularly used to estimate the rotor position because of its good comprehensive performance such as low cost, high reliability and sufficient precision.

However various possible faults may happen in a hall effect sensor such as flaws in the core (corrosion, cracks, residual magnetic fields and core breakage), changes in the bias current, change in the magnetic properties of the ferrite core due to temperature variations, changes in the orientation of the induced magnetic field in the sensor (due to mechanical shocks or other reasons). Any of these faults may result to breakdown of the hall effect sensor in BLDC motor.

There are a few research work on hall effect sensor failure of motor. Signal analysis, model based and knowledge based methods are the main techniques for fault diagnosis in the BLDC or PMSM drives[1]~[8].

In the signal analysis based methods, fault is detected and identified through comparison of the BLDC motor signals with the ideal situation. Therefore there is no need for an accurate dynamic model of the motor. Signal analysis fault diagnosis methods are generally slower than other two methods. Fault diagnosis in model based methods is quite fast and can be used for online fault detection; however the exact mathematical model of the BLDC motor is needed.

Model based observers are popularly used in fault diagnosis systems; however fault detection residuals are influenced not only by system faults but also by disturbances and model uncertainties.

Since uncertainties, disturbances or noise in system are unknown and cannot be modelled mathematically; therefore there is always a difference between the mathematical model results and actual system performance that can generates false fault detections in model based systems.

In expert systems, fuzzy or neural network controllers are developed based on the knowledge of system. The knowledge is collected and analysed either through an experienced engineer with a thorough understanding of the system, or via comprehensive study of the system dynamics through a validated simulation model.

In this paper, the hall sensor output fault-tolerant control strategy based on signal analysis and the proposed compensation logic is suggested.

2 Conventional Hall Sensor Fault-Tolerant Strategy

To rotate the BLDC motor, the stator windings should be energized in a sequence. It is important to know the rotor position in order to understand which winding will be energized following the energizing sequence.

Rotor position is sensed using Hall effect sensors embedded into the stator. Most BLDC motors have three Hall sensors embedded into the stator on the non-driving end of the motor.

Whenever the rotor magnetic poles pass near the Hall sensors, they give a high or low signal, indicating the Nor S pole is passing near the sensors. Based on the combination of these three Hall sensor signals, the exact sequence of commutation can be determined.

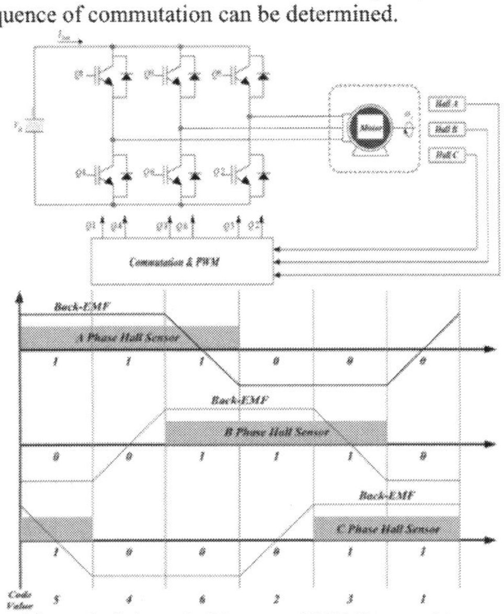

Figure 2. Schematic Diagram of BLDC motor drive

Hall Effect sensors are mounted inside the BLDC motor with 120 electrical degree phase difference to detect rotor position. Output of each sensor is high (logic '1') for 180 electrical degree and is low (logic '0') for the next 180 degree with respect to rotor position.

Each Hall Effect signal of BLDC motor has specific value at each instant of time with respect to permanent magnet rotor position. Electronic commutation is done by decoding the position sensor signals.

Decoding Hall Effect signals result to choose the proper voltage space vector for switching of the three phase Voltage Source Inverter (VSI) drive of BLDC motor.

Fig. 2 shows the schematic diagram of 3 phase Y – connected BLDC motor and three phase VSI drive and Fig. 3 shows the output signal according to hall sensor configurations.

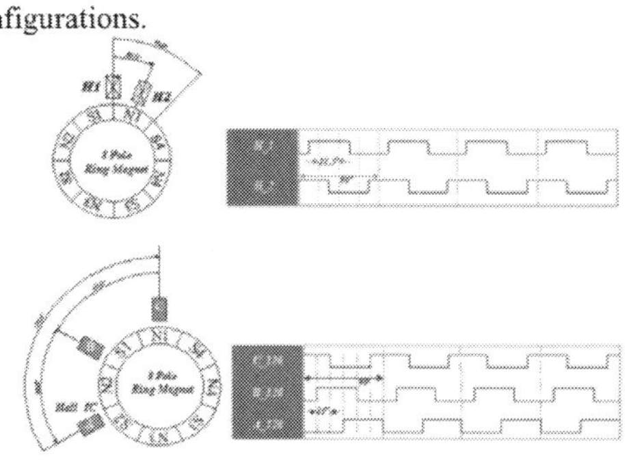

Figure 3. Output signal according to hall sensor configurations

In general, fault tolerant control systems are responsible to detect and identify the hall sensor fault condition and fault safety logic is used to isolate the faulty section and take the appropriate remedial action to keep the system working under the fault condition.

The conventional fault diagnosis methods are signal analysis, model based analysis and knowledge based analysis.

As mentioned above, in signal based method, fault is detected and identified through comparison of the ideal signals. Therefore motor dynamic model isn't needed but fault diagnosis is slow. And in model based analysis, fault is detected and identified through mathematical model & model observer. So motor dynamic model is needed but fault diagnosis is fast.

The proposed method is signal based analysis using a compensation signal for reconfiguration and it has two steps for checking fault conditons.

Step 1. Pre-Fault condition

detection and identification of the falut.

Step 2. Post-Fault condition

isolation of the falut & take action for motor operation

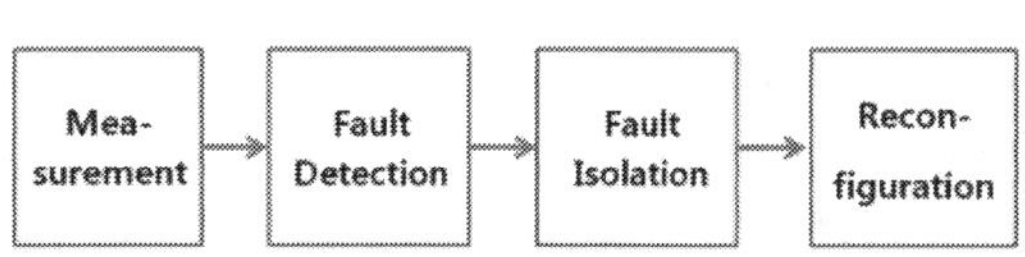

Figure 4. Hall Sensor Fault-Safety Strategy

3 The proposed Hall Sensor Fault-Tolerant Strategy

As mentioned above, the basic concept is the same with the conventional signal based method. So the operation rule check of hall sensor signal and switching sequence is shown in Fig 5 and Table 1.

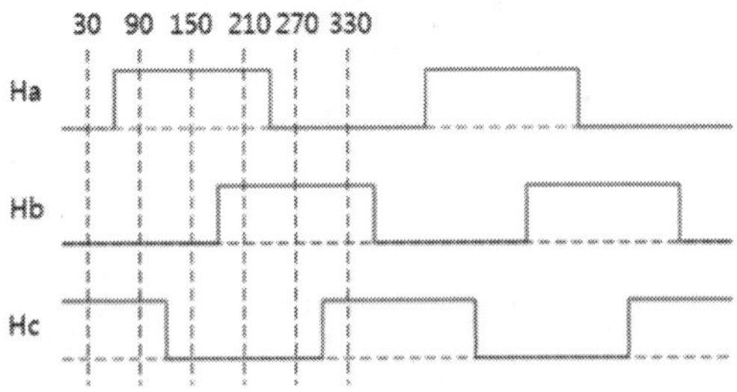

Figure 5. BLDC commutation diagram (Hall Sensor I/F)

Tablel 1. Hall Sensor Output Signal Decoding Rules

Rotor Angle*	Ha	Hb	Hc	Conducting Switches	Hf
30~90	1	0	1	S1, S6	5
90~150	1	0	0	S1, S2	4
150~210	1	1	0	S3, S2	6
210~270	0	1	0	S3, S4	2
270~330	0	1	1	S5, S4	3
330~30	0	0	1	S5, S6	1

The related equation for operation rule check is the following that :

$$H_f = w_a * H_a + w_b * H_b + w_c * H_c \quad (1)$$

here, w_n weight value (n : 4, 2, 1)

In normal condition, the output of H_f has the sequential value according to the rotor position. But in the fault condition, the output of H_f has not the normal sequential value because one of three has constant "0" or "1" despite of motor rotation like Fig 6.

In case of one hall seonsor fault, it occurs that XOR of three hall sensors and the decoding rules of hall sensor output signal are changed.

The proposed method is shown in Fig 6. The fault detection is executed by equation (1) and XOR logic pulse train based on three hall signal output.

If the fault condition is occurred, the XOR logic duty is non symmetric and it can be detected as a fault. The reconfiguration signal is made by XOR logic and the other two hall sensor output like Fig 6.

4 Simulation Results

In order to verification, Powesim is used to simulate the proposed hall sensor fault-tolerant strategy.

The proposed fault-detection and fault-safety logic is programmed by C-block function and the mentioned algorithm has the following main block :

1) hall sensor output decoding block
2) hall sensor output fault-detection block
3) hall sensor output fault-isolation block
4) hall sensor output reconfiguration block

Simulation condition has one hall fault condition of three hall sensors under motor operation at 3000 rpm and the proposed alogrithm does not allow start-up and near zero speed operation of a BLDC motor.

Because it is impossible to measure the pulse length on the other two hall sensors when the motor does not rotating.

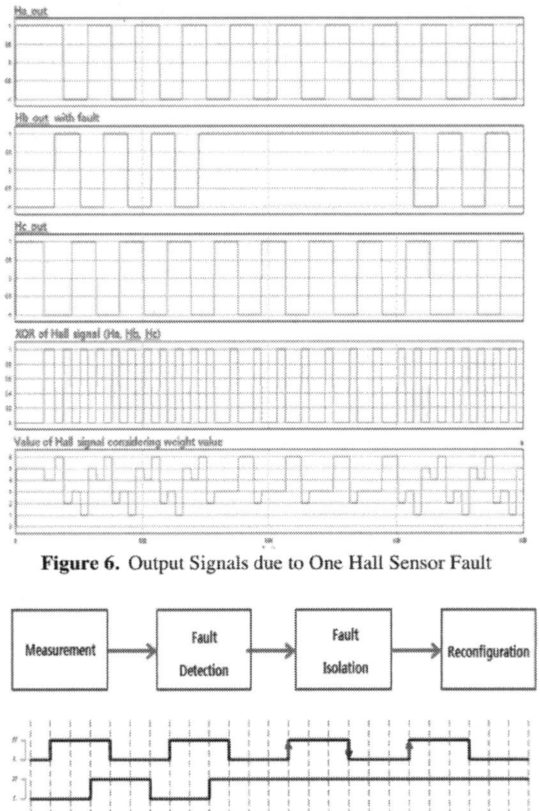

Figure 6. Output Signals due to One Hall Sensor Fault

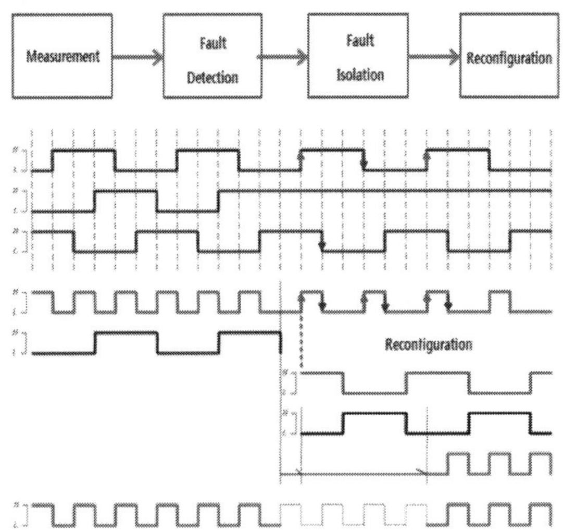

Figure 7. Proposed Hall Sensor Fault-Safety Strategy

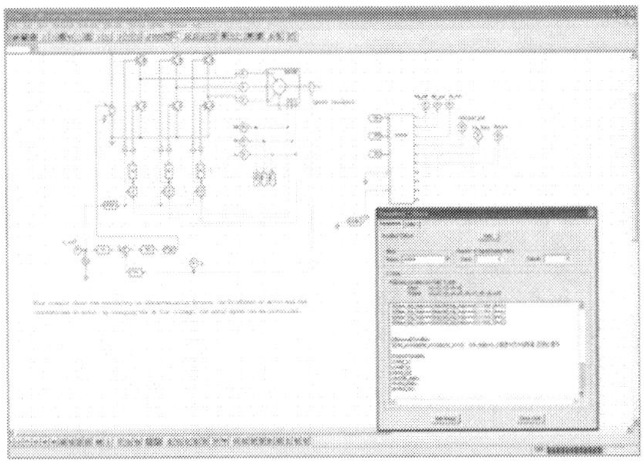

Figure 8. Simulation Schematics based on PowerSim

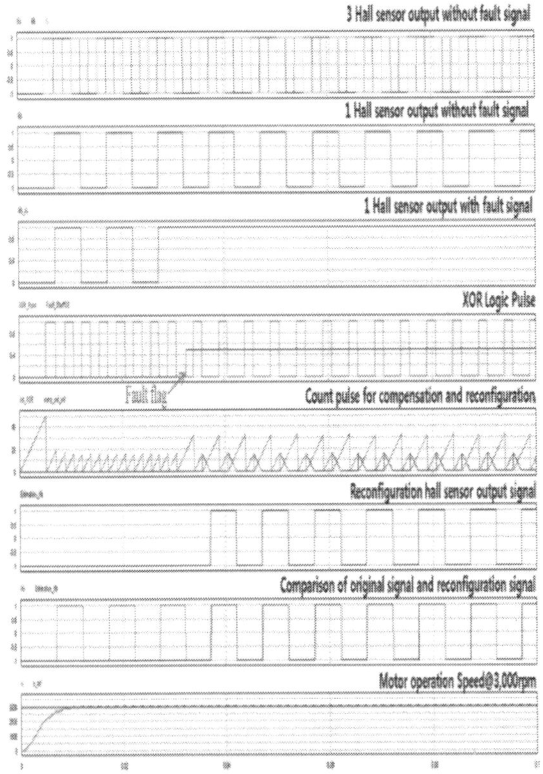

Figure 9. Simulation results

Fig. 9 shows the simulation results. As mentioned above, the operation rule check is executed to detect the fault condition. When fault is occurred by one of three hall sensor, the other two hall sensor output and count pulse train are used to rebuild the proper signal.

The proposed method can execute fault-detection and reconfiguration within 1 electrical cycle like Fig. 9 and therefore it has no interference for motor operation.

Conclusion

In this paper, the hall sensor output fault-tolerant control strategy based on signal analysis and the proposed compensation logic is suggested. The main logic that allows the operation of BLDC motor drive with one faulty hall position sensor is proposed. Fault condition is detected when the hall sensor holds continuously at the low or high level.

The proposed algorithm was verified by PSIM. As a result, it is shown that the proposed method can execute fault-detection and reconfiguration within 1 electrical cycle and therefore it has very short of reaction time and no interference for motor operation.

But the proposed method does not allow start-up and near zero speed operation of a BLDC motor, because it is impossible to measure the pulse length on the other two hall sensors when the motor does not rotating.

Acknowledgement

This project is supported and funded by Daegu-Kyoungbuk Institute for Regional Program Evaluation in 2015. Special thanks for STARGROUP INDUSTRY LTD in South of Korea.

References

1. X.-Q, “Fault detection & diagnosis of pm dc m-otor based on parameter estimation & neural net-work” , IEEE transactions on Industrial Elec., Vol.47, no. 5, pp.1021-1030, 2000
2. S. Mondal, “Unknown input high gain observer for fault detection & isolation of uncertain system” , Engineering Letters, vol.17, no.2, pp.121-127,2009
3. A. Tashakori, “A Simple Fault Tolerant Control System for Hall Effect Sensors Failure of BLDC Motor,” IEEE 8th Conference on industrial Electronics and Applications, 2013
4. N. Rajendran, “A Control Reconfiguration Strategy for Hall Sensor FTC in BLDC Motor,” International Journal of Scientific & Engineering Research, Vol.5, Issue 4, April, 2014
5. Song Ziyou, “Rule-Based Fault diagnosis of Hall Sensors and Fault-Tolerant Control of PMSM” Chinese Journal of Mechanical Engineering, Vol.26, Issue 4, 2013
6. Z. Wang, R. Schittenhelm, M. Borsdorf, and S. Rinderknecht, “Application of augmented observer for fault diagnosis in rotor systems,” Engineering Letters, vol. 21, no. 1, pp. 10–17, 2013.
7. Y.-S. Jeong, S.-K. Sul, S. Schulz, and N. Patel, “Fault detection and faulttolerant control of interior permanent-magnet motor drive system for electric vehicle,” IEEE Transactions on Industry Applications, vol. 41, no. 1, pp. 46–51, 2005.
8. L. Wang, J. Liu, and X. Wu, “Fault analysis on driving motors of lunar rover wheels,” in Proceeding of the International Conference on Electrical Machines and Systems, ICEMS 2011, (Beijing, China), Aug 2011.

A Novel Design of Needle Aspiration Biopsy Monitoring Instrument (NAOMI) Tested on a Low Cost Chest Phantom

Rino Ferdian Surakusumah[1], Muhammad Qurhanul Rizqie[1], Dwichayo Herlambang Prabowo Hidayat[3], Eko Supriyanto[2], and Muhammad Haikal Satria[2]

[1]*Faculty of Biosciences and Medical Engineering, Universiti Teknologi Malaysia, Johor Bahru, 81310 Malaysia*
[2]*IJN-UTM Cardiovascular Engineering Center, Institute of Human Centered Engineering, Universiti Teknologi Malaysia, Johor Bahru, 81310 Malaysia*
[3]*Institut Teknologi Bandung, Bandung, West Java, 40132 Indonesia*

Abstract. Needle biopsy is a medical intervention method for taking a lung tissue sample that suspected as a cancer. The disadvantage is the physicians directly visualize the anatomical structures in an open surgery for lung cancer biopsy procedure. There is a need to develop an instrument that may help the physician to guarantee the accuracy and efficiency while performing needle aspiration biopsy. Therefore, a needle aspiration biopsy monitoring instrument or named as NAOMI is proposed. It consists of a microcontroller system, an IMU sensor, an ultrasonic ranging module, a bluetooth module, and a 9V lithium battery. The experimental testing consist of performance testing, functional testing using chest phantom, and user acceptances. The results showed that the NAOMI improve the accuracy and efficiency while performing the needle biopsy operation.

1 Intoruction

Needle aspiration biopsy is a one of medical intervention method for taking a lung tissue sample that suspected as a cancer from a medical imaging procedure. It uses a long needle or known as biopsy needle inserted through the chest wall to the target of lung tissue sample. This method is used if the abnormal lung tissue is located close to the chest wall. A computed tomography (CT) scan, an ultrasound, or fluoroscopies are usually used to guide the needle to the abnormal tissue [1].

The disadvantage of these mentioned methods is the physicians directly visualize the anatomical structures in an open surgery for lung cancer biopsy procedure. The physician mentally recreates the spatial relationship between the surgical tools and anatomy based on the single or multiple 2D images. The outcome is highly dependent upon the physicians skills and experiences to mentally reconstruct a 3D scene, which in turn leading to inter and intra-observer variability. Accuracy and efficiency are predominant doubt in these cases. Besides, existing Computed Tomography (CT) Fluoroscopy also exposes both physician and patients to high doses of radiation, the longer time the procedure the greater damage may incurred [2], [3].

From this background, there is a need to develop an instrument that may help the physician to guarantee the accuracy and efficiency while performing needle aspiration biopsy. Therefore, a needle aspiration biopsy monitoring instrument or named as NAOMI is proposed. It is integrated with an image guided navigation system and aims to obtain orientation and depth information of the biopsy needle. It may shorten the learning period required for minimally invasive procedures and reduce the variability of the outcome, narrowing the gap between exceptional and standard practice without exposing to any doses of radiations. They may enable new minimally invasive procedures allowing physicians to perform procedures that were previously considered too dangerous.

2 Material and Method

2.1 Concept

In the last few years, there are some research that were conducted to enhance the accuracy of lung needle biopsy operation. In 2013, Zhou et al. [4], developed a robotic biopsy system that has capability to adapted to the patient breathing motion pattern. It obtained several images from the patient lung in breathing cycle then combined with a camera that recorded the patient skin surface to recognized breathing phase. In 2014, Surakusumah, RF et al. [5], [6], developed a novel design of flexible bronchoscopy for lung bronchoscopic biopsy using soft actuator mechanism. In 2015, Yaffe, D et al. [7], developed a novel technique for CT-guided transthoracic biopsy of lung lesions. It used a coaxial CT-guided TNAB while a 22-gauge guide wire was used to locate the lesion of the biopsy accurately. Based on authors knowledge, there was no study that improved the accuracy using orientation combined with depth measurement approach. Therefore, a Needle Aspiration Biopsy Monitoring Instriment or named as NAOMI that can guide a physician to know orientation and depth of a biopsy needle is proposed.

Based on study from Fadzil et al. [8] on 2015, orientation can be measured using Inertial Measurement Unit (IMU) sensor. It is relatively small and does not require sensor setup in different distances compared to other accelerometer system. Based on Dey, R et al. [9] who studied about expansion measurement of digital chest, an ultrasonic sensor was used to detected distance from any object to the surface of human body. Thus, an IMU and an ultrasonic sensors was intergrated to the NAOMI.

From the information in Haydock, MD et al. [10], there is a requirement for medical instruments to have evaluation of user acceptances. It is related to the ergonomic design concept of the medical instruments. So, the NAOMI design followed the ergonomic concept and be evaluated qualitatively by the possible users such as ordinary person, radiologist, and medical doctors. To measured the performance of the NAOMI, a lung nodule phantom that developed by Yusof NSM et al. [11] was used.

2.2 Design and fabrication

Design of the NAOMI consists of embedded system, and casing. The embedded system integrate a microcontroller system (ARDUINO Mini Pro), an IMU sensor (The InvenSense MPU-6050), an ultrasonic ranging module (HC-SR04), a bluetooth module (HC-06), and a 9V lithium battery. The IMU sensor obtained pitch and yaw position datas of the NAOMI with 1o resolution. The ultrasonic ranging module obtained the distance between the NAOMI front side (base point of the biopsy needle) and patient chest surface with 3mm resolution. A bluetooth module bridge the data transfer from the NAOMI to a 2D X-ray Lung Biopsy Software. In order to powered the NAOMI embedded system, the 9V lithium battery was connected to RAW pin of the microcontroller system that contain a feature to accepted an unregulated power supply. The NAOMI embedded system schematic is shown in Fig. 1. The fabrication of the PCB and component setup was conducted.

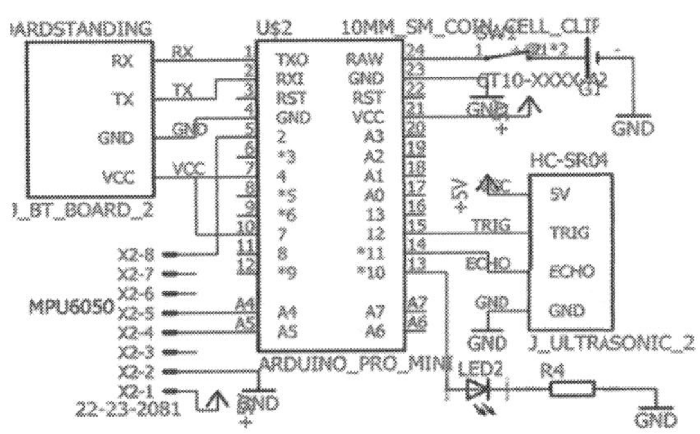

Figure 1. The NAOMI embedded system schematic.

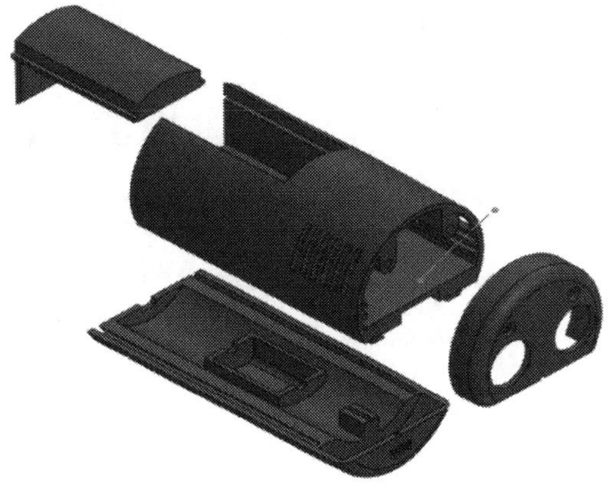

Figure 2. The NAOMI casing 3D design

The NAOMI casing was designed using rapid prototyping (Solidworks) software. It has cylindrical shaped with 50mm diameter and 105 mm length size in order to provide an optimum handling for the user. A placement of biopsy needle was designed in such a way that the user is able to replace the biopsy needle. As a note, this design only suitable for NAC-1825B-BX type of biopsy needle. The NAOMI casing 3D design is shown in Fig. 2. It was fabricated using 3D printer machine (MakerBot) with Acrylonitrile Butadiene Styrene (ABS) material.

2.3 Experiment testing

The testing of the NAOMI consisted of performance, functional, and user acceptance. The performance consisted of orientation and depth measurement. Orientation measurement was conducted using a camera that visualized the rotation in row and pitch of the NAOMI in front of an angle measurement sheet. As comparations, the NAOMI was mounted to a motorized orientation measurement device that gave the orientation data from the output of motors and also the orientation data which transferred to the 2D X-Ray Lung Biopsy software. The depth was measured using the camera that visualized the distance between the NAOMI to the object in front of a milimeter ruler. The depth data that transferred to the 2D X-Ray Lung Biopsy software was used for the comparation. The NAOMI performance testing setup is shown in Fig. 3.

In conjuction with 2D X-Ray Lung Biopsy software and using a lung nodule phantom as an object experiment, a functional system was tested. After an optimum orientation and a depth data of targeted lung nodule was collected by the software, the NAOMI mounted with a biopsy needle was prepared by the user. The tip of the biopsy needle was positioned at a hole of a red tape that indicated the starting point of needle insertion. The NAOMI transfered the biopsy needle orientation and depth information to the software. The user adjusted the NAOMI orientation in such a way that the actual NAOMI orientation is similar with the orientation that visualized in the software While holding the NAOMI, the user inserted the biopsy needle into the

DOI: 10.1051/matecconf/20165901008

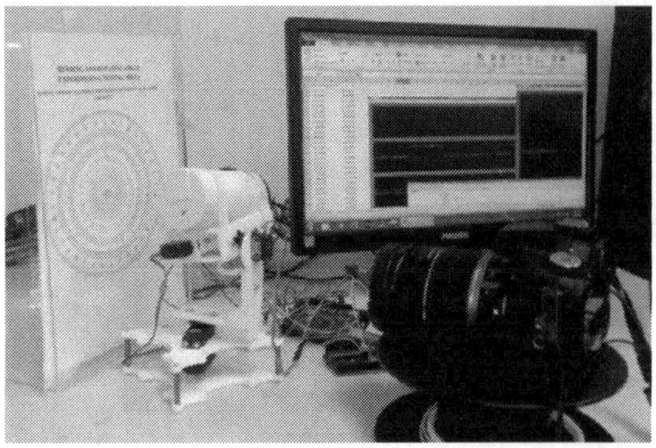

(a)

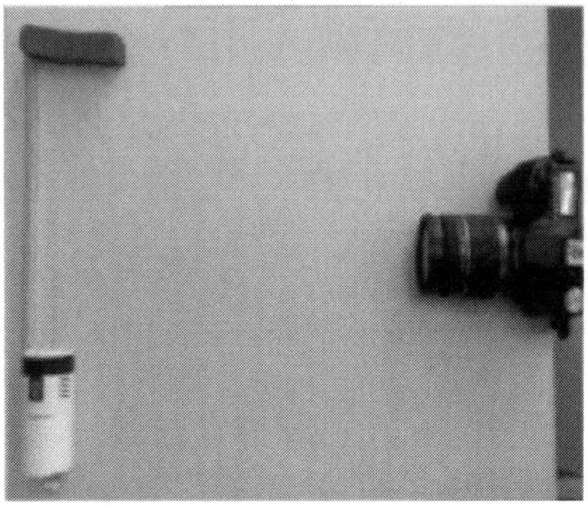

(b)

Figure 3. The NAOMI performance testing setup: (a) Orientation, (b) depth

lung nodule phantom until the depth reach the target that visualized in the software. After the biopsy needle reach the target, a sample taker needle was inserted through the biopsy needle in such a way that it reached the lung nodule and a sample will be taken out from the biopsy needle. Because of the lung nodule phantom provided different material of the lung nodule compared to the other tissue and by some numbers of experiment, the data of how accurate and fast the NAOMI guide the biopsy needle to take the tissue sample can be measured. The NAOMI functional testing setup is shown in Fig. 4.

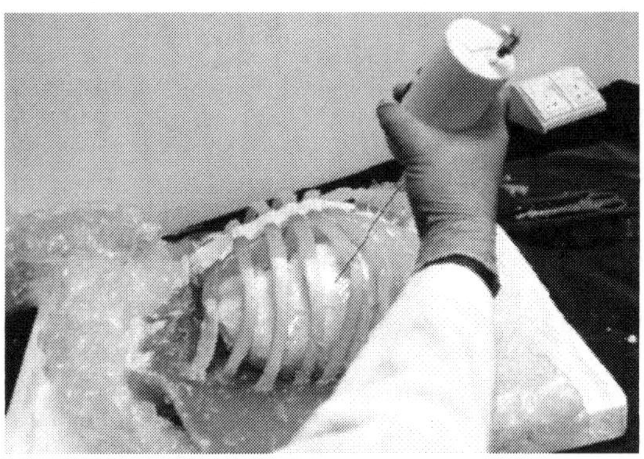

Figure 3. The NAOMI functional testing setup

For obtained the evaluation of user acceptances and ergonomic value of the NAOMI, a questionnaire was designed and distributed to 5 ordinary persons, 5 radiologists, and 5 medical doctors. The questionnaire consists of user personal information such as experiences in radiology and pulmonology, and procedural information such as evaluation in prototype demo, ergonomic design, and accuracy.

3 Result and discussion

Fig. 3 and Fig. 4 shows the final NAOMI prototype, which has a total length of 105mm and a diameter of 50mm. The lower part consists of a biopsy needle placement box (BNPB) that fit to the NAC-1825B-BX main housing part. The BNPB positioned about 32mm from the back surface of NAOMI in order to set the difference between biopsy needle tip and sample-taker needle position of 3mm. This is designed so that tissue sample cutting may be operated effectively. To combine the lower part and the upper part, a specific-shape connector is designed in both connecting surface of the part. This prototype configuration offered the optimum in user-friendliness and needle biopsy operation.

3.1. Performance result

The graph of orientation measurement experimental result is shown in Fig. 5. The graph consists of testing number of different angle position of pitch and yaw and different measurement methods which are from camera, motorized orientation sensor, and NAOMI depth sensor. The pitch measurement refer to 0o-350o line while the yaw measurement refer to 350o-0o. Using mean square error (MSE) analysis, the test show an optimum accuracy with 0.063 error.

The graph of depth measurement experimental result is shown in Fig. 6. The graph consists of testing number of different distance from front side of the NAOMI until the chest surface in milimeters. Using mean square error (MSE) analysis, the test show an optimum accuracy with 10 error.

3.2 Functional result

From 10 trials of the lung needle biopsy operation using NAOMI in conjuction with the 2D X-ray lung needle biopsy software, the result shows that there was not more than 1 minutes with 100% successful rate in obtaining the tissue sample at the first trial

3.3 User acceptance

Meetings with physicians provided important feedback concerning functional procedure and power handling issue of the NAOMI. The physicians was asked to operate an in-vitro lung needle biopsy procedure with the chest phantom without use NAOMI and use NAOMI in conjunction with 2D X-ray Lung Biopsy software. The NAOMI is improve the accuracy of the physician to conduct needle biopsy operation. It is lightweight and has a depression in the handle during operation. This was introduced after learning that the way physician conducted the lung neddle biopsy procedure without using NAOMI and with using NAOMI in conjunction with 2D X-ray Lung Biopsy software. The goal was to provide the NAOMI as ergonomic as well so the physicians would feel comfortable using the device.

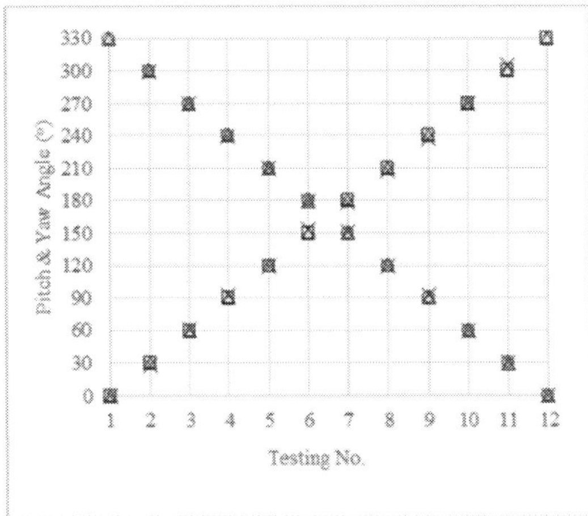

Figure 5. Orientation Measurement Experimental Result

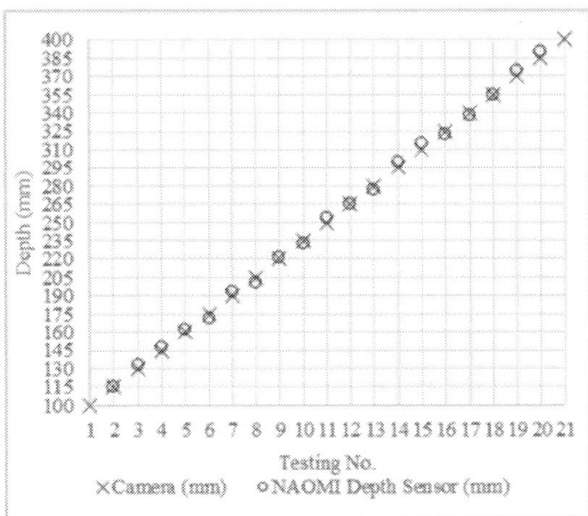

Figure 6. Depth Measurement Experimental Result

3.4 Discussion

Related to the performance of the NAOMI, based on the depth measurement result, the error was quite significant. It is expected because of the ultrasonic ranging module resolution and the orientation between the NAOMI and the chest surface. A higher resolution distance sensor may improve the accuracy. The orientation measurement already performed an optimum accuracy in determining the position of the NAOMI.

Related to the functional measurement and the user acceptances, the NAOMI in conjuction with 2D X-ray Lung Biopsy Software is able to guide the physiscian to orientate the biopsy needle in such a way that has optimum accuracy to take the tissue sample. It also able to measure the depth of the biopsy needle. However, during the operation, a skill and a human ability factor of the physician still importantly needed. When the physician is not well-trained enough for stabilizing the orientation of the NAOMI until it reach the target, the target will be possibly different. An equipment that able to fix the position of the biopsy needle or the use of robot may solve the issue. The lower part of the NAOMI may also redesign in such a way that it fit to other type of biopsy needle..

4 Conclusion

A novel design of Needle Aspiration Biopsy Monitoring Instrument (NAOMI) was developed for monitoring the lung needle biopsy by providing the orientation and depth information in in-vitro study. With considering the ergonomic design and using the NAOMI in conjuction with 2D X-Ray Lung Biopsy software, it is expected to help the physicians to improve accuracy and efficiency while performing the needle biopsy operation. In the future, further study on optimizing the design, accuracy, and the experiment using cadaver and also the clinical testing will be conducted to justify the development of the NAOMI.

Acknowledgement

This research was supported by the Ministry of Science, Technology and Innovation (MOSTI) Malaysia Grant No. 02-01-06-SF1191 (R.J130000.7909.4S091).

References

1. L. G. Collins, C. Haines, R. Perkel, and R. E. Enck, "Lung cancer: Diagnosis and management," Am. Fam. Physician, vol. 75, no. 1, pp. 56–63, 2007.
2. Y. Y. Cao, N. Fan, F. Xing, L. Y. Xu, Y. J. Qu, and M. Y. Liao, "Computed tomography-guided cutting needle pleural biopsy: Accuracy and complications," Exp. Ther. Med., vol. 9, no. 1, pp. 262–266, 2015.
3. C. M. Ogilvie, J. T. Torbert, J. L. Finstein, E. J. Fox, and R. D. Lackman, "Clinical utility of percutaneous biopsies of musculoskeletal tumors.," Clin. Orthop. Relat. Res., vol. 450, no. 450, pp. 95–100, 2006.
4. Y. Zhou, K. Thiruvalluvan, L. Krzeminski, W. H. Moore, Z. Xu, and Z. Liang, "CT-guided robotic needle biopsy of lung nodules with respiratory motion - Experimental system and preliminary test," Int. J. Med. Robot. Comput. Assist. Surg., vol. 9, no. 3, pp. 317–330, 2013.

5. M. Automation, R. F. Surakusumah, D. Ekashanti, O. Dewi, E. Supriyanto, and M. Engineering, "Development of a Half Sphere Bending Soft Actuator for Flexible Bronchoscope Movement," pp. 120–125, 2014.
6. A. M. Faudzi, R. F. Surakusumah, D. E. O. Dewi, I. N. A. M. Nordin, and M. R. M. Razif, "Development of Flexible Bronchoscope Device Using Soft Actuator," in Medical Imaging Technology, Springer Singapore, 2015, pp. 223–241.
7. D. Yaffe, M. Koslow, H. Haskiya, and D. Shitrit, "A novel technique for CT-guided transthoracic biopsy of lung lesions: improved biopsy accuracy and safety," Eur. Radiol., vol. 25, no. 11, pp. 3354–3360, 2015.
8. D. E. Octorina Dewi, M. Mohd. Fadzil, A. `Athif Mohd. Faudzi, E. Supriyanto, and K. W. Lai, "Position Tracking Systems for Ultrasound Imaging: A Survey," in Medical Imaging Technology Reviews and Computational Applications, W. K. Lai and E. D. Octorina Dewi, Eds. Singapore: Springer Singapore, 2015, pp. 57–89.
9. R. Dey, U. Thakur, L. Sunny, L. D'Almeida, and K. Chakravarty, "E-Health and Bioengineering Conference (EHB), 2015," E-Health and Bioengineering Conference (EHB), 2015. pp. 1–4, 2015.
10. M. D. Haydock, A. Mittal, C. F. Wilkes, D. H. Lim, E. Broadbent, and J. A. Windsor, "Interaction Between Objective Performance Measures and Subjective User Perceptions in the Evaluation of Medical Devices: a Case Study," Int. J. Technol. Assess. Health Care, vol. 31, no. 05, pp. 297–303, 2015.
11. M. Q. Rizqie, N. S. M. Yusof, R. F. Surakusumah, D. E. O. Dewi, E. Supriyanto, and K. W. Lai, "Review on Image Guided Lung Biopsy," in Medical Imaging Technology, Springer Singapore, 2015, pp. 41–56.

Quaternion-based pseudo kalman filter for wearable inertial/magnetic sensor applications

Jung Keun Lee

Department of Mechanical Engineering, Hankyong National University, Anseong 17579, Korea

Abstract. This paper deals with orientation estimation using miniature inertial/magnetic sensor comprised of a tri-axial rate gyro, a tri-axial accelerometer, and a tri-axial magnetometer. Particularly, a novel quaternion-based pseudo Kalman filter (KF) is proposed by modifying an indirect KF, in order to maximize the computational efficiency and implementation simplicity. In the proposed pseudo KF, time-update process for prediction is based on the quaternion itself, while measurement-update process for correction is performed through the quaternion error. Experimental tests were conducted to verify performance of the proposed algorithm in various dynamic conditions. By designing the pseudo KF structure, matrix operations required in a typical KF are simplified. For instance, the proposed KF does not require the evaluation of the a priori and a posteriori error covariance matrices. Thus, the proposed algorithm achieves higher computational efficiency even than a typical indirect KF, without sacrificing estimation accuracy. Due to its high efficiency, the proposed algorithm can be suitable for battery-powered and low cost processor-based wearable inertial/magnetic sensor applications.

1. Introduction

With the advances in micro-electro-mechanical systems (MEMS) technology, miniature inertial/magnetic sensors have emerged for wearable motion capture technology [1,2]. An inertial/magnetic sensor is typically comprised of a tri-axial rate gyro, a tri-axial accelerometer, and a tri-axial magnetometer, and is used to estimate accurate 3D orientation in an unconstrained environment (e.g., outdoors), associated with physical medicine and rehabilitation applications [3,4].

In terms of the orientation representation, Euler angles [5], quaternion [1,2,6,7], and direction cosine matrix [8] are common. Among them, the quaternion can arguably be considered as the most popular representation for 3D orientation estimation due to its singularity-free aspect and relatively small number of dimensions (i.e., four).

When quaternion is chosen as an orientation representation for a Kalman filter (KF) structure, there are mainly two approaches depending on how the quaternion is represented in the KF state. One approach is the direct KF which designates the quaternion itself as the state. The other approach is an indirect KF which does the quaternion error as the state [9].

With regard to the quaternion-based direct KF, Sabatini [6] proposed an extended KF by applying a first-order Taylor-Mac Laurin expansion to compute the Jacobian matrix, due to nonlinear nature of measurement model. Lee and Park [1] proposed a linear KF that is comprised of a quaternion measurement step and a KF step. The two steps have feedback relationship and enable the KF have a minimum-order structure (i.e., fourth order).

The advantage of an indirect KF over a direct KF is that the state dimension is smaller and its response is fast. Suh [10] proposed a quaternion-based indirect KF with adaptive estimation of external acceleration. Additionally, Suh *et al.* [9] proposed another indirect KF that can discard deteriorated pitch and roll information contained in magnetometer signals.

This paper proposes a quaternion-based pseudo Kalman filter by modifying an indirect KF, in order to maximize the computational efficiency. The proposed KF directly uses the quaternion for time update and then indirectly estimate the quaternion though quaternion error in measurement update. By doing so, the proposed algorithm achieves higher computational efficiency even than a typical indirect KF, without sacrificing estimation accuracy.

2. Method

Sensor Modelling and Quaternion Error. The coordinate transformation of a 3×1 vector **x** between the sensor frame *S* and the inertial frame *I* is

$$^{S}\mathbf{x}={}^{S}_{I}\mathbf{R}\,{}^{I}\mathbf{x} \quad (1)$$

where the left superscripts *S* and *I* of $\mathbf{x}$'s imply that the corresponding vectors are expressed in the sensor and inertial frame coordinates, respectively, and ${}^S_I\mathbf{R}$ is the rotation matrix of the inertial frame *I* with respect to the sensor frame *S*. The rotation matrix ${}^S_I\mathbf{R}$ (or simply $\mathbf{R}$ for convenience) can be expressed in terms of the orientation quaternion $\mathbf{q}=[q_0 \quad \mathbf{e}^T]^T$ [10], i.e.,

$${}^S_I\mathbf{R}=\mathbf{R}(\mathbf{q})=(q_0^2-\mathbf{e}^T\mathbf{e})\mathbf{I}_3+2\mathbf{e}\mathbf{e}^T-2q_0[\mathbf{e}\times]$$
(2)

where q_0 is the scalar part and $\mathbf{e}=[q_1 \quad q_2 \quad q_3]^T$ is the vector part of the quaternion. The 3×3 matrix $[\mathbf{a}\times]$ denotes a skew-symmetric matrix function of $\mathbf{a}$, the so-called cross-product matrix. The proposed algorithm estimates the quaternion error $\mathbf{q}_e \in R^3$. Now, the algorithm uses $\tilde{\mathbf{q}}_e$ to denote a small error in quaternion estimate $\hat{\mathbf{q}}$ so that

$$\mathbf{q}=\hat{\mathbf{q}}\otimes\tilde{\mathbf{q}}_e$$
(3)

where $\tilde{\mathbf{q}}_e=\left[1 \quad \mathbf{q}_e^T\right]^T \in R^4$. The violation of the quaternion constraint (i.e., unity norm) is allowed under the assumption of small error.

Sensor signals from the gyro (*G*), the accelerometer (*A*), and the magnetometer (*M*) are modeled, respectively, as follows [1]:

$$\mathbf{y}_G={}^S\boldsymbol{\omega}+\mathbf{n}_G$$
(4.a)

$$\mathbf{y}_A={}^S\mathbf{g}+{}^S\mathbf{a}+\mathbf{n}_A$$
(4.b)

$$\mathbf{y}_M={}^S\mathbf{m}+\mathbf{n}_M$$
(4.c)

where $\boldsymbol{\omega}$ is the angular velocity; $\mathbf{g}$ is the gravity; $\mathbf{a}$ is the body acceleration; $\mathbf{m}$ is the local magnetic field vector; $\mathbf{n}$'s are the measurement noises which are assumed to be uncorrelated and zero-mean white Gaussian. It is natural that every sensor reading is inherently expressed with respect to the sensor reference frame *S*. By considering Eq. (1), ${}^S\mathbf{g}$ and ${}^S\mathbf{m}$ can be rewritten as $\mathbf{R}(\mathbf{q})^I\mathbf{g}$ and $\mathbf{R}(\mathbf{q})^I\mathbf{m}$, respectively. Note that ${}^I\mathbf{g}$ and ${}^I\mathbf{m}$ are known vectors. The proposed algorithm does not take sensor biases into consideration. In Eq. (4.b), the body acceleration ${}^S\mathbf{a}$ is modeled as a first order low-pass filtered white noise process as in [11] i.e.

$${}^S\mathbf{a}_t=c_a\,{}^S\mathbf{a}_{t-1}+\boldsymbol{\varepsilon}_t$$
(5)

where c_a is a constant to be set and $\boldsymbol{\varepsilon}_t$ is the acceleration model error.

Time Update Process. With regards to the state equation of quaternion error, the time derivative of error, $\dot{\mathbf{q}}_e$, can be expressed as [10]

$$\dot{\mathbf{q}}_e \approx \mathbf{A}_e\mathbf{q}_e+\mathbf{w}_e$$
(6)

where $\mathbf{A}_e=-[\mathbf{y}_G\times]$ and the zero mean white noise $\mathbf{w}_e=-\frac{1}{2}\mathbf{n}_G$ with covariance of

$$\mathbf{Q}_e=E\left[\mathbf{w}_e\mathbf{w}_e^T\right]=\frac{1}{4}\mathbf{R}_G.$$
(7)

$\mathbf{R}_G$ is set equal to $\sigma_G^2\mathbf{I}_3$ where the gyro noise variance σ_G^2 is the gyro noise variance. By time-discretizing (6), we obtain the time process equation for quaternion error, i.e.,

$$\mathbf{q}_{e,t}=\mathbf{A}_{t-1}\mathbf{q}_{e,t-1}+\mathbf{w}_{t-1}$$
(8)

where $\mathbf{A}_{t-1}=\exp(\mathbf{A}_{e,t-1}\Delta t)\approx\mathbf{I}+\mathbf{A}_{e,t-1}\Delta t$ and Δt is the time stepsize. Process noise covariance matrix $\mathbf{Q}_{k-1}$ is then defined by

$$\mathbf{Q}_{t-1}=E\left[\mathbf{w}_{t-1}\mathbf{w}_{t-1}^T\right]=\int_{\tau=0}^{\Delta t}\exp(\mathbf{A}_e\tau)\mathbf{Q}_e\exp(\mathbf{A}_e^T\tau)\,d\tau \qquad (9)$$

By applying $\mathbf{A}_e=-\mathbf{A}_e^T$ and discarding high order terms, $\mathbf{Q}_{t-1}$ in Eq. (9) is approximated as

$$\mathbf{Q}_{t-1}=\mathbf{Q}_e\Delta t+\frac{1}{2}(\mathbf{A}_e\mathbf{Q}_e-\mathbf{Q}_e\mathbf{A}_e)\Delta t^2.$$
(10)

In comparison to the *quaternion error process* in (8), quaternion process equation is derived from quaternion strapdown integration equation, i.e.,

$$\mathbf{q}_t=\mathbf{q}_{t-1}+\Delta t\dot{\mathbf{q}}_{t-1}.$$
(11)

The time-derivative of the quaternion $\dot{\mathbf{q}}_t$ is [2]:

$$\dot{\mathbf{q}}=\boldsymbol{\Omega}(\boldsymbol{\omega})\mathbf{q}$$
(12)

where

$$\boldsymbol{\Omega}(\boldsymbol{\omega}) = \frac{1}{2}\begin{bmatrix} \mathbf{0} & -\boldsymbol{\omega}^{T} \\ \boldsymbol{\omega} & -[\boldsymbol{\omega}\times] \end{bmatrix}.$$

(13)

Since $\boldsymbol{\omega}$ is unavailable, Eq. (11) is rewritten using the actual gyro output, $\mathbf{y}_G\ (= \boldsymbol{\omega} + \mathbf{n}_G)$. Then, this leads to the time process equation for quaternion:

$$\mathbf{q}_t = \boldsymbol{\Phi}_{t-1}\mathbf{q}_{t-1} + \mathbf{w}_{\mathbf{q},t-1}$$

(14)

where

$$\boldsymbol{\Phi}_{t-1} = \mathbf{I} + \Delta t\,\boldsymbol{\Omega}(\mathbf{y}_{G,t-1})$$

(15)

$$\mathbf{w}_{\mathbf{q},t-1} = -\frac{\Delta t}{2}\begin{bmatrix} -\mathbf{e}^{T} \\ [\mathbf{e}\times] + q_0\mathbf{I} \end{bmatrix}\mathbf{n}_G.$$

(16)

Measurement Update Process. From Eq. (3), we have

$$\mathbf{R}(\mathbf{q}) = \mathbf{R}(\tilde{\mathbf{q}}_e)\mathbf{R}(\hat{\mathbf{q}})$$

(17)

where $\mathbf{R}(\tilde{\mathbf{q}}_e) \approx \mathbf{I} - 2[\mathbf{q}_e\times]$ [9]. In order to insert the body acceleration model Eq. (5) into the accelerometer model Eq. (4.b), we need to define the error of the predicted acceleration as

$${}^{S}\mathbf{a}_{\varepsilon,t}^{-} = {}^{S}\mathbf{a}_{t}^{-} - {}^{S}\mathbf{a}_{t}$$

(18)

where the minus superscript denotes the *a priori* (or predicted) estimate, while the plus superscript in Eq. (18) below denotes the *a posteriori* (or corrected) estimate [11]. In Eq. (18), the predicted body acceleration of the current time step ${}^{S}\mathbf{a}_t^{-}$ is

$${}^{S}\mathbf{a}_{t}^{-} = c_a\,{}^{S}\mathbf{a}_{t-1}^{+}.$$

(19)

By inserting Eqs. (17) ~ (19) to the accelerometer model Eq. (4.b) and by inserting Eq. (17) to the magnetometer model Eq. (4.c), we can derive the measurement model equation:

$$\mathbf{z}_t = \mathbf{H}_t\mathbf{q}_{e,t} + \mathbf{v}_t$$

(20)

where the measurement vector $\mathbf{z}_t$, observation matrix $\mathbf{H}$, and measurement noise $\mathbf{v}$ are

$$\mathbf{z}_t = \begin{bmatrix} \mathbf{y}_A - \mathbf{R}(\hat{\mathbf{q}})^{I}\mathbf{g} - c_a\,{}^{S}\mathbf{a}_{t-1}^{+} \\ \mathbf{y}_M - \mathbf{R}(\hat{\mathbf{q}})^{I}\mathbf{m} \end{bmatrix}$$

(21)

$$\mathbf{H} = 2\begin{bmatrix} [\mathbf{R}(\hat{\mathbf{q}})^{I}\mathbf{g}]\times \\ [\mathbf{R}(\hat{\mathbf{q}})^{I}\mathbf{m}]\times \end{bmatrix}, \text{ and}$$

(22)

$$\mathbf{v}_t = \begin{bmatrix} -{}^{S}\mathbf{a}_{\varepsilon,t}^{-} + \mathbf{n}_A \\ \mathbf{n}_M \end{bmatrix}$$

(23)

, respectively. Then, the measurement noise covariance matrix $\mathbf{M}_t$ is defined by

$$\mathbf{M}_t = E\left[\mathbf{v}_t\mathbf{v}_t^{T}\right].$$

(24)

Since ${}^{S}\mathbf{a}_{\varepsilon,t}^{-}$ is uncorrelated to $\mathbf{n}_A$, Eq. (24) can be divided into

$$\mathbf{M}_t = \boldsymbol{\Sigma}_{acc} + \boldsymbol{\Sigma}_A$$

(25)

where $\boldsymbol{\Sigma}_{acc}$ is the covariance matrix of the body acceleration vector and is set as $\boldsymbol{\Sigma}_{acc} = 3^{-1}c_a^2\left\|{}^{S}\mathbf{a}_{t-1}^{+}\right\|^2\mathbf{I}$, and $\boldsymbol{\Sigma}_A$ is the covariance matrix of the accelerometer's measurement noise and is set as $\sigma_A^2\mathbf{I}$ (similarly with $\boldsymbol{\Sigma}_G$) where σ_A^2 is the accelerometer noise variance [11].

Summary of Pseudo Kalman Filter. The procedure of the proposed pseudo KF can be summarized as follows.

1) Compute the *a priori* quaternion estimate:

$$\hat{\mathbf{q}}_t^{-} = \boldsymbol{\Phi}_{t-1}\mathbf{q}_{t-1}^{+}.$$

(26)

2) Compute the *a* posteriori quaternion error estimate:

$$\mathbf{q}_{e,t}^{+} = \mathbf{Q}_{t-1}\mathbf{H}^{T}(\mathbf{H}\mathbf{Q}_{t-1}\mathbf{H}^{T} + \mathbf{M}_t)^{-1}\mathbf{z}_t.$$

(27)

3) Compute the *a* posteriori quaternion estimate:

$$\hat{\mathbf{q}}_t^{+} = \hat{\mathbf{q}}_t^{-} \otimes \tilde{\mathbf{q}}_{e,t}^{+}.$$

(28)

4) Compute the body acceleration estimate:

$${}^{S}\mathbf{a}_t^{+} = \mathbf{y}_{A,t} - \mathbf{R}(\hat{\mathbf{q}}_t^{+})^{I}\mathbf{g}.$$

(29)

Note that, in the proposed pseudo KF, time-update process for prediction is based on the quaternion itself, while measurement-update process for correction is performed through the quaternion error. Thus, Eq. (26) of Step 1 is from Eq. (11) based on the quaternion process while Eq. (27) of Step 2 is from Eq. (10) associated with the quaternion error.

In Step 2, the observation matrix $\mathbf{H}$ and the measurement vector $\mathbf{z}_t$ are functions of the *a priori* quaternion estimate $\hat{\mathbf{q}}_t^-$. In our method, the *a priori* error covariance matrix (usually denoted by $\mathbf{P}_t^-$) is set to the process noise covariance matrix $\mathbf{Q}_{t-1}$. By doing so, $\mathbf{Q}_{t-1}\mathbf{H}^T(\mathbf{H}\mathbf{Q}_{t-1}\mathbf{H}^T + \mathbf{M}_t)^{-1}$ in Eq. (27) is no other than the Kalman gain (usually denoted by $\mathbf{K}_t$) and the proposed KF does not require the evaluation of the *a posteriori* error covariance matrix (usually denoted by $\mathbf{P}_t^+$). Also, it is notable that the measurement vector $\mathbf{z}_t$ plays a role as residual (also known as measurement innovation). The overall structure of the proposed algorithm is illustrated in Fig. 1.

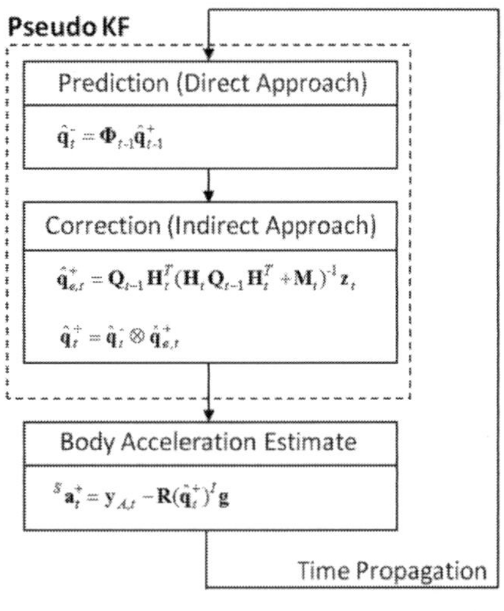

Figure 1. Flowchart of the proposed pseudo Kalman filter

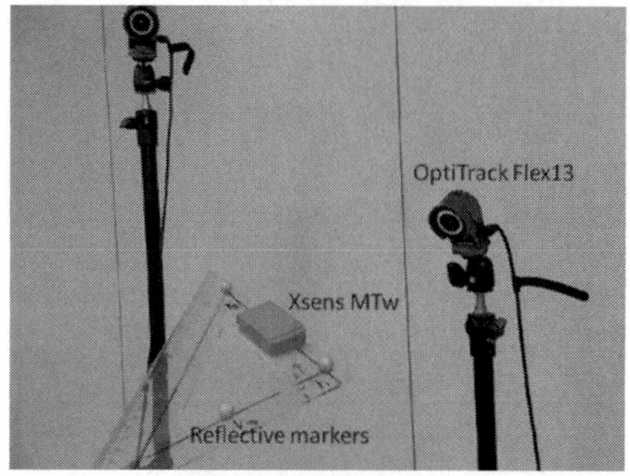

Figure 2. Experimental setup for sensor measurement

3. Experimental Results

Test Setup and Various Test Conditions. To verify the proposed algorithm, an MTw inertial/magnetic sensor (from Xsens technologies B.V., Netherlands) was used. It provided measurement inputs to the proposed algorithm at 120 Hz sampling rate. Also, in order to check the orientation estimation accuracy, OptiTrack Flex13 3D optical tracking system (from NaturalPoint, Inc. USA) was used with the same sampling rate. The orientation from the optical tracker was used as a truth reference for the subsequent orientation estimation tests (see Fig. 2).

To study the proposed algorithm's performance under various dynamic conditions, three different tests in terms of the magnitude and duration of the body accelerations were performed. All tests involved rotating the sensor manually in a random manner by hand. Test A was carried out with rarely accelerated movements: the averaged magnitude of the reference body acceleration, $\overline{\|\mathbf{a}_{ref}\|}$, of 1.16 m/s^2 with a maximum of 5.16 m/s^2. Test B was performed with a considerable amount of body accelerations: $\overline{\|\mathbf{a}_{ref}\|}$ of 6.36 m/s^2 with a maximum of 18.94 m/s^2. Test C involved much higher body acceleration magnitudes than Test B: $\overline{\|\mathbf{a}_{ref}\|}$ of 23.05 m/s^2 with a maximum of 87.66 m/s^2.

For each of the above tests, the orientation was estimated using three different estimation methods. Method 1 is the extended quaternion-based direct KF proposed by Sabatini (2006) [6]. Method 2 is the quaternion-based indirect KF proposed by Suh (2010) [10]. Method 3 is the proposed pseudo KF algorithm which is a sort of modified indirect KF. The estimation accuracy and computational cost were investigated.

Results of Estimation **Accuracy and Computational Efficiency**. In Test A, all of the three methods produced accurate results (i.e., <2°) since the test condition was desirable for inertial sensing-based orientation estimation (see Table 1).

In Test B, the estimation accuracies were decreased due to the higher body accelerations. However, all of the three methods have body acceleration compensating mechanism. This protects the algrotihms from severe deterioration of estimation accuracy (see Table 1 and Fig. 3). Particularly, Method 1 outperformed Methods 2 and 3.

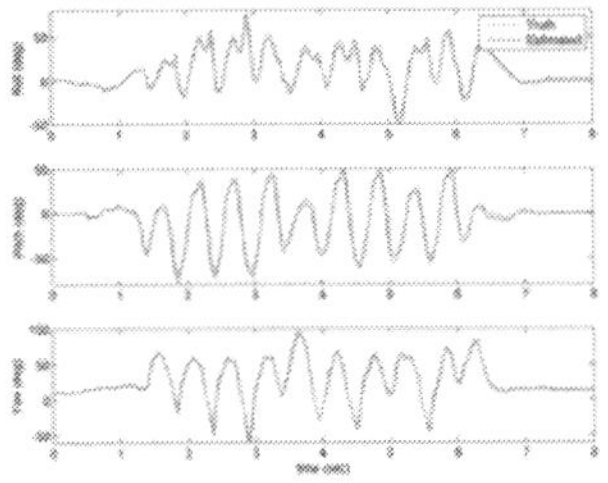

Figure 3.. Results of Test B from Method 3: (dotted) truth orientations, (solid) estimated orientations

In Test C, which involves the most severe condition (see Fig. 4.a for the magnitude of the applied body acceleration), all methods could not produce less than 4° accuracy in average. It was expectable due to the prolonged exposure to the accelerated conditions as well as the acceleration changes that were repetitively applied. However, Method 3 (the proposed method) performed the best among the three methods, which may imply the excellence of the acceleration model of the proposed method (see Table 1 and Fig. 4.b). With regard to the accuracy of acceleration estimation that is through the orientation estimation, the proposed method showed fairly accurate estimation, i.e., average root mean square error (RMSE) of 1.70 m/s^2 (see Fig. 4.a). In overall (based on the three tests), it can be said that the three methods have a similar degree of accuracy in orientation estimation.

In terms of the computational efficiency, the computation times for each time step for the conventional methods (Methods 1 and 2) and the proposed method (Method 3) on a 3.6 GHz personal computer (Intel® Core™ i7 processor) are given in Table 2. The proposed method was about two times faster than the conventional methods, due to the simplified pseudo KF procedures.

Table 1. Root mean square errors of orientations (°)

		Roll	Pitch	Yaw	Average
Test A	Method1	0.82	0.99	3.66	1.82
	Method 2	0.85	1.15	1.77	1.26
	Method 3	1.79	1.77	1.08	1.55
Test B	Method 1	1.09	0.96	1.48	1.18
	Method 2	2.59	2.37	4.90	3.29
	Method 3	3.22	3.92	2.67	3.27
Test C	Method 1	5.76	8.06	5.18	6.33
	Method 2	6.45	8.23	5.62	6.77
	Method 3	5.58	3.81	4.85	4.75

Table 2. Computation time per one time step.

	Method 1	Method 2	Method 3
Computation time (μs)	294	218	103
Ratio (%)†	288	214	100

† Ratio = Method x / Method 3

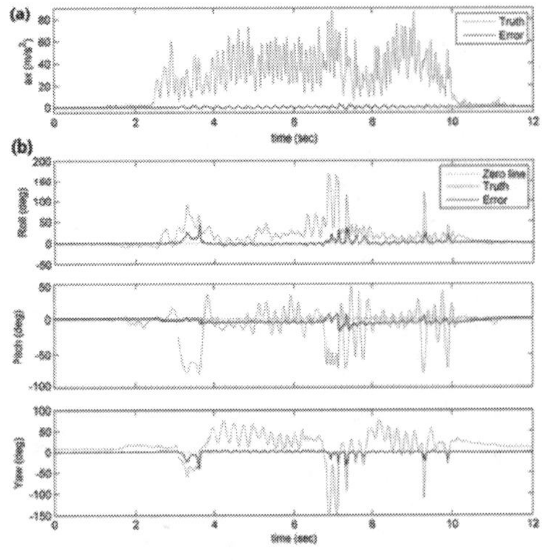

Figure 4.. Results of Test C from Method 3: (a) (dotted) truth acceleration norm (solid) estimated acceleration nor error; (b) (dotted) truth orientations (solid) estimated orientation error

4. Discussions and Conclusions

It is important to note that the proposed pseudo Kalman filter is realized only because the Kalman state is 'small' quaternion error. In measurement update procedure of a typical KF, *a posteriori* state estimate is determined by optimally weighting *a priori* state estimate and measurement vector depending on their reliabilities (represented by their covariance matrices). In the proposed method, the *a priori* state estimate is not used. This modification is possible because measurement vector in the proposed method is small and thus can play a role as residual vector. As a result, the simplified KF structure is achieved.

With regard to body acceleration compensating mechanism, concepts of acceleration compensation of Methods 1 and 2 are different from that of Method 3. Methods 1 and 2 switch the measurement noise covariance matrix of the accelerometer depending on the dynamic condition so that weight of accelerometer signals are adjusted compared to weight of gyro signals. This paper does not make claims that the compensation mechanism in the proposed method is superior to those of Methods 1 and 2, based on the result of Test C. However, the orientation estimation algorithm should have a proper acceleration compensation scheme to maintain an acceptable accuracy level in dynamic conditions.

This paper proposed a novel quaternion-based pseudo Kalman filter by modifying an indirect KF, in order to maximize the computational efficiency and implementation simplicity. The proposed algorithm achieves higher computational efficiency than a typical indirect KF, without sacrificing estimation accuracy. Due to its high efficiency, the proposed algorithm can be suitable for battery-powered and low cost processor-based wearable inertial/magnetic sensor applications.

Acknowledgment

This research was supported by Basic Science Research Program through the National Research Foundation of Korea (NRF) funded by the Ministry of Science, ICT & Future Planning (NRF-2015R1C1A1A02036373).

References

1. J. K. Lee, E. J. Park, Minimum-order Kalman filter with vector selector for accurate estimation of human body orientation, IEEE. T. Robot 25 (2009) 1196-1201.
2. J. K. Lee, E. J. Park, A fast quaternion-based orientation optimizer via virtual rotation for human motion tracking, IEEE. T. Biomed. Eng 56 (2009) 1574-1582.
3. J. K. Lee, G. T. Desmoulin, A. H. Kahn, E. J. Park, Comparison of 3D spinal motions during stair-climbing between individuals with and without low back pain, Gait Posture 34 (2011) 222-226.
4. J. K. Lee, E. J. Park, 3D spinal motion analysis during staircase walking using an ambulatory inertial and magnetic sensing system, Med. Biol. Eng. Comput 49 (2011) 755-764.
5. E. Foxlin, Inertial head-tracker sensor fusion by a complementary separate-bias Kalman filter, IEEE. Virt. Real. Ann 185-194 (1996).
6. A. M. Sabatini, Quaternion-based extended Kalman filter for determining orientation by inertial and magnetic sensing, IEEE. T. Biomed. Eng 53 (2006) 1346-1356.
7. E. R. Bachmann , X. Yun, A. Brumfield, Limitations of attitude estimation algorithms for inertial/magnetic sensor modules, IEEE. Robot. Autom. Mag 14 (2007) 76-87.
8. D. Roetenberg, H. J. Luinge, C. T. Baten, P. H. Veltink, Compensation of magnetic disturbances improves inertial and magnetic sensing of human body segment orientation, IEEE. T. Neur. Sys. Reh 13 (2005) 395-405.
9. Y. S. Suh, Y. S. Ro, H. J. Kang, Quaternion-based indirect Kalman filter discarding pitch and roll information contained in magnetic sensors, IEEE. T. Instrum. Meas 61 (2012) 1786-1792.
10. Y. S. Suh, Orientation estimation using a quaternion-based indirect Kalman filter with adaptive estimation of external acceleration, IEEE. T. Instrum. Meas 59, 12 (2010) 3296-3305.
11. J. K. Lee, E. J. Park, S. N. Robinovitch, Estimation of attitude and external acceleration using inertial sensor measurement during various dynamic conditions, IEEE T Instrum. Meas 61 (2012) 2262-2273.

Research on Proximity Magnetic Field Influence in Measuring Error of Active Electronic Current Transformers

Weijiang Wu[1,2], Yan Xu[3], Shihai Yang[1,2], Shufeng Lu[1,2], Xia Xiao[3]

[1]*Electric Power Research Institute, Jiangsu Electric Power Company, Nanjing 211103, China*

[2]*Key Laboratory of Electric Energy Metering, State Grid Corporation, Nanjing 210019, China*

[3]*School of Electrical and Electronic Engineering, Huazhong University of Science and Technology, Wuhan 430074, China*

Abstract. The principles of the active electronic current transformer (ECT) are introduced, and the mechanism of how a proximity magnetic field can influence the measuring of errors is analyzed from the perspective of the sensor section of the ECT. The impacts on active ECTs created by three-phase proximity magnetic field with invariable distance and variable distance are simulated and analyzed. The theory and simulated analysis indicate that the active ECTs are sensitive to proximity magnetic field under certain conditions. According to simulated analysis, a product structural design and the location of transformers at substation sites are suggested for manufacturers and administration of power supply, respectively.

1 Introduction

In recent years, there has been a wave of construction of smart grids. Electronic instrument transformers, which are an indispensable part of an intelligent substation, have been developed. New types of electronic instrument transformers will be widely used in the power grid system because of their advanced sensing mechanism, simple insulation structure, immunity to magnetic saturation, wide dynamic range, excellent frequency response, and ability to adapt to developments in automatic protection [1-4].

At present, electronic current transformers (ECTs) are classified into active ECTs and passive ECTs. Active ECTs are ECTs that have a primary current sensor or primary converter with a power supply. Passive ECTs refer to ECTs having a primary current sensor or primary converter without a power supply. Active ECTs with Rogowski coil are the most practical and have been widely applied after many years development research. Rogowski coils used to be applied in high pulse current measuring and large direct current measuring [5-9]. Owing to their measuring output power being small, Rogowski coils are not able to drive metering and protection devices in traditional substations [10]. Hence, their applications in metering and protection devices at substation sites are restricted. With the rapid development of smart grid and widely launch of intelligent substations, the shift to digitization in the power system made present-day metering and protection devices not need large power drive any more. Rogowski coils applied in current measurement in intelligent substations adapt to the development tendency and they are in extensive application at present stage [11].

In analyzing tests and runtime conditions so far, active ECTs, at intervals, have shown excessive errors in installation testing at substation sites, although they have passed type tests and delivery tests. Existing research into real error characteristics under online conditions is insufficient. The disparity between the error characteristics of active ECTs in the laboratory and at the substation site severely affects metering credibility. As a consequence, it is necessary to research error characteristics of active ECTs under the two conditions to determine a congruent relationship of transformer error characteristics and to provide a rationalization proposal for the application of active ECTs.

This paper focuses on impacts created by proximity magnetic field on active ECTs. The impacts of three-phase proximity magnetic field are simulated and analyzed by magnetic field simulation software. Several suggestions are given with regard to conclusion from simulation analysis.

2 Principle of Active Electronic Current Transformers and Sources of Error

Principle of Rogowski Coils. Rogowski coil, based on non-magnetic materials as skeleton, consists of a helical coil of wire with the lead from one end returning through the centre of the coil to the other end. The Rogowski coil with a winding of uniform density wraps around the current-carrying conductor when measuring the current value. The schematic of a Rogowski coil is illustrated in Fig. 1. The shape of skeleton is toroidal and the shape of section can be rectangle or circular. According to Ampere circuital theorem, the relationship between the current in line $I(t)$ and magnetic flux density that pass through turn area is

$$B_i = \mu_0 H_i = \frac{\mu_0 I(t)}{2\pi r} \quad (1)$$

where r is the distance between centerline of a coil and centerline of the current wire, and H_i is magnetic field intensity passing through the coil per turn.

When the area of the coil per turn S is small, the relationship between the induced voltage $e(t)$ and the current $I(t)$ is

$$e(t) = -n\frac{d\phi}{dt} = -nS\frac{dB_i}{dt} = \frac{\mu_0 nS}{2\pi r}\frac{dI(t)}{dt} \quad (2)$$

where n is number of turns in a Rogowski coil.

Thus it can be seen that the output voltage of a Rogowski coil is proportional to the time derivative of the busbar current. The measurement of busbar current can be implemented through integrating processing after measuring the output voltage of the Rogowski coil.

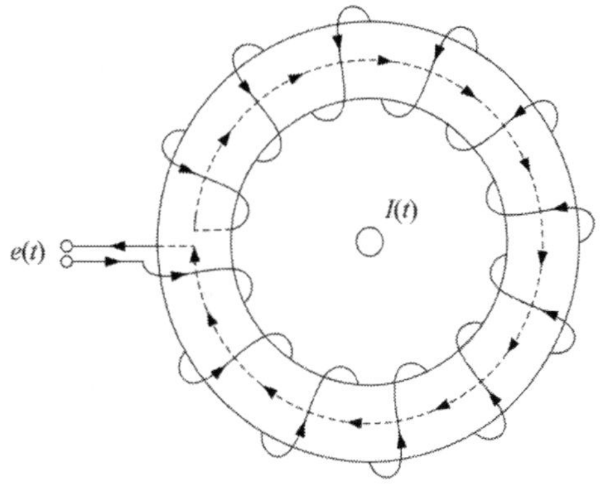

Figure 1. Schematic of sensor section in a Rogowski coil

Sources of Error Created by Proximity Magnetic Field. Many factors influencing the error of Rogowski coils are controllable after many years research. However, the research on proximity magnetic field influence in error of active electronic current transformers rarely gets enough attention.

The sensor of a Rogowski coil indirect measures the current through measuring the induced voltage from variable magnetic field created by the current. However, the measuring objects at the environment of the substation site tend to be three-phase current in simultaneous operation. The magnetic field created by proximal current busbar becomes disturbance magnetic field, which passes through coil section of the Rogowski coil with magnetic field created by measured current busbar and induces corresponding disturbance voltage. The disturbance voltage is superimposed on ideal measurement signals, which produces error to the measurement result. This kind of disturbance magnetic field influences active ECTs especially medium-voltage active ECTs greatly, and is a significant reason that active ECTs exhibit excessive errors at substation sites.

According to Eq. 2, the output voltage of a Rogowski coil is proportional to the time derivative of the busbar current. When the external disturbance magnetic field reaching the coil, the values of magnetic flux density of all coils B_i are no longer equivalent. Proceeding time integration to both side of Eq. 2, the busbar current can be expressed as follows:

$$I = -\frac{2\pi r}{\mu_0 nS}\int e(t)dt = \frac{2\pi r}{\mu_0 nS}\left(S\sum_{i=1}^{n} B_i\right) = \frac{2\pi r}{\mu_0 n}\sum_{i=1}^{n} B_i \quad (3)$$

And the Ampere circuital theorem is still valid even having the external disturbance magnetic field. Hence, the Ampere circuital theorem can be correspondingly expressed as follows:

$$I = \frac{1}{\mu_0}\oint_L \vec{B}\,d\vec{l} \quad (4)$$

Comparing Eq. 3 with Eq. 4, the algorithm principle of a Rogowski coil measuring current is that the line integral of the magnetic flux density around the coil skeleton circle in Ampere circuital theorem is equivalent to the average value of magnetic flux density of all coils multiplying by the perimeter of coil skeleton. The two are completely equal when the coils are symmetric even and the number of turns n approaches infinity. However, the number of turns is limited and the winding is uneven in reality. The distinction between the two is the origin of measurement error of Rogowski coils. In practical application, some uneven air gap tends to exist at outgoing line side. The PCB Rogowski coils are symmetric even while the number of turns is less leading to winding discontinuity. Both designs can bring error. The error can be eliminated through calibration in the ideal situation but can also be uncertainty error at substation sites with outside disturbance magnetic field.

3 Analysis of Error Influence of Proximity Effect of Active Electronic Current Transformers

With respect to active ECTs, different environmental settings around transformers can influence magnetic field distribution and consequently influence error. This section of this paper analyzes the impact of a proximity magnetic field using electromagnetic field analysis simulations. The effect degrees of proximity magnetic field in all situations are analyzed and some suggestions and measures to reduce the impact are given.

Three-phase Proximity Magnetic Field Simulation with Invariable Distance. Three-phase active ECTs are in simultaneous operation at intelligent substations. The magnetic field created by the adjacent phase current is disturbance magnetic field to ECT at its own phase. In this condition, the environment of active ECTs is nonideal. The magnetic field distribution of its own phase is changed leading to measurement accuracy of the ECT influenced to a certain degree, especially at special situation like three-phase imbalance.

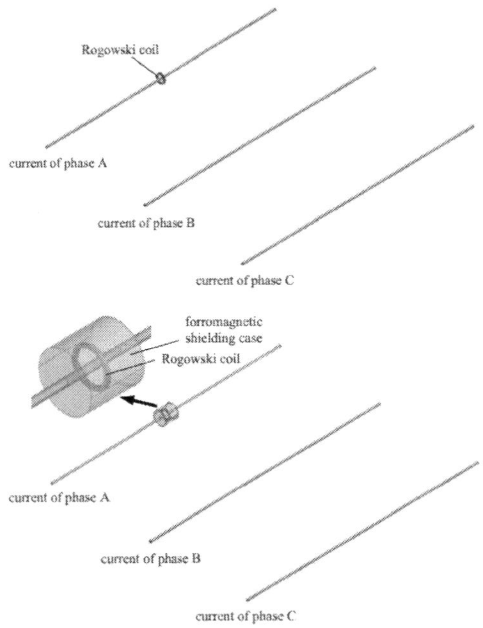

(a)(b)

Figure 2. Model of simulation of (a) three-phase proximity magnetic field (b) three-phase proximity magnetic field with ferro-magnetic shield

Taking 110kV high voltage line as an example, the model of three-phase proximity magnetic field 3D simulation is built, which is illustrated in Fig. 2(a). Considering the magnetic field created by current of phase A and phase C are counteracted at the position of phase B, this paper aims at research on current of phase B and phase C influencing an active ECT of phase A. The simulation applies SOLID97 3-D magnetic solid elements and INFIN111 3-D infinite solid elements. The Rogowski coil specific dimensions of phase A are modelled as product dimensions: the inner diameter of Rogowski coils is 140 mm, and the external diameter is 170 mm. The thickness of Rogowski coil is 15 mm, which means cross-section of the coil is a square. The radius of current busbar through the coil is 17 mm, and the electric current density is set as 1.2 A/mm^2. The total length of current busbar is set as double height of ECTs 4.68 m, which can be regard as infinite length and straight line comparing with the coil dimensions. Distance between proximity phases is set as 2 m according to the requirement of 110 kV intelligent substations. The Rogowski coil and three current busbar are placed in the air, and it can be considered that nearby devices are infinitely far apart from them (i.e., other devices will not create impacts). In postprocessing, the magnetic flux density in the position of diameter being equal to 155 mm (mean value of inner diameter and external diameter) integrate along the located circle. Compared with current set value, the theoretical errors of calculated current value after integral calculation in different conditions can be obtained.

In the industry, some active ECTs manufacturers applied ferro-magnetic shield around the Rogowski coil sensors in their active ECT products. Considering ferro-magnetic shield being able to decrease the impact of ambient magnetic field, this paper compares simulation of three-phase proximity magnetic field with ferro-magnetic shield against simulation of three-phase proximity magnetic field without ferro-magnetic shield to get contrastive analysis of error influence quantities. The model of ferro-magnetic shield is a cylinder with the external diameter being 240 mm, the inner diameter being 236 mm and length being 240 mm. The relative permeability of ferro-magnetic shield μ is 400, the same as the relative permeability of cast iron. The model of simulation of three-phase proximity magnetic field with ferro-magnetic shield is illustrated in Fig. 2(b).

Fig. 3(a) illustrates a magnetic flux density distribution pattern of the Rogowski coil in the ideal condition of the phase A current being 100%I_N without the impact of disturbance magnetic field from proximity phases. By contrast, Fig. 3(b) illustrates a magnetic flux density distribution pattern of the Rogowski coil in the condition of three-phase current all being 100%I_N. And Fig. 3(c) illustrates a magnetic flux density distribution pattern of the Rogowski coil in the same condition as Fig. 3(b) but with ferro-magnetic shield. From Fig. 3(a) to Fig. 3(c), it can be concluded that applying ferro-magnetic shield around the Rogowski coil can reduce the disturbance from proximity magnetic field and make the measured magnetic field distribution closer to ideal situations.

DOI: 10.1051/matecconf/20165901010

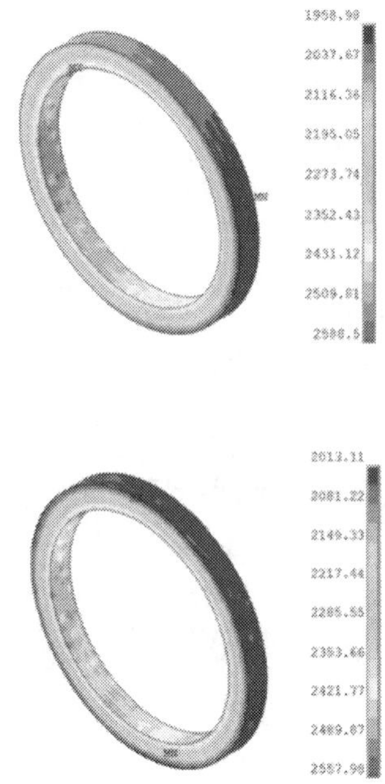

(a) (c) (b)

Figure 3. Magnetic flux density distribution pattern of the Rogowski coil (a) in the ideal condition (b) with proximity magnetic field disturbance (c) with ferro-magnetic shield

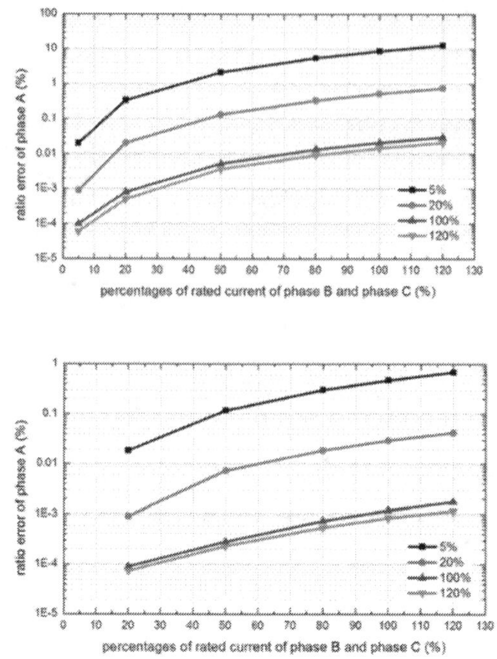

(a) (b)

Figure 4. The Rogowski coil error of phase A (a) without ferro-magnetic (b) shield with ferro-magnetic shield

Based on IEC standard [12], error of ECTs is requested within limits of error at four percentages of rated current: $5\%I_N$, $20\%I_N$, $100\%I_N$ and $120\%I_N$. Testing department conducts the accuracy test and error calibration to ECTs also at the four percentages of rated current. Hence, this paper considers the situation of phase A at four percentages of rated current influenced by three-phase proximity magnetic field. The current values of phase B and phase C stay the same all along the simulations and the variation range is from $5\%I_N$ to $120\%I_N$. Current error of phase A in the conditions of not applying ferro-magnetic shield and applying ferro-magnetic shield are demonstrated in Fig. 4(a) and Fig. 4(b), respectively. The current error of phase A is 0 when applying ferro-magnetic shield and the current of phase B and phase C are at $5\%I_N$. So the simulation data of phase B and phase C at $5\%I_N$ is not shown in Fig. 4(b).

With regard to 0.2 class ECTs, the limits of ratio error at $5\%I_N$, $20\%I_N$, $100\%I_N$ and $120\%I_N$ are 0.75%, 0.35%, 0.2% and 0.2%, respectively. Only the maximum ratio error caused by a three-phase proximity magnetic field being maintained better than 0.075%, 0.035%, 0.02% and 0.02% correspondingly can the influence be considered negligible. The simulation results indicate that the impact of the three-phase proximity magnetic field increases with the current value of phase B and phase C increasing. Under the same conditions, error of active ECTs caused by a three-phase proximity magnetic field reduces dramatically after adding a ferro-magnetic shield. Without ferro-magnetic shield, the impacts of current of phase A at $100\%I_N$ and $120\%I_N$ influenced by three-phase proximity magnetic field can be considered negligible. While the impacts at $20\%I_N$ and $5\%I_N$ influenced by three-phase proximity magnetic field make active ECTs exhibit excessive errors seriously. With ferro-magnetic shield, the impacts of current of phase A at $20\%I_N$, $100\%I_N$ and $120\%I_N$ influenced by three-phase proximity magnetic field can be considered negligible. And the impacts at $5\%I_N$ influenced by three-phase proximity magnetic field still make excessive errors phenomenon possibly.

Combining the results of Fig. 4(a) and Fig. 4(b), the following conclusion can be reached.

1) Under the condition of three-phase current equilibrium, active ECTs are less influenced by three-phase proximity magnetic field.

2) Under the condition of minor three-phase current imbalance, the influence by three-phase proximity magnetic field to active ECTs should not be ignored. Applying ferro-magnetic shield outside active ECTs can decrease the influence and guarantee the accuracy of measurement.

3) Under the condition of major three-phase current imbalance (e.g. there are $5\%I_N$ and $120\%I_N$ in three phases simultaneously), the disturbance of three-phase proximity magnetic field is not able to eliminate even though applying ferro-magnetic shield.

Three-phase Proximity Magnetic Field Simulation with Variable Distance. Distance between proximity phases of 110kV and above voltage classes is small, which is 2 m generally. The possibility of corresponding current in the condition of major three-phase current imbalance is also small. Hence, the impact of three-phase proximity magnetic field is relative small. With regard to 10kV voltage class, the corresponding ECTs are installed in power distribution cabinets with small distance between phases. In this condition, the disturbance of three-phase proximity magnetic field can change magnetic field distribution much more and creates worse impacts. Active ECTs probably exhibit excessive errors even though the three-phase current is balanced. Changing the distance between phases as a variable in the model of Fig. 2(a) and Fig. 2(b), and the distance is chosen from 0.2 to 2.0 m. The different relative permeability of ferro-magnetic shield μ is chosen from 400 to 1000 in the new simulation. The three-phase current is set as $100\%I_N$ under the condition of three-phase current equilibrium to simulate active medium-voltage ECTs influenced by three-phase proximity magnetic field.

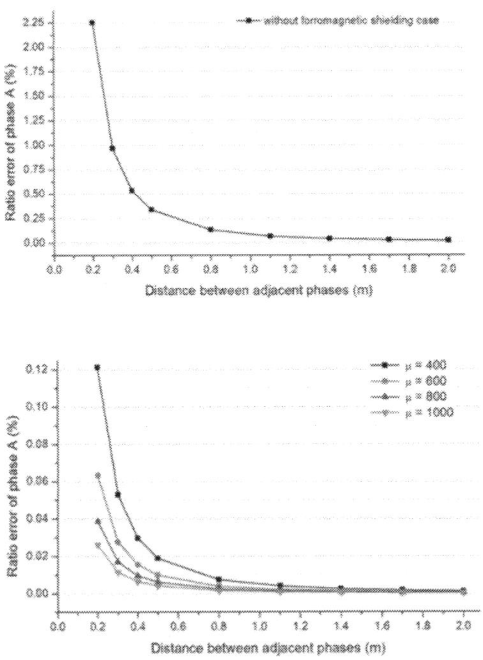

(a)(b)

Figure 5.. The error of active medium-voltage ECTs influenced by three-phase proximity magnetic field (a) without ferro-magnetic shield (b) with ferro-magnetic shield

The simulation results are shown in Fig. 5(a) and Fig. 5(b). The impacts of three-phase proximity magnetic field increase with the decrease of distance between phases. To active medium-voltage ECTs without ferro-magnetic shield, the ratio error exceeds the limit of error 0.2% when distance between phases reduces to approximate 0.7 m. At this distance point, the impacts of three-phase proximity magnetic field have already made active ECTs exhibit excessive errors. To active medium-voltage ECTs with ferro-magnetic shield, the impacts of three-phase proximity magnetic field decrease with the increasement of relative permeability of ferro-magnetic shield. But under the condition where relative permeability of ferro-magnetic shield $\mu \leq 1000$ and the distance between phases is 0.2 m, the impacts of three-phase proximity magnetic field still cannot be neglected. Combining the simulation data, conclusions can be reached as follows.

1) Under the condition where relative permeability of ferro-magnetic shield $\mu \geq 400$ and the distance between three-phases current lines is equal or greater than 0.5 m, the maximum ratio error of transformers that is caused by three-phase proximity magnetic field can be maintained at a value that is better than +0.02%.

2) Under the condition where relative permeability of ferro-magnetic shield $\mu \geq 800$ and the distance between three-phases current lines is equal or greater than 0.3 m, the maximum ratio error of transformers that is caused by three-phase proximity magnetic field can be maintained at a value that is better than +0.02% as well.

Combining all the results of three-phase proximity magnetic field simulations, three suggestions are given as follows.

1) With regard to active ECTs with Rogowski coil, it is suggested that manufacturers use ferro-magnetic shield to reduce disturbance magnetic field. The relative permeability of ferro-magnetic shield is suggested around 400 for active high-voltage ECTs while it is suggested above 1000 for active medium-voltage ECTs.

2) Under the condition of major three-phase current imbalance (e.g. the railway grid has major three-phase current imbalance because of articulated phase in its operation mode), the additional error of active ECTs should be taken into consideration.

3) The impacts of three-phase proximity magnetic field increase with the decrease of distance between phases. It is suggested to guarantee the distance between phases more than 0.3 m in power distribution cabinet for medium-voltage ECTs.

4 Conclusion

Error of active ECTs with Rogowski coil at substation site is influenced by proximity magnetic field. For high-voltage transformers, the influence of proximity magnetic field mainly occurs under the condition of three-phase current imbalance; for medium-voltage transformers, the influence of proximity magnetic field can still occur under the condition of three-phase current equilibrium. The influence is a significant reason that triggers off a disparity between the error characteristics of active ECTs in the laboratory and at the substation site. Applying ferro-magnetic shield with high relative permeability in the design structure of active ECTs with Rogowski coil and guaranteeing distance between three-phase current lines can reduce the influence of proximity magnetic field.

References

1. B. Djokic, E. So, IEEE Trans. Instrum. Meas., 54 (2005) pp. 479-482
2. A. Ehsan, M. Richard, IEEE Trans. Instrum. Meas., 56 (2007) pp. 753-759
3. G. Crotti, D. Gallo, D. Giordano, C. Landi and M. Luiso, IEEE Trans. Instrum. Meas., 63 (2014) pp. 2349-2357
4. X. Xiao, Y. Xu and Z. Dong, Sensors, 15 (2015) pp. 7125-7135
5. I. A. Metwally, IEEE SENSORS J., 13 (2013) pp. 538-547
6. I. A. Metwally, IEEE Trans. Instrum. Meas., 59 (2010) pp. 353-360
7. Y. Liu, F. Lin, Q. Zhang, and H. Zhong, IEEE SENSORS J., 11 (2011) 123-130
8. E. Suomalainen, J. K. Hällström, IEEE Trans. Instrum. Meas., 58 (2009) pp. 1054-1058
9. I. A. Metwally, IEEE Trans. Instrum. Meas., 62 (2013) pp. 2303-2313
10. P. Orr, P. Niewczas, C. Booth, G. Fusiek, A. Dysko, F. Kawano, T. Nishida and P. Beaumont, IEEE SENSORS J., 13 (2013) pp. 2053-2054
11. M. H. Samimi, A. Mahari, M. A. Farahnakian, and H. Mohseni, IEEE SENSORS J., 15 (2015) pp. 651-658
12. Instrument Transformers-Part 8: Electronic Current Transformers, document IEC 60044-8, Geneva, Switzerland, (2002)

MATEC Web of Conferences **59**, 01011 (2016) DOI: 10.1051/matecconf/20165901011
ICFST 2016

Fabrication of a capacitive relative humidity sensor using aluminum thin films deposited on etched printed circuit board

Jacqueline Ann L. Lee, and Ivan B. Culaba

Department of Physics, Ateneo de Manila University, Loyola Heights, Quezon City, Philippines, 1108

Abstract. A capacitive humidity-sensing device was created by thermal evaporation of 99.999% aluminum. The substrate used for the coating was etched double-sided printed circuit board. The etched printed circuit board serves as the dielectric of the capacitor while the aluminum thin films deposited on either side serve as the plates of the capacitor. The capacitance was measured before and after exposure to humidity. The device was then calibrated by comparing the readings of capacitance with that of the relative humidity sensor of the Vernier LabQuest2. It was found that there is a linear relationship between the capacitance and relative humidity given by the equation C=1.418RH+29.139 where C is the capacitance and RH is the relative humidity. The surface of the aluminum films is porous and it is through these pores that water is adsorbed and capillary condensation occurs, thereby causing the capacitance to change upon exposure to humidity.

1 Introduction

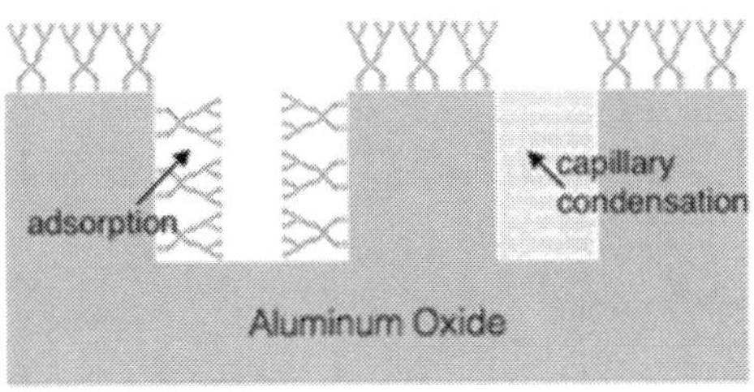

Figure 1. Adsorption and capillary condensation on the film.

Humidity sensors are used for numerous applications in various fields. Capacitive humidity sensors use changes in the capacitance reading to indicate the change in the moisture content of the air. In these sensors, dielectric thin films are deposited on a substrate. When there is a change in the humidity, the moisture in the air is taken in by the thin film. This changes the dielectric constant and indicates, in the change of capacitance, the change in the humidity. Metal oxide thin films are also used in creating these capacitive humidity sensors. The metal oxide thin films are made to be porous using various techniques such as glancing angle deposition wherein the substrate is rotated while the film is being deposited [1]. These films sense the changes in humidity through its pores, which are filled with water molecules, and thus its dielectric constant and capacitance changes. This is due to capillary condensation in the pores and adsorption of water on the surface of the film [2], which causes the electrical properties of the device to change as shown in the change in capacitance. A schematic for capillary condensation and adsorption is shown in Figure 1.

The other methods for fabricating these sensors are quite complex and involve complicated processes to create the porous surface needed. The novelty of the method presented in this paper is that it makes use of a simpler method of creating the porous structures. It is a one step process where the film has to simply be deposited at a normal incidence to the substrate. No intermediate steps are needed to ensure the formation of a porous film.

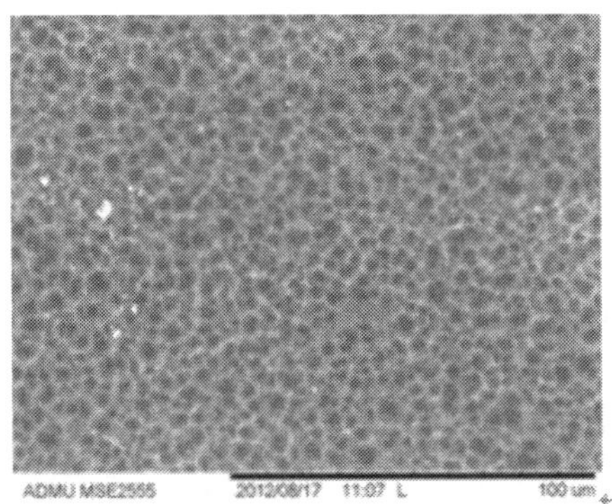

Figure 2. (a) SEM image of etched PCB. From: E.D. Dumayaca, N.N. Garcia, I.B. Culaba. *Proceedings 30th Physics Congress of the Samahang Pisika ng Pilipinas*. 22 (2012)

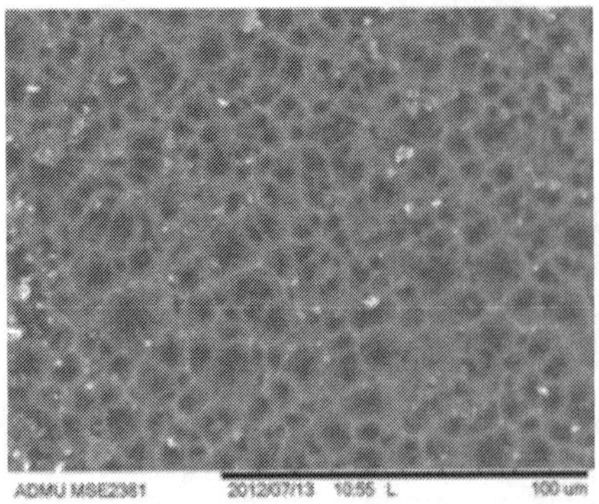

Figure 2. (b) SEM image of aluminum coated etched PCB. From: E.D. Dumayaca, N.N. Garcia, I.B. Culaba. Proceedings *30th Physics Congress of the Samahang Pisika ng Pilipinas*. 22 (2012)

From a study on the morphology of aluminum thin films deposited on etched printed circuit board, it was found that the surface of the etched printed circuit board was porous. After coating the surface with aluminum, it was found that the porous surface of the printed circuit board was retained [3] as shown in figures 2 (a) and 2 (b). The use of the already porous surface of etched printed circuit board, as the substrate for the coatings, produces a porous film and allows an easier method for constructing the capacitive humidity sensor.

In this paper, aluminum thin films deposited on etched printed circuit board is used as a capacitor. The aluminum thin films serve as the plates of the capacitor while the etched printed circuit board serves as the dielectric between the plates. This capacitor was used as a relative humidity sensor and its response to humidity was studied and compared to a commercially available sensor.

2 Methodology

Preparation of Substrates. To create the capacitor, the printed circuit board (PCB) is prepared. This will serve as the substrate on which the aluminum thin film will be deposited. First, double sided printed circuit board is cut into 1.27cm x 10.16cm (0.5" x 4.0") pieces and holes were drilled onto each end. One of the holes is drilled 1.75 cm from one end and the other is drilled 0.50 cm from the opposite end as shown in figure 3. This will allow bolts to be attached and it will be used to measure the capacitance.

Then the cut pieces of printed circuit board were submerged in ferric chloride ($FeCl_3$) to remove the copper on the surface. Once the copper was removed, the etched printed circuit board was rinsed with distilled water and was subjected to ultrasonic cleaning using the Fisher Scientific Ultrasonic Cleaner FS30H. They were further cleaned using isopropyl alcohol and Kimwipes®. Lastly, the etched PCB was masked with tape as shown in figure 3. The unmasked areas will be coated with aluminum while the masked areas will remain uncoated.

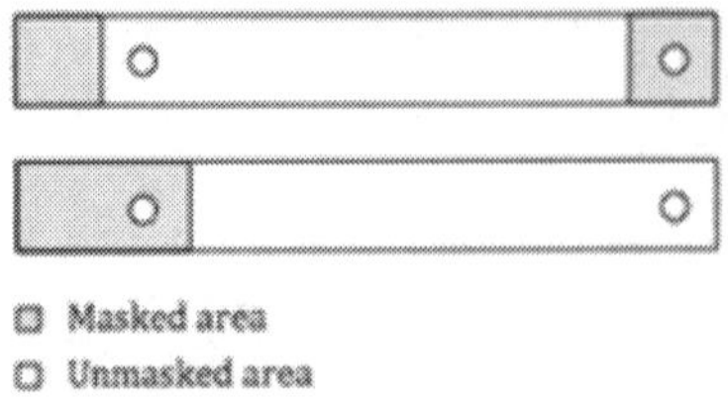

Figure 3. Masked areas and position of holes on the etched PCB

Coating Substrates with Aluminum. The Kinney vacuum system was used for coating the etched PCB. The substrates were placed on a rotating substrate holder inside the vacuum chamber. To coat the etched printed circuit board, 2 tungsten coils were attached to the current sources inside the vacuum chamber. On each coil, 20 pieces on 99.999% aluminum wires were attached. The coating was done when the pressure of the chamber reached 2.5 x 10^{-5} mmHg. To start the coating process, the first coil was heated up by applying a current. The current was gradually increased until it reached 60 amperes and the first side of the etched printed circuit board was exposed to the aluminum vapor for 3 minutes. The thickness of the film was then measured using a thickness monitor. After this, the substrate holder is rotated to prepare the second side for coating. The same procedure is applied to the second coil so that the second side is coated.

Assembling the sensor. Each aluminium coated PCB was fitted with nuts, bolts and washers to serve as contact points for the capacitor as shown in figure 4(a).

Four of these capacitors are connected in parallel to form the humidity sensor as shown in figure 4(b). The capacitance is then measured using an AC-2000 Digital Multi-meter.

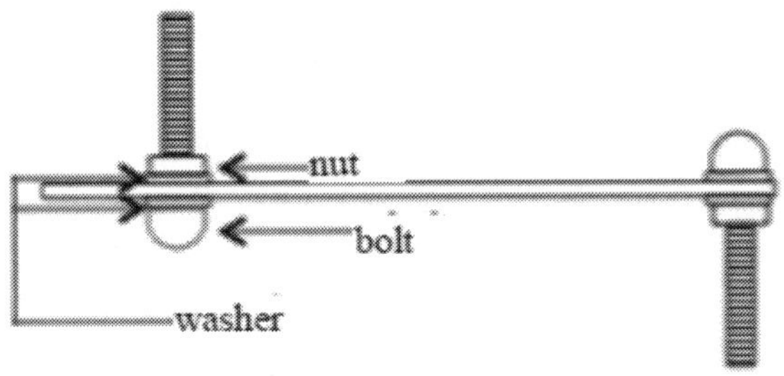

Figure 4. (a) Capacitor assembly

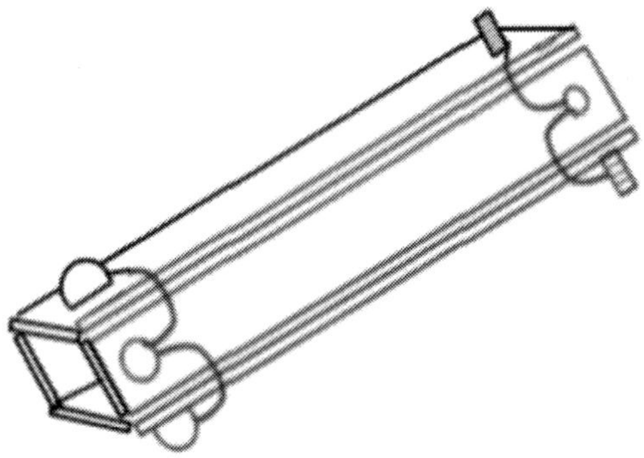

Figure 4. (b) Relative humidity sensor

Calibrating the Sensor. The sensor is calibrated by simultaneously measuring the capacitance of the sensor, using an AC-2000 Digital Multi-meter, and the relative humidity, using the relative humidity sensor of the Vernier LabQuest2. Both sensors are placed inside the same container and the humidity inside the container is decreased as capacitance and relative humidity are collected.
The capacitance is then plotted against relative humidity and the equation of the curve is obtained.

3 Results and Discussion

The thickness of the films for each side of the capacitor was measured to be 440Å and 480Å.

The capacitance readings obtained over time as the humidity is changed is shown in figure 5 while the change of relative humidity over time is shown in figure 6. It can be seen from figures 5 and 6 that there is a similar decreasing trend for both the capacitance versus time graph and the relative humidity versus time graph.

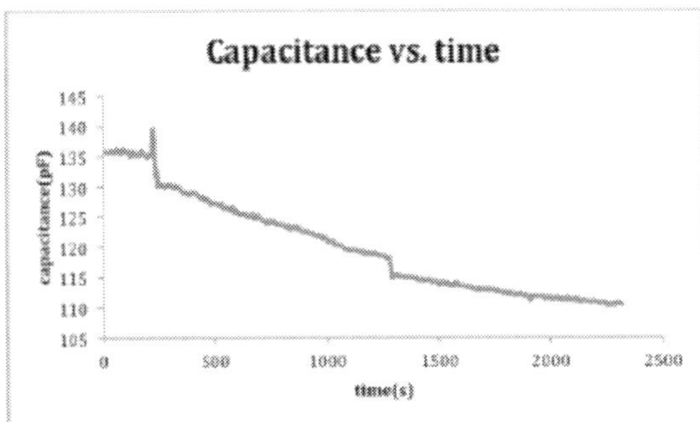

Figure 5. Graph of capacitance over time.

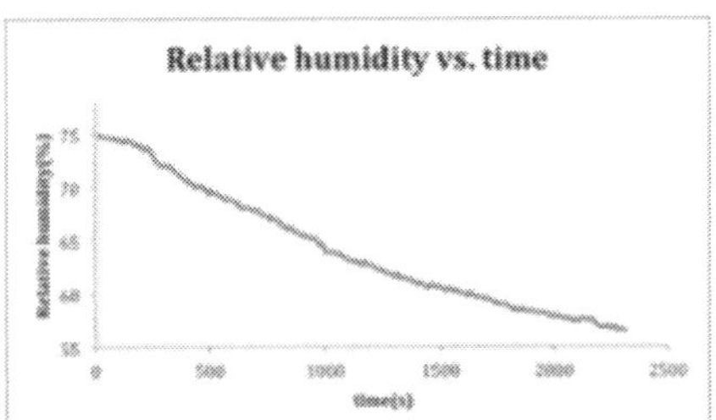

Figure 6. Graph of relative humidity over time.

Comparing the capacitance and relative humidity in figure 7 shows that there is a linear relationship between the relative humidity and the capacitance of the device. By obtaining the best-fit line (eq. 1) of the data points where C is the capacitance and RH is the relative humidity, the capacitance measured from the device can be easily converted to the relative humidity value and vice versa.

$$C = 1.418RH + 29.139. \quad (1)$$

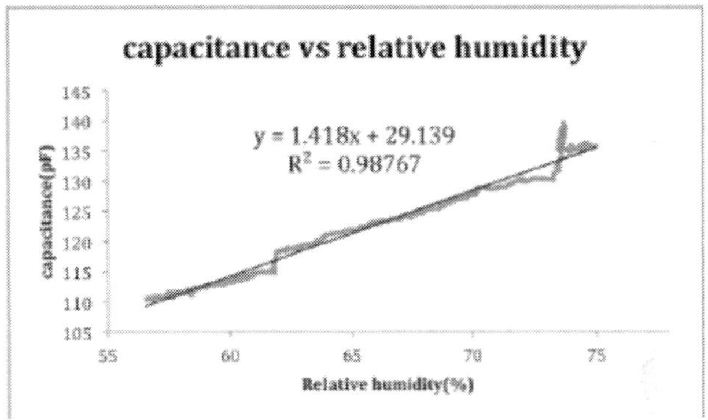

Figure 7. Graph of capacitance vs. relative humidity

When the alumimum coating is exposed to the air, an oxide layer is formed. The oxide layer enables the water molecules in the air to be adsorbed on the surface [4] as shown figure 8. The pores of the film are also filled by the water molecules through capillary condensation as shown previously in figure 1. This changes the electrical properties of the aluminum films and thus, changes the measured capacitance depending on the amount of adsorbed water. Water can also be taken in by the porous etched PCB and cause a change in the dielectric constant, and thus, the measured capacitance. As the relative humidity is increased, the amount of water molecules that that are adsorbed on the film surface increases and the amount of water molecules that enter the pores of the dielectric increases, thereby increasing the capacitance reading.

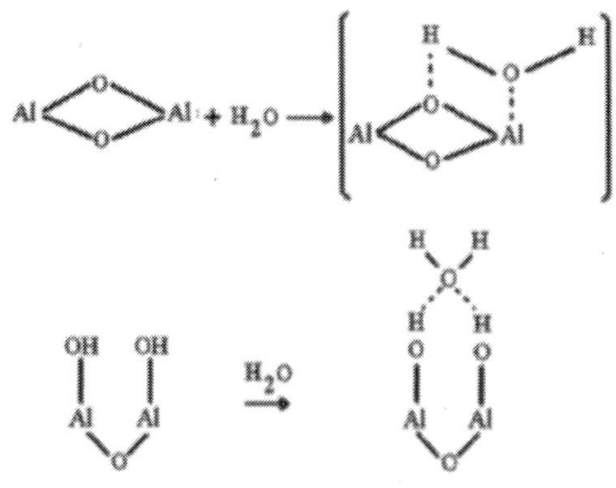

Figure 8. Adsorption of water on aluminum oxide.

DOI: 10.1051/matecconf/20165901011

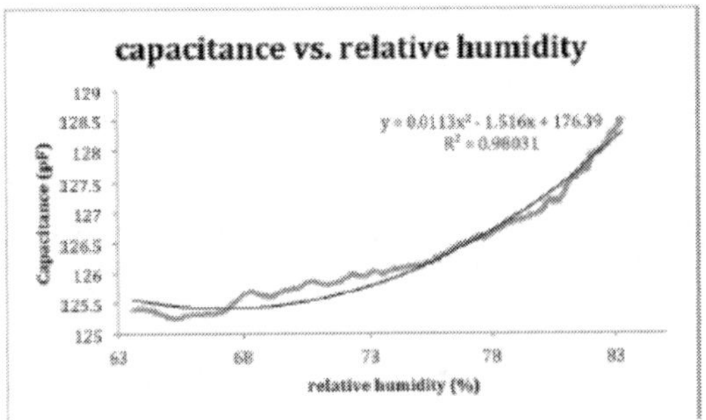

Figure 9. Graph of capacitance vs. relative humidity for the aged capacitor

After aging the sensor for three months, it was found that it no longer retained its original sensing properties. The capacitance no longer varied linearly with relative humidity as shown in figure 9. There is very little increase in capacitance as the relative humidity was increased. There was, approximately, only a 3pF change in the capacitance as the relative humidity was increased from 63% to 83% for the aged sensor compared to the, approximately, 25pF change in capacitance as the humidity was changed from 56% to 75% for the non-aged sensor. This could be attributed to contaminants in the pores that prevent the adsorption of water and capillary condensation; thus, hindering the sensor from responding to the humidity changes in the environment.

4 Conclusion

A capacitive relative humidity sensor was fabricated by depositing aluminum thin films on either side of etched double-sided printed circuit board. Exposing the sensor to humidity causes water molecules to fill the pores of the film and the dielectric. Water molecules are also adsorbed in the aluminum oxide surface, which causes a r

change in the capacitance reading. By comparing the readings of capacitance of the non-aged device with that of the relative humidity sensor of the Vernier LabQuest2, the device was calibrated. It was found that there is a linear relationship between the capacitance and relative humidity given by the equation $C=1.418RH+29.139$ where C is the capacitance and RH is the relative humidity.

References

1. J. J. Steele, G. A. Fitzpatrick, M. J. Brett, IEEE Sensors J. 6 (2007) 24-27
2. O. K. Varghese, C. A. Grimes, J. Nanosci. Nanotech. 3 (2003) 277-293
3. E.D. Dumayaca, N.N. Garcia, I.B. Culaba, Proceedings 30th Physics Congress of the Samahang Pisika ng Pilipinas. (2012.)
4. T. Morimoto, M. Nagao, F. Tokuda, J. Phys. Chem. 73 (1969) 243-248
5. V. Timár-Horváth, L. Juhász, A. Vass-Várnai, G. Perlaky, Microsyst. Technol. 14 (2008) 1081-1086
6. F. Casanova, C. E. Chiang, C. P. Li, I. V. Roshchin, A. M. Ruminski, M. J. Sailor, I. K. Schuller, Nanotechnology. 19 (2008)
7. T. Morimoto, M. Nagao, F. Tokuda, Bull. Chem. Soc. Jpn. 44 (1971) 1282-1288
8. J. J. Steele, G. A. Fitzpatrick, M. J. Brett, IEEE Sensors J. 7 (2007) 955-956
9. J. Li, Y. Liu, M. Tang, J. Li, X. Lin, IET Micro Nano Lett. 7 (2012) 1097-1100
10. H. Farahani, R. Wagiran, M. N. Hamidon, Sensors, 14 (2014) 7881-7939

Lipase Based Biosensors for Triglyceride Determination

Nurul Huwaida Rosli[1], Zainiharyati Mohd Zain[1] and Nor Monica Ahmad[2]

[1]*Faculty of Applied Sciences, Universiti Teknologi MARA, 40450 Shah Alam, Malaysia*

[2]*Faculty of Applied Sciences, Universiti Teknologi MARA, 47000 Kuala Pilah, Malaysia*

Abstract. A review of methods development in lipase based biosensor for triglyceride determination was briefly discussed. This review focuses on the basic principle of triglyceride biosensor that includes performances of triglyceride biosensor such as limit of detection, response time, and optimization.

1 Introduction

Triglycerides (TG) can be generated by esterification process of three hydroxyl (-OH) groups of glycerol with three molecules of fatty acids that produce an ester as product [1]. Triglyceride acts is important role in metabolisms as energy source and also as a dietary fat transporter. Fig. 1 shows the molecular structure of triglyceride:

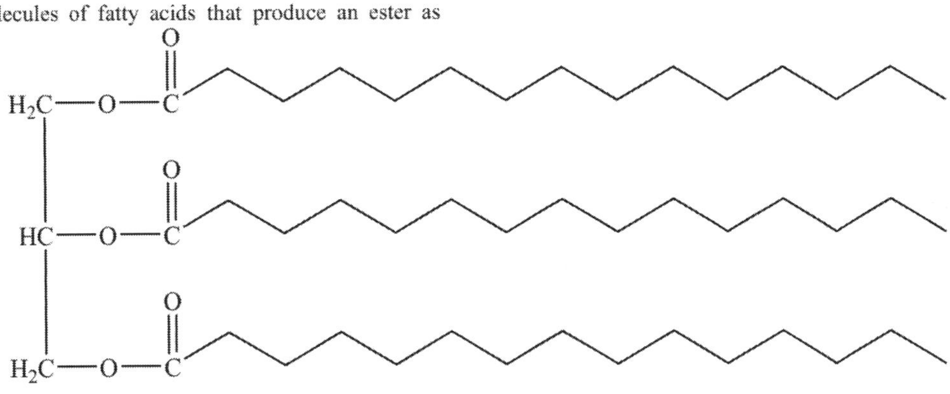

Figure 1. Molecular structure of Triglyceride (TG)

Determination of triglycerides (TG) is crucial since its high concentration could lead to hyperlipidemia. Therefore, ensuring the level of triglyceride in our body in normal range, women (35–135 mg/dl) and men (40–160 mg/dl) is a need [2]. Apart from coronary diseases such as heart attack and hyperlipidemia, are also associated to TG several disorders such as liver obstruction, diabetes mellitus and endocrine [3,4]. Hence, increased health awareness among community and stringent regulatory laws make the estimation of triglycerides content in food has become important nowadays. A biosensor is a device that enables the identification and quantification of an analyte of interest from a sample matrix such as food, water, blood and urine [5]. The most common device used for TG determination is enzymatic amperometric triglyceride biosensors. Normally, triglyceride biosensors are based on interactions of three enzymes, lipase, glycerol kinase (GK) and glycerol-3-phosphate oxidase (GPO). One of general strategies has been used for electrochemical sensing of Triglyceride is by measuring oxygen consumption or the amount of hydrogen peroxide produced by the enzymatic reactions [6]. The biosensor chemical reactions involved in the principle of this method re as equation (1) below; lipase hydrolyses triglycerides to glycerol and fatty acids [7]:

$$TG + H_2O \xrightarrow{\textbf{Lipase}} \text{Glycerol} + 3 \text{ free fatty acids} + H^+ \quad (1)$$

However, there are works that used single enzyme (lipase) for Triglyceride determinations (Solanki *et al.*, 2009) and (Pauliukaite *et al.*, 2011) [7,8]. This methods are more

 DOI: 10.1051/matecconf/20165901012

preferable than multi-enzyme becaus MATEC Web of Conferences
e single enzyme methods are less time consuming and inexpensive.

2 Triglyceride (TG) biosensor

Triglyceride biosensor has been studied by other researchers using various properties and method. Most of the studies focus on surface of electrode, to determine triglyceride at optimum level. Table 1 shows recent studies on triglyceride biosensor.

There are several types of working electrode that have been used for triglyceride biosensor, such as Platinum (Pt), gold (Au), carbon (C), and indium tin oxide (ITO), most of this electrode were stable as base for working electrode. Screen printed carbon electrode (SPCE) has been used as working electrode in this study. Modification on surface, able the detection of hydrogen ions, (H^+) at optimum level. The H^+ ions produced during the enzymatic reaction, will be detected electrochemically on bioelectrode surface at low applied potential [9]. Furthermore, present researches has describes the improvement of TG biosensor by employing ionic liquid on the surface of SPCE which is relatively cheaper compared to other common working electrode such as glassy carbon electrode (GCE), Au and ITO [9].

Meanwhile, triglyceride can be detected by electrochemical or amperometric detection. Besides that, performance of triglyceride biosensor is measured by identifying its detection limit, linear range, response time, sensitivity and storage ability. Based on Table 1, Narang and Pundir (2011) shows a modest detection limit, nevertheless it shows wide linear range that are corresponding with triglyceride concentration that varies in the range between 30 and 550 mg/dL [14,15]. Furthermore, time taken for the electrode to response is faster and the life time is longest. Therefore, high performance of triglyceride biosensor must have lower detection limit. A linear working range that correspond with triglyceride concentration, fast response time (2 sec) and long life time (210 days) in which the biosensor can be used several time with constant performance.

3 Electrochemical characterization for Triglyceride biosensor

Phongphut *et al.,* (2013) state that cyclic voltammetry was used to optimize the electrochemical response of modified electrode, hence triglyceride detection was then performed by chronoamperometry at a constant applied potential [15]. Solanki *et al.,* (2009) shows cyclic voltammetric (CV) studies of ITO modified electrode and bioelectrode have been carried out in PBS containing $[Fe(CN)_6]^{3-/4}-$ at scan rate of 50 mV/s [8]. Cyclic voltammetric (CV) studies have been carried out in phosphate buffer solution (PBS) at potential range -1.0 to 1.0 V. Furthermore, calibration curve, response time and detection limit are then able to be estimated from the amperometric responses.

Pundir *et al.,* (2010) stated that, an amperometric triglyceride biosensor was applied in their research, as current produced from the cyclic voltammetry (CV) was optimum at potential 0.4 V, and at that potential has been selected for standardization of working conditions for triglyceride, determination in amperometry [13]. Therefore, cyclic voltammetry was used for triglyceride determination, in present work due to its rapid capability in measuring redox behavior over a wide potential range [16].

4 Parameters optimization for Triglyceride biosensor

There are several parameters that have been study to increase optimization level of triglyceride biosensor, such as effect of triglyceride biosensor towards applied potential, enzyme concentration, pH, and temperature. Phongphut *et al.,* (2013) stated that, the operational potential for triglyceride determination on Au electrode have been characterized by ranging the potential from 0.1 V to 0.6 V, and the optimum potential for biosensor was at 0.4 V, most of optimum potential was at lower potential since higher applied potential leads to the interferences [15]. Pundir., (2008) stated that, effect of pH towards biosensor have been studied, and optimum pH level for biosensor to response was at pH 6.5, while result that found by Narang and Pundir., (2011), the biosensor was found optimum at pH 7.5, most of triglyceride biosensor shows optimum pH between 6.5 to 7.5, after that it start decreasing [11,14]. Meanwhile, for optimum temperature, Narang *et al.,* (2013) state that, the changing in temperature will give an effect on current response of the biosensor and from their result, it reached at approximately 35 °C [1]. In contrast with result from Pundir., (2008), the biosensor shows the temperature was optimum at 25°C [11].

In our study, we incorporated ionic liquid (IL) that has a high ionic conductivity and well biocompatibility to enhance the electrochemical response. A single enzyme which is lipase with ionic liquid modified electrode as a base for tributyrin sensor was developed for determination of lipase activity. IL was used to ensure fast and easy electron transfer, since most of IL does not give harm towards enzyme [7]. Besides that, it presented as suitable compounds for biosensor because most of them seem to have the ability to solubilize proteins without denaturation [17]. The optimization parameters of SPCE based IL electrode was carried out, and pH 7, 30°C and 5 % of lipase enzyme loading resulted in optimum performance.

Table 1. Analytical properties of recent triglyceride (TG) biosensor

Authors	Type	Type of support for immobilization	Type of electrode	Detection Limit (mg/L)	Linear range (mg/L)	Response Time (s)	Storage stability
[7]	Electrochemical	MWCNT–RTIL ZnO(NPs)-CHIT	GCE	0.02	0-1603	NR	NR
[14]	Amperometric	NiO(NPs)-CHIT/ZnO-ZnHCF	Pt	200	500-6500	6	210 days
[1]	Amperometric	Au/PEDOT-PSS CeO_2/TCO	Au	100	500-7000	4	50% loss in 180 days
[15]	Amperometric	Cellulose acetate, CA	SPCE	7.9	0-531	30	
[9]	Electrochemical	C/Iridium(NPs)	TCO	0.60	100-599	5	40% loss in 30 days
[11]	Amperometric	CeO_2	Pt	180	180-3100	40	42 days
[10]	Electrochemical	Egg shell membrane PVA	SPCE	NR	0-8850	10	25 days
[8]	Electrochemical		ITO	328	500-5000	20	NR
[12]	Amperometric		Pt	247	500-2000	10	84 days
[13]	Amperometric		Pt	190	500-2000	2	70 days
							50% loss in 50 days

5 Conclusion

In conclusion, cyclic voltammetry is the fastest and simplest way for TG determination. Most of studies shows optimize conditions at pH (6.5-7.5) and temperature (25-35°C). Modification of working electrode improves TG biosensor performance.

Acknowledgment

Financial and technical support from The Ministry of Higher Education under Research Acculturation Grant Scheme (RAGS) 600-RMI/RAGS 5/3 (31/2013) and Universiti Teknologi MARA (UiTM) are highly acknowledged.

References

1. J. Narang, N. Chauhan, C.S. Pundir, Construction of triglyceride biosensor based on nickel oxide–chitosan/zinc oxide/zinc hexacyanoferrate film, International Journal of Biological Macromolecules 60 45– 51 (2013).
2. L.A. Caplan, A.J. Pesce, Clinical Chemistry, Theory, Analysis and Correlation, C.V. Mosby, St. Louis, 1989, 481.
3. R. Avramoglu, K. Basiciano, H.K. Adeli, Lipid and lipoprotein dysregulation in insulin resistant states, Clinica Chimica Acta 368 (2006) 1–19.
4. A. Vijayalakshmi, Y. Tarunashree, B. Baruwati, S.V. Manorama, B.L. Narayana, R.E.C. Johnson, N.M. Rao, Enzyme field effect transistor (ENFET) for estimation of triglycerides using magnetic nanoparticles, Biosens. Bioelectron. 23 (2008) 1708–1714.
5. S. Borgmann, A. Schulte, S. Neugebauer, W. Schuhmann. Amperometric Biosensors, (2011)
6. C.S. Pundir, J. Narang, Determination of triglycerides with special emphasis on biosensors: a review, International journal of biological macromolecules, 61, 379–89 (2013).
7. R. Pauliukaite, A.P. Doherty, K.D. Murnaghan, C.M. a. Brett, Application of room temperature ionic liquids to the development of electrochemical lipase biosensing systems for water-insoluble analytes, J. of Electroanalytical Chemistry, 656(1-2), 96–101 (2011).
8. P.R. Solanki, C. Dhand, A. Kaushik, A. Ansari a, K.N. Sood, B.D. Malhotra, Nanostructured cerium oxide film for triglyceride sensor, Sensors and Actuators B: Chemical, 141(2), 551–556 (2009).
9. S. Panky, K. Thandavan, D. Sivalingam, S. Sethuraman, U.M. Krishnan, B.G. Jeyaprakash, J.B.B. Rayappan,

Lipase immobilized on nanostructured cerium oxide thin film coated on transparent conducting oxide electrode for butyrin sensing, Materials Chemistry and Physics, 137(3), 892–897 (2013).
10. W.Y. Liao, C.C. Liu, T.C. Chou, Detection of triglyceride using an iridium nano-particle catalyst based amperometric biosensor, The Analyst, 133(12), 1757–63 (2008).
11. C.S. Pundir, Construction of an amperometric enzymic sensor for triglyceride determination, Sensors and Actuators B: Chemical, 133(1), 251–255 (2008).
12. J. Narang, Minakshi, M. Bhambi, C.S. Pundir, Determination of serum triglyceride by enzyme electrode using covalently immobilized enzyme on egg shell membrane, International Journal of Biological Macromolecules, 47(5), 691–5 (2010).
13. C.S. Pundir, B.S. Singh, J. Narang, Construction of an amperometric triglyceride biosensor using PVA membrane bound enzymes, Clinical Biochemistry 43 467–472 (2010).
14. Narang, C.S. Pundir, Construction of a triglyceride amperometric biosensor based on chitosan-ZnO nanocomposite film, International Journal of Biological Macromolecules, 49(4), 707–715 (2011).
15. Phongphut, a, C. Sriprachuabwong, Wisitsoraat, a, Tuantranont, a, S. Prichanont, P. Sritongkham: A disposable amperometric biosensor based on inkjet-printed Au/PEDOT-PSS nanocomposite for triglyceride determination, Sensors and Actuators B: Chemical, 178, 501–507 (2013).
16. P.T. Kissinger, W. Lafayette, W.R. Heineman, Cyclic voltammetry, (vol 60, no.9), 702–706 (1983).
17. K.S. Galhardo, R.M. Torresi, S.I.C. de Torresi, Improving the performance of a glucose biosensor using an ionic liquid for enzyme immobilization.On the chemical interaction between the biomolecule, the ionic liquid and the cross-linking agent.ElectrochimicaActa, 73, 123–128 (2012).

Integration of Hybrid PV/Wind Generation System Using Fuzzy MPPT in Grid Connected System for Remote Area

Soedibyo, Thohaku Abdul Hadi, Hadyan Perdana Putra , Heri Suryoatmojo , Mochamad Ashari

Electrical Engineering Department, Institut Teknologi Sepuluh Nopember, Surabaya-East Java. Indonesia

Abstract. Photovoltaic and wind are renewable energy resources that widely used and grow rapidly in fulfilling electricity demand. Powers from both technologies depend on sunlight intensity and wind speed. For small scale power generation, DC voltage from both technologies is low and requires step-up converter to raise DC voltage ratio before converted into AC voltage. To optimize this system, step-up converter must have high ratio and efficiency to a distance of wide voltage input. This paper proposed an operation simulation and arrangement of DC-DC converter along with DC-AC from hybrid source PV-Wind which integrated to grid utilities without using storage device. High Gain Integrated Cascade Boost (HGICB) is DC-DC converter that has quadratic voltage ratio and used in this research. Then DC link connected to Voltage Source Inverter (VSI) which interconnected with utility grid and controlled by current control method. The total installed capacity of hybrid source is 4.4 kW. Wind turbine uses PMSG along with full bridge rectifier. To maximize and stabilize the generated power, MPPT fuzzy is used. Result from the simulation shows that converter capable to maintain maximum power whether from PV and wind turbine which canalized to utility grid in various irradiation condition, wind speed, and grid load alteration.

1 Introduction

In this modern era, electricity becomes one of basic needs of mankind. However, there are still many places around the world which have no access to electricity. International agencies report that more than 20% of world's population have no access to electricity [Energy 2009]. Generation of electricity through renewable energy sources such as solar and wind could be potential options for these remote places. Solar and Wind are renewable energy sources which are easily found in nature and also kind of alternative energy that develop in a rapid speed. To convert energies from solar and wind, photovoltaic (PV) and wind turbine is used. PV and Wind turbine has a characteristic of power which depends on solar irradiation and wind speed based on geographical areas. This paper presented a control methodology of hybrid small scale PV/Wind energy system that use a fuzzy logic controller (FLC) to extract the maximum power point tracking (MPPT) in different solar irradiation and wind speed.

Fig. 1 A remote village in Nepal with small scale wind turbine (University of Canterbury, New Zealand, 2013)

This paper discuss simulation and analysis of the generation process and power conditioning unit (PCU) of Photovoltaic (PV) and Wind Turbine (WT) that is connected to the three phase low voltage electricity grid system (380V). The capacity of the sources used is 2.2kWp PV and 2.5kW PMSG (Permanent Magnet Synchronous Generator) - WT power rating. The Modelling of hybrid PV/Wind as well as integrated power electronics components in grid connected system is simulated using MATLAB/Simulink.

Fig. 2 A remote village with small scale photovoltaic (bravenewclimate.com)

2 Methodology

2.1 Photovoltaic, Wind Turbine, Grid System Modelling

This step is modelling small scale PV and Wind Turbine and its DC/DC converter to show the characteristic of power at different solar irradiation and wind speed in grid interconnecting system.

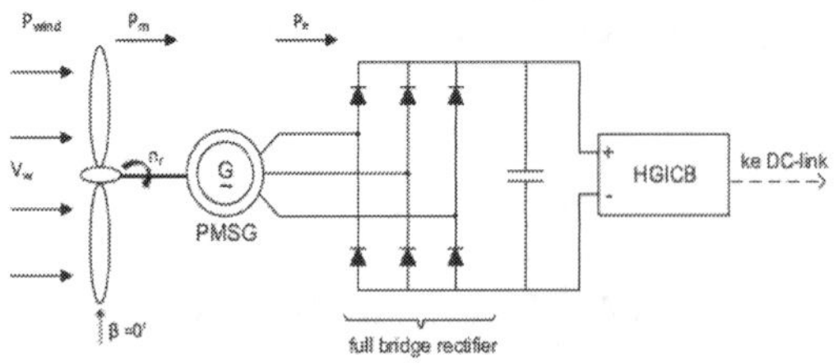

Fig. 3 Equivalent circuit of photovoltaic cell

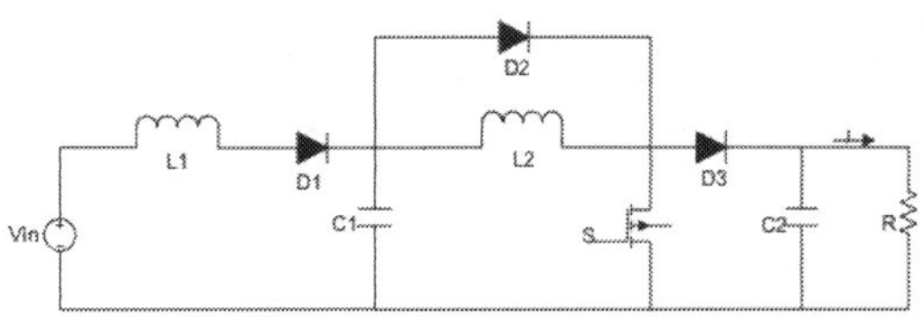

Fig. 4 Schematic of wind turbine system

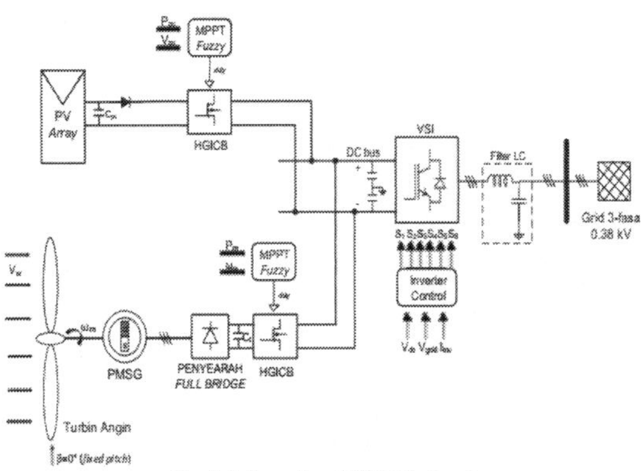

Fig. 5 Schematic of HGICB circuit

Figure 6 shows the schematic diagram of hybrid PV/Wind that connects to the grid through a common DC bus at the inverter. The solar PV with the capacity of 2.2 kWp and PMSG Wind Turbine with the capacity of 2.5 kW connects to the DC-DC converter that have a MPPT function. The inverter is transformerless, so the DC-DC converter use a High Gain Integrated Cascade Boost (HGICB) topology which have a high ratio to bring the output voltage of PV and WT to DC common bus voltage. The MPPT was applied to these HGICB converter using fuzzy logic controller (FLC) that achieve the maximum power from any irradiation and wind speed condition. The DC bus voltage is controlled by the inverter that it can in stable value. The inverter is VSI and use a fast decoupled current control to get synchronization with the utility grid. LC filter is use to damp the harmonic output voltage and current so its value must not exceed from the grid requirement.

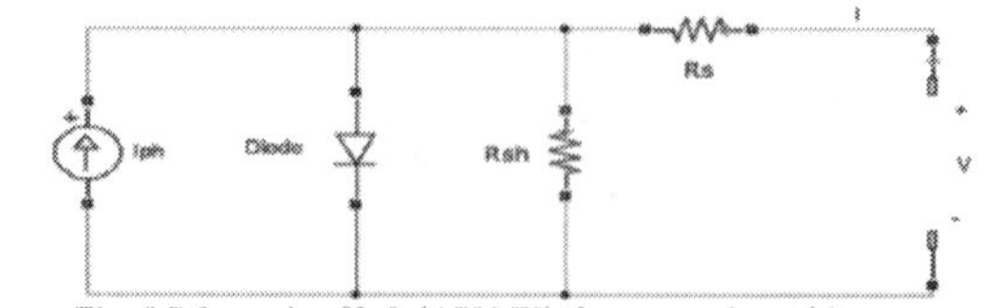

Fig. 6 Schematic of hybrid PV-Wind connected to grid system

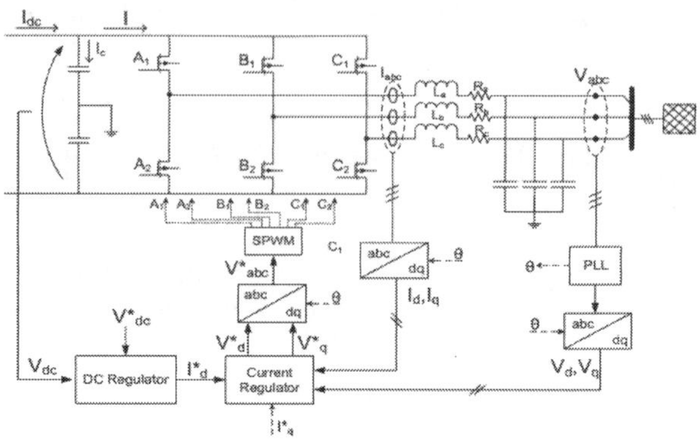

Fig. 7 Current control scheme of the VSI

2.2 Design a Fuzzy Logic Controller (FLC) to Achieve the Maximum Power

This step is designing a control of maximum power point tracking (MPPT) using fuzzy logic. Then show its work in different solar irradiation and wind speed. The MPPT was designed for two DC-DC converter both for solar PV and wind turbine. Figure and table below shows the fuzzy control scheme, membership function and rule base of MPPT control for PV and wind turbine respectively.

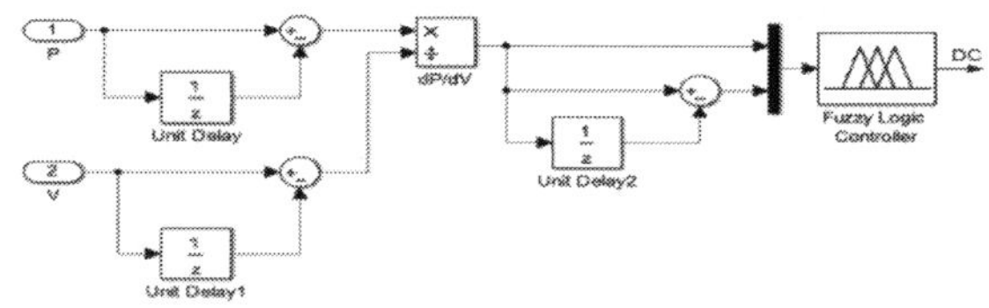

Fig. 8 PV MPPT fuzzy control scheme

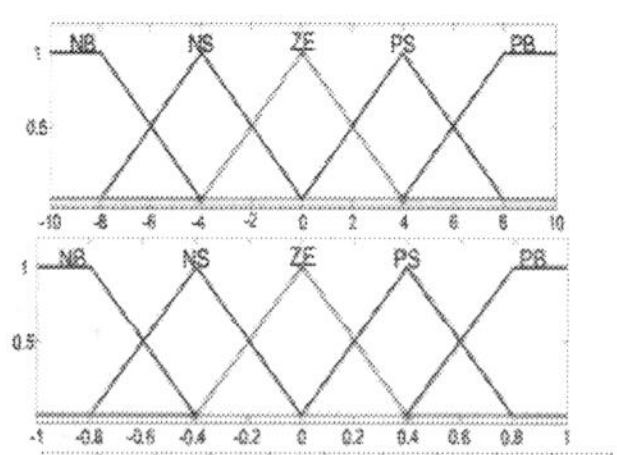

Fig. 9 Membership function of error (e), delta error (de) and duty cycle (duty) as a input – output FLC for PV MPPT

TABLE I
RULE BASE FUZZY LOGIC IN PV MPPT

de \ e	NB	NS	ZE	PS	PB
NB	ZE	ZE	PB	PB	PB
NS	ZE	ZE	PS	PS	PS
ZE	PS	ZE	ZE	ZE	NS
PS	NS	NS	NS	ZE	ZE
PB	NB	NB	NB	ZE	ZE

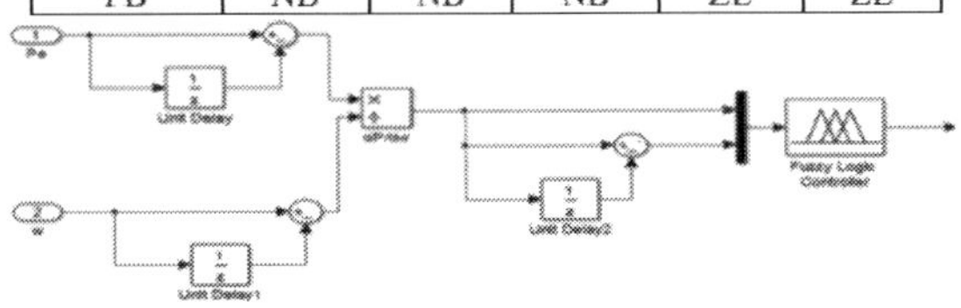

Fig. 10 wind turbine MPPT fuzzy control scheme

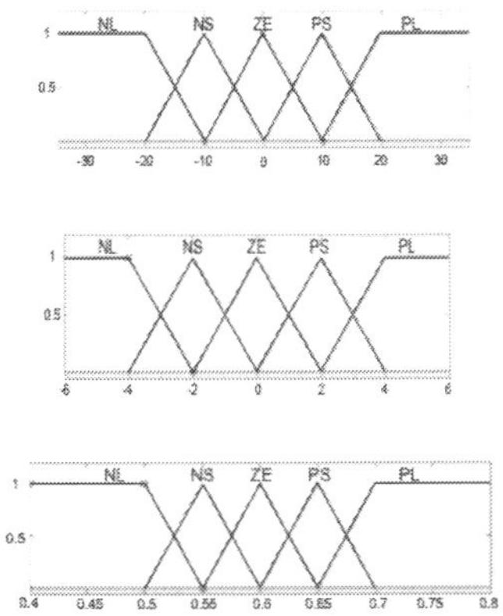

Fig. 11 Membership function of error (e), delta error (de) and duty cycle (duty) as a input – output FLC for WT MPPT

TABLE II
RULE BASE FUZZY LOGIC IN WIND TURBINE MPPT

de \ e	NL	NS	ZE	PS	PL
NL	NL	NL	NS	NS	ZE
NS	NL	NS	NS	ZE	PS
ZE	NS	NS	ZE	PS	PS
PS	NS	ZE	PS	PS	PL
PL	ZE	PS	PS	PL	PL

3 Result

The result shows that hybrid PV/Wind generation system can achieve the maximum power at different solar irradiation and wind speed. The maximum power is controlled by using fuzzy logic applied in its DC/DC Converter to set the duty cycle of the High Gain Integrated Cascade Boost (HGICB) converter. Output of HGICB converter is connected to Voltage Source Inverter (VSI) using voltage and current regulator. Thus, the DC-link voltage is kept constant at 650 V. Output of VSI is connected to 1 MVA grid.

Using fuzzy logic, the duty cycle of DC/DC converter is set to reach the maximum power. The input of fuzzy logic controller is error (e) and change in error (de). In PV system, the error is changed in power divide by change in voltage.

$$e = \frac{\Delta P}{\Delta V} = \frac{P(t)-P(t-1)}{V(t)-V(t-1)} \quad (1)$$

But in Wind turbine system using PMSG in variable speed-fixed pitch, the converter used are 3-phase full bridge uncontrolled rectifier and HGICB converter. The fuzzy logic design of wind turbine is different from PV system. The input of fuzzy logic is changed in mechanical power (Pm) divide by change in rotational speed (w).

$$e = \frac{\Delta Pm}{\Delta w} = \frac{Pm(t)-Pm(t-1)}{w(t)-w(t-1)} \quad (2)$$

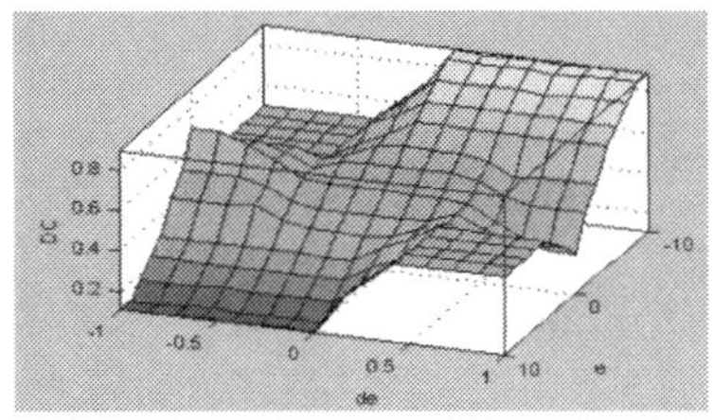

Fig. 12 A surface of fuzzy logic MPPT for PV system

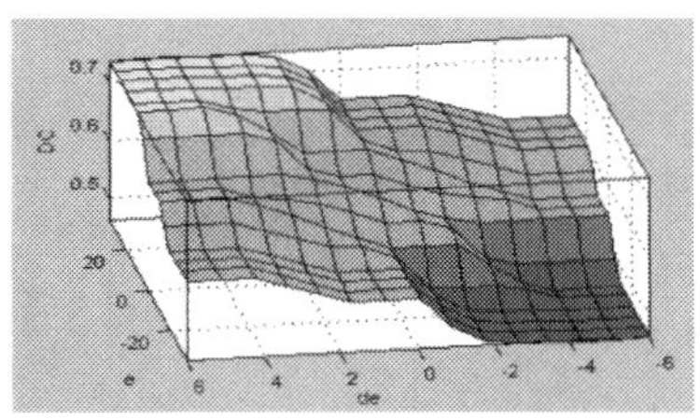

Fig. 13 A surface of fuzzy logic MPPT for Wind turbine system

There are two (2) scenarios to experimenting the Hybrid PV/Wind with the control scheme above. First is the hybrid system is tested to connects to the grid with constant load. The various combination of irradiation and wind speed was applied. The second is hybrid PV/Wind connects to the grid with the load profile with the high irradiation condition occur at daylight and high wind speed occur at daynight.

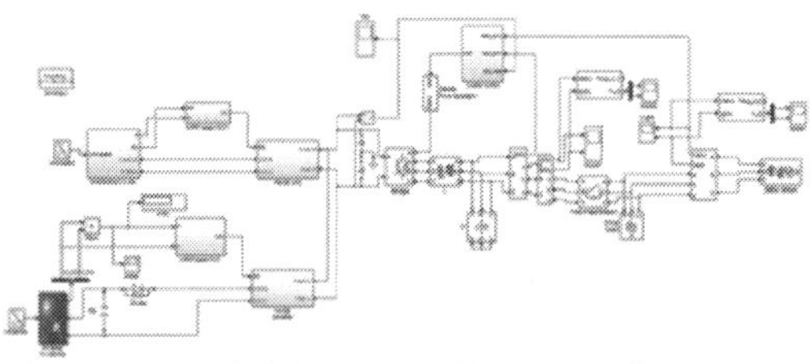
Fig. 14 Hybrid PV/Wind on 1MVA grid system Simulink model

As a result, the following figure below shows the power output from the Hybrid PV/Wind Turbine system that connects to the grid both active and reactive power. The common DC bus voltage stay in constant due to irradiation and wind speed changes. Irradiation was drop from $1kW/m^2$ to $0.4kW/m^2$ and wind speed changes from 9m/s to 4m/s.

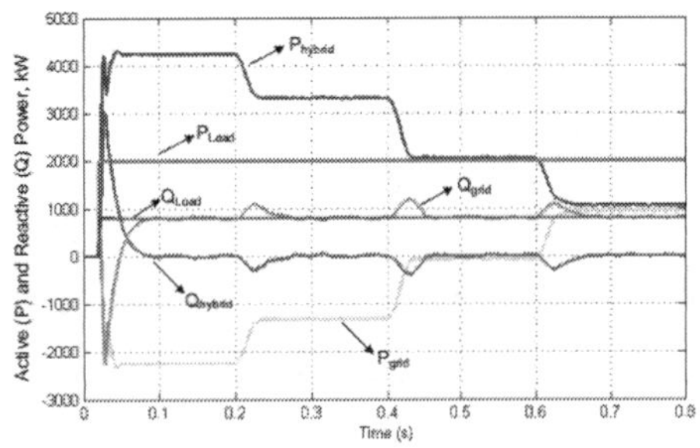

Fig. 15 Hybrid PV/Wind power delivered to the grid with constant load

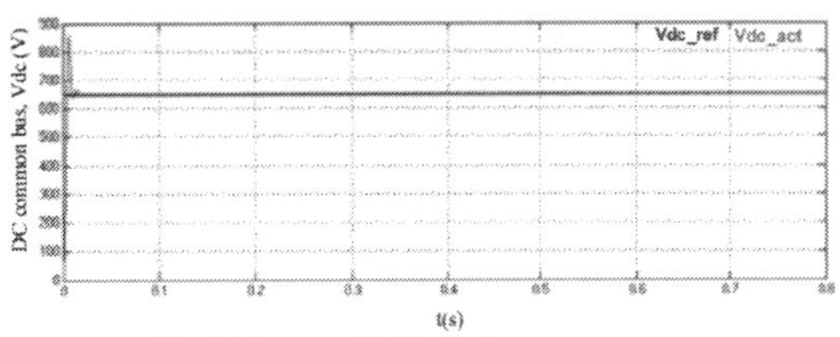

Fig. 16 Common DC bus voltage stay constant

Second, the hybrid PV/Wind was applied to the grid which has load profile, irradiation profile and wind profile. The blue line in the figure below shows the total amount power of the hybrid PV/Wind that injected to the grid as a result from irradiation and wind changes.

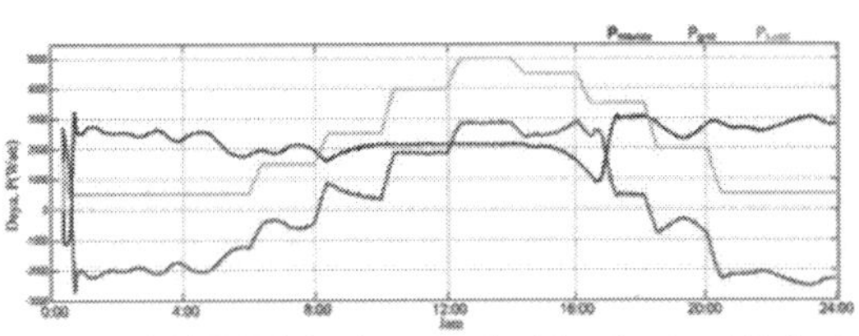

Fig. 17 Hybrid PV/Wind active power that injected to the grid in load profile condition

4 Conclusion

The conclusion from this proposed paper explained as follow:

1. The utilizing of HGICB in small scale hybrid system connected to grid could improve conversion effectiveness because it have high ratio conversion and wide input range.
2. Total installed capacity of hybrid system is 4.4 kW with DC voltage constant in 650 V without using storage device.
3. The utilizing Hybrid PV and wind turbine can reduce the power from existing conventional generating unit in utility grid system to supply the load.
4. The utilizing of fuzzy MPPT to arrangement of switching duty cycle in HGICB both in PV and Wind has error percentage low (<2%). It's a fast-reacting MPPT.
5. The utilizing of current control PWM inverter with synchronies reference theory could accelerate transient response time while hybrid system connected to grid utilities.

References

1. Abbasi, R. *Improvement of the Integration of a Grid-Connected Wind-Photovoltaic Hybrid System.* Electrical Engineering and Software Application (ICEESA) International Conference on, 2013
2. Soedibyo, Feby Agung P., Prof. M. Ashari, *Control Design of Photovoltaic BPSX-60 Using Fuzzy Logic Controller for Low Voltage Grid.* International Seminar on Intelligent Technology and Its Application, 2015
3. Narsa Reddy Tummuru, Mahesh K. Mishra. *Integration of PV/Battery Energy Conversion System to the Grid with Power Quality Improvement Features*, IEEE Transactions On Power Electronics, 978-1-4673-4569-9, 2013
4. Lakshman Rao S.P, Dr. Ciji Pearl Kurian, Dr. B.K.Singh, Athulya Jyothi V. *Simulation and Control DC/DC Converter for MPPT Based Hybrid PV/Wind Power System,* International Journal of Renewable Energy Reasearch Vol.4 No.3, 2014
5. Ping Yang, Jianping Xu, *A New Quadratic Boost Converter with High Voltage Step-up Ratio and Reduced Voltage Stress,* IEEE International Power Electronics and Motion Control Conference-ECCE Asia, 2012
6. Ahmed Mohamed Abdel Moniem El-Sebani. *Sensorless MPPT Technique for Grid Connected PMSG Wind Turbine*, Thesis of Electrical and Control Engineering Arab Academy For Science and Technology and Maritime Transport, Egypt, 2014

MATEC Web of Conferences 59, 02002 (2016) DOI: 10.1051/matecconf/20165902002
ICFST 2016

Analysis of the disturbance voltage and gpr in substation and mobile base station co-construction

Jie Long[1], Shu-Yan Jia[1], Shou-Ren Diao[2] and Lu Wang[3]

[1]*Jiaozuo Power Supply Company, State Grid Henan Electric Power Company, Jiaozuo, China*
[2]*Information Communications Branch, State Grid Hebei Electric Power Company, Shijiazhuang, China*
[3]*North China Electric Power University, Baoding, China*

Abstract. It is essential to consider the electrical safety protection distance when the mobile base station and the substation are building together. This paper makes an analysis from three aspects: the influence on the disturbance voltage of the secondary cable in substation when the station is struck by lightning; the influence on the base station grounding grid caused by the station switching operation and short circuit faults. According to the relevant national and international standards, the electrical parameters are calculated when the substation and mobile base station grounding grid connected in different ways and different distance. The conclusion of the paper provides a theoretical basis for engineering construction with much higher practical value. Meanwhile, it is of great significance to develop the win-win co-operation between the power and telecommunications operators and to explore the diversified businesses of electric power.

1 Introduction

In recent years, with the communications users growing, the construction of mobile base station increased year by year. For independent mobile base stations which need to supply power to some large base stations, the costs associated are growing. The co-construction of mobile base station and substation may solve the problem of independent transformers, and to provide the thought of deep cooperation for the electric power department and telecommunication. According to the current study, the effects of lightning current to the base station were studied when mobile base stations struck by the lightning [1-2], the transient response of lightning strike to the substation ground network were researched in the literature[3-5], and the electromagnetic interference of switch operation of substation to secondary cable and the substation short-circuit fault were also studied [6-8]. Meanwhile, the related issue of the co-construction has not been clearly researched at home and abroad. In this paper, disturbance voltage and GPR are analyzed in substation and mobile base station co-construction. The calculation results provide theoretical basis for the construction of mobile base stations and substations.

2 Model analysis

In this article, calculation models of the substation grounding grid, mobile base station grounding grid, base station tower and secondary cable shielding layer are established through CDEGS (Current Distribution, Electromagnetic Interference, Grounding and Soil Structure Analysis). The disturbance voltage and GPR are analyzed from three aspects. Firstly the substation and mobile base station grounding grid are connected in different ways and different distance. Then it is occurred that base station tower top is struck by lightning. Finally substation switch is operated and short circuit fault is appeared.

In this paper, practical operating 110 kV substation and mobile base station in a certain area of Hebei served as the analysis background. Among that, the specification of the substation grounding grid is 72 × 64.2m2. The buried depth of horizontal ground rod is 0.8m. Vertical grounding depth is 1.5m. The flat steel of 50×6mm2 and steel angle of 50×5×2500mm3 are used as level ground conductor and vertical grounding conductor separately. The vertical grounding system spacing of equalizing bring is 6m. The resistivity and permeability of the grounding conductor are 1.75×10-8 Ωm and 4π×10-7H/m. Soil resistivity of single layer used is 100 Ωm. KYJCE/B3-14×1.5 is adopted as the 40m target cable on both ends. Connected and unconnected model of substation and mobile base station grounding grid is shown in Figure 1

The mobile base station grounding grid is 3 × 3m2. Galvanized steel is used in ground network, grounding electrodes with 50×50×5mm3 steel angle and the connection and introduction using 40×4 mm2 flat steel. The high of base station tower reaches 45m. The base station tower model is shown in Figure 2.

DOI: 10.1051/matecconf/20165902002

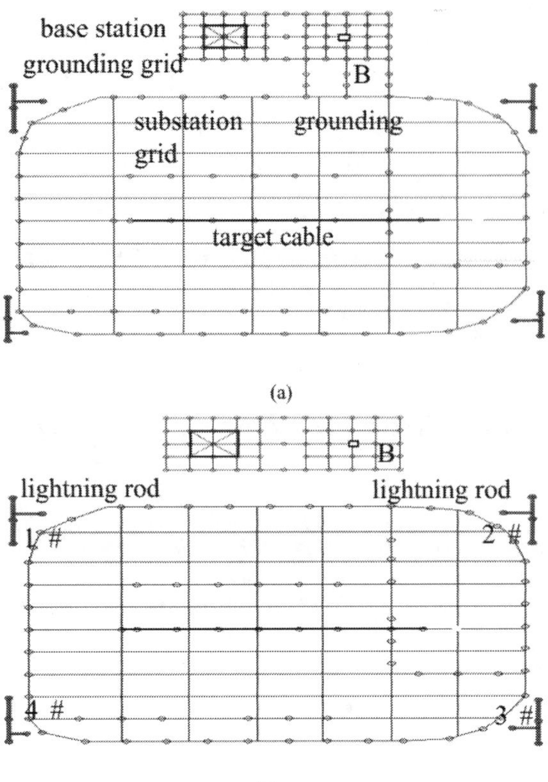

Figure 1. Grid model of substation and mobile base station:(a)Connected model,(b)Unconnected model.

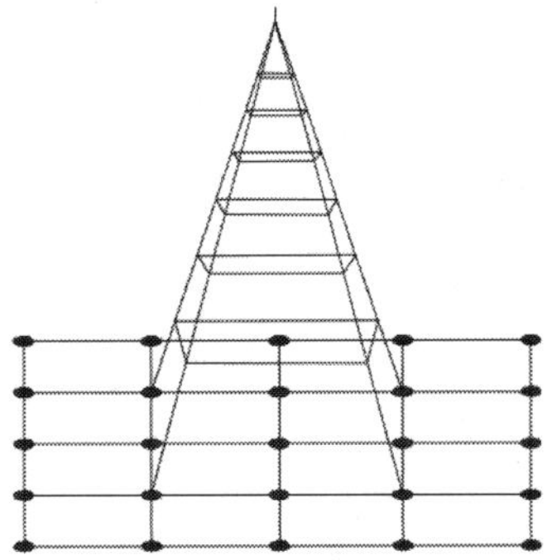

Figure 2. Station tower model

From the existing literature and experiments, lightning current amplitude in hebei region is mainly concentrated under 100kA. According to the most serious situation, 10/350s, 250kA double exponential lightning current model for the international standard IEC 62305-1in [9] is adopted. Lightning current expression is as follows:

$$I(t)=I(e^{-\alpha t}-e^{-\beta t}) \tag{1}$$

Among that, I=256107.3, a=2049.38, b=563758.3.

Transfer impedance expressions of secondary cable [10] is as follows:

$$Z_T = Z_d + j\omega(M_h \pm M_b) \tag{2}$$

For that, computational formulas of Scattering impedance Zd ,Small inductance Mh, Weave the inductance Mb are as follows:

$$Z_d \approx \frac{4}{\pi d^2 \sigma n C\cos\alpha}\frac{(1+j)d/\delta}{\sinh((1+j)d/\delta)} \tag{3}$$

$$M_h = \frac{2\mu_0 C}{\pi\cos\alpha}\left(\frac{b}{\pi D_m}\right)^2 \exp\left(-\frac{\pi d}{b}-2\right) \tag{4}$$

$$M_b = \pm\frac{\mu_0\left[ndh+\left(b-\frac{bh}{d}\right)\frac{(h+b)}{2}+B\right]A}{2\pi C(D_0+2d+h)\cos\alpha} \tag{5}$$

Of which: b=4πcos/C-nd, h=2bd/(b+d), B=n(d2-πd2/4), A=($\sqrt{v}$ +1)2cos(2α), v=C2sinα/(4πacosα).

The diameter of the insulation layer is D0=3.1mm. The diameter of each braided line is d=0.2mm. The number of weaving beam braid is C=24. The number of wires in weaving beam is n=6. The average radius of woven layer is a=7.8mm, the braided angle α=38.1°。

3 Calculation method and process

3.1 Lightning

In order to reduce the lightning current to zero, the duration of the lightning current is T=2500s. The sampling index is n=15. The sampling time interval is t=0.0763s. The Nyquist frequency is f=6.554MHz. Because the lightning current waveform rise is very steep, it leads to a wide spectrum of lightning current. However, the main frequency components are less than 50kHz. Therefore, it selects 422 frequency points, which are mainly low frequency components and a small amount of high frequency components to get the main frequency of lightning current. The lightning current waveform is shown in figure 3. The main calculation process is three steps. Firstly, the spectral density of the lightning current is calculated with the FFT. Based on the characteristics of its spectral density, the right amount of calculating frequency points is selected. Second, it is necessary to calculate the unit current of the cable shielding layer through the CDEGS, when the excitation source is given frequency points and the corresponding unit injection current. The unit current of the cable shielding layer and the amplitude spectrum of the corresponding lightning current are multiplied to obtain the total current of shielding layer. At last, it is obtained the disturbance voltage that the total current of shielding layer and the transfer impedance of the shielded cable are multiplied. And the time domain waveform of the disturbance voltage is calculated with IFFT.

3.2 Switching operation

This paper mainly studies the influence of switching operation in substation on the second equipment of mobile base station. The transient overvoltage caused by the station switching operation causes the uneven ground potential rise of the substation grounding grid and the base station grounding grid. The specific calculation process is similar to that of the lightning.

This paper analyzes the over voltage with the double exponential function[11]. The specific expression is as follows:

$$u(t) = AU_{p}(e^{t/\tau_1} - e^{t/\tau_2}) \tag{6}$$

Where, Up for peak value, A for wave coefficient, τ1, τ2 for time constant, the front time/half peak time: 250/2 500, τ1=-0.315, τ2=-62.500, A=1.104, Up is normalized

3.3 Short circuit fault

The most serious harm is the fault current in the ground when the short-circuit failure occurs in the substation, rather than the total short-circuit current. The calculation of fault current in the ground needs to consider the shunt coefficient of the ground grid. The expression of the shunt coefficient of the grounding grid is Ksg. That is:

$$K_{sg} = \frac{I_g}{I_0 - I_N} \tag{7}$$

Where I0 is the short-circuit current provided by the power supply. IN is the current flowing through the neutral point of the transformer or the generator. Iw is the total current flowing through the ground wire.

According to the actual data of the actual substation, the maximum short-circuit current is 11kA. And the shunt coefficient of the grounding grid is Ksg=0.5. So the fault current is Ig=5.5kA. The specific calculation process is similar to that of the lightning.

4 Computation results

Through the model of CDEGS, this paper analyzes the connection and non-connection of substation grounding grid and mobile base station grounding grid. The article gives an analysis of the grounding grid spacing in five respects: d1=10m、d2=20m、d3=30m、d4=40m、d5=50m. The calculation result of the disturbance voltage of the secondary cable in substation when the station is struck by lightning is shown in Figure 3. The GPR of point B caused by the station switching operation is shown in Figure 4. The power frequency transient voltage of point B with short circuit fault is shown in figure 5.

Three conclusions can be obtained from the analysis of the results of the calculation. The disturbance voltage of the secondary cable and GPR are maximum with the grounding grid connection and minimum distance. (1) The maximum disturbance voltage is 0.2772kV when the top of the base station is struck by lightning. (2) The maximum GPR of point B caused by the station switching operation is 0.5965kV.The maximum power frequency transient voltage of point B with short circuit fault is 3.3583kV.

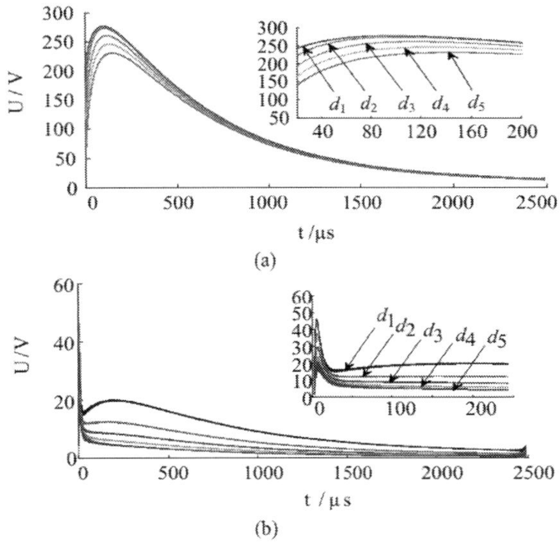

Figure 3. Disturbance voltage waveform of the cable shielding layer (different distance):(a)connection,(b)non-connection.

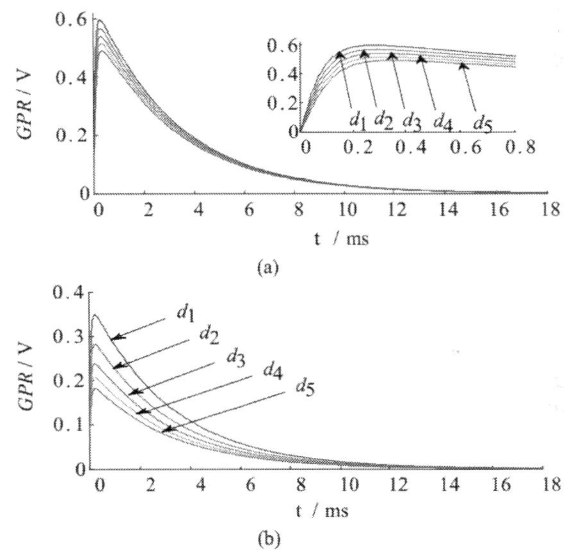

Figure 4. GPR of point B with switching overvoltage (different distance):(a)connection,(b)non-connection

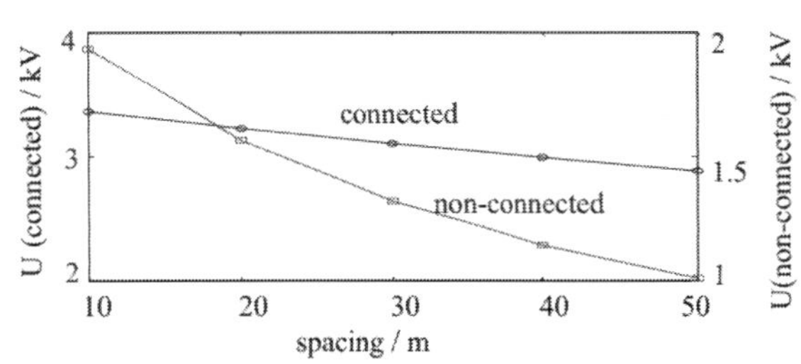

Figure 5. Power frequency transient voltage of point B with short circuit fault (different distance, connected, non-connected).

In this paper, the size of disturbance voltage of cable port with lightning striking the different lightning rod is calculated to verify the connection reliability of the grounding grid. The calculation results are shown in Table 1. Where, 1#: lightning striking the 1st lightning rod; 2#: lightning striking the 2st lightning rod. 3#: lightning striking the 3st lightning rod. 4#: lightning striking the 4st lightning rod. 5#: lightning striking the 5st lightning rod (grounding grid connection). 6#: lightning striking the 6st lightning rod (grounding grid non-connection). As shown in Table 1, the disturbance voltage of the cable with the grounding grid connected or non-connected is less than that of the cable when the lightning strikes each lightning rod. The disturbance voltage of the cable is less than 0.4kV when the lightning amplitude is lower than 100kV. Because the lightning current amplitude of Hebei area is mainly concentrated in 100kA, the disturbance voltage can meet the national standards. The maximum disturbance voltage of the cable when lightning strikes the base station is 0.15kV.

Table 1. Relationship between lightning amplitude and cable port disturbance voltage with different lightning point.

Lightning Amplitude /kA	Disturbance Voltage /kV					
	1#	2#	3#	4#	5#	6#
200	0.615	0.580	0.649	0.656	0.221	0.037
150	0.461	0.435	0.487	0.492	0.166	0.027
100	0.307	0.290	0.324	0.328	0.110	0.018
80	0.246	0.232	0.259	0.262	0.088	0.014
80	0.246	0.232	0.259	0.262	0.088	0.014

It is essential to consider the interference degree of the equipment when the electrical safety protection distance is calculated. According to the existing standard [12], the limiting value of the target cable disturbance voltage is U1=0.5kV when the station is struck by lightning. The limiting value of transient ground potential rise is U2=50kV with the point B caused by the station switching operation. The limiting value of the power frequency transient voltage with short circuit fault is U3=5kV.

Because the electromagnetic disturbance voltage of the connected grounding grid is less than that of the non-connected grounding grid, the protective distance of the non-connection mode is mainly analyzed. As shown in Table 2 the relationship between the cable safety length and different distance with grids non-connected is analyzed, when the lightning current amplitude is 250kA. The relationship between the operating voltage and distance of grounding grid non-connected is calculated. According to the amplitude of different operation voltage waveform, the corresponding grounding grid spacing is selected as shown in Table 3. The relationship between different short circuit grounding current and distance of grounding grids non-connected is analyzed as shown in table 4. The corresponding grounding grid spacing is selected by using different short-circuit current.

Table 2. Relationship between the cable safety length and different distance with grids non-connected.

Length /m	U1/kV				
	d1	d2	d3	d4	d5
200	0.6156	0.5801	0.6497	0.6569	0.2218
150	0.4617	0.4351	0.4872	0.4927	0.1663
100	0.3078	0.2901	0.3248	0.3285	0.1109
80	0.2462	0.2320	0.2599	0.2628	0.0887

Table 3. Relationship between the operating voltage and distance of grounding grid non-connected.

Length /m	U1/kV				
	d1	d2	d3	d4	d5
140	47.423	----	----	----	----
150	50.585	----	----	----	----
170	----	48.628	----	----	----
180	----	51.187	----	----	----
210	----	----	49.731	----	----
220	----	----	51.893	----	----
240	----	----	----	48.746	----
250	----	----	----	50.621	----
270	----	----	----	----	49.681
280	----	----	----	----	51.337

Table 4. Relationship between different short circuit grounding current and distance of grounding grids non-connected.

Length /m	U1/kV				
	d1	d2	d3	d4	d5
14	4.922	----	----	----	----
15	5.274	----	----	----	----
17	----	4.833	----	----	----
18	----	6.329	----	----	----
20	----	----	4.802	----	----
21	----	----	5.042	----	----
24	----	----	----	4.994	----
25	----	----	----	5.203	----
27	----	----	----	----	4.9605
28	----	----	----	----	5.1441

Through the analysis of the above, the longer the length of the cable safety, the greater the magnitude of the operating voltage, the greater the limitation of the short circuit current can be allowed with the greater the distance between the grounding grid.

5 Conclusions

This paper makes an analysis about the problem of electrical protection spacing when the mobile base station and the substation are built together. It establishes the calculation model of substation grounding grid, mobile base station grounding grid, base station tower and the secondary cable. The minimum disturbance voltage, minimum ground potential rise and power frequency transient voltage can be obtained by calculation with grounding grid non- connected and the largest spacing.

The program of 10 meters and non-connected grounding grid spacing can be used when the mobile base stations and substations are building together. The conclusion of the paper provides a theoretical basis for engineering construction with much higher practical value. Meanwhile, it is of great significance to develop the win-win co-operation between the power and telecommunications operators and to explore the diversified businesses of electric power.

References

1. P. Farid, B. Dawalibi, W. Ruan, S. Fortin, Proceedings of Internat-ional Conference on Electromagnetic Compatibility, (1995)
2. P. Li, Y.-H. Lv, Z.-W. Li, Chinese Journal of Radio Science **101**, 66 (2012)
3. L.-L. Li, Z.-Y. Xie, N. Li, Electrical Measurement & Instrumentation **2**, 37 (2008)
4. M.-L. Wu, X. Cui, High Voltage Engineering **6**, 1 (2003)
5. W. Ruan, S. Fortin, F. P. Dawalibi, Asia-Pacific International Conference, (2011)
6. C.-H. Zhang, Y. Wang, X.-M. Xie, Electrical Measurement &Instrumentation 10, 94 (2013)
7. G. Yu, J.-L. He, B.-C. Mao, Electro technical Journal **12**, 19 (2002)
8. B. Zhang, J.-P. Wu, H. Xiao, High Voltage Engineering, **3**, 720 (2012）
9. IEC, IEC 62305-1 Protection against lightning-Part 1： General principles, (Geneva: IEC, 2006)
10. H. Ma, J.-P. Ge, K. Wu, Wire and cable **6**, 1 (1993)
11. GB/T 17626-1 Electromagnetic compatibility-Testing and measurement techniques ， (China Standard Press, 2006)
12. GB/T 17626-1 Electromagnetic compatibility-Testing andmeasurement techniques, (China St-andard Press, 2006)

MATEC Web of Conferences 59, 02003 (2016) DOI: 10.1051/matecconf/20165902003
ICFST 2016

A New Method of Grounding Grid Design

Zhen-Bao Feng[1] , Chun-Sheng Wang[1], Ji Xia[2], Lu Wang[2]

[1]*Jiaozuo Power Supply Company, State Grid Henan Electric Power Company, Jiaozuo, China*
[2]*North China Electric Power University, Baoding, China*

Abstract. In order to equalize the electric potential distribution of the grounding grid surface, and improve the safety level of the grounding grid, method for optimal arrangement of conducts in the grounding grid is proposed in this paper. The calculation results about maximum touch voltages and mesh potentials show that the method can reduce touch voltage and equilibrium surface potential. It is shown that the design of the grounding grid is related to not only the soil environment, but also the rectangular shape. Comparing with other methods, the method proposed in the paper can reduce maximum touch voltage about 12%. This method also can be applied in a uniform and non-uniform soil, rectangular and square grounding grids. The optimize effect meets the related standards of the power industry.

1 Introduction

Grounding grid is an important guarantee of the safe operation in substation. Grounding grid performance has been the hot spot research for a long time. Ground grid can provide a common reference point for various electrical equipment. When lightning strike or malfunction occur in power system, the grounding grid can reduce the substation ground potential rising and discharge fault current rapidly. The performance of grounding grid is directly related to the personal safety and electrical equipment operating normally.

The earliest research is the theory of the arrangement grounding grid of the non - uniform soil which was put forward by Sverak [1]. Dawalibi had made some researches about optimal placement of the ring conductor in the grounding grid [2-3]. In 1986, Professor Chen Xianlu of Chongqing University proposed a optimization method of grounding grid in uniform soil. Professor Zeng Rong and He Jinliang researched on grounding grid performance in the high frequency situation. Professor Sima Wenxia, Wen Xishan arranged grounding grid ring conductor non-uniformly in uniform soil situation by means of Genetic Algorithm (GA). Professor Gao Yanqing studied on the optimum compression ratio of grounding grid in non-uniform soil situation. Cao Xiaobin, Hu Jinsong took the lead on the optimization design of the rectangular grounding grid [4-10].

The research shows that the index arrangement of the grounding grid is not the optimal layout scheme. Due to the rapid attenuation of the exponential function, it will lead to dense edge mesh when the grounding grid is optimized. In this paper, a new optimize method can be applied in the uniform and non-uniform soil condition, and it is suitable for the square and rectangular grounding grid.

There are many electrical parameters to evaluate the performance about substation grounding grid. The touch voltage is the most difficult to meet the requirements, so it is selected as the optimization goal to reach the safety requirements in this paper.

2 The establishment of simulation model

Firstly, it is considered that the square grounding grid optimization, assuming that a=b=100m, the square length divided into n-1 parts by N ring conductors, each part being numbered from left to right, defining Mi (i=1, 2….., n-1) for the No.i segment accounted for the proportion of total length.

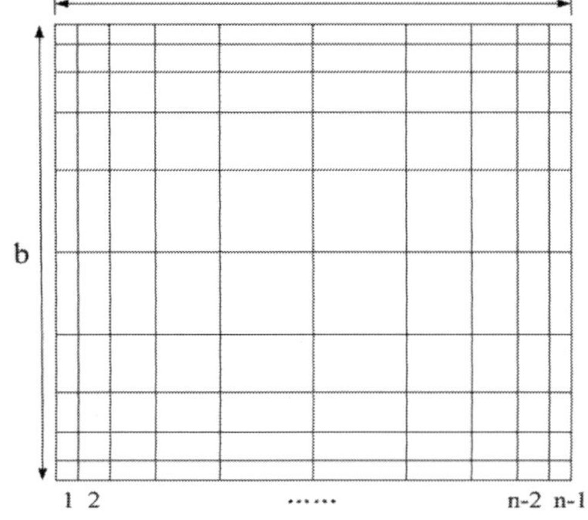

Figure 1. Unequally spaced grounding grid

When the maximum touch voltage of each mesh is equal, the percentage of each segment is as shown in

Table I. Of which, the percentage of the No.i segment is Mi. With the least square method, the fitting formula is shown in formula 1.

$$M_i = e^{-(\frac{i-n/2}{2})^2} \quad n=1, 2, \ldots, n-1 \qquad (1)$$

In this formula, only the denominator of the exponent is change, when the area of grounding grid and the number of ring conductors are changed. Therefore, the optimal fitting formula is shown in formula 2.

$$M_i = e^{-(\frac{i-n/2}{b})^2} \ ; \ n=1, 2, \ldots, n-1 \qquad (2)$$

Table 1. Optimal ground in different segment

n	i=1	i=2	i=3	i=4	i=5	i=6	i=7	i=8	i=9	i=10
1	1									
2	0.5	0.5								
3	0.3125	0.3751	0.3125							
4	0.2048	0.2952	0.2952	0.2048						
5	0.1327	0.2295	0.2755	0.2295	0.1327					
6	0.0824	0.1711	0.2465	0.2465	0.1711	0.0824				
7	0.0481	0.1199	0.2074	0.249	0.2074	0.1199	0.0481			
8	0.0261	0.0781	0.1622	0.2336	0.2336	0.1622	0.0781	0.0261		
9	0.0131	0.0469	0.1168	0.202	0.2425	0.202	0.1168	0.0469	0.0131	
10	0.006	0.0258	0.0772	0.1602	0.2308	0.2308	0.1602	0.0772	0.0258	0.006

In the fitting function b can been considered as a parameter. As long as the optimal value of b is obtained, the optimal layout scheme of the grounding grid can be received.

3 The influence of various factors on the optimal coefficient

3.1 The upper soil thickness

The relationship between the optimal parameter value b and the upper soil thickness h and the reflectional coefficient k is shown in figure 2.

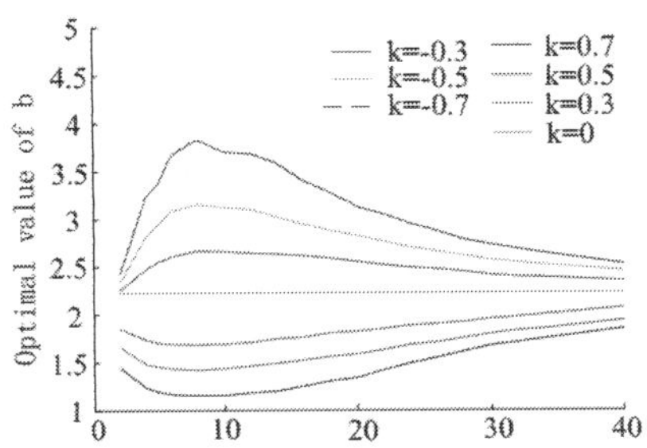

Figure 2. L=100 b-h curves

This paper makes a calculation from two situations: the reflectional coefficient k which is a fixed value and greater than zero and less than zero. In the situation of k<0, b increases with the increase of h. When h reaches a certain value (about 9 m), b reaches a maximum value, and then decreases with the increase of h. In the case of k>0, b decreases with the increase of h. When h reaches a certain value (about 7 m), b reaches a minimum value, and then increases with the increase of h.

3.2 Reflectional coefficient

Figure 3 shows that relationship between b and reflectional coefficient of double soil. When the upper soil thickness h is a constant coefficient, b decreases with the increase of k. That means the lower soil resistivity is larger and ground grid is more uniform, when the upper layer soil thickness is constant value.

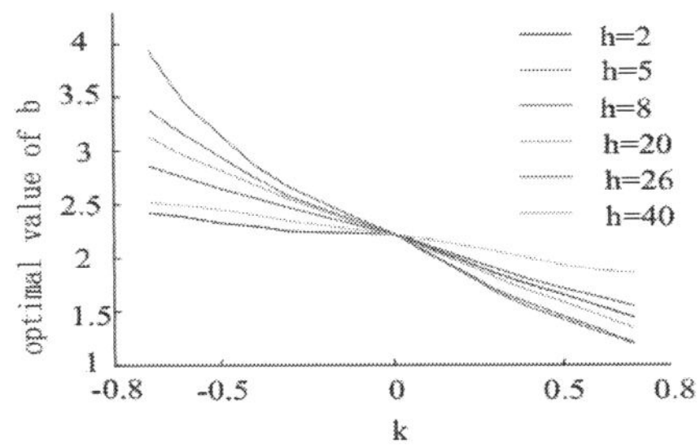

Figure 3. L=100m b-k curves

3.3 Grounding grid area

With k=-0.3, the relationship between b and side length of grounding grid L is shown in figure 4.

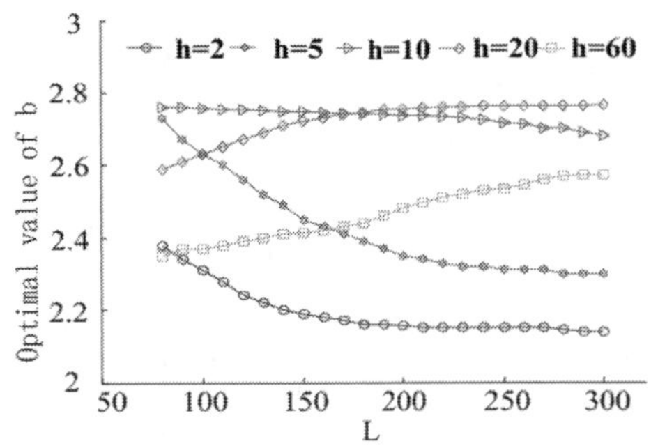

Figure 4. k= - 0.3 b-L curves

The upper layer soil in more than or less than 10m b-L curve shows different trends.

If the h<10m, the value of b decreases as the length increases. If the h>10m, the value of b increases as the length increases. When the length increased to 180m, the optimal b value tends to be stable.

According to the above results, the optimal fitting formula of b value should be divided into two situations: h⩾10m and h<10m.

3.4 Fitting formula for computing the optimal value of b

Due to the optimization parameters is related to soil thickness, reflectional coefficient and the area of the grounding grid, three parameters are set as independent variables with the neural network fitting method, and each data is considered as a three-dimensional coordinate point. Let b=f(h,k,L), and the expression is shown in formula 3.

$$b = a + c \times h + d \times L + e \times k \tag{3}$$

The corresponding coefficients of soil thickness of different grounding grids are shown in the following table 2.

Table 2. Coefficient values of fitting formula

h	a	c	d	e
$h < 10m$	2.1492	0.0526	-0.0015	-1.1351
$h \geq 10m$	2.4067	-0.0053	0.0007	-1.0298

3.5 The optimization effect inspection in non-uniform soil situation

The enclosed station grounding grid is 100×100m2. There are seven ring conductors in horizontal and vertical directions. Grounding depth is 0.5m. The ring conductor radius is 0.08m. The upper soil resistivity is 185.714Ω•m, and the thickness is 2 meters. The lower soil resistivity is 100Ω•m. The current is 1kA. The grounding grid is arranged respectively with the optimal compression ratio optimization method (method I) and the fitting function method (method II which is the method of this paper)

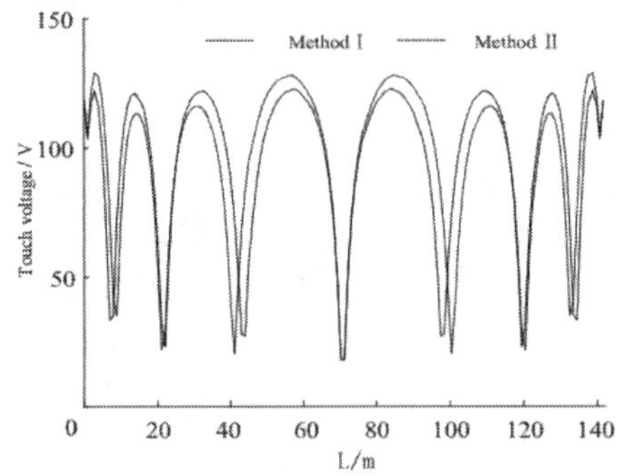

Figure 5. Touch voltage distribution on diagonal in non-uniform area

In the Figure 5, the touch voltage of edge mesh increases with method II, but the touch voltage of middle mesh reduces. And the peak touch voltage of the whole network is basically the same relative to the method I. Finally, it is balanced mesh touch voltage in grounding grid with method II.

4 To optimize the formula in the application of the rectangular grounding grid

4.1 Ensure each side of the rectangle section number

Ring conductor layout of rectangular grounding grid which is 100×60m is obtained by the fitting formula. The number of long segments is 8. The number of wide segments starts from 1 to 8 as shown Figure 6. The touch voltages with the different widths are shown in table 3.

Table 3. Max value, min value and D-value in rectangle

The segmentation number of width	The maximum value / V	The minimum value / V	The difference value / V
1	205.098	70.168	134.930
2	179.139	66.220	112.919
3	156.484	57.622	98.861
4	132.232	52.192	80.04
5	113.678	46.748	66.930
6	104.897	37.293	67.603
7	101.374	27.399	73.976
8	97.163	19.048	78.115

DOI: 10.1051/matecconf/20165902003

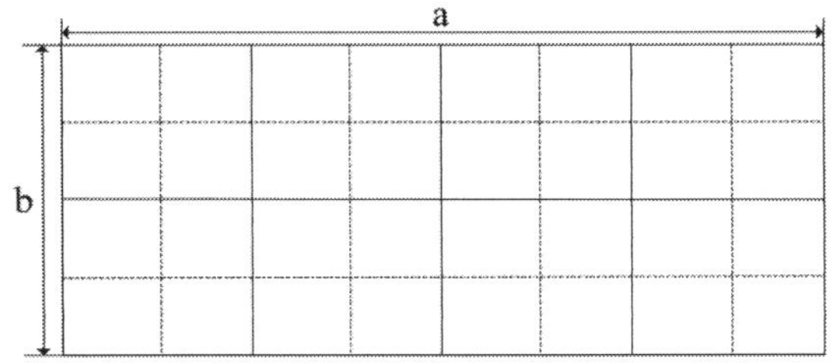

Figure 6. Rectangle grounding grid for simulation

When the maximum touch voltage is close to the minimum touch voltage , the optimization is the best.

The Table III shows that when the segmentation number of width is 5, maximum and minimum difference is minimal. When the aspect ratio is close to the ratio of segments, the optimization effect is best.

Table 4. Divided method of width in different rectangle

The length of the side a /m	The length of the side b /m	The segmentation number of length	The segmentation number of width
100	50	8	4
100	60	8	5
100	70	8	6
100	80	8	7
100	90	8	8

4.2 The optimization effect inspection in non-uniform soil

In a rectangular grounding grid, rectangular length is 120m and the width is 80m.The soil situation is same to III.

With the optimal compression ratio method(method I) and the fitting function method (method II which is the method of this paper), the touch voltage distribution on diagonal is shown in Figure 7. The touch voltage distribution of the total station with method II is shown in Figure 8. As shown in Figure 9, the touch voltage distribution of the total station with method I.

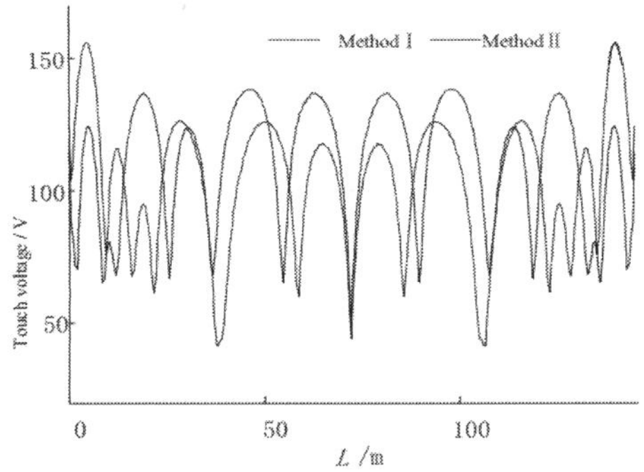

Figure 7. Diagonal touch voltage distribution in homogeneous soil with the method I and II

The method II reduces the marginal mesh size, and increase the middle mesh size relative to the method I. This arrangement can balance mesh touch voltage and reduce the maximum touch voltage .

After fitting, the maximum touch voltage is 155V with the method I. The maximum touch voltage is 138V with the method II. It is reduced by 12.1%.

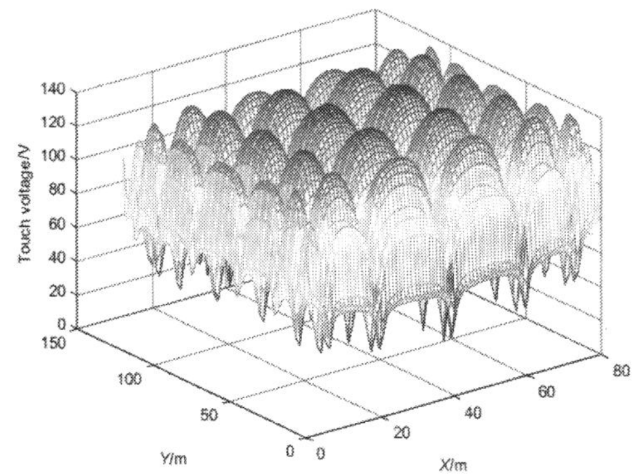

Figure 8. Touch voltage distribution of the total station with method II

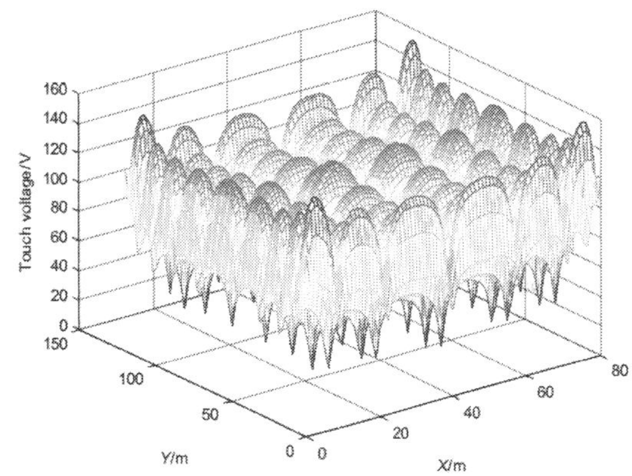

Figure 9. Touch voltage distribution of the total station with method I

5 Conclusion

By analyzing the influence of different reflectional coefficient, the area of grounding grid and the thickness of the upper soil on the optimal coefficient, this paper obtains the optimal formula of the optimal coefficient, and applies it to the optimization of the rectangular grounding grid in the non-uniform soil. The calculation results show that:

In view of the different reflectional coefficient, the upper layer soil thickness and the area of the grounding grid, the calculation formula of the optimal layout is obtained.

In the rectangular grounding grid, when the aspect ratio is close to the ratio of segments, the optimization effect is best.

Through the above results, this formula is also applicable for rectangular grounding grids and non-uniform soils. The maximum contact voltage is reduced by about 12%.

References

1. R. Zeng, X.-H. Gong, J.-L. He, IEEE Transactions on Power Delivery **23**, 667 (2008)
2. J G. Sverak, IEEE T-PAS **98**, 362 (1976)
3. F. Dawalibi, D. Mukhedkar, IEEE T-PAS **94**, 252 (1975)
4. F. Dawalibi, D. Mukhedkar, IEEE TEAS **94,** 262 (1975)
5. L. Huang, X. Chen, H. Yan, IEEE Trans on PWRD **10**,. 716 (1995)
6. W.-M. Sun, J.-L. He, R. Zeng, Y.-Q. Gao, Proceedings of the CSEE **20**, 15 (2000)
7. X.-L. Chen, E.-Y. Huang, Y. Huang, High Voltage Engineering **13**, 60 (1987)
8. Z.-Q. He, X.-S. Wen, K. Zhang, Electric Power Automation Equipment **28**, 27 (2008)
9. H. Xu, X.-S. Wen, L. Huang, High Voltage Engineering **31**, 63 (2005)
10. C.-J. Zhuang, R. Zeng, B. Zhang, High Voltage Engineering **34**, 893 (2008)

MATEC Web of Conferences **59**, 03001 (2016) DOI: 10.1051/matecconf/20165903001
ICFST 2016

Model Building of Photovoltaic Array with MPPT Function and Research on Single Phase Grid Connected

Li Zhengzhou

North China Electric Power University, School of Electrical and Electronic Engineering, 2nd Beinong Road, Huilongguan, Changping District, Beijing, China

Abstract. With the continued development of solar photovoltaic technology, research on distributed grid connected photovoltaic system has become a research focus in the field of photovoltaic grid power plant and the computer simulation technology is an effective technology means in the study. On the basis of the photovoltaic array output characteristic equation, the photovoltaic array maximum power control simulation model based on M function is established by using MATLAB/Simulink and the simulation model of single phase grid connected photovoltaic array is proposed. It overcomes the shortcomings of the process of building the model of the PV array by using Simulink component library and provides the basic guarantee for the realization of system simulation, guiding theory research and system design.

1 Introduction

With the world's energy shortage and environmental pollution becoming more and more serious, the energy and the environment become a major problem faced by mankind in the 21st century. The development and application of clean renewable energy have received more and more attention from all over the world. In the last twenty or thirty years, the solar photovoltaic (PV) technology has been developed continuously and photovoltaic has become one of the main ways of using solar energy. The development of solar photovoltaic generation system is of great theoretical and practical significance for the mitigation of energy and environmental issues, developing a broad photovoltaic power generation market and mastering the advanced technology related fields[1].

In the previous research on photovoltaic power generation system simulation, most of them are based on the quasi steady state theory to model the system components, to calculate the system power flow and to evaluate the long-term performance of the system. But in the study of the dynamic performance of photovoltaic grid-connected system, the model cannot reflect the transient change of the running state of the photovoltaic power station and the influence of the change on the power grid when the solar radiation intensity and environmental temperature changed. This needs to establish the dynamic simulation model of PV array.

PV array is the key component of distributed photovoltaic grid connected power system and its *I-V* characteristics are the nonlinear function of solar radiation intensity, ambient temperature and PV module parameters. To realize the dynamic simulation of PV power generation system, the first step is to solve the problem of how to simulate the *I-V* characteristics of PV array. Once established, the model can be used to simulate the input power of the system. But the model cannot follow the solar radiation intensity, the environment temperature and the PV array parameters in real time, so the system simulation cannot reflect the influence of the above parameters on the whole system performance. At present, the relevant work in this area has not been published in the literature. Although there are public literatures in this area abroad but the model is mainly aimed at specific PV modules and thus lack of general[2].

Based on the PV array output characteristic equation, a general simulation model of PV array maximum power control based on M function is established by using MATLAB/Simulink. The model can reflect the influence of environment temperature and solar radiation intensity on *I-V* characteristics in real time and consider the influence of PV module parameters on the output of the general model. The application example of the model in the simulation of single phase PV grid connected system is presented in this paper. The results show that it has a strong versatility and exhibits good dynamic characteristics.

2 Mathematical Model of PV array and Characteristics Analysis

PV array is the key component of the PV power generation system, and its output characteristics are greatly influenced by the external environment. In order to study the influence of the external environment change on its output characteristics, we need to establish a mathematical model. And In the MATLAB environment, the characteristics of the output under different environmental parameters need to be analyzed.

2.1 Establishment of Mathematical Model of PV Array

In practical applications, the PV modules are usually connected in series and parallel to a M×N PV array(M and N are the number of PV modules in series and parallel)[3].

In this paper, the mathematical model of the PV array for MATLAB modeling is as follows.

Under the condition of any solar radiation intensity R(W/m^2) and ambient temperature Ta(℃), the temperature Tc(℃) of solar cell is as follows:

$$T_C = T_a + t_C \times R \tag{1}$$

R is the total solar radiation on the inclined plane of the PV array and t_c is the temperature coefficient of the PV array module.

Under the reference condition, I_{sc} is short circuit current; V_{oc} is open circuit voltage; V_m and I_m are the current and voltage of peak point. When the voltage of the PV array is V, the corresponding point current is I.

$$I = I_{SC}\left(1 - C_1\left(e^{\frac{V}{C_2 V_{OC}}} - 1\right)\right) \tag{2}$$

$$C_1 = \left(1 - \frac{I_m}{I_{SC}}\right) e^{-\frac{V_m}{C_2 V_{OC}}} \tag{3}$$

$$C_2 = \frac{\left(\frac{V_m}{V_{OC}} - 1\right)}{\ln\left(1 - \frac{I_m}{I_{SC}}\right)} \tag{4}$$

Considering the change of solar radiation and temperature:

$$I = I_{SC}\left(1 - C_1\left(e^{\frac{V-DV}{C_2 V_{OC}}} - 1\right)\right) + DI \tag{5}$$

$$DI = \alpha \cdot \frac{R}{R_{ref}} \cdot DT + \left(\frac{R}{R_{ref}} - 1\right) \cdot I_{SC} \tag{6}$$

$$DV = -\beta \cdot DT - R_s \cdot DI \tag{7}$$

$$DT = T_C - T_{ref} \tag{8}$$

In the above formula, R_{ref} and T_{ref} are the reference values for the solar radiation and the PV array temperature and they are generally taken as 1kW/m^2 and 25℃.

α—the current variation of temperature coefficient under the reference sunlight (A/℃);

β—the temperature coefficient of voltage variation under the reference of the sunshine(V/℃);

R_s—series resistance of PV module (Ω)。

2.2 Characteristic Analysis of PV Array

PV array is the basic component of the solar PV power generation system. The characteristics of I-V and I-P are influenced by the sun light intensity, ambient temperature and parameters of PN junction photocell which presents a non-linear relationship. On the basis of the mathematical model of the PV array mentioned above, the simulation results are as follows by using the MATLAB environment to analyze the characteristics of the PV array[4].

The typical characteristics of PV arrays are shown in Figure 1, Figure 2. Figure 1 and Figure 2 show that P-V and I-V change with the change of the solar radiation. Figure 3 shows the law of P-V with the change of environment temperature.

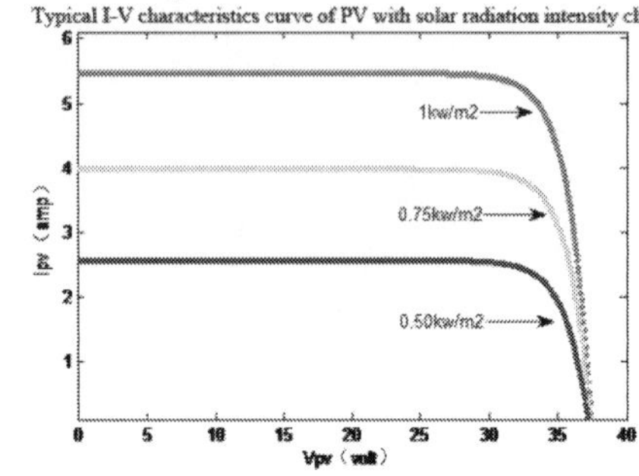

Figure 1 I-V Characteristics of Typical PV Arrays with the Variation of Solar Radiation Intensity

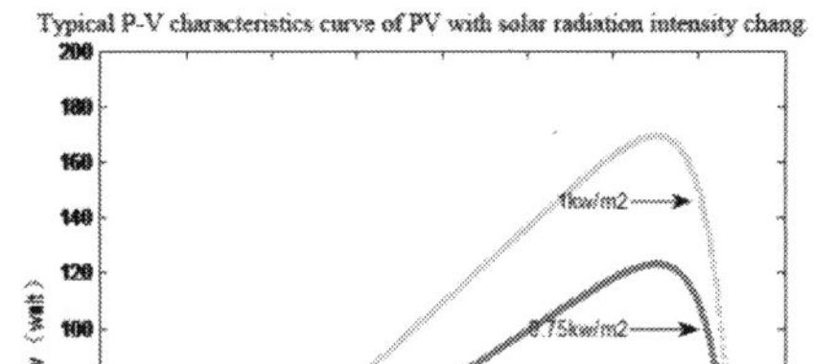

Figure 2 P-V Characteristics of Typical PV Arrays with the Variation of Solar Radiation Intensity

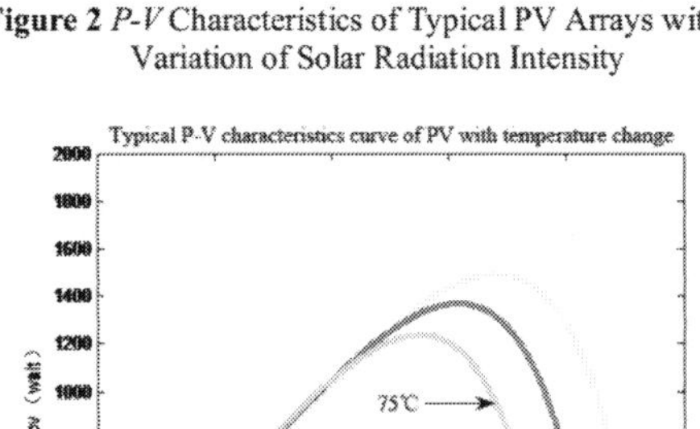

Figure 3 P-V Characteristics of Typical PV Arrays with the Variation of Temperature

3 Research on Maximum Power Tracking Control of PV Array

To improve the efficiency of PV power generation system and make full use of the energy generated by the PV array is the basic requirements of PV power generation system. Now the PV power generation system usually requires the output power of the PV array has always been the largest. It is that the system should be able to track the maximum power point of the PV array[5].

The output characteristic of PV array is nonlinear and its output is influenced by the intensity of solar radiation and ambient temperature and load. In the solar radiation intensity and environment temperature, PV array can work in different output voltage. But only in a given output voltage, the output power of PV array can reach the maximum and then the working point of PV array reaches the highest output power voltage curve which is known as maximum power point. Therefore, to improve the overall efficiency in the PV power generation system, an important way is to adjust the working point of the PV array in real time so that it always works in the vicinity of the maximum power point. This process is referred to as the maximum power point tracking (MPPT).

In this paper, a mathematical model of the maximum power point tracking of PV array is established based on the Newton iteration method for solving MPPT points.

The power of PV array in any solar radiation intensity and ambient temperature is as follows:

$$P = IV = \left(I_{sc}\left(1-C_1\left(e^{\frac{V-DV}{C_2V_{oc}}}-1\right)\right)+DI\right)V \tag{9}$$

By the extreme conditions of $dP/dV=0$, we can know:

$$I_{sc}\left(1-C_1\left(e^{\frac{V-DV}{C_2V_{oc}}}-1\right)\right)+DI-VI_{sc}e^{\frac{V-DV}{C_2V_{oc}}}/(C_2V_{oc})=0 \tag{10}$$

The formula above is a transcendental equation and we can use Newton method to iterative solution of the corresponding maximum power point of the voltage V_{max}.

$$V_{k+1}=V_k-\frac{P'(V_k)}{P''(V_k)} \tag{11}$$

$$=V_k-\frac{1+V_k\left(-I_{sc}\frac{C_1}{C_2V_{oc}}e^{\frac{V_k-DI}{C_2V_{oc}}}\right)}{\left(2+\frac{V_k}{C_2V_{oc}}\right)\left(-I_{sc}\frac{C_1}{C_2V_{oc}}e^{\frac{V_k-DI}{C_2V_{oc}}}\right)} \tag{12}$$

When $|V_{k+1}-V_k|<\varepsilon_1, V_{max}=V_{k+1}$.

In the above formula, V_{k+1} and V_k are the iteration values of $k+1$ and k respectively. ε_1 is iterative precision. $P'(V_k)$ and $P''(V_k)$ are the first and second order derivatives of V under the iteration of the k. Putting all the V_{max} into the equation (5), we can get I_{max}. Thus the maximum power P_{max} can be obtained by the following formula:

$$P_{\max}=V_{\max}\cdot I_{\max} \tag{13}$$

The simulation results show that it is able to track the maximum power point of the PV array stably and efficiently under the control of this method and quickly find a new job, keep the system stable, exhibit good dynamic characteristics at the same time in sunshine intensity, environment temperature and other parameters of the system disturbance.

4 General MATLAB Simulation Model of PV Array

A general simulation module of PV array with MPPT is established based on the above specific algorithm and simulation model in the MATLAB environment by using Simulink tools and combined with the preparation of the M function. The M function is used to solve the voltage and current of the maximum power point of the PV array in the corresponding arbitrary solar radiation intensity.

Figure 4 is the realization of the voltage of any PV cell by MATLAB Function to obtain the maximum power control voltage and to obtain the maximum current by using the $f(u)$ function in the MATLAB/Simulink and the formula (5) in the previous mathematical model.

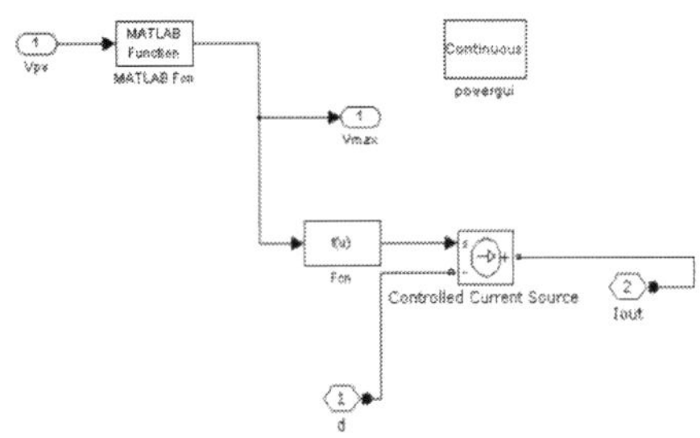

Figure 4 The Internal Structure of the General Simulation Model of PV Array

After the common simulation module of the PV array being written, it will be packaged. Figure 5 shows the appearance of the model and its internal package a number of parameters. The user can get the user interface which is shown in Figure 6 by clicking the icon in Figure 5 to facilitate the setting of the parameters thereby forming a PV array under different parameters.

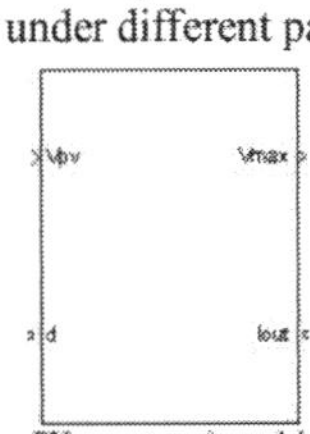

Figure 5 General Simulation Package Model of PV Array

The general simulation model of PV array with MPPT is the key component of the PV power generation system. The establishment of the model provides a basic guarantee for the research of the maximum power point tracking, guiding theory and system design of PV system.

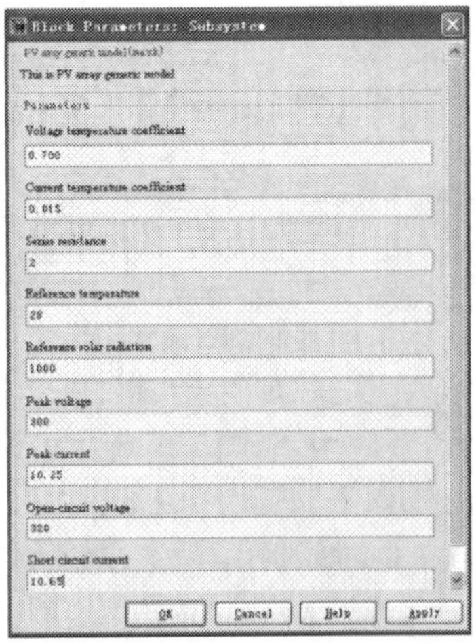

Figure 6 User Interface of General Simulation Model of PV Array

5 Application of General Simulation Model of PV Array in Grid Connected System

A general simulation model of the PV array is obtained which is used in the single phase PV grid connected system. The basic idea is to put the maximum current output of the PV array to the positive end of the inverter bridge. And the inverter bridge is triggered by the pulse trigger and the DC power is changed into AC so as to realize the grid connection with the single phase 220V. The single phase AC voltage and current and voltage amplitude after grid connection are measured by the measuring element and as the simulation results. The PV array is respectively provided with the simulation results when the solar radiation intensity is reduced from 1000W/m^2 to 800W/m^2 and 600W/m^2. DC-AC inverter module is the universal inverter bridge in MATLAB. MATLAB model of single-phase grid-connected system is shown in the following figure[6][7].

Based on the research above, we obtain the grid-connected voltage fundamental amplitude, single-phase grid connected voltage, grid connected current, voltage of the input terminal of the inverter and the output current of the PV array of PV grid connected system when solar radiation intensity is reduced from 1000W/m^2 to 800W/m^2 and 600W/m^2. We can see from the simulation results that the simulation model has the change of the PV array in real time simulation of different solar radiation intensity and the simulation results are as follows.

It can be seen from the simulation results that the inverter input voltage has been dropped from 268V to 265.6V and 264V and the drop range is within the range of 4V because of the control strategy adopted for the fixed modulation ratio when the solar radiation intensity is decreased from 1000 W/m^2 to 800 W/m^2 and 600 W/m^2 and t=1s, t=2s. The simulation result is shown in Figure 8. As can be seen from Figure 9, the PV array output current is reduced from 10.65A to 8.45A and 6.25A respectively. Figure 10 shows the change of grid connected voltage and current with the change of solar radiation. As can be seen from the figure that the grid connected voltage amplitude is constant and the grid connected current decreases with the decrease of the solar radiation intensity when the solar radiation changes. And the current is gradually reduced from the 22A to 12A.

The simulation results show that the the Matlab simulation model of single-phase PV grid-connected system in this paper can output current and voltage quickly and accurately and with the changes in the intensity of solar radiation, the output current of inverter can also change correspondingly.

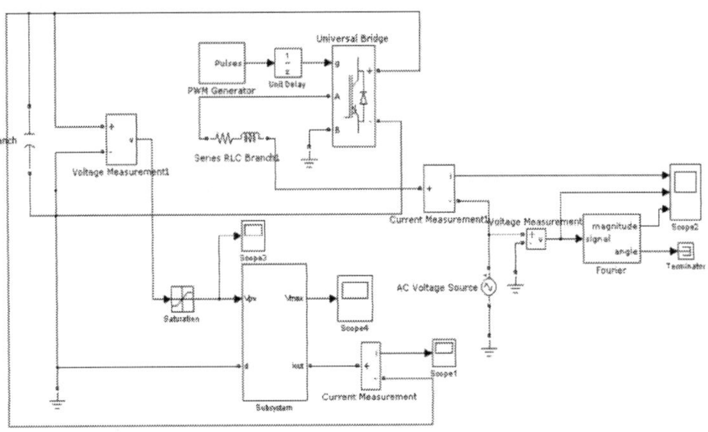

Figure 7 MATLAB Simulation Model of Single Phase PV Grid-connected System

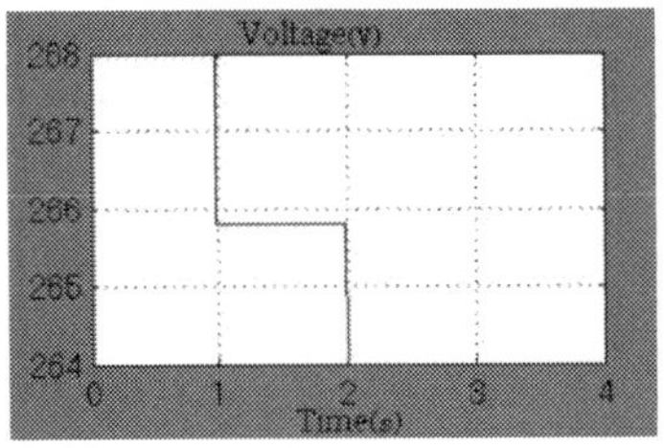

Figure 8 Relationship Between the Output Voltage of PV Array and the Solar Radiation Intensity

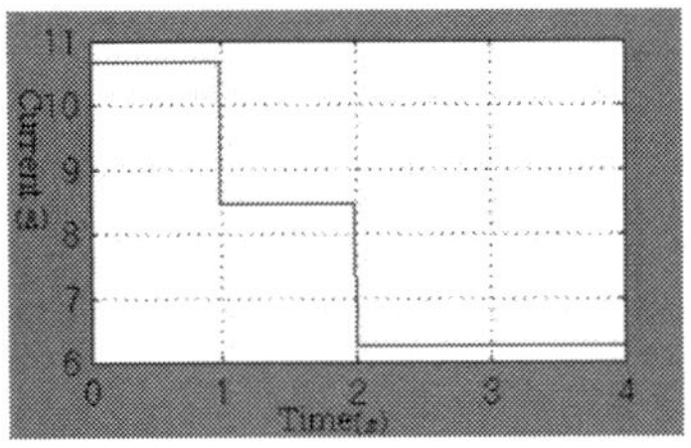

Figure 9 Relationship Between the Output Current of PV Array and the Solar Radiation Intensity

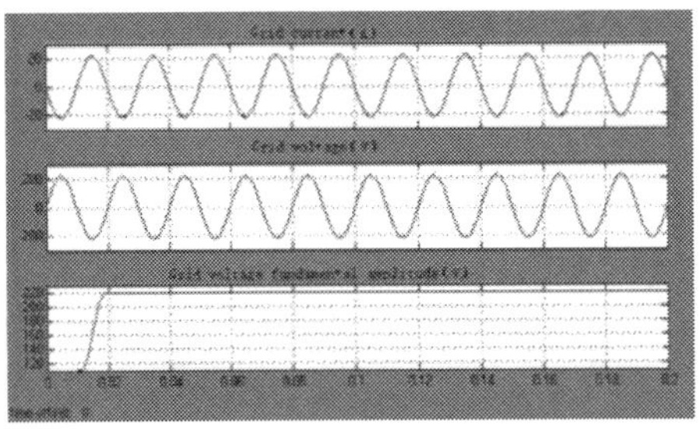

a) The Grid-connected Voltage and Current Curve of the Inverter When R=1000 W/m^2

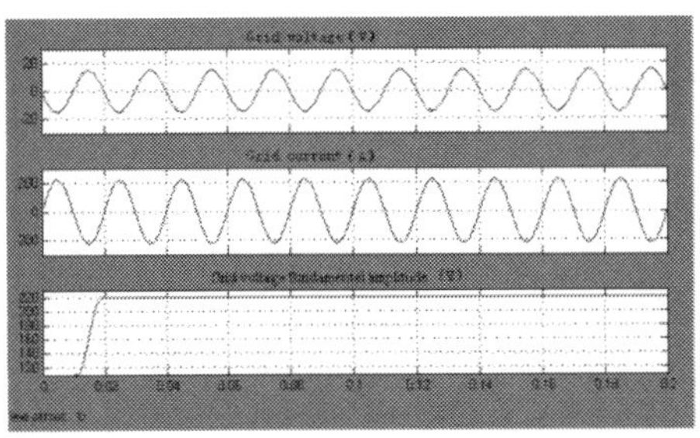

b) The Grid-connected Voltage and Current Curve of the Inverter When R=800 W/m^2

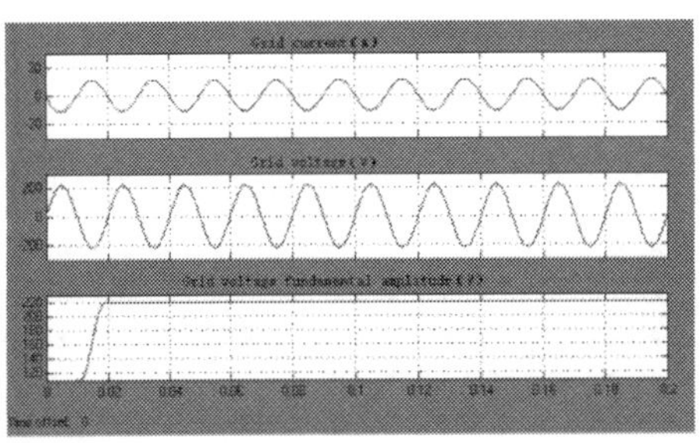

c) The Grid-connected Voltage and Current Curve of the Inverter When R=600 W/m^2

Figure 10 The Change of Grid Connected Voltage and Current with the Change of Solar Radiation

6 Conclusions

In this paper, the MATLAB is used to establish the PV array simulation model which is based on the existing mathematical model of PV array. The theoretical analysis of simulation results shows that the simulation model can accurately reflect the output characteristics of PV array. Based on the theoretical research, the general simulation model of PV array with MPPT function is established by using Matlab/ Simulink. This model can realize the maximum power tracking control using the M function in Simulink and can simulate the maximum voltage and current output of PV array quickly and accurately under different solar radiation intensity and ambient temperature. The Matlab simulation model of single-phase PV grid-connected in this paper can output current and voltage of grid-connected system quickly and accurately and with the changes in the intensity of solar radiation, the output current of inverter can also change correspondingly. It has important theoretical effect and practical value to understand and get the characteristics of PV arrays and to guide the design of practical systems.

References

1. Li Junfeng, Wang Sicheng, Zhang Sicheng etc. China PV Development Report of 2007, China Environmental Science Press,2007
2. A.ZAHEDI. Development of An Electrical Model For A PV/Battery System for Performance Prediction. Renewable Energy, 1998, 15(1): 531-534
3. Mao Meiqin, Yu Shijie, Su Jianhui, Versatile Matlab Simulation Model for Photovoltaic Array with MPPT Function. J. Journal of System Simulation, 2005,5:1248-1251
4. Zhou Dejia, Zhao Zhengming, Wu Libo. Analysis of the Characteristics of Solar Photovoltaic Cell Array based on Simulation Model. Journal of Tsinghua University (NATURAL SCIENCE EDITION) , 2007,47(7): 1109-1117
5. Wu Libo, Zhao Zhengming, Li Jianzheng etc. Research on the Stability of the Maximum Power Point Tracking Algorithm in the Single Stage Photovoltaic Grid Connected Inverter System. J. Chinese Journal of Electrical Engineering, 2006,26(6)73-77
6. Liu Shu, Liu Jianzheng, Zhao Zhengming etc. Single Stage Photovoltaic Grid Connected System Based on Improved MPPT Algorithm. Journal of Tsinghua University,2005,45(7):873-876
7. Li Jing, Xu Honghua, Zhao Hai xiang. Dynamic Modeling and Simulation Analysis of Grid-connected Photovoltaic Power Plant. Automation of Electric Power System. 2008, 32(24): 8-10

MATEC Web of Conferences **59**, 03002 (2016) DOI: 10.1051/matecconf/20165903002
ICFST 2016

A New Hybrid Bathroom System Based on Energy Saving Concept

Bo-wen Cui

Shandong University of Science and Technology, Mining and Safety engineering institute,ShanDong,QingDao,266590)

Abstract: Based on the characteristics of hot water supply in bathroom, this article proposes a new hybrid energy hot water supply system. The programmable logic controller(PLC) as the master controller was adopted in this system, which could automatically detect and storage main thermal physical of the system, such as temperature, water level, solar radiation intensity, power consumption and so on. The active thermal utilization technology of solar energy, air-source heat pump technology, solar energy intensive natural ventilation technology and low temperature hot water floor radiant heating technology were organically integrated in this system, which has the advantages of energy conservation and environment protection, high automation, safe and reliable operation, etc. At the same time, it can make good use of electric power cost between on-peak and off-peak, and promote the optimal allocation of power resources and reduce the cost, which can achieve the goal of intelligent control and energy saving.

1 Introduction

With the continuous improvement of living standards, people's demand for quality is also getting higher and higher, in order to relax and shape a good personal image, bath has become an indispensable activity in people's lives. Comfortable bath can ease the pressure, can provide a stable supply of hot water and a good heating and ventilation system is an important guarantee for the realization of the bathroom. At present, the solar hot water supply technology is becoming more and more mature, has been widely used, but it is highly dependent on environmental conditions, the stability and reliability is poor, often need to combine with the stable heat source. Boiler hot water supply system combustion of coal and other fossil fuels, emissions of SO2, CO2 and other waste gas, serious pollution and energy consumption. The traditional bathroom heating system usually use gas or coal boiler as heat source, rely on the installed on the wall of the indoor heating radiator heating device. The heating efficiency of the bathroom heating system is relatively low, and the heated air is concentrated in the upper part of the room, and the discomfort of the head and the feet is cool. In addition, most of the traditional bathroom adopt simple natural ventilation technology and its effect is poor, lead to indoor air pollution, hypoxia and make people feel serious discomfort.

In this article, we come up a heat pump water heater[1], which drived by motor, use the vapor compression refrigeration cycle with air as the heat source and provide hot water. In the system we combine solar energy utilization technology and heat pump technology for hot water supply[2], which can bring into full play to the advantages of energy conservation and achieve all weather reliable operation of the system. In recent years, low-temperature floor radiant heating system is a new heating type, it takes the whole ground as a heat sink, when convective heat transfered between floor and indoor air, people can have a comfortable feeling- feet warm head cool, then it can use the low-grade heat source as heat source, it is a kind of ideal heating system to reduce the energy consumption and improve the comfort of the building.

In part 1, we introduce the composition of hybrid energy-saving bathroom system; In part 2 we introduce the working principle and performance analysis of the system; In part 3 we introduce the innovation and application in other fields.

2 Composition of Hybrid Energy-saving Bathroom System

Aiming at the fixed opening time, large water consumption, stable discharge temperature and moderate washing temperature in colleges and enterprises, we construct a new hybrid energy-saving bathroom system, which comprises a solar assisted air-source heat pump hot water subsystem, a low temperature hot water floor radiant heating subsystem and a solar energy intensive natural ventilation subsystem. Solar assisted air-source heat pump hot water subsystem[3] consisted the plate type solar heat collector, heat exchanger, heat storage tank, circulating water pump and other components; Air-source heat pump unit, heat storage water tank, circulating water pump and other parts constitute air-source heat pump system. Low temperature hot water floor radiant heating system constituted by the air-source heat pump, thermal storage tank, circulating water pump and a heating coil. The solar energy intensive natural ventilation system composed by a transparent cover plate, an air channel, a heat absorbing plate, a heating accumulator and an insulating board.

Figure 1. The system model in physical map

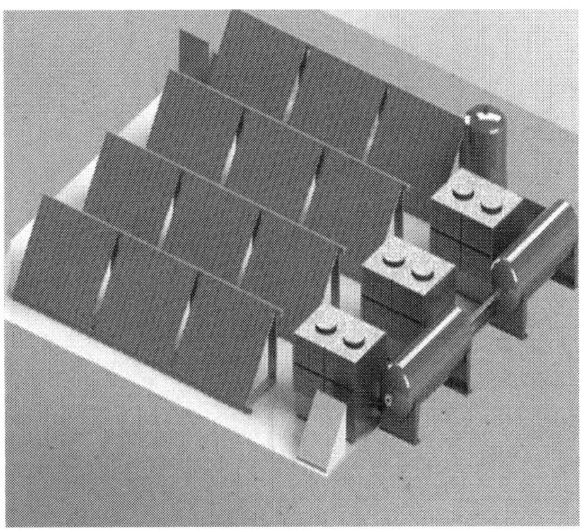

Figure 2. The solar energy, air source heat pump system

3 Working Principle and Performance Analysis of the System

3.1 Hot water heating system

Aiming at the characteristics of the solar collector's large area and low energy conversion rate, the system adopts the combination of air-source heat pump system and the solar heat collection as a supplement[4]. In order to prevent freezing in winter, the working medium in solar collector subsystem is selected as propylene glycol solution, transfered the collected heat to the storage tank through heat exchanger. The system operated by temperature cycle control mode, using two temperature sensors and one temperature controller, one of the temperature sensor is installed in the heat collector, which is close to the outlet of propylene glycol, the other installed in the storage tank bottom near the outlet. The circulating water pump started when the temperature of the propanediol is higher than the heat storage water tank and the temperature difference controller achieved the starting temperature difference. When the difference is lower than the temperature controller set, the water pump will stop running, repeatedly like this, the water temperature in the heat storage tank is increasing continuously. In the setting time, if the temperature of the heat storage water tank can't reach the set, which the bath water supply, the electromagnetic valve will be opened, and the air source heat pump unit is auxiliary heating.

The air-source heat pump heating subsystem use heat storage water tank temperature control mode, when the heat storage water tank water temperature is lower than the temperature of water bath start followed by the circulating water pump and the heat pump unit, until the water temperature in the water tank reaches the set value, and then turn to stop heat pump unit and a water circulating pump. When the outdoor temperature in winter is very low, air source heat pump heating efficiency is greatly reduced, and at the set time of insufficient heat. At this point the system by thermal power plant auxiliary steam heating mode of operation, to open the solenoid valve until two heat storage water temperature in the water tank reaches the set temperature of bath water. Combined with the state of peak and valley electricity policy, the heat pump can be in the off peak period (23:00 - 7:00 the next day) run, this will not only ease the peak power demand and supply gap, and promote the optimal allocation of power resources, is a "cut peak and fill valley" win-win strategy.

3.2 Bathroom heating system

Beforethe period when people use their bathroom, namely at the off-peak period, the running air-sourceheat pump heats th heat storage water tank for floor heating, and when the temperature of water reach the limit, the air-source-heat pump shuts down. When it needs heat, hot water circulating pump is switched on. The hot water in the heat storage water tank is transported into the floor heating coil, heating the floor in the bathroom and the space above the floor by radiation and convection and then returning to the heat storage water tank. When the temperature of the hot water hit the floor level, the air-source heat pump is switched on.

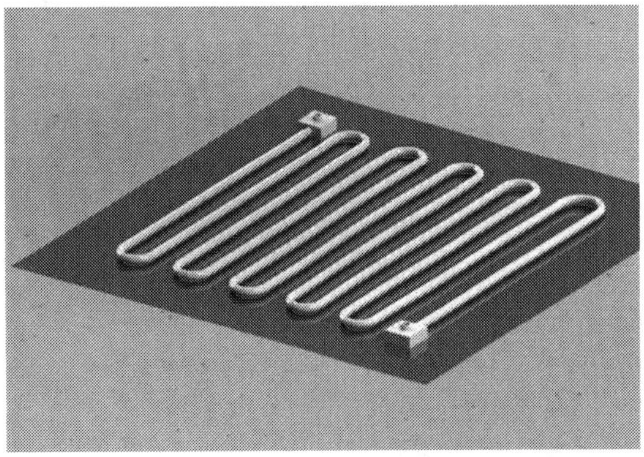

Figure 3. The heating coil diagram

3.3 Bathroom ventilation system

During the daytime , the sunshine penetrates the transparent glass cover with radiation from it absorbed by absorber plate, heating the gas in the air pipes, producing the temperature difference between the inside and the outside of the air pipes and resultant larger hot-pressure. Then the gas is flowing up along the pipes, thus the air in the bathroom is pushed into the air pipes through air outlet. The outdoor fresh air flows into the indoor space through air inlet, thus strengthening the indoor air ventilation and in the same time, the remaining part is stored in the heat storage plate for repeating the circulation in the evening.

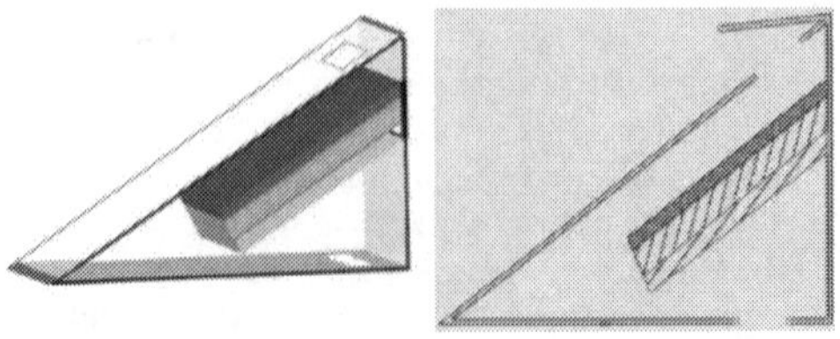

Figure 4. The solar ventilation system

4 Innovation and Application

(1) The hot water subsystem uses air-source heat pump system as main method and solar energy collection is supplemented by the hot water supply;

(2) The system uses the off-peak period to heat, promoted the electric power resources optimized disposition, which is a " peak load shifting " double-wins strategy;

(3)By using the solar energy and the solar radiation to heat the air, the system can strengthen natural ventilation technology and produce differential pressure to strengthen the natural ventilation cycle, which will provide comfort and the effective use of solar energy resources.

Reference:

1. Kong Xiangqiang,An Peng,Liu Xiaoxiao,Li Xinling,Li Ying. Development of Experimental Platform with Hybrid Energy Hot Water Supply System for College Bathrooms [P]. National utility model patent, ZL 201220300696.6. 2012.
2. Yang Lei, Zhang Xiaosong. Performance simulation of multiple heat source solar heat pump water heater system[J]. Acta Energiae Solaris Sinica,2011,32(1):120-126.
3. Tie Yan, Luo Huilong. Techno-economic analysis of solar water heating system in conjunction with air-source heat pump [J]. refrigeration and air-conditioning,2009,9(4):31-32
4. Yuan Jianshan, Zhang Guangshun, Zhang Jinfeng. Handbook of solar water heating systems [M].Beijing: Chemical Industry Press,2010.

Parameters extraction for perovskite solar cells based on Lambert W-function

Junyu Ge, Manlin Luo, Wanli Pan, Na li and Wei Peng
School of Physics and Electronics, Hunan University, Changsha 410082, China

Abstract. The behaviors of the solar cells are decided by the device parameters. Thus, it is necessary to extract these parameters to achieve the optimal working condition. Because the five-parameter model of solar cells has the implicit equation of current-voltage relationship, it is difficult to obtain the parameters with conventional methods. In this work, an optimized method is presented to extract device parameters from the actual test data of photovoltaic cell. Based on Lambert W-function, explicit formulation of the model can be deduced. The proposed technique takes suitable method of selecting sample points, which are used to calculate the values of the model parameters. By comparing with the Quasi-Newton method, the results verify accuracy and reliability of this method.

1 Introduction

Compared with other energy resources, solar energy is sustainable and far more abundant than our projected energy needs as a species. Thus, it is believed as the largest source of electricity in the future[1]. Solar energy can be used in photothermal conversion, photochemical conversion and photovoltaic conversion. Using the photovoltaic effect, solar cells convert light into electric current, which is convenient for energy storage and transport. This advantage makes solar cell a promising electrical device. Perovskite solar cell has become one of the most widely used solar cells due to its high power conversion efficiency and low material costs[2]. A new category of perovskite solar cell has been put into use these years, with remarkable rapid progress made in energy conversion efficiency. In 2014, its energy conversion efficiency has increased to a confirmed data of 17.9%. The unique quality of perovskite solar cell contributes to the outstanding performance[3].

In order to do the quantitative analysis of solar cells, it is significant to utilize a photovoltaic model which can represent the electrical characteristics of the solar cells accurately. Among different solar cell models used to describe the I-V relationship, single and double diode models are used practically. The single-diode model is now most widely used to describe solar cells' behavior. It is the simplest model with only five parameters and satisfied with most of the applications[4]. The five parameters are listed as follows: the series photocurrent (I_{ph}), the resistance(R_s), the shunt resistance (R_{sh}), the ideality factor (n) and the saturation current(I_0). This single-diode model is used to evaluate solar cells' performance, optimize device structure and fabrication process as well.

There are different methods to extract model parameters, such as genetic algorithm, differential evolution method and particle swarm optimization[5]. Generally, these algorithms require much time and computational cost to get high accuracy[6]. Herein, we present a method based on Lambert W-function to extract perovskite solar cells' parameters[7]. Due to the simple calculation process, this approach has high accuracy and calculation speed.

The key factor for precise result and efficient computation is choosing points. Points selected around short-circuit current (I_{sc}), open-circuit voltage (V_{oc}) and max power point have significant effects for the outcomes. In this paper, distinction between different point selection methods is studied. Moreover, the initial values of the parameters have no impact on the results[8].

2 Methods

All the work is based on a single-diode model here. The single-diode model is depicted in Figure 1. It is well known that the output current of this model can be described as[9]:

$$I = I_{ph} - I_0(\exp(\frac{V+IR_s}{nV_{th}})-1) - \frac{V+IR_s}{R_{sh}}, \quad (1)$$

DOI: 10.1051/matecconf/20165903003

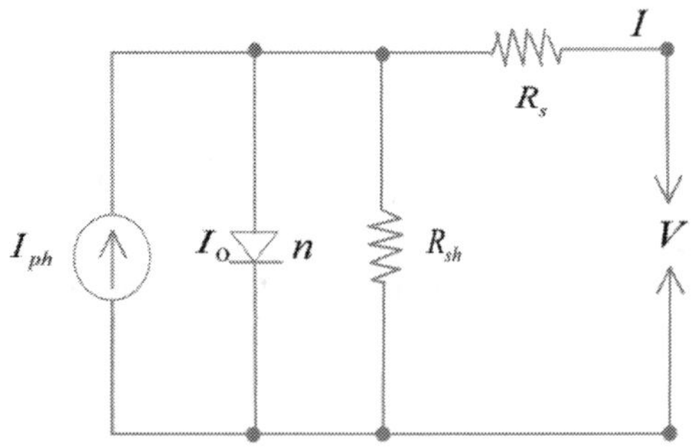

Figure 1. Single-diode model of solar cells.

The V_{th} is given by $V_{th}(T)=\frac{kT}{q}$, where q is the electron charge (1.60217646×10^{-19} C), k is the Boltzmann constant (1.3806503×10^{-23} J/K) and T is the Kelvin temperature of the p-n junction[10].

The Eq.(1) is an implicit non-linear transcendental function which has no direct general analytical solution. Lambert W-function is used to obtain the explicit formulation of the model, and this function can be described as:

$$W(x)\exp[W(x)]=x \tag{2}$$

"W(x)" represents the usual short-hand notation for the principal branch of Lambert W-function.

In our work, $R_{sh}>>0,\frac{V+I_{RS}}{R_{sh}}\approx 0$. In this way, the Eq.(1) can be written as:

$$I=I_{ph}-I_0[\exp(\frac{V+R_sI}{nV_{th}})-1] \tag{3}$$

Then,

$$\exp\left(\frac{R_sI}{nV_{th}}+\frac{V}{nV_{th}}\right)=-\frac{I}{I_0}+\frac{I_{ph}+I_0}{I_0} \tag{4}$$

Independent variable I is replaced by x, then the solution of Eq. (4) can be expressed as[11]:

$$\exp(ax+b)=cx+d \tag{5}$$

Variables are defined as follows:

$$\begin{cases}\frac{R_s}{nV_{th}}=a\\ \frac{V}{nV_{th}}=b\\ \frac{1}{-I_0}=c\\ \frac{I_{ph}-I_0}{I_0}=d\end{cases} \tag{6}$$

Multiply $-\frac{a}{c}\exp(-ax-\frac{ab}{c})$ at both sides of the equation, then x can be calculated by:

$$x=-\frac{d}{c}-\frac{1}{a}W[-\frac{a}{c}\exp(\frac{bc-ab}{c})] \tag{7}$$

Substitute a, b, c, d with Eq.(6):

$$I=\frac{R_{sh}(I_{ph}+I_0)-V}{R_s+R_{sh}}-\frac{nV_{th}}{R_s}W(X), \tag{8}$$

Here,

$$X=\frac{R_sR_{sh}I_0}{nV_{th}(R_s+R_{sh})}\exp\left(\frac{R_{sh}(R_sI_{ph}+R_sI_0+V)}{nV_{th}(R_s+R_{sh})}\right).$$

Integral-equation method is also available for the formulation simplification. But differential-equation method is utilized here for the simpler computation . The differential dV/dI can be expressed as:

$$\frac{dV}{dI}=-R_s+\frac{nV_{th}R_{sh}}{V+(R_s+R_{sh})I-nV_{th}-R_{sh}(I_0-I_{ph})} \tag{9}$$

This equation can be greatly simplified with algebraic method, and then the differential dV/dI can be given as below:

$$\frac{dV}{dI}=-C_1-\frac{C_2}{-V-C_3I-C_4} \tag{10}$$

It is not an implicit equation anymore, and the main advantage is speed. The number of unknown parameters is reduced from five (R_s, R_{sh}, n, I_{ph}, I_0) to four (C_1,C_2,C_3,C_4) , which are defined by the following expressions:

DOI: 10.1051/matecconf/20165903003

$$\begin{cases} C_1 = R_s, \\ C_2 = nV_{th}R_{sh}, \\ C_3 = R_s + R_{sh}, \\ C_4 = nV_{th} + R_{sh}(I_0 + I_{ph}) \end{cases} \quad (11)$$

Thus, the value of the five parameters can be easily obtained by solving the Eq. (11).

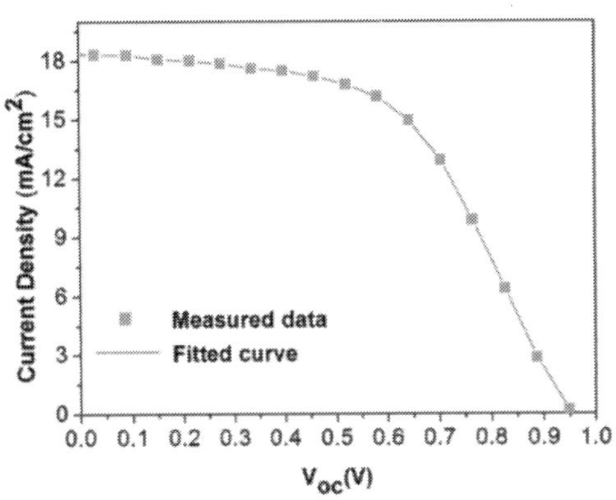

Figure 2. I – V characteristics of perovskite solar cells.

The curve fitting is performed with practical measured I-V data, this I-V curve of a perovskite solar cell was ploted in Figure 2. Moreover, this set of data is fitted with eleven-degree polynomial function in MATLAB. Four point were chose from the I-V curve to get the values of R_s, R_{sh}, n, I_{ph} and I_0 according to Eq.(11). Calculating the coefficients of algebraic equation has simplified the process of obtaining higher quality results. The exact expressions of I_{ph} and I_0 can be obtained by combining Eqs.(12) and (13)[11]:

$$I_{ph} = \frac{C_4 - C_2/(C_3 - C_1)}{C_3 - C_1} \quad (12)$$

$$I_0 = \frac{(I_{ph} - I)R_{sh} - (V + IR_s)}{R_{sh}(\exp((V + IR_s)/nV_{th}) - 1)} \quad (13)$$

3 Result and discussion

A method suitable for obtaining parameters has been developed. In the extraction process, it is found that the calculation results were closely related to the four sample points, these points are chose to solve the equations.

Furthermore, tests were carried out for comparison. First, points were only selected from the left side or right side of the max power point respectively. We failed to get reasonable results.

Then, two groups of points were chose from the I-V curve for parameters extraction, the positions of these points are shown in Figure 3. Two points from each side of the max power point were selected. But the distances between any two adjacent points of the Group1 are shorter than that of the Group2. The calculation results were shown in Table 1.

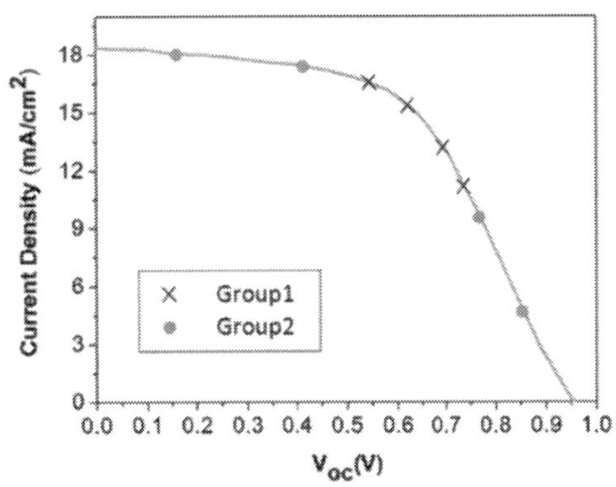

Figure 3. The selected points in the I-V curve for parameters extraction.

In order to estimate the accuracy of our method, the parameters were also extracted by Quasi-Newton method as Table1 shown. When comparing the extracted results of three groups, considerable error can be found in the results of Group2. The value of R_s in Group2 is about 20 Ω lower than others, and the difference of I_0 is more than an order of magnitude. By comparison, disparities between Group1 and Quasi-Newton methods are unremarkable.

Table 1. Results of parameters extraction by Lambert W-function.

	R_s(Ω)	R_{sh} (Ω)	n	I_{ph} (A)	I_0 (A)
Group 1	93.08366	3622.54335	2.23511	0.00169	9.26446×10^{-8}
Group 2	73.35790	3308.57911	2.18622	0.00170	1.07354×10^{-7}
Quasi-Newton methods	91.00027	3623.95822	2.23999	0.00169	9.30074×10^{-8}

According to the slight variation between Group1 and Quasi-Newton methods, the results which solved by the four points of Group1 are regarded as the true parameters in the perovskite solar cells. This conclusion was proved to be reliable after repeating this process. The accuracy of result is largely dependent on the position of selected points. In order to get accurate presentation, the calculation points should be chose on both sides of the max power point with an appropriate distance between each others. When the starting point and the ending point are chose nearer the short-circuit current and the open-circuit voltage respectively, the values of parameters exhibit higher quality, and show a better fit to the consequence of Quasi-Newton method.

4 Conclusion

In this paper, an efficient method has been proposed to extract all the parameters of a solar cell's mathematical model. Based on a single I-V curve under the one-diode photovoltaic model, parameters can be easily calculated. With the help of Lambert W-function, the explicit analytic expression for I-V relationship can be obtained. This expression is directly used to determine the values of the parameters through the selecting points form fitted curve. It is shown that our method can be easily used to analyze perovskite solar cells. The result accuracy of this method is as high as that of the numerical methods. But the computational process is simpler, and the computational speed is faster.

Acknowledgments

The authors acknowledge support from the Fundamental Research Funds for the Central Universities of China and the Hunan University Scientific research Promotion Plan for Undergraduate Students.

References

1. L. Jingshan, I. Jeong-Hyeok, M. T. Mayer, S. Marcel, N. Mohammad Khaja, P. Nam-Gyu, et al., Science, **345**, 1593-1596 (2014)
2. M. Liu, M. B. Johnston, and H. J. Snaith, 09/19/print Nature, **501**, 395-398 (2013)
3. M A. Green, A Ho-Baillie and H J. Snaith, Nature Photon.,**8**, 506-514 (2014)
4. C. Zhang, J. Zhang, Y. Hao, Z. Lin, and C. Zhu, J. Appl.Phys., **110**, 064504 - 064504-7 (2011)
5. W. Peng, Y. Zeng, H. Gong, Y. Q. Leng, Y. H. Yan, and W. Hu, Micro. Nano Lett., **8**, 86-89 (2013)
6. L. Xue, L. Sun, W. Huang, and C. Jiang, *ICMTMA*, 306-309 (2011)
7. J. Ding and R. Radhakrishnan, Sol. Energ. Mat. Sol. C., **92**, 1566-1569 (2008)
8. D. Bari, N. Wrachien, A. Cester, G. Meneghesso, *IRPS*, **470(9)**, 5E.6.1-5E.6.7 (2011)
9. Y. Chen, X. Wang, D. Li, R. Hong and H. Shen, Appl. Energy, **88**, 2239–2244 (2011)
10. K. Ishaque, Z. Salam, H. Taheri and A. Shamsudin, in *IAPEC*, 2011 IEEE, 10-15(2011)
11. J. Ding, Univ. Sci. Technol. China (2007)

Long – term performance evaluation of a fixed and solar follow – up systems with modified astronomical positioning in Polish conditions

Artur Bugała[1], Grażyna Frydrychowicz – Jastrzębska[1], Zbyszek Zbytek[2], Jacek Dach[3], Damian Janczak[3]

[1]*Institute of Electrical Engineering and Electronics, Poznan University of Technology, Poznań, Poland*
[2]*Industrial Institute of Agricultural Engineering, ul. Starolecka 31 60-963 Poznan, Poland*
[3]*Institute of Biosystems Engineering, Poznan University of Life Sciences, Poznań, Poland*

Abstract. The paper presents a physical construction of a measurement stand consisting of a photovoltaic module operating in a fixed optimal configuration and in a solar tracking with a modified astronomical positioning. In order to reduce the electric energy consumption by two direct current linear motors system is positioned three times a day during winter months. Prepared energy balance, using annual measurements, confirmed the legitimacy of using two - axis Sun tracking units for central Poland. Basing on the daily and monthly electricity production and solar irradiance for both photovoltaic units a mathematical formula with proposed correction factors is presented to calculate with good accuracy monthly electric energy generation for a fixed photovoltaic installation. The results of calculations were compared with measurements showing good correlation.

1 Introduction

The efficiency of photovoltaic conversion depends, on the one hand, on parameters of the receiver, on the other - on the availability of solar radiation on its surface. It is possible to increase the amount of solar radiation on the surface of the receiver. Its value is determined by external factors like: geographical location of the PV module, the period of operation (during the day and year) and meteorological conditions, particularly cloud cover and composition of the atmosphere described by a transparency factor as well as random shading of photovoltaic cells.

Negative impact of most external conditions can be reduced by choosing appropriate spatial orientation of the receiver, including azimuth and elevation angle to the ground. The simulation carried out for Belgrade (44°47' N), with similar latitude to Poznań (52°25' N), where four different seasonal and two different biannual scenarios were considered, has shown that monthly, seasonal and yearly tilt angle optimization leads to an increase of electric energy from photovoltaic modules by 15,42 %, 13,55 % and 5,98 %, respectively [1].

Among many works available in the literature, concerning the legitimacy of using various types of Sun tracking systems, exist mainly those relating to uniaxial horizontal or vertical systems, considered as units with less complexity. The problem of a single axis positioning of the photovoltaic modules was analyzed in publications [2, 3, 4, 5, 6], where authors presented different projects and executions of the follow - up systems for different geographical locations. The investigation conducted by G. C. Lazaroiu et al. [7] using a fixed and single – axis Sun tracker, during May and June 2014, showed the increase by 12 % - 20 % of the produced energy by using the Sun tracker. The growth is maximized during "clear sky" and the ending hours of the day.

The analysis of selected works allows to state that increased complexity in the design and implementation of dual - axis control algorithm affect the growth of the electric power generation [8, 9], compared to systems with low complexity level [10].

The comparison of two - axis follow - up and fixed systems, performed in AGH Laboratory of Automation, Robotics, and Photovoltaic Systems (Cracow, Poland), showed that azimuth and elevation angle adjustment, in the range of 0° - 350° and 0° - 70°, leads to 35 % increase in electric energy during the year [11, 12].

In the foreign literature there are a few publications comparing two - axis trackers and fixed units in a different time scale, e.g. for Texas (USA) [13], for Denizli [14] and Kahramanmaras (Turkey) [15], Almaty (Kazakhstan) [16], Gwangju (South Korea) [17], Montreal (Canada) [18], Bengaluru (India) [19].

In 2013 Ch. Okoye and S. Abbasoglu presented the results of measurements of electric energy generated by three photovoltaic modules in fixed (inclination angle $\beta = 36^0$), horizontal ($\beta = 0^0$) and two – axis tracking configurations. In case of using mobile construction the highest percentage energy gain, comparing to fixed installation, was recorded for June (58 %), the lowest for January (28 %) [20].

The influence of cloud cover and the contribution of the diffuse solar radiation on the electrical parameters of solar cells in high latitudes (Montreal, Canada) was investigated by G. Quesada et al. [21], showing that some tracking strategies can lead to decrease in electric energy production up to 25 %, comparing to fixed and horizontal panels, especially on cloudy days.

A limited number of publications, describing the legitimacy of using 2 – axis Sun tracking systems for Polish climatic conditions (Central Europe), where the contribution of diffuse radiation in total solar radiation reaches even 70 %, forces to present long – term measurements in order to perform complete energy and economic balance, that is also common and important in case of other renewable sources of energy like biogas [22][23].

2 Two – axis follow – up system

A two - year comparative analysis of the operation of a photovoltaic system consisting of a polycrystalline PV module in a stationary and 2 – axis Sun tracking configuration was performed using a measurement stand showed in figure 1.

Figure 1. A photovoltaic system consisting of the fixed and the 2 - axis tracking construction.

Table 1 presents the parameters of the single photovoltaic module and DC/AC microinverter used in the comparative analysis.

Table 1. Parameters of the photovoltaic modules and microinverters

Photovoltaic module	Value	DC/AC converter	Value
Maximum power	210 W_p	Nominal input power	240 W
Voltage at max. power point	29.64 V	Maximum DC voltage	44 V
Current at max. power point	6.98 A	Minimum DC voltage	20 V
Open circuit voltage	35.94 V	Range of MPPT	23 V-35 V
Short circuit current	7.6 A	Maximum input current	12 A

In the project the magnetic incremental encoders are used to describe the current position of the solar receiver in a horizontal and vertical axis. Two Hall sensors measure the voltage at the electrode terminals perpendicular to the magnetic field line. The current magnet displacement is correlated with the magnetic field intensity. The resolution of used direct current linear motors is 380 imp./mm. The control unit used in the tracking system allows to connect external sensors e.g. to measure the wind speed and direction, snow cover, temperature or weight on the photovoltaic module plane. In the case of high wind speed the solar receiver is positioned to horizontal position. Snow coverage implies vertical position with the inclination angle to the ground $\beta = 90°$, which allows to reduce the shading effect.

In order to determine the incidence angle of solar radiation on a plane tilted at β angle to the ground, a relationship between vector **a** $\{a_x, a_y, a_z\}$ - perpendicular to the surface of the photovoltaic module – and **b** $\{b_x, b_y, b_z\}$ – identical with direct solar radiation – should be described. A scalar product of two vectors equals cos θ.

A unit vectors can be determined using relationships (1) - (3) [24]:

$$a_x = \cos\alpha\cos\beta \qquad b_x = \cos\alpha'\cos\beta' \tag{1}$$

$$a_y = \sin\alpha\cos\beta \qquad b_y = \sin\alpha'\cos\beta' \tag{2}$$

$$a_z = \sin\beta \qquad b_z = \sin\beta' \tag{3}$$

where: α, β – azimuth and inclination angle of the photovoltaic module, α' – Sun azimuth angle, β' – Sun elevation angle.

Sun elevation and azimuth angle are defined by the relation [24]:

$$\beta' = \arcsin(\sin\varphi\sin\delta + \cos\varphi\cos\delta\cos\omega) \tag{4}$$

$$\alpha' = \arcsin(\frac{\cos\delta\sin\omega}{\cos\beta'}) \tag{5}$$

Taking into account equations (1) – (5) formula describing the incidence angle of solar radiation on a plane tilted at β angle to the ground has the following form [25]:

$$\begin{aligned}\cos\theta &= \sin\delta\sin\varphi\cos\beta - \sin\delta\cos\varphi\sin\beta\cos\gamma + \\ &+ \cos\delta\cos\varphi\cos\beta\cos\omega + \cos\delta\sin\varphi\sin\beta\cos\gamma\cos\omega + \\ &+ \cos\delta\sin\beta\sin\gamma\sin\omega\end{aligned} \tag{6}$$

For considered spatial orientation (photovoltaic module plane is facing south) the relation (6) can be expressed by the following formula:

$$\cos\theta = \sin\delta\sin(\varphi - \beta) + \cos\delta\cos(\varphi - \beta)\cos\omega \tag{7}$$

Time between the local noon and the sunrise (H_{sr}) and sunset (H_{ss}) were calculated using formula [26]:

$$Hsr = -Hss = \arccos(-\tan\varphi\tan\delta)/15 \tag{8}$$

The hour of the day when the Sun reaches its highest altitude can be described using the relation [27]:

 DOI: 10.1051/matecconf/20165903004

$$LN = 12:00 - \{229{,}2\{0{,}000075 + 0{,}00187\cos\omega - \\ + 0{,}03208\sin\omega - 0{,}0146\cos 2\omega - \\ + 0{,}0409\sin 2\omega\}\}/60 \quad (9)$$

Local time LT for an analyzed longitude can be expressed by the following relation [26]:

$$LT = GMT + Long./15 \quad (10)$$

where: GMT - Greenwich Mean Time, moment the Sun crosses the Greenwich meridian.

3 Stationary photovoltaic configuration

Fixed photovoltaic unit, analyzed with 2 – axis follow – up construction, consists of polycrystalline module installed with calculated elevation angle $\beta = 37°$ in south direction. The Liu – Jordan isotropic model, which describes solar radiation on tilted surface using direct and diffuse radiation for a horizontal plane, was used to determine the elevation angle [25]:

$$G_\beta = G_b \cdot \frac{\sin\delta(t)\cdot[\sin\varphi\cdot\cos\beta - \cos\varphi\cdot\sin\beta\cdot\cos\gamma]}{\sin\delta(t)\cdot\sin\varphi + \cos\delta(t)\cdot\cos\varphi\cdot\cos\omega(t)} + \\ \frac{\cos\delta(t)\cdot[x_0(\varphi,\beta,\omega) + x_1(\varphi,\beta,\gamma,\omega) + x_2(\beta,\gamma,\omega)}{\sin\delta(t)\cdot\sin\varphi + \cos\delta(t)\cdot\cos\varphi\cdot\cos\omega(t)} + \\ + G_d\cdot(\frac{1+\cos\beta}{2}) + (G_b + G_d)\cdot\rho_o\cdot(\frac{1-\cos\beta}{2}) \quad (11)$$

$$x_0 = \cos\varphi\cdot\cos\beta\cdot\cos\omega(t) \quad (12)$$

$$x_1 = \sin\varphi\cdot\sin\beta\cdot\cos\gamma\cdot\cos\omega(t) \quad (13)$$

$$x_2 = \sin\beta\cdot\sin\gamma\cdot\sin\omega(t) \quad (14)$$

where: G_b, G_d – direct and diffuse solar radiation for a horizontal surface; G_β - total solar radiation for a plane tilted at β angle; φ - latitude angle; $\delta(t)$ - declination angle; $\omega(t)$ - hour angle; ρ_o - reflectance factor.

The calculations were conducted using Matlab simulation software. The aim was to verify the total value of radiation power density in a one - degree step. For each new value of the β angle, the insolation value was determined for particular hours and was added on a daily and yearly basis to a two – dimensional array. The calculations were repeated until the highest value corresponding to the amount of electricity was obtained. The values of total, direct and diffuse solar radiation at selected spatial orientations were obtained from a typical metrological year TMY2 for the city of Poznań (Poland) [28].

4 Results

The two - year measurements of electric power and cumulative energy, AC voltage and current, irradiance and insolation on a photovoltaic module plane in a stationary and follow – up configuration were conducted in natural solar radiation.

The value of electric energy produced by a photovoltaic module, dependent on the registered insolation, efficiency of a photovoltaic conversion, area of a PV module and value of electric energy losses (e.g. temperature losses, shading effect) may be defined by the relation:

$$E_{rz} = E\cdot S\cdot\eta\cdot\frac{100\% - \sum P_{S\%}}{100\%} \quad (15)$$

where: E – insolation on a horizontal photovoltaic module plane, S – area of a PV module plane, $P_{s\%}$ – power losses in analyzed system, η – efficiency of a photovoltaic conversion.

A modified relationship, presented in a general form in [29], describing monthly electric energy production by a fixed photovoltaic unit, using determined correction factors for Polish climatic conditions presented in table 2, has the following form:

$$E_{rz,fixed} = (\sum_{i=1}^{12} E_{fix,i}\cdot k_i)\cdot P_m\cdot[1 - \frac{1}{100}(\sum P_{s\%})] \quad (16)$$

where: E_{fix} – insolation on a fixed photovoltaic module plane, P_m – maximum power of a photovoltaic module, k_i – average value of determined correction factor.

In addition, monthly electric energy produced by a 2 – axis Sun tracker has been described using a modified formula [30]:

$$E_{rz,track} = n\cdot\sum_{i=1}^{i=12} E_{tr,i}\cdot\sum P_s\cdot P_{oc,sc}\cdot FF\cdot|1 - \frac{1}{\sum P_s}| \quad (17)$$

where: E_{tr} – insolation on a 2 – axis tracking plane, n – number of PV modules, $P_{oc,sc}$ – apparent power, FF – fill factor.

Results of measurements, conducted between July 2014 and June 2015, were compared with values determined using formula (16) and (17).

On the basis of a thirty – second measurements of irradiance, conducted by means of prepared microprocessor system and a silicon sensor, monthly insolation E_{fix} and E_{tr} was determined on a photovoltaic module plane in a fixed and tracking configuration.

Average correction factors calculated for 12 months of the year, on the basis of 24 – month measurements conducted between 07.2013 - 06.2015, describe the relationship between monthly electric energy generated by a photovoltaic module in a fixed configuration and value determined using general formula (15), for central Poland.

Table 2 presents calculated correction factors of monthly electric energy for a fixed photovoltaic unit.

Table 2. Correction factors of the electric energy determined on the basis of 24 - month measurements

Month number	I	II	III	IV	V	VI
Value	0.93	0.93	0.83	0.87	0.88	0.84
Month number	VII	VIII	IX	X	XI	XII
Value	0.73	0.78	0.93	0.88	0.77	0.73

Figure 2 shows the value of annual electricity production for a fixed photovoltaic installation, on the basis of real measurements conducted between July 2014 - June 2015 and calculations using modified mathematical relationship (16) with relative error analysis, including determined correction factors and without correction.

Figure 3 shows the value of annual electricity production for a 2 – axis Sun tracking construction, on the basis of real measurements conducted between July 2014 - June 2015 and calculations using mathematical formula (17).

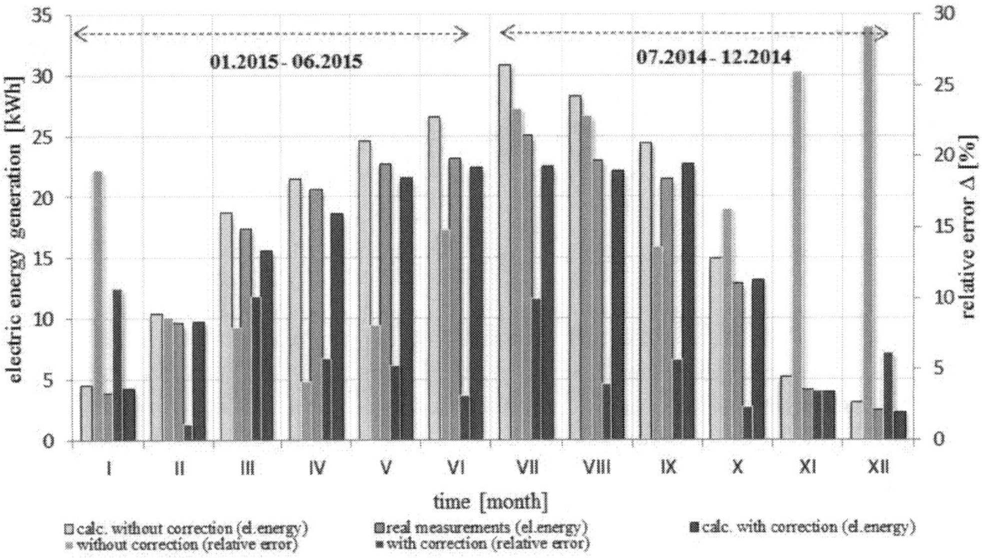

Figure 2. Energy differences between calculated without correction, with proposed monthly correction factors and measured values of electric energy for a fixed system including relative error analysis.

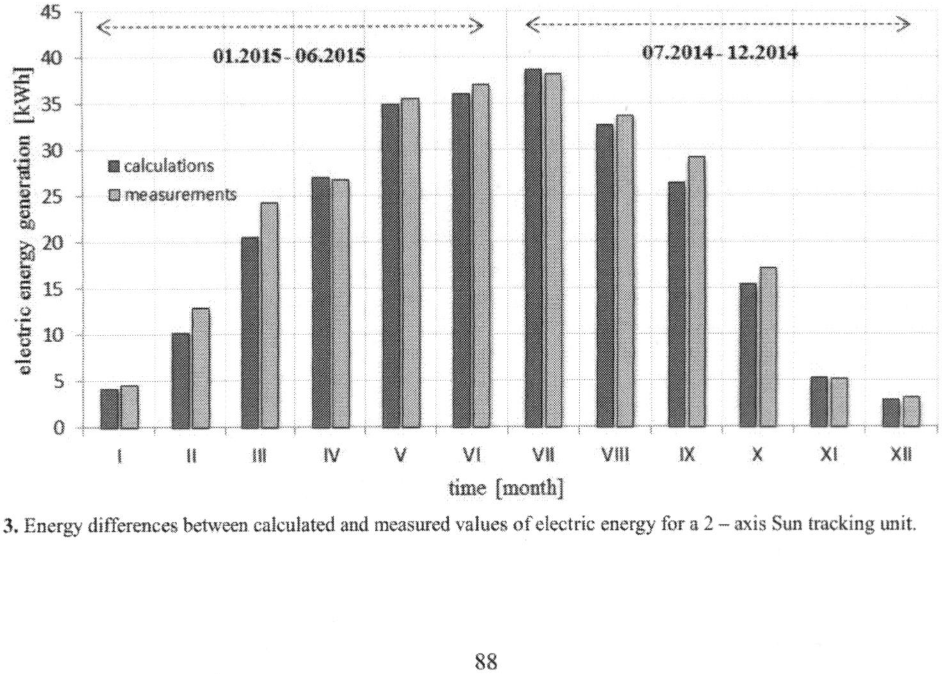

Figure 3. Energy differences between calculated and measured values of electric energy for a 2 – axis Sun tracking unit.

5 Discussion

The comparative analysis for 24 months proved the legitimacy of using 2 - axis follow – up systems in photovoltaics for the climatic conditions of central Poland.

Energy gain from the proposed algorithm was 56 %, 61 %, 53 %, 47 %, 36 % and 34 % gross, and 43 %, 47 %, 40 %, 32 %, 23 % and 16 % net in May, June, July, August, September and October 2014, respectively.

The proposed mathematical relationship for a stationary PV system, based on the monthly insolation and the proposed 12 correction factors using 2 - year measurements, allows to estimate the value of monthly electric energy with sufficient accuracy. The application of average correction factors allowed to reduce the percentage relative error from 23.32 %, 22.81 %, 13.6 %, 16.26 %, 25.93 %, 29.09 % to 9.92 %, 3.95 %, 5.64 %, 2.31 %, 3.41 %, 6.11 % in July, August, September, October, November, and December 2014. The lowest value of the Δ% parameter, in case of a fixed photovoltaic unit, was observed in February 2014, where introduced correction decreased the percentage error more than four times from 6 % to 1.39 %. A similar situation occurs in the following measurement year, where the value of the percentage relative error Δ%, between measured and calculated amount of electric energy in February 2015 is reduced from 8.63 % to 1.06 %, which is presented in figure 2.

The biggest differences between a measured value of the electric energy produced by the fixed photovoltaic module, and one determined on the basis of the relation without correction, are observed in June, July and August. A high temperature of silicon solar cells during summer months, without sufficient ventilation, may cause the decrease in voltage at the maximum power point, leading to significant differences in electric energy. This situation does not occur at such a level in the case of well ventilated solar module in tracking unit (figure 3).

A mathematical formula describing monthly electric energy production by the 2 - axis follow - up unit allows high accuracy prediction using values of insolation on a surface at a particular spatial orientation and the electrical parameters of a solar module, which is confirmed in figure 3. Presented formula does not require to use correction factors.

Calculations based on the generalized mathematical relationship (15), without monthly correction, lead to significant forecast errors.

References

1. M. Despotovic, V. Nedic, Comparison of optimum tilt angles of solar collectors determined at yearly, seasonal and monthly levels, Energ. Convers. Manage. **97**, 121 – 131 (2015)
2. N. Bansal, A. Chaudhari, M. Mishra, J. Jadav, Fabrication of a cost effective automatic dual/single axis active solar tracker with built in inverter designed for grid connectivity and standalone systems, IJEAT **3**, 159 – 162 (2014)
3. Z. Li, X. Liu, R. Tang, Optical performance of vertical single-axis tracked solar panels, Renew. Energy **36**, 64 – 68 (2011)
4. G. Li, R. Tang, H. Zhong, Optical Performance of Horizontal Single-Axis Tracked Solar Panels, Energy Procedia **16**, 1744 – 1752 (2012)
5. I. Sefa, M. Demirtas, I. Colak, Application of one - axis sun tracking system, Energ. Convers. Manage. **11**, 2709 – 2718 (2009)
6. T. Tudorache, L. Kreindler, Design of a solar tracker system for PV power plants, Acta Polytech. Hung. **1**, 23 – 29 (2010)
7. G. Lazaroiu, M. Longo, M. Roscia, M. Pagano, Comparative analysis of fixed and sun tracking low power systems considering energy consumption, Energ. Convers. Manage. **92**, 143 – 148 (2015)
8. B. Khadidja, K. Dris, A. Boubeker, S. Noureddine, Optimisation of a Solar Tracker System for Photovoltaic Power Plants in Saharian region, Example of Ouargla, Energy Procedia **50**, 610 – 618 (2014)
9. H. Mousazadeh, A. Keyhani, A. Javadi, H. Mobli, K. Abrinia, A. Sharifi, A review of principle and sun - tracking methods for maximizing solar systems output, Renew. Sustainable Energy Rev. **13**, 1800 – 1818 (2009)
10. Y.B. Gebremedhen, Determination of Optimum Fixed and Adjustable Tilt Angles for Solar Collectors by Using Typical Meteorological Year data for Turkey, Int. J. Renew. Energy Res. **4**, 924 – 928 (2014)
11. J. Chojnacki, J. Teneta, Ł. Więckowski, Development of PV systems and research studies on photovoltaics at the AGH University of Science and Technology in Krakow, 22nd European Photovoltaic Solar Energy Conference, Milan, Italy, 3 – 7 September 2007, 3049 – 3052
12. J. Teneta, Ł. Więckowski, Dissemination of Polish and world achievements in photovoltaics in the education process at high level - II edition, 178 – 189
13. W. Grady, H. Thomas, Predicting the Impact of Panel Orientation and Tracking on PV Daily Energy Harvest, IEEE Energytech Conference, Cleveland, USA, 21 – 23 May 2013, 1 – 3
14. S. Kivrak, M. Gunduzalp, F. Dincer, Theoretical and experimental performance investigation of a two - axis solar tracker under the climatic condition of Denizli, PE. **2**, 332 – 336 (2012)
15. S. Yilmaz, H. Ozcalik, O. Dogmus, F. Dincer, O. Akgol, M. Karaaslan, Design of two axes sun tracking controller with analytically solar radiation calculations, Renew. Sustainable Energy Rev. **43**, 997 – 1005 (2015)
16. S. Sadyrbayev, A. Bekbayev, S. Orynbayev, Z. Kaliyev, Design and Research of Dual - Axis Solar Tracking System in Condition of Town Almaty, J. Scien. Res. **17**, 1747 – 1751 (2013)
17. J. Byeong – Ho, P. Ju – Hoon, K. Seung – Dai, K. Jong – Ho, Performance Evaluation of Dual Axis Solar Tracking System with Photo Diodes, International Conference on Electrical Machines

and Systems, Busan, Korea, 26 - 29 October 2013, 414 – 417
18. M. Mehrtash, G. Quesada, Y. Dutil, D. Rousse, Performance Evaluation of Sun Tracking Photovoltaic Systems in Canada, 20th Annual International Conference on Mechanical Engineering - ISME, Shiraz, Iran, 16 – 18 May 2012, 1 – 4
19. R.R. Rao, H.R. Swetha, J. Srinivasan, S.K. Ramasesha, Comparison of performance of solar photovoltaics on dual axis tracker with fixed axis at 13°N latitude, Curr. Sci. **10**, 2087 – 2093 (2015)
20. Ch. Okoye, S. Abbasoglu, Empirical investigation of fixed and dual axis Sun tracking photovoltaic system installations in Turkish Republic of Northern Cyprus, J. Asian Scien. Res. **5**, 440 – 453 (2013)
21. G. Quesada, L. Guillon, D. Rousse, M. Mehrtash, Y. Dutil, P. Paradis, Tracking strategy for photovoltaic solar systems in high latitudes, Energ. Convers. Manage. **103**, 147 – 156 (2015)
22. J. Dach, P. Boniecki, J. Przybył, D. Janczak, A. Lewicki, W. Czekała, K. Witaszek, P.C. Rodríguez Carmona, M. Cieślik, Energetic efficiency analysis of the agricultural biogas plant in 250 kWe experimental installation, Energy **69**, 34 - 38 (2014)
23. W. Czekała, K. Kozłowski, J. Dach, P. Boniecki, A. Lewicki, D. Janczak, K. Jóźwiakowski, T. Piechota, Energy Conversion from Biomass to Hydrogen and Methane, 4th International Conference on Materials Engineering for Advanced Technologies, London, England, 27 – 28 June 2015, 654 - 657
24. A. Binkiewicz, The possibility of increasing the solar radiation flux using tracking systems, PE. **4**, 18 – 20 (2011)
25. G. Jastrzębska, A. Bugała, Modeling the distribution of solar radiation on two-axis tracking plane for the photovoltaic conversion, Energies **8**, 1025 – 1041 (2015)
26. F.M. Al – Naima, R.S. Ali, A.J. Abid, Solar tracking system. Design based on GPS and Astronomical Equations, IT-DREPS Conference & Exhibition, Amman, Jordan, 29 – 31 May 2013, 1 – 6
27. F.M. Al – Naima, R.S. Ali, A.J. Abid, Design and implementation of a smart dual axis sun tracker based on astronomical equations, European Conference on Renewable Energy Systems (ECRES), Antalya, Turkey, 17 – 28 September 2012
28. https://cms.transport.gov.pl (Ministry of Infrastructure and Development)
29. A. Bugała, G. Frydrychowicz – Jastrzębska, Energy balance for tracking systems in photovoltaics for local urban conditions - part II, PUT Academic J. **79**, 63 – 69 (2014)
30. A. Bugała, G. Frydrychowicz – Jastrzębska, The UNIPEDE method of assessing electric energy generation costs in photovoltaic systems, PE. **12**, 68 – 71 (2014)

Investigating the heat transfer phenomena of CO_2-EGS in the reservoir by experiment verification

Bo-Yen Shih[1], Jui-Ching Hsieh[2], Hsin-Jung Huang[3], David T.W. Lin[1], Chung-Neng Huang[1] and Chi-Chang Wang[4]

[1]*Institute of Mechatronic System Engineering, National University of Tainan, No.33, Sec. 2, Shu-lin St., Tainan City 700, Taiwan (R.O.C.)*
[2]*Industrial Technology Research Institute, Green Energy and Environment Research Laboratories, No.195, Sec. 4, Chung-Hsing Rd., Hsinchu City 31040, Taiwan (R.O.C.)*
[3]*Department of System and Naval Mechatronic Engineering, National Cheng Kung University, No.1, University Rd., Tainan City 70101, Taiwan (R.O.C.)*
[4]*Department of Mechanical and Computer-Aided Engineering, Feng Chia University, No.100, Wenhwa Rd., Taichung City 40724, Taiwan (R.O.C.)*

Abstract. The purpose of this study is to find the heat transfer phenomena of CO_2-EGS in the reservoir. The heat transfer model conjugated with the Brinkman model is used. This numerical model is validated by the experiment of supercritical CO_2. The heat transfer coefficient of experiment is derived from the thermal resistance method of comparison between numerical model and experiment. Further, the heat transfer coefficients with different operating conditions are build in this study. This study will provide the better combination of operating conditions for the improved heat extraction.

1 Introduction

Recently, the geothermal exploration obtains more efficiency for the technological breakthroughs, especial in the enhanced geothermal system. The geothermal temperature of deepest stratum reaches 6000 °C but power generation only need the depth of 1 km to 10 km, where geothermal temperature is 300°C. In earlier, the technology of exploitation is restricted in the extracting hot underground water directly for generating electricity. It is the reason for the few regions are feasible to develop the geothermal system. International Energy Agency indicates the geothermal projects can improve the exploration technique and decrease the cost of geothermal system. The geothermal system will be a larger utilization in future.

Nowadays, the deep geothermal power generation technology common with enhanced geothermal system (EGS) and complex energy extraction geothermal system (CEEG) are proposed for the higher efficiency of geothermal system. Water is first applied on the EGS as the working fluid because of the high latent heat, the specific enthalpy and the cheap cost. However, a number of disadvantages have to overcome, the main ones are mineralization. These will bring the pipeline fouling. Moreover, CO_2 may be the practicable working fluid for the heat extraction of EGS. The researches of EGS proliferate recently from CO_2-EGS is first proposed by Brown from 2000[1]. There are many benefits in the CO_2-EGS, such as reducing the pumping power for the thermal siphon phenomena, hardly combining with minerals in the surface of the pipes, reducing CO_2 emissions for the geological storage, reducing cost of the maintenance and improving the efficiency of the heat extraction. In order to obtain more fully understand the CO_2-EGS, Zhang et al. compare with the heat transfer performance between CO_2-EGS and water-EGS systems [2]. Fouillac et al. discuss the sequestration benefits of CO_2 [3]. The deep-saline aquifers the advance of CO_2 sequestration [4] is presented by Rosenbauer et al.. Ueda et al. [5] and Wan et al. [6] propose the interaction of fluid and rock. A series of studies of CO_2-EGS, such as heat transmission [7], sequestration of CO_2 [8], and production behavior [9] are proposed by Pruess et al. from 2006. In addition, the chemical changes of the granite and sandstone in CO_2 injection [10], the thermosiphon of CO_2 [11], the thermal convection of CO_2 in the vertical tube [12] and miniature tube [13] are presented, respectively.

However, the above researches have some limits. We difficult to know actual circumstances of the heat transfer phenomena for few comparative studies for the heat extraction. Therefore, the purpose of this study proposes an integrated method which combines the experiment. The results can prove reliability of numerical model whereby also can observe the heat transfer phenomena of CO_2-EGS in the horizontal tube.

DOI: 10.1051/matecconf/20165903005

2 Methods

2.1 Experiment

This study investigates the efficiency of the heat extraction of supercritical CO_2 at different pressure, mass flow rate. The efficiency of the heat extraction depends on the pressure and mass flow rate. Therefore, the pressure of experimental system is assumed as 7.5, 9, 10, 11 and 12.5 MPa, and the mass flow rate is 0.00082 and 0.00109 kg/s.

To simulate the reservoir temperature, the wall of test section is heated to 200 °C as the circumstance of the reservoir. The reservoir assumes as filling the silica-base particles, the average size of particle is 1.54 mm. The schematic diagram of the experimental system is shown in figure 1. It contains of a CO_2 cylinder, high pressure pump, pre-heated bath, cooling bath, test section, differential pressure gage, flow meter, and data logger. The photograph of experimental system is shown in figure 2. The beginning of the experimental procedure injects CO_2 into the pre-heater bath and reaches the supercritical status. Then the supercritical CO_2 flows into the test section until the experimental steady. The test section is composed of the stainless steel, ten calibrated thermocouples (T-type) and the differential pressure gage, are inserted into the test section to record experimental data. The inside, outside diameter and length of the test section are 29.9, 55.0 mm, and 235.0 mm, respectively. The temperature, pressure, and mass flow rate are measured and analysis to obtain the heat extraction.

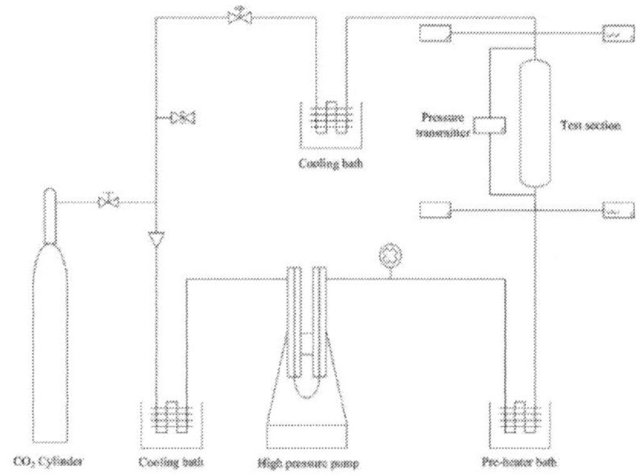

Figure 1. The schematic diagram of experimental apparatus.

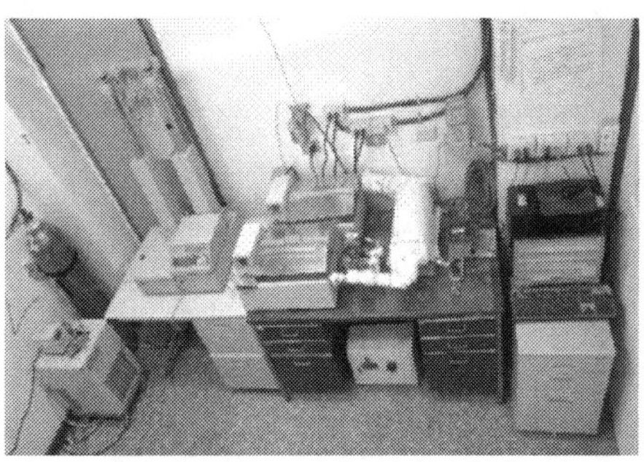

Figure 2. The photograph of experimental apparatus.

2.2 Simulation

A simulated model is assumed as the part of reservoir of EGS. This model is simulated based on the experimental test section as shown in figure 3. In this model, a 3D porous media flow model combines with heat transfer model established by finite element method – COMSOL multiphysics package. The Brinkman model is presented as the porous media. The material of test section is made of stainless steel. The properties of supercritical CO_2 are interpolate function based on the National Institute of Standards and Technology (NIST) standard reference database 69.

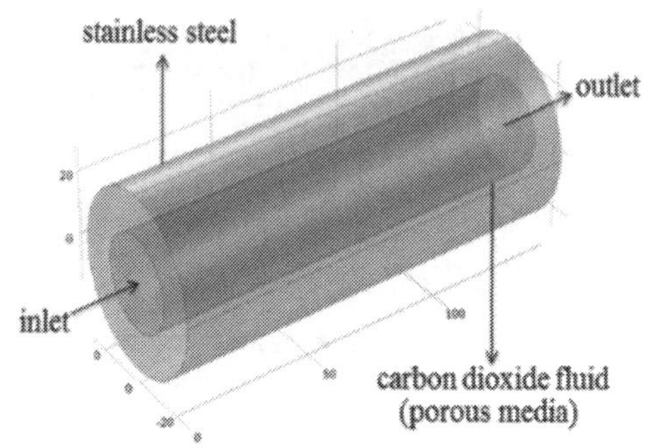

Figure 3. The schematic diagram of test tube.

The governing equations of non-porous medium are described as follows:

Continuity equation is:

$$\frac{\partial \rho}{\partial t} + \nabla \cdot (\rho u) = 0 \tag{1}$$

Momentum equation is:

$$\rho \frac{\partial u}{\partial t} + \rho (u \cdot \nabla) u = \nabla \cdot \left[-pI + \mu(\nabla u + (\nabla u)^T) - \frac{2}{3}\mu(\nabla \cdot u)I \right] + F \tag{2}$$

here, μ is viscosity; u is velocity; ρ is fluid density; p is pressure and F is forced term.

The governing equations of the Brinkman model are described as follows:

Continuity equation is:

$$\frac{\partial}{\partial t}(\varepsilon \rho_f) + \nabla \cdot (u \rho_f) = Q_{br} \tag{3}$$

Momentum equation is:

$$\frac{\rho_f}{\varepsilon_p}\left(\frac{\partial u}{\partial t} + (u \cdot \nabla)\frac{u}{\varepsilon_p}\right) = -\nabla P_f + \nabla \cdot \left[\frac{1}{\varepsilon}\left\{\mu(\nabla u + (\nabla u^T)) - \frac{2}{3}\mu(\nabla \cdot u)I\right\}\right] - \left(\frac{\mu}{\kappa} + \frac{Q_{br}}{\varepsilon^2}\right)u + F \tag{4}$$

here, ε is porosity; κ is permeability; Q_{br} is mass force and the subscript label f means fluid and p is media.

The Fourier's law of porous medium are described as follows:

$$\frac{\partial}{\partial t}\left[(1-\varepsilon)\rho_p C_{p,p} T_p\right] - (1-\varepsilon)\nabla\cdot\left(k_p \nabla T_p\right) = 0 \quad (5)$$

Energy balance equation is:

$$\frac{\partial}{\partial t}\left[\varepsilon\rho_f C_{p,f} T_f\right] + \nabla\cdot\left(\rho_f C_{p,f} D T_f\right) - \varepsilon\nabla\cdot\left(k_f \nabla T_f\right) = 0 \quad (6)$$

here, T is temperature; C_p is specific heat; k is thermal conductivity and D is Darcy flow.

The relationship between heat extraction, temperature difference and thermal resistance is defined as follows:

$$\mathrm{Q} = \frac{\Delta T}{\mathrm{Rth}} \quad (7)$$

$$\mathrm{Rth} = \frac{1}{hA} \quad (8)$$

here, Q is the heat extraction, ΔT is the temperature difference between the inlet and outlet of test tube, Rth is the thermal resistance od test tube, and A is the surface area of the heat exchange in the test section.

The heat transfer coefficient of numerical model can be obtained from the thermal resistance. Therefore, the heat transfer coefficient can be derived from the validation between experiment and numerical model.

3 Results and Discussion

This study processes the experimental verification of the heat extraction under the various pressure and mass flow rate. The comparison of the heat extraction between simulation and experiment with the different pressure (7.5, 9, 10, 11 and 12.5 MPa) and mass flow rate (0.00082 and 0.00109 kg/s) is presented.

Through figure 4 & 5, the heat extraction increases as the mass flow rate increases. We can find that the maximum heat extraction is 39.22 W occurred at 9 MPa. It is better than that of 7.5 and 12.5 MPa, and reaches the highest value at 0.00109 kg/s. We observe that the heat extraction at 7.5 to 9 MPa increases 44.93 %, at 9 to 10 MPa decreases 8.08 %, then decreases 15.64 % from 10 to 11 MPa, final decreases 4.42 % from 11 to 12.5 MPa. The nonlinear variation of heat extraction is shown as the pressure variation. The major reason is that the pressure depends on the thermal properties of the supercritical CO_2. The error of heat extraction are less than ±1.9% It can proof that the simulation is reliable.

Through figure 6 & 7, the outlet temperature increases as the mass flow rate decreases. We can find that the maximum outlet temperature is 50.90 °C occurred at 7.5 MPa, which is better than other pressures, and reached the highest value at 0.00082 kg/s. We observe that the outlet temperature at 7.5 to 9 MPa decreases 9.52 %, at 9 to 10 MPa increases 3.21 %, then increases 2.09 % from 10 to 11 MPa, final increases 2.7 % from 11 to 12.5 MPa. The analysis show that the error outlet temperature are less than ±1.7%. It can proof that the simulation is very reliable under the variable pressure and mass flow rate.

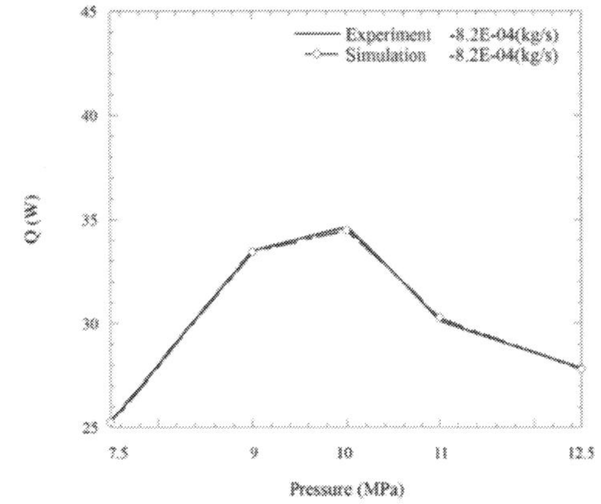

Figure 4. The comparison of heat extraction between experiment and simulation (mass flow rate is 0.00082 kg/s).

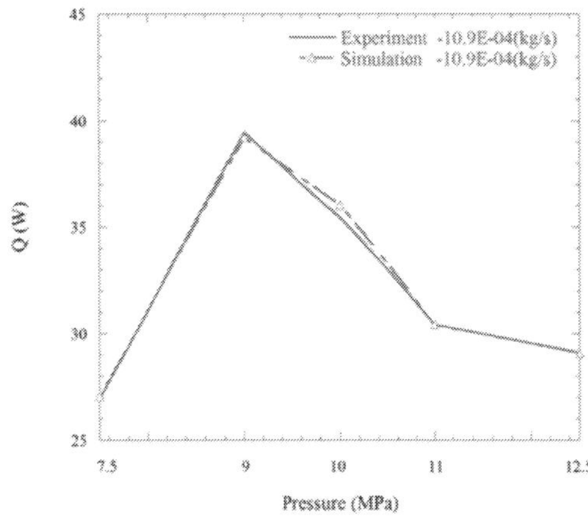

Figure 5. The comparison of heat extraction between experiment and simulation (mass flow rate is 0.00109 kg/s).

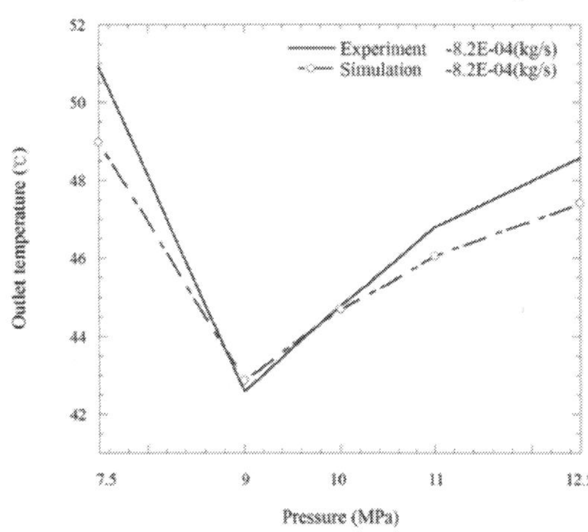

Figure 6. The comparison of outlet temperature between experiment and simulation (mass flow rate is 0.00082 kg/s).

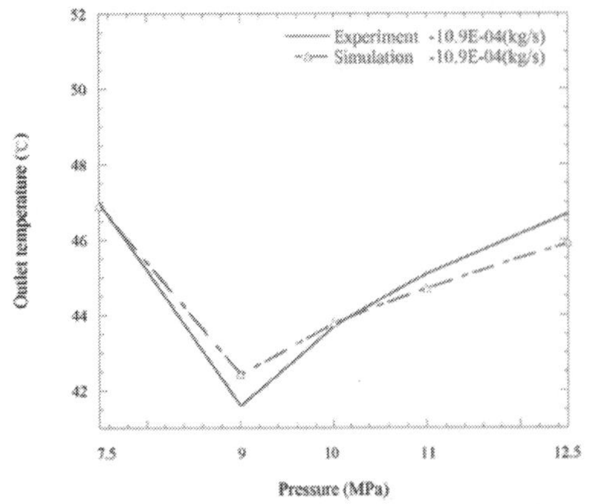

Figure 7. The comparison of outlet temperature between experiment and simulation (mass flow rate is 0.00109 kg/s).

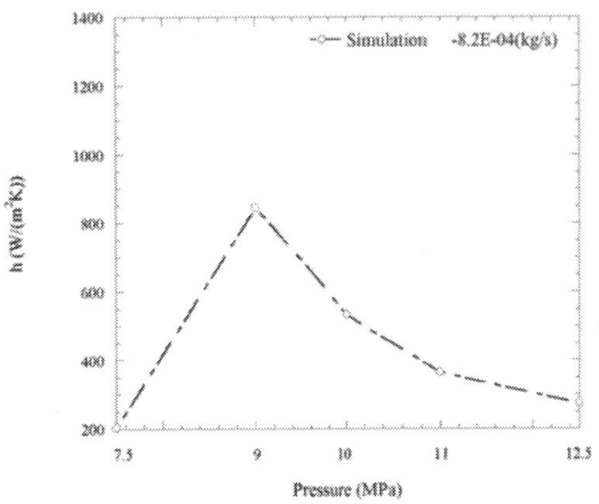

Figure 8. The profiles of heat transfer coefficient with different pressure and mass flow rate (mass flow rate is 0.00082 kg/s).

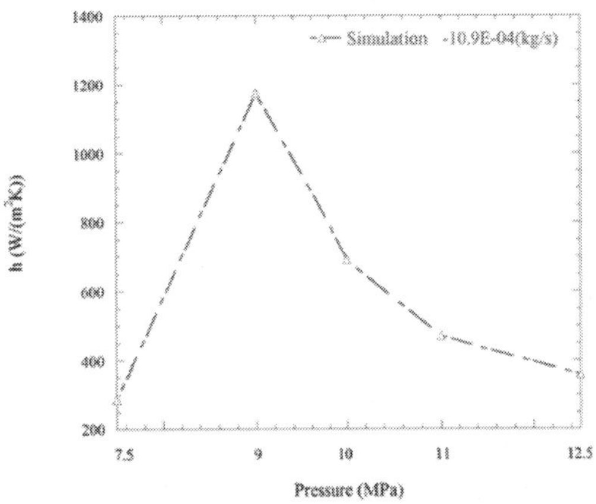

Figure 9. The profiles of heat transfer coefficient with different pressure and mass flow rate (mass flow rate is 0.00109 kg/s).

Figure 8 & 9 shows the heat transfer coefficient of simulation. It derives from the definition of thermal resistance. Through the validation results of numerical model, the heat transfer coefficient can be applied as the actual circumstances of experimental system. We also observe that the heat transfer coefficient increases as the mass flow rate increases. We can find that the maximum heat transfer coefficient is 1174.1 $W/m^2 \cdot k$ occurred at 9 MPa, which is better than that of at 7.5 and 12.5 MPa, and reaches the highest value in 0.00109 kg/s. We observe that the heat transfer coefficient at 7.5 to 9 MPa significantly increases to 311.4 %, at 9 to 10MPa decreases 41.14 %, then decreases 32.02 % from 10 to 11 MPa, final decreases 23.95 % from 11 to 12.5 MPa. In addition, it figures out the complex phenomena of porous flow of supercritical CO_2 to study the actual circumstances of the heat extraction.

4 Conclusion

This study is to investigate the actual circumstances of the heat transfer phenomena of CO_2-EGS. The results show that the error of the heat extraction between simulation and experiment is very small. The heat transfer coefficient of this experimental system can derives from the comparison between the experiment and simulation. It observe that the best heat extraction close to 9 MPa. In addition, the complex phenomena of the porous flow of supercritical CO_2 is shown in the numerical model and proof the nonlinear tendency of the heat extraction with various pressure and mass flow rate. These results can validate that the experimental system is very reliable and reduces the cost of actual test of CO_2-EGS. This study will provide the better combination of operating conditions for the improved heat extraction.

Acknowledgements

The financial support provided to this study by the Ministry of Science and Technology of the Republic of China under Contract No. MOST 104-3113-M-024 -001 is gratefully acknowledged.

References

1. D. Brown, Geothermal Reservoir Engineering **25**, 233 (2000)
2. F.Z. Zhang, P.X. Jiang, R.N. Xu, Applied Thermal Engineering **61**, 236 (2013)
3. C. Fouillac, B. Sanjuan, S. Gentier, C.L Isabelle, Carbon Capture and Sequestration **3**, 3 (2004)
4. R.J. Rosenbauer, K. Tamer, J.L. Palandri, Fuel Processing Technology **86**, 1581 (2005)
5. A. Ueda, K. Kato, T. Ohsumi, T. Yajima, H. Ito, H. Kaieda, R. Meycalf, H. Takase, Geochemical Journal **39**, 417 (2005)
6. Y. Wan, T. Xu, K. Pruess, Geothermal Reservoir Engineering **36**, SGP-TR-191 (2011)
7. K. Pruess, M. Azaroual, Geothermal Reservoir Engineering **31**, 386 (2006)
8. K. Pruess, Geothermics **35**, 351 (2006)
9. K. Pruess, Energy Conversion and Management **49**, 1446 (2008)
10. L. Lihui, S. Yuko, B. Greg, Y. Nakamichi, H. Energy Conversion and Management **44**, 1399 (2003)

11. A.D. Atrens, G. Hal, R. Victor, Geothermics **39**, 161 (2010)
12. P.X. Jiang, Y.J. Xu, R.F Shi., S. He, J.D. Jackson, Applied Thermal Engineering **24**, 1255 (2004)
13. S.M. Liao, T.S. Zhao, International Journal of Heat and Mass Transfer **45**, 5025 (2002)

Distribution and risk evaluation of heavy metal in the soil of typical grassed swale

Binghui Li and Chaohong Tan

Beijing Engineering Research Center of Sustainable Urban Sewage System Construction and Risk Control, Beijing University of Civil Engineering and Architecture, 100044, Beijing, China

Abstract. The concentration and distribution of Pb, Zn, Cu and Cr in the surface soil of three grassed swales located in the city of Beijing, Jiaxing and Shenzhen were studied in this paper. Evaluations on the risk of heavy metals were conducted based on the second level limit of soil environmental quality standard, using different indexes (i.e., single item, comprehensive items and potential ecological risk indexes). Results show that the concentration of Pb, Zn, Cu and Cr in all the soil samples is within the limit of second level of soil environmental quality standard of China, except the soil samples of the grassed swale from Shenzhen city. The result also shows that the concentration of the four kinds of heavy metals in the surface soil of grassed swales is linearly correlated with the depth of soil sampling points. In summary, ecological risk of heavy metals in the three grassed swales is slight.

1 Introduction

Grassed swale is a kind of engineering infrastructure with grass planted in the surface ditch. The purpose of this infrastructure is to collect and reuse of rain water. When the rain water flows through the swale, the pollutants in the runoff will be removed by the process of precipitation, filtration, infiltration, absorption and biodegradation. The grassed swale is generally applied in impervious road side and city park, etc. [1]. As a kind of typical runoff pollutants, heavy metals are trapped and accumulated in the surface soil of grassed swale with a potential risk to human health. Relevant investigation has been made in the surface soil of grassed swale in some countries. Beijing, Jiaxing and Shenzhen are the first batch of sponge cities in China. These three cities all experienced of water logging problem [2], thus rainwater utilization constructions in these cities are more common than elsewhere in China. As a kind of typical urban rainwater utilization engineering facilities easy for construction, grassed swales are applied in park and resident area in these three cities [3]. Heavy metal content is studied and risk assessment is carried out in the surface soil of three typical grassed swales located in Beijing, Jiaxing, and Shenzhen city. The result can be regarded as scientific basis for the prevention and control of heavy metal pollution in grassed swales.

2 Materials and methods

To ensure the typical of soil sample, the collection method of Simon T [4] is used in this study. Research results show that the top 30 cm soil is most vulnerable to heavy metal pollution [5]. Depth of grassed swale planting soil is also less than 30 cm in China according to our research.

2.1 Site description

Three typical grassed swales respectively locate in the Olympic Forest Park (116°23'3"E, 40°1'3"N, Beijing), Sunny Bay district (120°42'43"E, 30°44'57"N, Jiaxing) and Niushan park (113°57'11"E, 22°45'14"N, Shenzhen). The length of the three swales is all about 150 m. Beijing Olympic Forest Park was builted in 2008. It covers an area of 680 hectares with 478 hectares of green area, and 67.7 hectares of water surface. The annual average rainfall of Beijing is 600 mm. Jiaxing Sunny Bay district was built in 2012. The average annual rainfall of Jiaxing is 1169 mm. NiushanPark, a bright and high tech Industrial Park, is located in the southeast of Shenzhen City, which covers an area of 570 thousand square meters. The park was put into use in 2011. Shenzhen belongs to the subtropical monsoon climate, with an average annual rainfall of 1933 mm.

2.2 Sample Collection

Soil samples were collected from the top soil layer (30 cm) in the swales during the period from October to December in 2015. The distribution of sampling points of each grassed swale is shown in Figure 1. The three sampling points are arranged at both ends and the middle of each grassed swale. The soil samples of each sampling point were collected by 0-10cm, 10-20cm and 20-30cm depth using a stainless steel sampler. 27 soil samples were collected in each grassed swale. The samples were collected and stored in clean plastic bags with record of data, sampling points and other information. Then the samples were sent to the laboratory for further test.

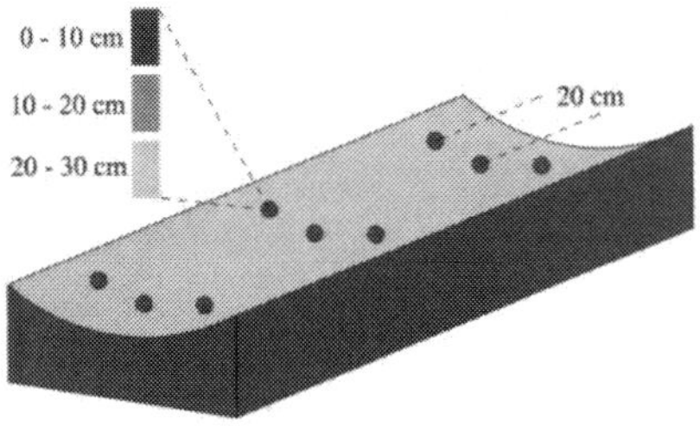

Figure 1. The distribution of sampling points.

2.3 Tests and Analyses

The collected soil samples (no less than 500 g) were air dried in natural condition indoor. Stones and debris in each sample were removed by hand. With the use of an agate mortar, the soil was sieved through a 2 mm nylon screen, and then filtered through a 0.15 m nylon screen[6]. Each tested sample is a mixture of samples from the same soil layer.

0.1 g soil was weighted by an analysis balance and dropped in a PTFE digestion tank. 5 ml Nitrate, 2 ml hydrogen peroxide, and 3 ml hydrofluoric acid were then added to the tank. All reagents are GR. Level. The tank was then put in the Topwave microwave for digestion. The tank was first heated to 150 degrees and kept for 20 min, then heated to 250 degrees and kept for 30min. When the tank was cooled down, the liquid in it was poured into 50 ml PTFE crucible. The liquid was heated on an electric hot plate (200 degrees) to remove the redundant acid in it. The final volume of liquid sample is 0.5 ml. The liquid was transferred to a 50 ml volumetric flask and set to 50 ml with deionized water. The contents of heavy metals were analyzed by flame atomic absorption spectrophotometer using Z-2010 HITACHI flame atomic absorption spectrometry. The correlation coefficient of the standard curve was no less than 0.9995.

10.0 g soil was weighted with an analysis balance and put into 50 ml high type beaker, adding 25 ml water. After sealing the container, the sample was stirred using the blender for 5 min, and the soil pH was measured with the FE20-K pH measurer after 2 h . This method was followed by NY/T 1377-2007 standard of China. Microsoft Excel 2003 and SPSS 17.0 software were used for data analyses.

2.4 Evaluation of heavy metal pollution

2.4.1 Evaluation criteria

The second level limit value of soil environment quality standard of China was applied as the evaluation criteria in this paper. This level of soil quality is basically does not cause damage and pollution to the plant and environment. The standard value of different metals is listed in Table 1.

Table 1. Second level limit metal value of soil environment quality standard [7] (mg/kg)

Metal/pH	pH<6.5	6.5<pH<7.5	pH>7.5
Pb	250	300	350
Zn	200	250	300
Cu	50	100	100
Cr	150	200	250

2.4.2 Evaluation Methods

Single factor index, comprehensive items and potential ecological risk indexes were applied to the risk evaluation. The single factor index method evaluates the ratio of tested value and relative standard limit value of pollutant. The calculation formula is as follows:

$$P_i = C_i / S_i \tag{1}$$

In the formula: P_i represents to the single factor pollution index. C_i represents to the content value of soil. S_i represents to the criteria value. This evaluation method can evaluate the risk of pollution of single heavy metal in soil. Nemero comprehensive pollution index method is based on the single factor index. The calculation formula is as follows:

$$P = \sqrt{[(P_{imax}^2 + P_{iavg}^2)/2]} \tag{2}$$

In the formula, P represents to the comprehensive pollution index. P_{imax} represents to the largest single factor index value. P_{iavg} represents to the average single factor index value. Table 2 shows the relation of P value with soil quality.

Table 2. Comprehensive pollution index criteria [8]

Level	Index Value	Soil quality
1	250	300
2	200	250
3	50	100
4	150	200

Potential ecological risk index takes heavy metal content and its ecological effects, environmental effect and toxicological effect together into consideration. The calculation formula is as follows:

$$RI = \sum_{i=1}^{n} = \sum_{i=1}^{n} T_r^i \times P_r^i \tag{3}$$

RI represents to the potential ecological risk index value. E_r^i represents to the single potential metal ecological risk value. T_r^i represents to the coefficient of single metal (Pb=5, Zn=1, Cu=5, Cr=2[4]). P_r^i represents to the single factor pollution index value. If the value of *RI* is less than 150, the ecological risk is slight. If the value of *RI* is between 150 and 300, the ecological risk is moderate. If the value of *RI* is beyond 600, then the ecological risk is serious [9].

3 Results and discussion

3.1 Heavy metal contents

The concentration of heavy metal in different level of soil is shown in Figure 2, 3 and 4. The maximum content of Cr, Cu, Zn and Pb in Olympic Forest Park is 126.3 mg/kg, 85.4 mg/kg, 230.3 mg/kg and 58.8 mg/kg, respectively. The maximum content of Cr, Cu, Zn and Pb in Sunny Bay district is 158.4 mg/kg, 85.2 mg/kg, 213.5 mg/kg and 23.6 mg/kg, respectively.

All the maximum values occur in the top 10 cm soil layer without exceeding the evaluation criteria, which actually means there is no heavy metal pollution in these two grassed swales caused by the four kind of heavy metals examined. The maximum content of Cr, Zn and Pb in Niushan Park is 124.6 mg/kg, 151.7 mg/kg and 144.6 mg/kg, respectively. Similarly, the maximum content of Cr, Zn and Pb occur in the top 10 cm soil layer without exceeding the evaluation criteria. Concentration of Cu in Niushan Park grassed swale is between 138.6-167.3 mg/kg, which greatly exceed the evaluation criteria and shows a heavy pollution condition. Data from Figure 4 also shows that basically, the metal concentration decrease along with the depth of soil. This phenomenon is due to the process of precipitation, filtration, infiltration and absorption by the grassed swale.

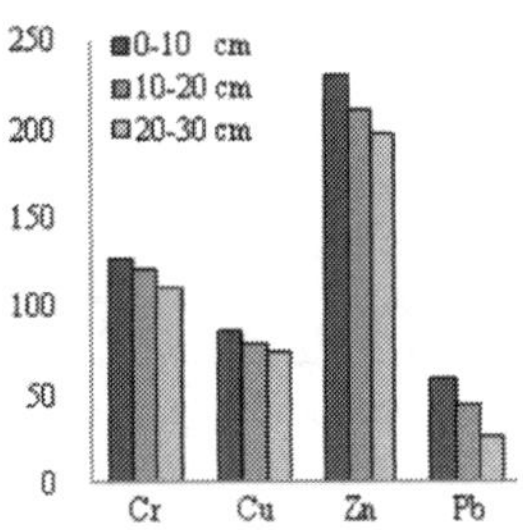

Figure 2. Concentration of Heavy metal in the soil of Olympic Park (mg/kg)

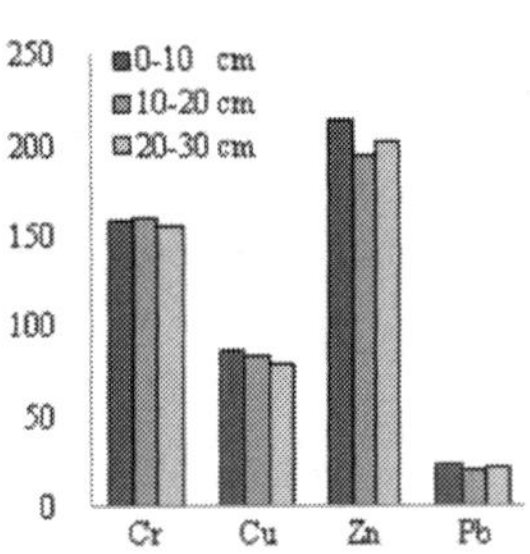

Figure 3. Concentration of Heavy metal in the soil of Sunny Bay (mg/kg)

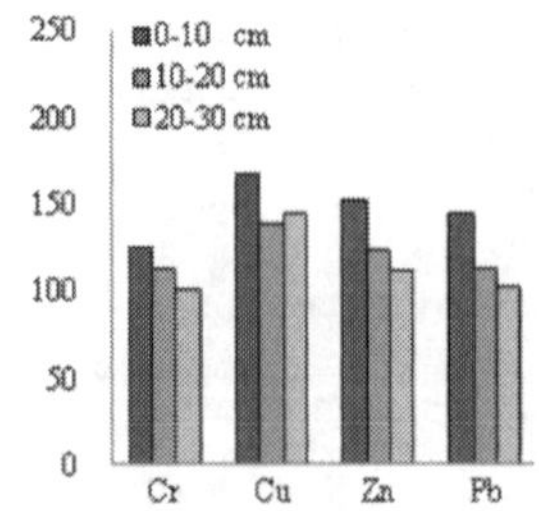

Figure 4. Concentration of Heavy metal in the soil of Niushan Park (mg/kg)

The background value of Pb, Zn, Cu and Cr of soil in Beijing is 24.6 mg/kg, 57.5 mg/kg, 18.7 mg/kg and 29.8 mg/kg, respectively [10]. The concentration of heavy metal all exceed this background value in the soil sample from Beijing. This indicates that the possibility of soil pollution still exits. According to the review of soil concentration in Beijing, Guo et al. believes that the main kind of heavy metal pollution in soils of Beijing is Pb, Zn, Cu and Cr, based on the background value [11]. In Jiaxing and Shenzhen city, there is no heavy metal background value data of soil until now. The static analysis results show that each of the heavy metal significantly associated with the other three metals at 0.01 level, so there is the possibility that the metal come from the same source. During the test we recognized that pH in the soil of Shenzhen grassed swale is less than 6, indicating that the soil is acidic, which is a vulnerable condition for heavy metal migration [12]. Besides, the annual rainfall in Shenzhen is larger than the other two cities, leading to a greater risk of heavy metal pollution. Besides pH, the organic matter in soil may also have an impact on the distribution of heavy metal in soil. The concentration of Zn in Shenzhen soil is significantly less than the other two cities, mainly because of the larger rainfall and the more chance of migration of Zn.

The standard of soil in China is founded in the last century, which is not strict enough. During the collection of soil sample, we found that the distribution of heavy metal in soil is not only associated with rainfall but also with the maintenance measure of grassed swale, such as irrigation and fertilization. Although the date of construction of these three grassed swale is different, they all perform well until now. Thus, the possibility of heavy metal pollution in grassed swale is low in short time. However, the long time performance of grassed swale in China is still unknown and need to be further investigated.

The concentration of Pb in Sunny Bay district is significantly less than the other two park. Probable reason for this is that the mobile population is larger in park than residential areas. The flush of rainwater in the surface soil is another factor of heavy metal distribution in soil, which can bring heavy metal in soil back into the rainwater. Compared with the results of heavy metal in soil of grassed swale besides traffic roads, heavy metal concentrations in these areas is lower, showing that road traffic is an important source of heavy metals in grassed swale soil. The concentration of Pb in the soil of Niushan Park is higher than the othe two areas. Laterite soil may be the cause of high concentration of Pb in the soil of ShenZhen.

3.2 Risk evaluation of heavy metal

The evaluation results of comprehensive index are showed in Table 3. Single factor index value of each kind of metal from Olympic Forest Park and Sunny Bay district is less than 1, which means a clean soil environment. Nemero comprehensive index value of this two areas is 0.7 or 0.8, within the range of level 2, which is a still clean state. Single factor index value of Cu in Niushan Park is between 2.9 and 3.3, with a comprehensive index value between 2.1 and 2.5. This shows that soil of grassed swale in Niushan Park is moderately polluted by Cu. The evaluation results of potential ecological risk is shown in Table 4. Potential ecological value of three areas is between 5.8 and 21.9, less than 40, which means that the ecological risk is slight.

Table 3. Evaluation results of Nemero Comprehensive pollution index criteria

Depth/Site	Olympic Park, Beijing	Sunny bay district, Jiaxing	Niushan Park, Shenzhen
0-10 cm	0.8	0.8	2.5
10-20 cm	0.7	0.7	2.1
20-30 cm	0.7	0.7	2.2

The evaluation results of potential ecological risk is shown in Table 4. Potential ecological value of three areas is between 5.8 and 21.9, less than 40, which means that the ecological risk is slight. In fact, the ecological risk value is far low than 150. However, other kinds of heavy metal, such as Cd, Mn and Ni also can cause potential risk. This should be investigated in the following research.

Table 4. Evaluation results of potential ecological risk

Depth/Site	Olympic Park, Beijing	Sunny bay district, Jiaxing	Niushan Park, Shenzhen
0-10 cm	7.6	6.9	21.9
10-20cm	6.5	6.9	18.7
20-30cm	5.8	6.9	18.5

Conclusions

The concentration of the four heavy metal (Pb, Zn, Cu and Cr) in the surface soil of three grassed swales decrease with the soil depth. All the metal concentration in soil samples is in the second level limit of soil environment quality standard of china, except the Cu concentration of soil samples from Shenzhen city. The ecological risk of the soil in the three grassed swales examined is in low risk condition. Grassed swale has a significant performance without causing soil heavy metal pollution, at least in short time. However, the long term performance of grassed swale is still unknown and further study is needed. Distribution of heavy metal in soil of grassed swale is the result of natural rainfall and manual maintenance.

Acknowledgements

This work was financially supported by Monitor and Risk Assessment on Heavy Metal Pollution in Stormwater Treatment Facilities Based on Low Impact Development (LID), a National Geographic Society project (GEFC18-14). This work was also financially supported by Beijing outstanding talent project for excellent youth team (2015000026833T0000). The authors would like to express their gratitude for the assistance.

References

1. Zhang Wei, et al. The Application of Glassed Swales in Urban Stormwater Utilization System. J. Water Supply and Drainage, **32**, 8 (2001).
2. Chen Youyun. Analysis of Waterlogging in Beijing and Shenzhen city. J. Water Resources Development Research, **1**,3 (2013).
3. Li Haiyan, et al. Progress of Grassed Swale for Runoff Disposal. C. Annual Meeting of Chinese Society for Environment Science, Chendu, China, (2014).
4. Simon T. Ingvertsen, et al. Assessment of existing roadside swales with engineered filter soil: I. Characterization and lifetime expectancy. J. Journal of Environmental Quality, **41**, 6 (2012).
5. Batjargal,Tserennyam, et al. Assessment of metals contamination of soils in Ulaanbaatar, Mongolia. J. Journal of Hazardous Materials, **184**, 3 (2010).
6. CHEN Hao, et al. The Monitoring Process of Heavy Metals in Soils and its Quality Control Measures. J. Environmental Monitoring in China, **26**, 5, (2010).
7. Ministry of Environmental Protection of the People's Republic of China. Environmental quality standard for soils. GB 15618, (1995)
8. Xu Zhengqi, et al. Calculation of Heavy Metals' Toxicity Coefficient in the Evaluation of Potential Ecological Risk Index. J. Environmental Science & Technology, **31**, 2, (2008).
9. Lars Hakanson. An ecological risk index for aquatic pollution control.asedimentological approach. J. Water Research, **14**, 8, (1980).
10. CHEN Tong-bin. Background Concentrations of Soil Heavy Metals in Beijing. J. Environment Science, **25**, 1, (2004).
11. Guo guang-hui, LEI mei, QIAO Peng-wei. Spatial Distribution of Heavy Metals in Soils during Urban Development of Beijing. J. Journal of Environmental Engineering Technology, **5**, 5, (2015).
12. Ramakrishnaiah H, Somashekar R.K. Heavy metal contamination in roadside soil and their mobility in relations to pH and organic carbon. J. Soil and Tillage Research, **11**, 5, (2002).

NanoCrystalline Cellulose isolated from oil palm empty fruit bunch and its potential in cadmium metal removal

Yong Hui Lim[1,2], Irene Mei Leng Chew[3], Thomas Shean Yaw Choong[1], Mei Ching Tan[2] and Khang Wei Tan[2]

[1]*Faculty of Engineering, University Putra Malaysia, 43400 UPM, Serdang, Selangor Darul Ehsan, Malaysia*
[2]*Faculty of Engineering, Technology and Built Environment, UCSI University, Cheras, 56000 Kuala Lumpur, Malaysia.*
[3]*School of Engineering, Monash University Sunway Campus, 46150 Bandar Sunway, Selangor Darul Ehsan, Malaysia.*

Abstract. NanoCrystalline Cellulose (NCC) was isolated via ultrasonic cavitation assisted acid hydrolysis method. Characterization was done using Dynamic Light Scattering (DLS) together with Scanning Electron Microscope (SEM) imaging to double prove the existence of NCC. DLS measures length of 236.6 nm with width of 34.40 nm, supported by SEM which showed NCC a rod-like shaped particle with large surface area and high porosity. It was then attempted for heavy metal cadmium ion (Cd^{2+}) removal from aqueous solution. The pH implication to the rate of Cd^{2+} adsorption was investigated by varying the solution to pH 4, pH 7 and pH 10 over a duration of 120 minutes. The removal efficiency was analyzed using Atomic Absorption Spectroscopy (AAS) resulting in pH 7 being the most favorable for Cd^{2+} removal.

1 Introduction

The growth of environmental issues has called forward to the search of new materials that are sustainable, biodegradable and environmental friendly [1]. Cellulose by far is the most abundant and inexhaustible natural resources which come with various benefits which has gain a lot of attention from researchers. Cellulose is available in various sources such as plant fibers and also lignocellulosic biomass. The biocompatibility towards human body and non-toxic nature were its winning points [2]. Cellulose can be decomposed into nanocrystalline cellulose (NCC) [3], which is a highly emerging bio-nanomaterial that offers many advantages i.e. high surface area, high aspect ratio, high specific strength together with stiffness, environmental benefits and low in cost [4].

Malaysia is an agricultural based tropical country with many industrial crops mainly the oil palm, paddy rice and sugarcane [5]. Malaysia the second largest oil palm producer with total oil palm plantation area up to 5.39 million hectares in 2014 (MPOB 2014). According to Malaysia Palm Oil Board (MPOB), about 22% of oil palm empty fruit bunch (EFB) was produced after fresh fruit bunch (FFB) processed in a palm oil mill. EFB contains high fibrous materials and packed with lots of lignocellulosic materials which can be source for cellulose. In 2013, it was reported that 8 million tonnes (dry weight) of EFB were produced annually making it a suitable raw material for the isolation of NCC.

Extensive reviews suggested that NCC has great potential in fields of medicine, optics, energy and composite materials. However, the applications of NCC in the following fields are still remain in the research stage. The negatively charged hydroxyl group on its surface giving opportunities for surface modification to its functionalization. Isolation of NCC via sulphuric acid hydrolysis method caused a negatively charge sulphate group to attach on its surface also acting as a stabilizer to suspense NCC in liquid such as water [6]. This suggests an opportunities for its application in drug delivery and waste water treatment in many industries.

Heavy metal pollution is one of the serious matters due to its toxicity not only towards human but also towards the ecosystem. Various heavy metal ions such as lead (Pb), cadmium (Cd), chromium (Cr), copper (Cu), nickel (Ni), zinc (Zn) and mercury (Hg) are product of industrial wastewaters mainly from industry that processes metal plating, mining, and paint [7]. Substantial amounts of heavy metals was introduced into the atmosphere, aquatic and environment, leading to the increase of heavy metal exposure [8]. Heavy metals cannot be degraded by their own. They tend to accumulate themself in living organisms leading to various diseases and disorder. Lead and cadmium are considered the most toxic heavy metals ions which impose serious health issues [9]. High exposure from Lead can potentially lead to encephalopathy, cognitive impairment, behavioural disturbances, kidney damage, anaemia and toxicity to the reproductive system. Cadmium causes nephrotoxic effects. With its long term exposure, it may cause bone damage [8]. Based on World Health Organization (WHO), the permissible limit in waste water for lead is 0.015 mg/L and cadmium is 0.01 mg/L [9].

Biosorption of heavy metals from aqueous solution is a promising process to remove contaminants from aqueous effluent [10]. Compared to conventional treatment, biosorption is low in cost, high efficient, recyclable, and possibility in the metal recovery [7]. A previous research had studied the feasibility of using NCC in waste water treatment, and the outcome is encouraging [9].

In the current study, NCC was isolated from oil palm biomass i.e. EFB using ultrasonic cavitation assisted acid hydrolysis. Sample characterization was done using Dynamic Light Scattering (DLS) measurement together with Scanning Electron Microscope (SEM). The extracted NCC was then employed to remove Cadmium ion from an aqueous solution. The mechanism of removal and the effect of pH against the performance of adsorption were investigated

2 Methodology

2.1 Preparation of samples

The pretreatment, bleaching and acid hydrolysis process to isolate NCC from Palm oil empty fruit bunch (EFB) via a conventional method was published in previous work [11]. Briefly, EFB was dried and grinded using a household grinder before pass screening through a sieve of 300 μm. The sieved EFB was first washed using distilled water at a temperature of 50°C for 2 hr to remove dust and impurities together with waxy substances on the surface followed with oven-drying at 70°C for 24 hr. The oven-dried EFB was then treated with 4% NaOH solution at temperature of 80°C for 6 hr and washed thoroughly using distilled water until pH 7 was reached. This process is known as alkaline treatment, repeated twice using the same setting and then proceeded with oven-drying for 24 hr at 70°C temperature. 30% hydrogen peroxide (H_2O_2) solution with few drops of glacial acetic acid was further applied on the alkaline-treated EFB for 7 hr as the delignification or bleaching process. After washing it thoroughly until pH 7 was reached, it was proceed to oven-drying for 24 hr at 70°C.

2.2 Isolation of nanocrystalline cellulose

Bleached EFB was dispersed in 60% sulphuric acid (H_2SO_4) solution at a constant temperature of 45°C for 2 h. The acid used to hydrolyze the amorphous region of the cellulose, isolating only the crystalline region cellulose, which is less susceptible to acid attack. Once the reaction completed, cold distilled water was introduced to quench the reaction, bringing the acid hydrolysis to a stop. The post-reaction liquid was then repeatedly centrifuged for 15 min at 7,000 rpm, to obtain raw NCC suspension. This process was repeated till an inseparable suspension is formed. The suspension was filtered and washed with distilled water until pH 7 was reached. The filtrate was then dispersed in distilled water.

2.3 Ultrasonic cavitation

NCC suspended in distilled water was subjected to ultrasonic cavitation treatment using KUDOS ultrasonic bath model SK7210HP at power of 100% with frequency of 53 kHz for 15 min. Ultrasonic cavities created by ultrasonic bath further break down the crystalline cellulose into nano-size known as NCC. The NCC produced was left dispersed in water for further characterization and application as described next.

2.4 Dynamic light scattering measurement

Particles size of NCC were measured using Dynamic Light Scattering (DLS) method with Malvern Zetasizer Nano ZS as described in [1]. The measurement was repeated twice to obtain the average value for z-average.

2.5 Cadmium metal ions removal

Heavy metal ions was removed from aqueous solution by adsorption via NCC as described in [9] with slight modification. In this study, a 500 mL standard solution of 1 mg/L cadmium was mixed with 1 mL of NCC suspension and stirred at 120 rpm for 2 hr. The pH of the solution is adjusted using NaOH and HCl solution to obtain pH 4, pH 7, and pH 10. Samples were collected every 20 min and analyzed using Atomic Absorption Spectroscopy (AAS, Pelkin Elmer AAnalysis 400) to determine the percentage of cadmium ions removal.

3 Results and Discussion

3.1 Isolation of nanocrystalline cellulose

The EFB was pre-treated with mechanically disintegration, washing and drying prior to alkaline treatment and delignification process, before the last stage cellulose isolation. Theoretically, lignin, hemicellulose, wax and oils on the surface of cellulose can be effectively removed via alkaline treatment which helps to swell the cellulose, defibrillates the cellulose, and exposing the crystallites [12]. Delignification or bleaching process further the alkaline treatment and gives a bleaching effect on the treated EFB. It can be observed from the colour change i.e. from brown to white, indicating a decrease in percentage of hemicellulose and lignin from EFB [13]. After that, isolation of NCC driven by sulphuric acid hydrolyse the amorphous region due to its high reactivity via cleavage of the cellulose glycosidic bonds, and the introduction of hydrophilic sulphate (-OSO_3H) groups to the crystalline cellulose [14]. This suggest the formation of negatively charged NCC that induce electrostatic repulsion forces among NCC particles making the aqueous suspension stable [15]. The NCC usually range in length from 100 nm to 3000 nm for many cellulosic materials and individualized by ultrasonic cavitation. The dimensions of nanocrystalline cellulose are affected by the source of cellulose and hydrolysis conditions [16].

3.2 Dynamics Light Scattering measurement

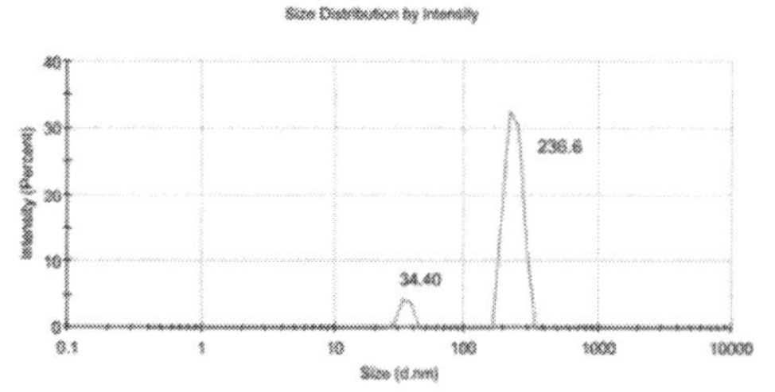

Figure 1. Particle size analysis using Dynamic Light Scattering Measurement using Malvern Zetasizer, Nano ZS Series.

DLS is a noninvasive method of measuring particle dispersed in a medium. Particles in dispersion will undergo Brownian motion and DLS measures the translational diffusion coefficient of the particle in the dispersion [2]. Figure 1 shows the results of DLS which indicates two size distribution intensity represented by the two peaks. An average size, z-average diameter was obtained as 561.0 nm. Peak 1 recorded particle size of 236.6 nm and peak 2 recorded particle size of 34.40 nm which can be interpret to the corresponding length and the width of the NCC. NCC has a rod-like shaped as described by various reviews, with length ranging from 100 nm to 3000 nm and width ranging from 3 nm to 30 nm using imaging technique of measurement such as atomic field microscope (AFM) and transmission electron microscope (TEM). Based on Stoke-Einstein theory, larger particle moves slower in a medium as compared to smaller particle which diffuses at a greater rate [17]. In this case, if NCC diffuses across the aqueous medium in a horizontal orientation against the light source, DLS would probably interpret it as a smaller particle due to its higher diffusion speed. Likewise, if NCC diffuses across the aqueous medium in a vertical orientation against the light source, DLS would most probably interpret it a bigger particle due to its slower diffusion rate as a result of higher resistance received toward the moving direction. Concisely, the orientation of particles scattered the light in different angle outlined two peaks of information that reflect the NCC diffusion orientation. The DLS measurement proved the existence of the NCC and a Scanning Electron Microscope (SEM) imaging was done to further support the result.

3.3 Scanning Electron Microscope

The morphology of the extracted NCC was determined using SEM imaging. NCC was dried before proceeding to SEM imaging. Figure 2 shows rod-like shaped NCC on the bottom right while some rod-like NCC clustered or bundled on at the top left corner of the image, in line with a previous study [18]. The length and width was confirmed. As from the SEM image, NCC is found to have large surface area and the bundles which looks like the stacking of the NCC, produced network-structure. The network structure also found to be the result of NCC in overcoming the repulsion of the surface negative charges which led to formation of self-assembled porous networks [18]. This phenomena brings to the high porosity properties of NCC that would benefit in the heavy metal removal. High surface area and high porosity properties can provides more active sites for the adsorption and desorption ions.

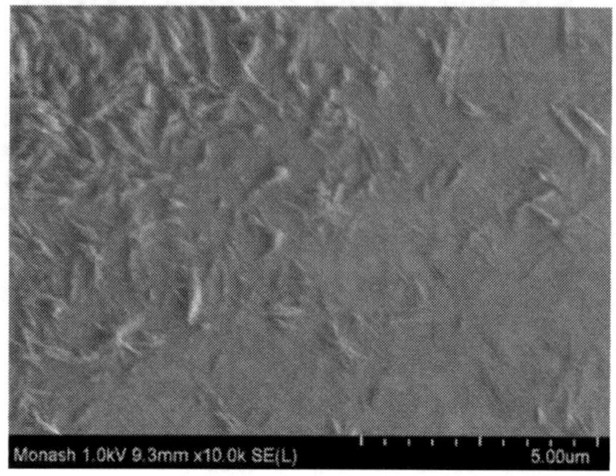

Figure 2. SEM imaging of NCC at 10,000 X magnification.

3.4 Cadmium removal

A similar research was reported [9] by functionalizing NCC into sodic NCC (NaSCNCs). The NaSCNCs acts as nanoadsorbent in removing Pb^{2+} and Cd^{2+}, and it is able to regenerate with NaCl solution. These suggested that the chemically modified cellulose could potentially achieve efficient adsorption capacity of heavy metal ions as stated by [19]. The cellulose can adsorb metal ions via ion exchange which contributed from the active sites present on them. NCC as reported earlier was found to have great surface area exposed and high porosity making it more susceptible to absorb more metal ions.

Solution pH plays a major role in heavy metal removal as it affects the activity of the functional groups of the sorbent and the metal ions [20]. As in this case, which is the hydroxl (OH) group on the surface of NCC and the Cd^{2+}. Figure 3 shows the comparison of the cadmium removal rate at pH value of 4, 7 and 10. It shows highest rate of removal at pH 7 which is almost the same as reported in [20] and [21]. As expected in lower pH 4, there are lesser ligands available for metal ions to binding due to the high concentration of hydrogen ions (H_30^+, hydronium) that suggest competition between hydronium and metal ions for the free binding sites. With increase of pH, lesser hydronium present, deprotonate the binding sites on the surface making more ligand groups available for metal ion bonding. The electrostatic attraction between the hydroxide group and the free electron-pairs from cadmium ions enhanced [21]. The increase of pH also indicated that the surface became negatively charged, making cations adsorption (which in this case the Cd^{2+}) more favorable [22].

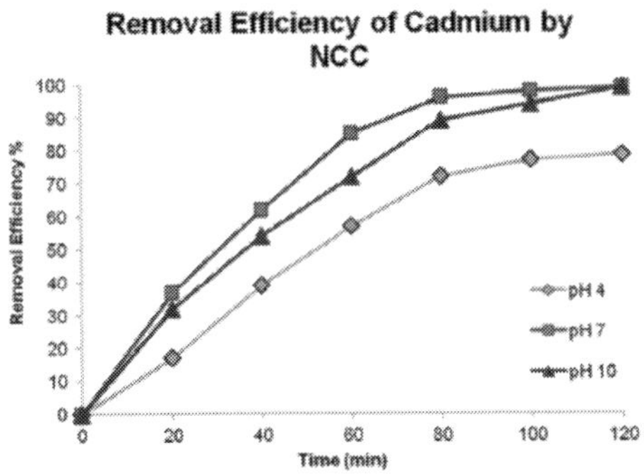

Figure 3. Cadmium metal ions removal efficiently by NCC.

4 Conclusion

NanoCrystalline Cellulose (NCC) was successfully isolated from palm oil empty fruit bunch (EFB) using ultrasonic cavitation assisted acid hydrolysis method. The Dynamic Light Scattering (DLS) measurements gave two peaks readings at 34.40 nm and 236.6 nm which interpreted as the diameter and length measurement respectively. The unique properties of NCC isolated from acid hydrolysis using sulphuric acid, which is the sulphate bonding on NCC surface promote the adsorption of cadmium in the application of heavy metal ions removal. As the result pH effect, it was found that NCC showed optimal cadmium ions removal rate at pH 7, which is at neutral state.

4 Acknowledgement

The authors are thankful to the Ministry of Higher Education, Malaysia (MoHE) for providing financial support (ERGS) on this research. Credit also goes to University Putra Malaysia, UCSI University Malaysia and Monash University Malaysia Campus for various infrastructures provided for advanced analysis work in this research field.

References

1. Khouri, S., Shams, M., & Tam, K. C. (2014). Determination and prediction of physical properties of cellulose nanocrystals from dynamic light scattering measurements. *Journal of Nanoparticle Research*, **16 (7)**. http://doi.org/10.1007/s11051-014-2499-7.
2. Boluk, Y., & Danumah, C. (2014). Analysis of cellulose nanocrystal rod lengths by dynamic light scattering and electron microscopy. *Journal of Nanoparticle Research*, **16** **(1)**. http://doi.org/10.1007/s11051-013-2174-4.
3. Klemm, D., Kramer, F., Moritz, S., Lindström, T., Ankerfors, M., Gray, D., & Dorris, A. (2011). Nanocelluloses: A new family of nature-based materials. *Angewandte Chemie - International Edition*, **50** **(24),** 5438–5466. http://doi.org/10.1002/anie.201001273.
4. Abdul Khalil, H. P. S., Davoudpour, Y., Islam, M. N., Mustapha, A., Sudesh, K., Dungani, R., & Jawaid, M. (2014). Production and modification of nanofibrillated cellulose using various mechanical processes: a review. *Carbohydrate Polymers*, **99**, 649–65. http://doi.org/10.1016/j.carbpol.2013.08.069.
5. Hosseini, S. E., & Wahid, M. A. (2013). Feasibility study of biogas production and utilization as a source of renewable energy in Malaysia. *Renewable and Sustainable Energy Reviews*, **19**, 454–462. http://doi.org/10.1016/j.rser.2012.11.008.
6. Moon, R. J., Martini, A., Nairn, J., Simonsen, J., & Youngblood, J. (2011). *Cellulose nanomaterials review: structure, properties and nanocomposites. Chemical Society reviews* **40.** http://doi.org/10.1039/c0cs00108b.
7. Demirbas, A. (2008). Heavy metal adsorption onto agro-based waste materials: A review. *Journal of Hazardous Materials*, **157** **(2-3)**, 220–229. http://doi.org/10.1016/j.jhazmat.2008.01.024.
8. O'Connell, D. W., Birkinshaw, C., & O'Dwyer, T. F. (2008). Heavy metal adsorbents prepared from the modification of cellulose: A review. *Bioresource Technology*, **99** **(15)**, 6709–6724. http://doi.org/10.1016/j.biortech.2008.01.036.
9. Yu, X., Tong, S., Ge, M., Wu, L., Zuo, J., Cao, C., & Song, W. (2013). Adsorption of heavy metal ions from aqueous solution by carboxylated cellulose nanocrystals. *Journal of Environmental Sciences*, **25** **(5)**, 933–943. http://doi.org/10.1016/S1001-0742(12)60145-4.
10. L.Y.Lee, X.J.Lee, P.C.Chia, K.W.Tan and S.Gan (2014) Utilization of Cymbopogon citratus (lemon grass) as biosorbent for the sequestration of nickel ions from aqueous solution: Equilibrium, kinetic, thermodynamics and mechanism studies, *Journal of the Taiwan Institute of Chemical Engineers* **45**, 1764-1772. doi:10.1016/j.jtice.2014.02.002.
11. Shanmugarajah B., Kiew P. L., Chew M. L., Chong S. Y., Tan K.W. (2015) Isolation of nanocrystalline cellulose (NCC) from palm oil empty fruit bunch (EFB): Preliminary result on FTIR and DLS analysis. Chemical Engineering Transactions, 45, 1705-1710. http://doi:10.3303/CET1545285.
12. Abraham, E., Deepa, B., Pothan, L. a., Jacob, M., Thomas, S., Cvelbar, U., & Anandjiwala, R. (2011). Extraction of nanocellulose fibrils from lignocellulosic fibres: A novel approach. Carbohydrate Polymers, **86(4)**, 1468–1475. http://doi.org/10.1016/j.carbpol.2011.06.034
13. Lu, P., & Hsieh, Y.-L. (2012). Preparation and characterization of cellulose nanocrystals from rice straw. Carbohydrate Polymers, **87(1),** 564–573. http://doi.org/10.1016/j.carbpol.2011.08.022
14. Hu, T. Q., Hashaikeh, R., & Berry, R. M. (2014). Isolation of a novel, crystalline cellulose material from the spent liquor of cellulose nanocrystals

(CNCs). Cellulose, 3217–3229. http://doi.org/10.1007/s10570-014-0350-z

15. Mueller, S., Weder, C., & Foster, E. J. (2014). Isolation of cellulose nanocrystals from pseudostems of banana plants. RSC Advances, **4(2),** 907. http://doi.org/10.1039/c3ra46390g
16. Sheltami, R. M., Abdullah, I., Ahmad, I., Dufresne, A., & Kargarzadeh, H. (2012). Extraction of cellulose nanocrystals from mengkuang leaves (Pandanus tectorius). Carbohydrate Polymers, **88(2),** 772–779. http://doi.org/10.1016/j.carbpol.2012.01.062
17. Sharma, M., & Yashonath, S. (2007). Size dependence of solute diffusivity and Stokes-Einstein relationship□: effect of van der Waals interaction. Diffusion Fundamentals, **7**, 1–15.
18. Tang, Y., Yang, S., Zhang, N., & Zhang, J. (2013). Preparation and characterization of nanocrystalline cellulose via low-intensity ultrasonic-assisted sulfuric acid hydrolysis. Cellulose, **21(1),** 335–346. http://doi.org/10.1007/s10570-013-0158-2
19. Kamel, S., Hassan, E. M., & El-Sakhawy, M. (2006). Preparation and application of acrylonitrile-grafted cyanoethyl cellulose for the removal of copper (II) ions. *Journal of Applied Polymer Science*, **100 (1)**, 329–334. http://doi.org/10.1002/app.23317.
20. Wei, W., Kim, S., Song, M.-H., Bediako, J. K., & Yun, Y.-S. (2015). Carboxymethyl cellulose fiber as a fast binding and biodegradable adsorbent of heavy metals. *Journal of the Taiwan Institute of Chemical Engineers*, **57**, 1–7. http://doi.org/10.1016/j.jtice.2015.05.019.
21. Gurgel, L. V. A., & Gil, L. F. (2009). Adsorption of Cu(II), Cd(II) and Pb(II) from aqueous single metal solutions by succinylated twice-mercerized sugarcane bagasse functionalized with triethylenetetramine. *Water Research*, **43 (18)**, 4479–4488. http://doi.org/10.1016/j.watres.2009.07.017.
22. Guo, H., Zhang, S., Kou, Z., Zhai, S., Ma, W., & Yang, Y. (2015). Removal of cadmium(II) from aqueous solutions by chemically modified maize straw. *Carbohydr Polym*, **115,** 177–185. http://doi.org/10.1016/j.carbpol.2014.08.041.

Evaluation of low degree polynomial kernel support vector machines for modelling Pore-water pressure responses

Nuraddeen Muhammad Babangida[1], Muhammad Raza Ul Mustafa[1], Khamaruzaman Wan Yusuf[1], Mohamed Hasnain Isa[1] and Imran Baig[2]

[1]*Department of Civil & Environmental Engineering, Universiti Teknologi PETRONAS, 32610 Bandar Seri Iskandar, Perak Darul Ridzuan, Malaysia.*
[2]*Department of Electrical and Computing Engineering, Dhofar University, Salalah, Oman*

Abstract. Pore-water pressure (PWP) is influenced by climatic changes, especially rainfall. These changes may affect the stability of, particularly unsaturated slopes. Thus monitoring the changes in PWP resulting from climatic factors has become an important part of effective slope management. However, this monitoring requires field instrumentation program, which is resource and labour expensive. Recently, soft computing modelling has become an alternative. Low degree polynomial kernel support vector machine (SVM) was evaluated in modelling the PWP changes. The developed model used pore-water pressure and rainfall data collected from an instrumented slope. Wrapper technique was used to select input features and k-fold cross validation was used to calibrate the model parameters. The developed model showed great promise in modelling the pore-water pressure changes. High correlation, with coefficient of determination of 0.9694 between the predicted and observed changes was obtained. The one degree polynomial SVM model yielded competitive result, and can be used to provide lead time records of PWP which can aid in better slope management.

1 Introduction

Climatic factors, such as rainfall and evaporation play an important role in pore-water pressure (PWP) changes, which in turn is vital to slope studies. In tropical region, climatic conditions are typically characterized by variable rainfall, frequent temperature variations and high evapotranspiration. Changes in climatic factors are mostly accompanied by changes in PWP, and sometimes leading to excessively high levels of PWP. High levels or positive PWP levels may lead to failure in slopes [1, 2]. Because of such changes, it is important to monitor PWP changes, particularly to rainfall. Damiano and Mercogliano [3] recommended that for effective monitoring of hydrological behaviour of slope, field in-situ records of PWP amongst other factors should be monitored. Evidently the knowledge of PWP will aid in making objective judgment with regard to slope management.

Pore-water pressure monitoring typically entails field instrumentation of the slope in question [4, 5]. This field instrumentation is expensive, time consuming and requires expertise. Modelling of PWP responses can provide an alternative to the instrumentation. Although it will not fully replace the instrumentation but it can provide an enhancement in that few lead time responses can be obtained. PWP modelling has been carried out using several variations of artificial neural networks (ANN) [6–10]. Albeit the ANN models boosted a remarkable correlation between observed and predicted records of PWP, ANN algorithm has some flaws that makes it a less preferable modelling algorithm these days. ANN is comparatively more complex, can get stuck in local optima and prone to over-fitting during training. The support vector machines (SVM), unlike ANN, is less complex as it has a simple geometric interpretation and its solution is always global [11]. SVM has been shown to outperform ANN in many engineering fields, as evident in some studies [12–14].

One of the most challenging aspects to the use of SVM is deciding which kernel to use. Different kernels suit different problems. In most cases literature do point the way as to which kernel to use, seldom, this may not be the case, as SVM usage is lacking in some studies. SVM was used by Lamorski et al. [15] in developing a point pedotransfer function, they compared a radial basis function (RBF) kernel and a linear kernel, and found the RBF more suitable to their data. Zhu et al. [16] found the sigmoid kernel most suitable for infrared data. The polynomial kernel has the advantage of behaving as two different kernels in one (polynomial and linear), since at a point during parameter optimization, it can assume the form of linear kernel, if the right parameter search space is selected. Furthermore a low degree polynomial has the advantage of fast training, while preserving the nonlinear capabilities of a nonlinear kernel. This study therefore aims to evaluate the capability of SVM with low degree polynomial kernel in modelling the nonlinear responses of PWP to rainfall.

2 Support Vector Machines Theory

Support vector machines has established itself as a robust method of modelling in the field of machine learning. Unless one has a valid reason for using another method, SVM should be the first technique to be employed.

Support vector machines, developed by Vapnik [17], originates from the frame work of statistical learning theory. Originally applied to classification problems where outputs are hard limited and then later extended to regression problems, and termed as support vector regression (SVR). In SVR, one seek an estimator of some real valued targets. Kecman [18] explain that the problem may be formulated as below. Given a training data sets : $\mathbf{x}_1 y_1, \mathbf{x}_2 y_2 ... \mathbf{x}_m y_m : \mathbf{x} \in \ ^n, y \in \ $, where input $\mathbf{x}$ are n dimensional vector, and y are real valued ($\mathbb{R}$) responses. The SVR function that represents the optimal separating hyper-plane may be introduced via linear regression, as shown in Eq. 1.

$$f(\mathbf{x}, \mathbf{w}) = \mathbf{w}\mathbf{x} + b \tag{1}$$

With $\mathbf{w}$, being a weight vector and b a bias. Typically a loss function to measure the approximation error is required. Vapnik introduced (Eq. 2) a loss function of the form;

$$|y - f(\mathbf{x}, \mathbf{w})|_\varepsilon = \begin{cases} 0, & |y - f(\mathbf{x}, \mathbf{w})| \le \varepsilon \\ |y - f(\mathbf{x}, \mathbf{w})| - \varepsilon, & otherwise \end{cases} \tag{2}$$

This function is insensitive to errors within a certain ε zone, a zone within which all predicted errors are tolerable and are considered as 0. The objective of the SVM algorithm is to find an estimator of the real valued targets, by minimizing the empirical risk and at the same time maximizing the width of the insensitive zone $\|\mathbf{w}\|^2$. This is formulated as Eq. 3, under the constraints in Eq. 4.

$$R = \frac{1}{2}\|\mathbf{w}\|^2 + C\left(\sum_{i=1}^{m}\xi + \sum_{i=1}^{m}\xi^*\right) \tag{3}$$

$$|y - f(\mathbf{x}, \mathbf{w})|_\varepsilon - \varepsilon \quad \xi \quad \mathbf{x}_{\mp} \tag{4a}$$

$$|y - f(\mathbf{x}, \mathbf{w})|_\varepsilon - \varepsilon \quad \xi^* \quad \mathbf{x}_{-} \tag{4b}$$

$$\xi, \xi^* \ge 0 \tag{4c}$$

The ξ^*, ξ are slack variables that indicates distance of miss classified points from the insensitive zone, C, is a regularization parameter that needs to be tuned, along with ε. The optimization problem in Eq. 3 is solved first by forming a primal variable Langrangian. This is then readily solved in the dual space. A standard optimization algorithm is used to solve the resulting Hessian matrix. Ultimately, learning results in the Langrange multipliers α, and the number of free α is equal to number of support vectors (SV). The obtained α values are used to find the optimal weights $\mathbf{w}$ and optimal bias b, which are in turn used to determine the optimum separating hyper-plane or estimator of Eq. 1.

For nonlinear regression problems, where no linear separating plane exist. Inputs are mapped to higher dimensional space. This is done by use of kernel functions. Kernel maps input features from a given dimension to a higher dimensional feature space. With this, nonlinearly separable problems can be readily solved as linear problems in high dimensional space. Basic kernel functions include, the Linear, Polynomial, Sigmoid and Gaussian. Only basic explanation of SVM formulation is offered here. Kecman [18] provides, detailed explanation on SVM formulations.

2.1 Polynomial Kernel

Polynomial kernel, given in Eq. 5, is less frequently used, compared to say RBF kernel. It generally does not give better accuracy. However, low degree polynomial (d ≤ 3) trained and tested in similar fashion as RBF perform only slightly less, but their training and testing time is much faster, and they can also be trained using linear SVM methods [19]. Additionally, polynomial kernels are found to perform better in natural language predictions [20].

$$K(\mathbf{x}_i, \mathbf{x}_j) = (\gamma\mathbf{x}_i, \mathbf{x}_j + r)^d \tag{5}$$

Where, γ, r and d, are kernel parameters that need to be tuned.

3 Methodology

3.1. Field data

The data used was part of the data from a slope instrumentation program, setup to monitor the PWP responses in a slope. The slope is located in Universiti Teknologi PETRONAS (UTP) in Malaysia. The slope was instrumented with tensiometers (fitted with transducers and data logger) and rain gauge. Part of the data from the crest at 0.6 meter depth was used for this study. Its properties are shown in Table 1. The highest water table around the area was found to be about 1.4 meters below the ground level (slope toe), as monitored from four monitoring wells sited around the slope area. Results from undisturbed sample of the slope material, indicated that the slope soil material is sandy clay in the top one meter, to clay from about two meters.

3.2 Input Features Determination

The selection of model representative attributes constitutes great challenge in soft computing modelling. Selecting the optimum number of attributes or input features require a good knowledge of the system to be modelled. The knowledge of the physical system of PWP has been documented in literatures.

Table 1. Data Properties

	Training (3577) 30/10/14 – 31/1/15		**Test set (2832) 1/2/15 – 31/3/15**	
	PWP (kPa)	**Rain (mm)**	**PWP (kPa)**	**Rain (mm)**
Mean	-9.5	0.2	-14.5	0.1
Max	-3.5	46.0	-5.5	26.5
Min	-34.1	0.0	-27.7	0.0
SD	3.3	1.6	3.9	1.2
Var	10.8	2.6	14.9	1.4
Skew	-3.4	16.0	-0.5	15.8

Rahardjo et al. [5] showed that five antecedent records of rainfall are needed to model the PWP responses to rainfall. Mustafa et al. [10], showed that five antecedent PWP records will be needed to develop a good model of PWP. A combination of both of these input features is considered herein. Therefore five antecedent records of each of PWP and rainfall were considered for analysis and determination of the needed features. However, only a subset of these may be relevant.

The state of the art input feature selection involves the use of either, the filter method or wrapper method and to a lesser extent, the embedded method, which has comparatively limited applications [21, 22]. Filter method aims to find the relevant features, outside the framework the algorithm, by evaluating the relevance of the features using a statistical measure. The wrapper method aims to evaluate prediction power of the features within a chosen algorithm. Although filter method is computationally cheaper, it provides a broad selection of features, which are not specific to any algorithm. On the other hand, the wrapper method yields features which are algorithm and data specific. In practice this is the most ideally sought option [23].

3.3 Wrapper method of Input determination

The wrapper method requires the selection of a search algorithm and evaluation function. The search algorithm explores the feature search space, using operators tailored to the search engine, which significantly aid in how the search space is manipulated by the search algorithm. The evaluation function, employs the use of a learning scheme, which generally happens to be the algorithm, which the features are to be used within to develop the model in question. A learning scheme is used as the feature evaluation algorithm. By using a learning scheme, K-fold cross validation (CV) may be used to estimate the accuracy of the learning scheme for any selected set of sub features, this way the selected sub features can be evaluated. An evaluation measure such as root mean squared error (RMSE), coefficient of determination (R^2), mean absolute error (MAE), etc. are used as accuracy measure of the CV. In essence this serves as the objective function of the search engine.

Several search algorithms, such as evolutionary algorithm (EA), greedy search etc. may be used as the search engine with its objective set as the performance of the evaluation function. The search engine is tasked with exploring the feature search space. Detailed use of the wrapper method can be found in the literature of [23]. Chandrashekar and Sahin [21] explained some search methods as heuristic search algorithms, methods such as that of genetic algorithm (GA) used as a search engine for feature subsets selection, since these methods do not follow the usual methods of forward or backward selection. They went on to explain that the chromosomes of the GA can be used to represent if a feature is selected or not. With the objective function being set as the performance of the prediction algorithm, which uses the selected chromosomes as input features. In this study such method will be used with GA being employed as the search algorithm. The GA uses the result to determine which amongst the selected feature subsets has less relevance to the predictor algorithm. Upon that the GA creates a better subset by manipulation within it, aided by the following operators: single point crossover, bit-flip mutation, generational replacement with elitism. The wrapper attribute selection will be implemented using Waikato environment for knowledge analysis (WEKA) [24].

3.4 SVM model Development

Basic SVM model development may be implemented with four basic steps as follows;

• Identifying the kernel to be used in the model. Although, many studies choose to use the RBF on the basis of being the most common. It is wise to compare the performance of few kernels or consult literature on similar data.

• Identify the kernel parameters to be optimized in addition to the SVM meta-parameters (C, ε). Use a suitable optimization technique, such as the GA, Particle swamp optimisation PSO), grid search etc. to find the optimum of the meta-parameters.

• Use the optimum parameters to train the model on part of the data (training data).

• Test the model on unseen data sets (test set), to evaluate its generalization performance. Appropriate accuracy measures are used for the evaluation. The SVM within the wrapper algorithm, is implemented using WEKA's library SVM [25] and, model development is implemented using library for support vector machines (LIBSVM) [26]

4 Results and Discussions

4.1 Data Analysis

The wrapper method was employed in this study to determine the optimal subset of input features. Evolutionary algorithm was used as the search engine in the wrapper method. The feature space with 11 sets of features (chromosomes) was exploited through manipulation within the GA to select subsets. SVM was used as the evaluation function, with the data evaluated over fivefold grid search. The performance of the CV was evaluated using root mean squared error (RMSE), and is shown in Table 2.

 DOI: 10.1051/matecconf/20165904003

Table 2. Root mean squared error (RMSE) of cross validation results, using different chromosome combinations- as implement with the wrapper algorithm.

	Current Fitness		Global Fitness	
	Best	**Least**	**Best**	**Least**
RMSE	0.0267	0.0272	0.0267	0.1085
CC*	2, 4,7,9	2,3,4,6,7,9	2,4,7,9	11

$*U_{t-1}=2,\ U_{t-2}=3,\ldots U_{t-5}=6,\ r_t=7,\ r_{t-1}=8\ldots\ r_{t-5}=12$

Where U represents PWP, and r is rainfall. The subscript indexes, are lag positions of the variables. E.g. U_{t-1} represents one antecedent (30 minutes lag) PWP.

The RMSE of best feature combinations (Chromosome combination—CC), that appears in the fivefold CV showed lowest values for combination of two, four, seven and nine. Thus the result of the wrapper input determination, showed Eq. 6 as an excellent set of parameters for modelling PWP.

$$U_t = f\left(U_{t-1}, r_{t,t-2}\right) \qquad (6)$$

Although other feature combination yielded similar result to this 3 features, however, they have more number of features and that makes them a less preferable choice for the model build. Fewer features with the same fitness level will ensure a simpler, less complex model.

4.2 SVM Calibration and Testing

The training data was scaled between 0 and 1. Base on the objective of the study, the polynomial kernel was selected. The input features obtained from the wrapper input determination were used as the input for the model development. Grid search with fivefold CV was used to search the optimum model meta-parameters. The grid search was carried out in three stages. First, a coarse grid search, with the following range $10^{-3} \le C \le 10^3$, $10^{-4} \le \varepsilon \le 10^1$, $10^{-3} \le \gamma \le 10^3$, $1 \le d \le 3$, and $0 \le r \le 5$, was carried out, and then followed by two steps of finer grid search. Table 3 shows the obtained meta-parameters for the model. The combination of parameters from Table 3, gave the best of results during a fivefold CV.

The calibrated parameters from the grid search were used to train the SVM model, which was then tested on the test set. The result of the calibration and testing is shown in Table 3. The results presented in Table 3 are results of analysis carried out using scaled data sets. The polynomial kernel yielded very good results as can be seen from the performance measures in Table 3. A slightly better performance was obtained during the test stages, compared to results from the calibration stages. This can be attributed to data variability. The variance of the data in the test stages is higher than that in the calibration stage, making the training data much smoother, and thus better result in the calibration stage. This is consistent with effects of data variability [27].

Table 3 Performance metrics (for calibration and testing) of the developed Polynomial kernel Model

C=1, ε=0.001, γ=0.03, r=0, d=1

Calibration (3577 sets)			Test (2832 sets)		
RMSE	**MAE**	**R^2**	**RMSE**	**MAE**	**R^2**
0.0219	0.0115	0.9590	0.0222	0.0101	0.9694

Figure 1. Shows a scatter plot from the calibration of the model. The fivefold CV calibration result was compared with the observed training result. The accuracy measure obtained was good, with R^2 of 0.9845, and low values of MAE and RMSE as well. The model fit line indicates the best fit attained by the model (not the perfect fit line which indicates, perfect agreement), since unlike the perfect fit line, the model fit shows the correlation extent, and what is important is to have all points on the line. The calibration results shows how good the developed Model did in data training. However, in machine learning, a model may yield good approximation result, during training but may fail to yield even close to that result on test data. What is vital is for the model to perform well on unseen data, and that is the generalization power of the model.

Thus the generalization ability of the Model was tested, and the estimation errors are shown in Table 3. The estimation errors obtained are low, and very close to the approximation errors from the fivefold CV. The generalization performance of the model was further brought to light in the individual event based Model predictions, shown in the scatter plot of Figure 2. The result obtained has an R^2 value of 0.9694, indicating not only a strong agreement, but near perfect correlation. This is an indication of good generalization ability of the Model.

The relationship between the observed and predicted values, is again shown in a better picture, in the time series plot of the two events (shown in Figure 3). The Model mimicked the reality, the complex trend, including several points where a rainfall event has pushed the PWP to high values. Rainfall events push PWP to extreme values, and most often, these are the points that tend to be difficult predict.

Generally in time series modelling, predicted values tend to lag the observed, such lag was not observed here, indicating the robustness of the Model. Nevertheless, some points were difficult to predict, and are unfortunately mis-estimated. Some of these points have a sudden unprecedented high rise in negative PWP. These are considered as outliers, as there was no prior event that could trigger such rise. However, no analysis of outlier is offered in this study. The most pronounced of such points are clearly shown in Figure 2. They have deviated the most from the model fit line, indicating less correlation between the Model predictions and observed records.

Support vector machines with polynomial kernel, has mimicked the nonlinear responses of PWP to rainfall events quite well as indicated by some measures explained above. The results are competitive even in comparison to the neural network model trained with eight input features and nine months of data [6].

DOI: 10.1051/matecconf/20165904003

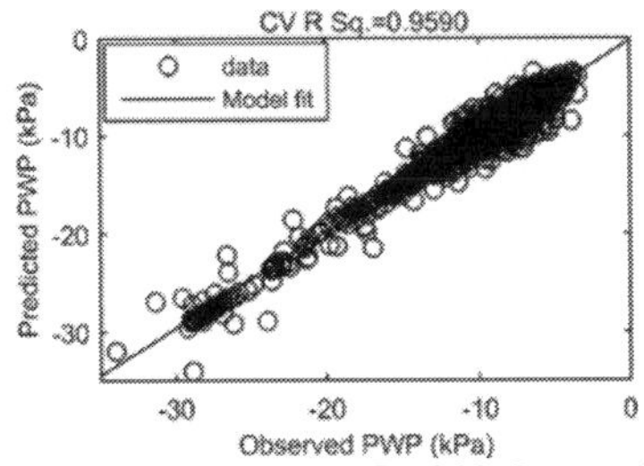

Figure 1. Scatter plot of the fivefold Cross validation and observed PWP, with Line of fit, indicating model fitness to observed data

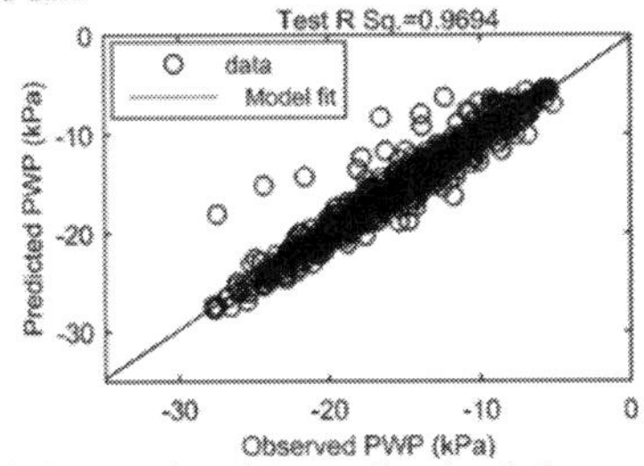

Figure 2. Scatter plot of the predicted and observed PWP, with Line of fit, indicating model fitness to observed data

5 Conclusions

A simple one degree polynomial kernel, has performed well in predicting the complex nonlinear behaviour of PWP, with high correlation recorded between observed and predicted records. This is aided with the excellent choice of input features. As it aided in obtaining a comparatively simple model (three inputs features), and that is what the low degree polynomial is all about simplicity. The low degree Polynomial kernel SVM model, thus can be readily employed to model PWP responses. However, this by no means constitute polynomial kernel as the best kernel to be used for PWP modelling, as this study did not compare results of the polynomial Model with other kernels. In fact a linear kernel Model, could perform nearly as good, since a linear kernel is equivalent to a polynomial with $r = 0$, $d = 1$, and $\gamma = 1$. In the calibration of this Model only γ was different (0.03). Therefore the developed model is quite close to a linear kernel model. Thus there is the need to check other basic kernels to ensure the best choice.

In addition to its proven advantages, such as fast implementation and training, the result of the developed Model are competitive and the low degree polynomial kernel SVM, with just one degree can serve as an excellent option in modelling PWP responses.

Acknowledgement

This study is financed by Ministry of Education Malaysia under fundamental research grant scheme (FRGS) of cost centre no 0153AB-I61. The first author is thankful to UTP for study sponsorship, under the graduate assistantship scheme.

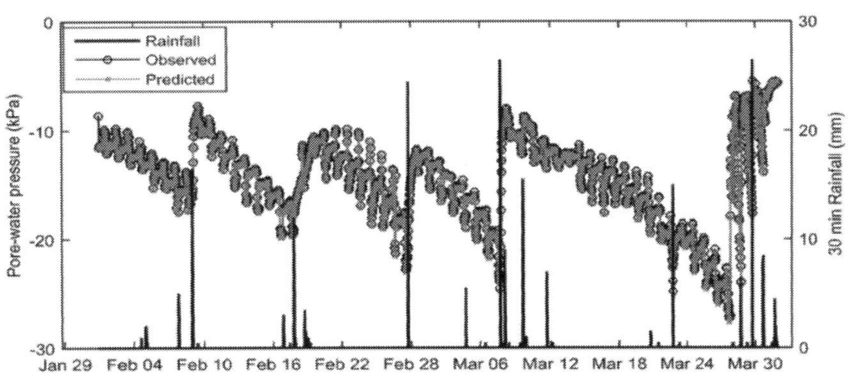

Figure 3. 30 minutes time series plot, showing comparison between observed and Model predicted PWP. With the Model prediction closely following the observed, including records immediately after rainfall event

References

1. E. E. Alonso, N. M. Pinyol, and a. Yerro, "Mathematical Modelling of Slopes," Procedia Earth Planet. Sci., vol. **9**, pp. 64–73 (2014)
2. H. F. Yeh and C. H. Lee, "Soil water balance model for precipitation-induced shallow landslides," Environ. Earth Sci., vol. **70**, no. 6, pp. 2691–2701 (2013)
3. E. Damiano and P. Mercogliano, "Potential Effects of Climate Change on Slope Stability in Unsaturated Pyroclastic Soils," in *Landslide Science and Practice*, vol. **4,** C. Margottini, P. Canuti, and K. Sassa, Eds. Berlin: Springer (2013)
4. H. Rahardjo, A. Satyanaga, and E. C. Leong, "Characteristics of Pore-Water Pressure Response in Slopes During Rainfall," in *3dr Asian Conference on Unsaturated soils-UNSAT-ASIA*, pp. 493–498. (2007)
5. H. Rahardjo, E. C. Leong, and R. B. Rezaur, "Effect of antecedent rainfall on pore-water pressure distribution characteristics in residual soil slopes under tropical rainfall," *Hydrol. Process.*, vol. **22**, no. 4, pp. 506–523, (2008)
6. M. R. Mustafa, R. B. Rezaur, H. Rahardjo, M. H. Isa, and A. Arif, "Artificial neural network modeling for

spatial and temporal variations of pore-water pressure responses to rainfall Artificial neural network modeling for spatial and temporal variations of pore-water pressure responses to rainfall," Adv. Meteorol., vol. **2015**, pp. 1–39, (2015)
7. M. R. Mustafa, R. B. Rezaur, M. H. Isa, and H. Rahardjo, "Estimation of soil pore-water pressure variations using a thin plate spline basis function," in *WIT Transactions of the Built Environment*, vol **137** pp. 615–624 (2014)
8. M. R. Mustafa, R. B. Rezaur, S. Saiedi, H. Rahardjo, and M. H. Isa, "Evaluation of MLP-ANN Training Algorithms for Modeling Soil Pore-Water Pressure Responses to Rainfall," J. Hydrol. Eng., vol. **18**, no. January, pp. 50–57, (2013)
9. M. R. Mustafa, R. B. Rezaur, M. H. Isa, S. Saiedi, and H. Rahardjo, "Effect of antecedent conditions on prediction of pore-water pressure using artificial neural networks," Mod. Appl. Sci., vol. **6**, no. 2, pp. 6–15, (2012)
10. M. R. Mustafa, R. B. Rezaur, H. Rahardjo, and M. H. Isa, "Prediction of pore-water pressure using radial basis function neural network," Eng. Geol., vol. **135–136**, pp. 40–47, (2012)
11. C. J. C. Burges, "A Tutorial on Support Vector Machines for Pattern Recognition," Data Min. Knowl. Discov., vol. **2**, no. 2, pp. 121–167, Jun. (1998)
12. R. Noori, Z. Deng, A. Kiaghadi, and F. T. Kachoosangi, "How Reliable Are ANN , ANFIS , and SVM Techniques for Predicting Longitudinal Dispersion Coefficient in Natural Rivers ?," J. Hydraul. Eng., pp. 1–8 (2015)
13. W. Zhao, T. Tao, and E. Zio, "Parameters tuning in support vector regression for reliability forecasting," Chem. Eng. Trans., vol. **33**, pp. 523–528 (2013)
14. Z. A. Zakaria and A. Shabri, "Streamflow Forecasting at Ungaged Sites Using Support Vector Machines," *Appl. Math. Sci.*, vol. **6**, no. 60, pp. 3003–3014 (2012)
15. K. Lamorski, C. Sławiński, F. Moreno, G. Barna, W. Skierucha, and J. L. Arrue, "Modelling soil water retention using support vector machines with genetic algorithm optimisation.," Scientific World Journal., vol. **2014** (2014)
16. B. Zhu, Z. Cheng, and H. Wang, "A kernel function Optimisation and Selection Algorithm Based on Cost Function Maximization," in *IIEEE International Conference on Imaging Systems and Techniques (IST), 2013*, no. **2**, pp. 259–263 (2013)
17. V. N. Vapnik, *The Nature of Statistical Learning Theory*, vol. **8**, no. 6. New York, New York, USA: Springer, (1995)
18. V. Kecman, *Learning and Soft Computing*. London, England: MIT press, (2001)
19. Y. Chang, C.-J. Hsieh, K.-W. Chang, M. Ringgaard, and C. Lin, "Training and Testing Low-degree Polynomial Data Mappings via Linear SVM," J. Mach., vol. **11**, pp. 1471–1490, (2010)
20. S. Yaman and J. Pelecanos, "Machines for Speaker Veri fi cation," IEEE Signal Process. Lett., vol. **20**, no. 9, pp. 901–904, (2013)
21. G. Chandrashekar and F. Sahin, "A survey on feature selection methods," Comput. Electr. Eng., vol. **40**, no. 1, pp. 16–28, (2014)
22. S. Maldonado and R. Weber, "A wrapper method for feature selection using Support Vector Machines," Inf. Sci. (Ny)., vol. **179**, no. 13, pp. 2208–2217, (2009).
23. R. Kohavi and R. Kohavi, "Wrappers for feature subset selection," Artif. Intell., vol. **97**, no. 1–2, pp. 273–324, (1997).
24. H. Mark, F. Eibe, H. Geoffrey, P. Bernhard, R. Peter, and I. H. Witten "The WEKA Data Mining Software: An Update," SIGKDD Explorations, vol **11**, no 1 (2009)
25. Y. EL-Manzalawy and V. Honavar, "WLSVM : Integrating LibSVM into Weka Environment," (2005). http://www.cs.iastate.edu/~yasser/wlsv
26. C. C. Chang, C. J. Lin "LIBSVM: A Library for Support Vector Machines," ACM Trans Intell Syst Technol (ACM TIST) vol **2**, pp. 1–27 (2011)
27. P. A. Kagoda, J. Ndiritu, C. Ntuli, and B. Mwaka, "Application of radial basis function neural networks to short-term streamflow forecasting," *Phys. Chem. Earth, Parts A/B/C*, vol. **35**, no. 13–14, pp. 571–581, (2010)

 DOI: 10.1051/matecconf/20165904004

Adsorptive Removal of Formaldehyde by Chemically Bamboo Activated Carbon with addition of Ag nanoparticle: Equilibrium and Kinetic

Wara Dyah Pita Rengga[1,2], Mahmud Sudibandriyo[1] and Mohammad Nasikin[1]

[1]*Department of Chemical Engineering, Universitas Indonesia, Kampus UI Depok, Depok 16424, Indonesia*
[2]*Department of Chemical Engineering, Universitas Negeri Semarang, Sekaran Gunungpati Semarang 50229, Indonesia*

Abstract. Carbon was prepared from dried waste bamboo (Dendrocalamus asper) using chemical activation with KOH. The carbon was prepared with the activating agent in a mass ratio of KOH and dried bamboo (3:1) at 800oC. Using impregnation technique, the bamboo-based activated carbon has developed with modified Ag nanoparticle (Ag-AC) to capture formaldehyde. The Ag-AC has characteristics of moderate surface area of 685 m2/g and average pore size of 2.7 nm. The adsorption equilibriums and kinetics of formaldehyde on Ag-AC measured. The influences of initial formaldehyde on adsorption performance have measured in a batch system. The equilibrium data were evaluated by isotherm models of Langmuir, Freundlich, and Temkin. The Langmuir model well describes the adsorptive removal of formaldehyde on Ag-AC in this study. Pseudo-first-order and pseudo-second-order kinetic equations were applied to test the experimental data. The pseudo-second-order exhibited the best fit for kinetic study.

1 Introduction

The quality of indoor air has become areas of concern since it related to human health, directly. Maintaining adequate indoor air quality through controlling ambient concentrations of formaldehyde is recommended, although the relationships between air quality and formaldehyde rarely determined. Formaldehyde with a high level can cause symptoms such as eye, nose, and throat irritation, chest pains, bronchitis, and prolonged formaldehyde has associated with reduced pulmonary function and asthma [1,2]. General methods are being used to purify the polluted air, including adsorption [3], membrane, absorption, plasma decomposition [4], and photocatalytic oxidation [5] techniques. Considering adsorption has a promising method for controlling low concentration of formaldehyde because of its simple technique and low energy requirement. Activated Carbon (AC) of bamboo (Dendrocalamus asper) is extremely porous material with a moderate surface area and the most widely used as adsorbent. The AC is used for capturing indoor air pollutants at ambient temperature is considered economically feasible. However, AC is a non-polar adsorbent that sometimes exhibits low selectivity with polar contaminants. In recent years, modified AC with metal nanoparticle was considered a potential adsorption and oxidation catalytic process due to increased capacity to formaldehyde. Chemically bamboo AC was produced by with the activating agent in a mass ratio of KOH and dried bamboo (3:1) at 800°C. Silver nanoparticles are impregnated on AC to enhance the adsorption capacity of removal formaldehyde. An interesting fact was that the Ag nanoparticles dispersed on commercial coal-based AC to effectively remove formaldehyde from air [6]. Silver also possesses high applicability for use in controlling indoor air due to its non-toxic feature [7].

In this study, AC was modified with silver nanoparticles (Ag-AC) to improve their adsorption capacities. It is important to determine the Ag-AC by submitting it for the adsorption of formaldehyde and evaluating the equilibrium isotherms and kinetics adsorption of formaldehyde.

2 Materials and Methods

2.1 Materials

The AC had conducted as Rengga's method [7]. The Bamboo Chars and KOH were mixed with 1:3 of weight ratio. The immersed bamboo chars in KOH solution was dehydrated until it formed the slurry. Subsequently, the slurry was activated in a tube-electronic heater at heating rate of 10°C/min in the presence of nitrogen gas with flowing rate of 200 mL/min. The activation process was continued up to temperature of 800°C and maintained for 1 h. The activated product was washed with HCl followed by rinsing with distilled water until the pH of the washing effluent was 6. The activated carbon was dried at 105 °C and grinded into 200 mesh prior to characterization and adsorption study.

2.2 Synthesis Adsorbent

Silver nanoparticle synthesized was conducted with reduction process similar to Gaedhi's method [8] in the previous paper [7]. As much 4 mL of 0.1 M $AgNO_3$ added to 1 L of an aqueous solution with 0.15 wt% of the soluble starch. The mixture vigorously stirred for 1 h under atmosphere. The pH of resulting solution adjusted to 8.0 by adding NaOH solution. The mixture was maintained at 50 °C until a change of the reaction solution became yellow. The next step, silver nanoparticles in solution was mixed with 107.9 g of activated carbon under magnetic stirring, resulting in the deposition of about 4 wt% of silver nanoparticles on AC. The AC-Ag was filtered then washed with distilled water and dried at 105 °C until its constant weight.

2.3 Batch equilibrium studies

The adsorption tests were performed to determine the addition amounts of Ag-AC (0.5 g) to capacity adsorption of formaldehyde at various gas-phase concentrations. The test was done into 10 mL glass vial equipped with a septum, and subsequently the vials was shaken at 25 °C for 24 h. The inlet and outlet concentrations of formaldehyde in the batch tests were collected using a 0.5 mL gas-tight syringe. Furthermore, the sample was analyzed immediately by using gas chromatography equipped with a flame ionization detector (GC/FID, Shimadzu GC-2014). In the experiments, procedures of kinetic tests were identical to equilibrium tests of adsorption. The samples were drawn at suitable time intervals. Each experiment continued until equilibrium conditions were reached when no further decrease in the formaldehyde concentration was measured. Initially on the Ag-AC did not contain any formaldehyde. Furthermore, formaldehyde adsorbed was calculated by using Eq. 1.

$$q_e = \frac{V(c_0 - c_t)}{m}$$

where qe is formaldehyde adsorbed at equilibrium [mg/g]; V is a volume of adsorbate [L]; m is the mass of modified adsorbent [g]. The value of c_0 is formaldehyde concentration initially, and c_t is formaldehyde concentration after adsorption at equilibrium time [ppm], respectively. The best-fit model determined for a system of formaldehyde adsorption, experimental data substituted for equilibrium isotherm model equations and kinetic models. Parameter values of equations were obtained using non-linear regression analysis. The correlation coefficient (R^2) of the fit these values, were also reported for each model.

3 Results and Discussion

The characteristics of pore are shown in Table 1. We calculated the mesoporosity of AC by BET and BJH methods with surface area and average size of pore of 685 m^2/g and 2.7 nm, respectively.

Tabel 1. The Characteristics of Pores

S_{BET} [m^2/g]	D_{BET} [nm]	$V_{meso\ BJH}$ [cm^3/g]	$V_{micro\ BJH}$ [cm^3/g]	V_{BJH} [cm^3/g]
685	2.7	0.2009	0.0336	0.2355

3.1 Isotherm Equilibrium

In isotherm equilibrium, formaldehyde removal increases with an increase in the initial formaldehyde concentration. Formaldehyde at equilibrium concentration was indicated the maximum of adsorption capacity is shown in Fig. 1. The adsorption process reaches an equilibrium state, occurs when the molecules distribute between gas and solid phase. The data of formaldehyde adsorption equilibrium were analyzed using Langmuir, Freundlich, and Temkin isotherms, based on the equilibrium of the sorption between the adsorbate and adsorbent. The Langmuir model describes a monolayer sorption with a homogenous distribution both energies and sites of sorption The Freundlich model describes the adsorption distribution energy for heterogeneous sites. The Temkin isotherm assumption is the adsorption heat of all the molecules in AC decreases linearly with coverage. In order, removal formaldehyde from indoor for optimization adsorption system is important to be done to establish the most appropriate correlation in equilibrium curve in Fig 1.

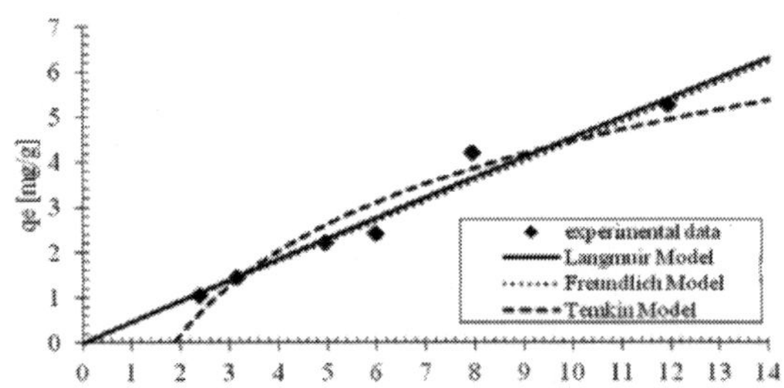

Figure 1. Representation of formaldehyde adsorption on g-AC for Langmuir, Freundlich, and Temkin Models

The Langmuir, Freundlich, and Temkin isotherm equations with their equation are presented in Table 2. In the Langmuir model, Q_m is the maximum monolayer coverage adsorption capacity [mg/g], and K_L is the Langmuir isotherm constant [m^3/g]. In the Freundlich model, the values of K_F and n are Freundlich isotherm constant and adsorption intensity. The value of $1/n$ is the heterogeneity of the adsorbent. The value of 1/n is higher than 1 for favorable adsorption process. In Temkin model, B = RT/b, R is the universal gas constant [8.314 J/K mol], and T is the absolute temperature [K]. The value of B_T is constant related to the heat of adsorption [m^3/g], and A_T is the equilibrium binding constant.

The isotherm constants of adsorption models and correlation coefficients are reported in Table 2. The adsorption of formaldehyde on Ag-AC for Langmuir isotherm is shown range studies with high correlation coefficient ($R^2 > 0.96$). Therefore, the conclusion of isotherm that Langmuir model is the best isotherm to predict the adsorption of formaldehyde on Ag-AC adsorbent. It reported that Ag-AC had a relatively suitable adsorption capacity of 157.223 mg/g in this work. The Langmuir model predictions for concentration formaldehyde of 50 mg/m^3 using commercial coal-based carbon is 15.8 mg/g [5] that was close to this study (21 mg/g). Confirmation of the experimental data to the Langmuir isotherm equation proves that the surface of the Ag-AC for the adsorption of formaldehyde consists of homogeneous sites. The results are in agreement with previous researchers to the adsorption of pollutants using AC prepared from bone char [9].

Table 2. Parameter of Isotherm Constant and Correlation Coefficient for Adsorption of Formaldehyde on Ag-AC

Model	Non-linear equation	Parameter	Value
Langmuir	$q_e = \frac{Q_m . K_L . C_e}{1 + K_L . C_e}$	Q_m [mg/g]	157.223
		K_L ([m^3/mg]	0.0030
		R^2	0.966
Freundlich	$q_e = K_F . C_e{}^{1/n}$	$1/n$	1.005
		K_F [mg/g][m^3/mg]$^{1/n}$	0.466
		R^2	0.962
Temkin	$q_e \left(\frac{RT}{b_T}\right) . \ln(A_T . C_e$	$\left(\frac{RT}{b_T}\right)$ [J/mol]	2.668
		A_T [m^3/mg]	0.536
		R^2	0.925

3.2 Adsorption kinetic

The kinetics of adsorption of formaldehyde on Ag-AC were investigated by using pseudo-first-order and pseudo-second-order models. The pseudo-first-order and pseudo second-order kinetic models of k_1, k_2, and R^2 are presented in Table 3. The value of k_1 is the adsorption rate constant (1/h) dan k_2 is the rate constant of second-order adsorption (g/mg.h). The values of q_e and q_t are the amounts of formaldehyde adsorbed at equilibrium and time [mg/g], respectively. It can see that the values of R^2 obtained did not trend consistently at low formaldehyde initial concentration for the pseudo-first-order model. The pseudo-second-order model predicts the behavior of the concentration range adsorption. Between the experimental and the calculated q_e values (Fig. 2) have a good agreement. Besides that the experimental q_e value agree with the estimated q_e values obtained from non-linear. Obtained that the correlation coefficient values $R^2 = 0.995$ for the second-order kinetic model, it is clear that the adsorption follows this model. The value of R^2 was close to unity, indicating that Ag-AC fitted well to this model for the adsorption of formaldehyde. Between these models, the criterion for their applicability is based on judgment on the respective R^2 with the high values indicates that the understudy kinetic of formaldehyde adsorption follows pseudo-second-order model. Agreed the results of the work performed by previous researchers for adsorption of formaldehyde pollutants using activated carbon prepared from bone char [9]. Wu et al. [10] reported that the pseudo second-order model were suitable for adsorption of adsorbate with low molecular weight on smaller adsorbent particles as described for this work.

Tabel 3. Kinetic Parameter for The Adsorption of Formaldehyde on Ag-AC

Kinetic Model	Non-linear equation	Parameter	Value
Pseudo-first-order	$q_t = q_e[1 - \exp(-k_1 . t)]$	k_1 [1/h] R^2	0.333 0.991
pseudo-second order	$q_t = . q_e^2 k_2 . t/(1 + q_e k_2 . t)$	k_2 [g/mg.h] R^2	0.00015 0.995

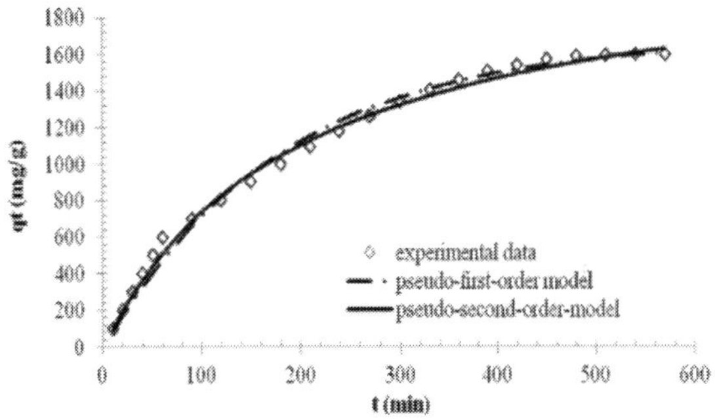

Figure 2. Pseudo-first-order and pseudo-second-order kinetics plot for the adsorption of formaldehyde on Ag-AC

4 Conclusions

The present investigation showed that low concentration formaldehyde by chemically activated carbon of bamboo modified with Ag nanoparticles is an effective adsorption. In the equilibrium condition, adsorption using modified activated carbon is appropriater with Langmuir isotherm model, as well as Freundlich and Temkin models. Isothermal equilibrium model is suitable Langmuir monolayer adsorption model, with adsorption capacity of 157.223 mg/g. For kinetics adsorption of formaldehyde using modified activated carbon, the uses of pseudo-second-order model is more appropriate rather than pseudo-first-order-model.

Acknowledgements

We gratefully acknowledge Directorate General of Higher Education, Indonesian Ministry of National Education, Universitas Negeri Semarang, and Universitas Indonesia to support and assist of this work. (Fundamental Research Grant No. 523/UN37.3.1/LT/2015).

References

1. C.M. Thompson, R.P. Subramaniam and R.C. Grafstrom: Toxicol. Appl. Pharmacol Vol. 233 (2008), p. 355
2. P. Chin, L.P. Yang and D.F. Ollis: J. Catal. Vol. 237 (2006), p. 29
3. S. Kumagai, K. Sasaki, Y. Shimizu and K. Takeda: Sep. Purif. Technol. Vol. 61 (2008), p. 398
4. L. Fang, G. Zhang and A. Wisthaler: Indoor Air Vol. 18 (2008), p. 375
5. N. An, Q. Yu, G. Liu, S. Li, M. Jia and W. Zhang: J. Hazard. Mater. Vol. 186 (2011), p. 1392
6. S. K. Shin and J. H. Song: J. Hazard. Mater. Vol. 194 (2011), p. 385
7. W.D.P. Rengga, M. Sudibandriyo and M. Nasikin: IJCEA Vol. 4,(2013), p. 334.
8. M. Ghaedi, M.N. Biyareh, S.N. Kokhdan, S. Shamsaldini, R. Sahraei, A. Daneshfar and S. Shahriyar: Mater. Sci. Eng Vol. C32 (2012), p. 725
9. A. Rezaee, H-A Rangkooy, A. Khavanin, A. Jonidi-Jafari, R.D.C. Soltani and A. Nili-Ahmadabadi: IJCEA Vol. 2 (2011), p. 423.
10. F.C. Wu, R.L. Tseng, S.C. Huang and R.S. Juang: Chem. Eng. J. Vol. 151 (2009), p. 1.

MATEC Web of Conferences 59, 04005 (2016) DOI: 10.1051/matecconf/20165904005
ICFST 2016

Copper Determination in Gunshot Residue by Cyclic Voltammetric and Inductive Coupled Plasma-Optical Emission Spectroscopy

Nurul'Afiqah Hashimah Mohd Hashim[1], Zainiharyati Mohd Zain[1] and Mohd Zuli Jaafar[2]

[1]*Faculty of Applied Sciences, Universiti Teknologi MARA, 40450 Shah Alam, Malaysia*

[2]*Universiti Teknologi MARA, 77300 Jasin, Malaysia*

Abstract. Analysis of gunshot residue (GSR) is a crucial evidences for a forensic analyst in the fastest way. GSR analysis insists a suitable method provides a relatively simple, rapid and precise information on the spot at the crime scene. Therefore, the analysis of Cu(II) in GSR using cyclic voltammetry (CV) on screen printed carbon electrode (SPCE) is a better choice compared to previous alternative methods such as Inductive Coupled Plasma-Optical Emission Spectroscopy (ICP-OES) those required a long time for analysis. SPCE is specially designed to handle with microvolumes of sample such as GSR sample. It gives advantages for identification of copper in GSR on-site preliminary test to prevent the sample loss on the process to be analyzed in the laboratory. SPCE was swabbed directly on the shooter's arm immediately after firing and acetate buffer was dropped on SPCE before CV analysis. For ICP-OES analysis, cotton that had been soaked in 0.5 M nitric acid was swabbed on the shooter's arm immediately after firing and kept in a tightly closed sampling tube. Gold coated SPCE that had been through nanoparticles modification exhibits excellent performance on voltammograms. The calibration was linear from 1 to 50 ppm of copper, the limit of detection for copper was 0.3 ppm and a relative standard deviation was 6.1 %. The method was successfully applied to the determination of copper in GSR. The Cu determination on SPCE was compared and validated by ICP-OES method with 94 % accuracy.

1 Introduction

Gunshot residue is classified into forensic microtraces category [1] which is valuable evidence in the forensic analysis [2]. This is due to the condensation of metallic particles from the primer of a firearm discharged that is characterized by their uniqueness of chemical content and morphology of GSR [3]. It provides very useful information for forensic analysts to obtain the information about the type of gun and ammunition used in a crime when proper methods are used [2,4]. A suitable method used in GSR analysis offers more reliable results in reducing the erroneous convictions [4].

Many GSR analyses were reported using different techniques separately for identification organic or inorganic compounds. The organic compounds that are usually obtained in GSR sample are methyl centralite, ethyl centralite and diphenylamine [5]. Inorganic components that had been detected as reported in the previous study are lead, barium [6–9] and antimony [6,7].

Analysis of organic compounds in GSR was implemented by a combination of solid-phase microextraction (SPME) and gas chromatography mass spectrometry (GC-MS) technique [10], desorption electrospray ionization mass spectrometry (DESI–MS) [5] and fourier transform infrared (FTIR) spectroscopy [11]. Analyses of inorganic compounds commonly used scanning electron microscope energy dispersive x-ray spectrometer (SEM-EDS) [12–15], sector field inductively coupled plasma mass spectrometer (SF-ICP-MS) [6], inductively coupled plasma atomic emission spectrometer analysis (ICP-AES), ion beam analysis particle induced x-ray emission (IBA-PIXE) [7], Raman spectroscopy [2], capillary electrophoresis [16], ion chromatography [17], differential pulse anodic stripping voltammetry (DPASV) [8], cyclic voltammetry [4] and abrasive stripping voltammetry (AbrSV) [18].

A fast and precise technique in GSR analysis is needed for identifying and discriminating firearm ammunition and GSR rapidly. Salles *et al.* in 2012 was proposed a fast electrochemical technique that can be miniaturized and applied in GSR analysis. Electrochemical analysis provides simple responses to different samples and is generally cheaper compared with other conventional spectroscopic methods such as atomic absorption spectrometry and mass spectrometry. Cyclic voltammogram (CV) is a selected method than the other electrochemical technique for GSR analysis. GSR was collected from the hands of the shooters before analyzing using CV and a gold microelectrode as the electrode [4].

Recently, analysis of GSR had been performed by voltammetry of microcrystals or abrasive stripping voltammetry (AbrSV) which one of an electrochemical technique. The metal deposited produces an analytically diagnostic signal in stripping voltammetry. It is a very suitable method for GSR analysis those contain organic components [19-20]. AbrSV technique had been implemented in GSR analysis based on an initial mechanical transfer of trace amounts of the GSR from the hand of the suspect directly to the surface of the working-electrode sensor strip without intermediate processing steps. The used of screen-printed carbon (SPE) as a working electrode in this abrasive technique is an alternative way to overcome the limitations related with the effective collection of residues and simplifying the overall electroanalytical protocol [18].

2 Experimental

Chemical and Materials. Acetate buffer (0.7 M, pH 4.6) and standard solutions of copper (1000ppm) were purchased from Fluka (St. Louis, MO). A range of 1 to 50 ppm of copper standard solution was made from the copper stock solution. Deionized water (18 M MΩ cm) from a NANOpure Diamond system (Thermo Scientific, Dubuque, IA) was used to prepare all solutions. Both SPCE and gold coated SPCE were purchased from Metrohm (The Netherlands) used for all measurements.

Instruments and Procedure. An Autolab PGSTAT 12 (Eco Chemie, The Netherlands) was used to perform electrochemical (cyclic voltammetry) measurements and an Inductively Coupled Plasma Optical Emission Spectrometry (ICP-OES) instrument (Perkin Elmer, Optima 2100DV, USA) was used for the quantification of copper.

GSR Sampling. GSR samplings were performed in enclosed areas at Science and Technology Research Institute for Defense (STRIDE) at Batu Arang, Selangor. Glock with 9 mm ammunition, Rifle Stryer AUG-A1 with 5.56 mm ammunition and Rifle Colt M16-A1 with 5.56 mm ammunition were used in this work. GSR triplicates sampling was carried out in two ways; gold coated SPCE was directly swabbing on the shooter's arm for voltammetric determination. Secondly, cotton swab that was soaked in 0.5 M of nitric acid was swabbed on the shooter's arm and kept in a tightly closed sampling tube.

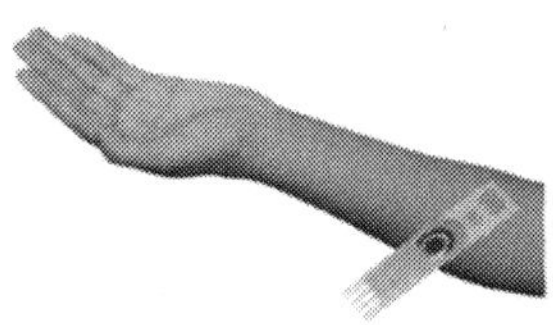

Figure 1. SPCE swabbed on the shooter's arm

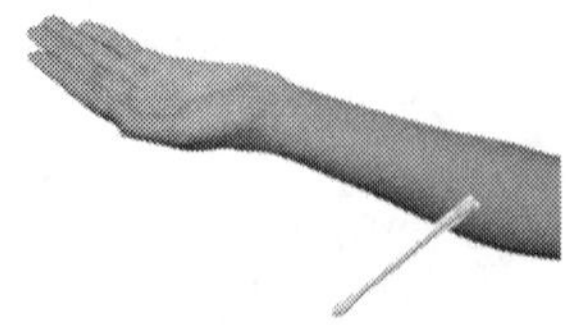

Figure 2. Soaked cotton swabbed on the shooter's arm

Cyclic Voltammetry (CV). Cyclic voltammetry was applied to characterize GSR electrochemical signatures. 3-electrode strip surface was immediately swabbed after firing on the shooter's arm. 20 µL of acetate buffer was dropped on SPCE and immediately analyzed using cyclic voltammetry on-site preliminary test. Scan potential of -100 to 100 mV and scan rate of 100 mVs^{-1} was set up as operating parameters.

Inductive Coupled Plasma-Optical Emission Spectroscopy (ICP-OES). GSR samples were collected by swabbing cotton soaked in 0.5 M of nitric acid on the shooter's arm. Then the swabbed cottons were soaked in 0.5 M of nitric acid overnight. Operating parameters were set up as follows: 1500 W of a RF power, 15 $Lmin^{-1}$ of a plasma flow rate, 0.2 $Lmin^{-1}$ of an auxiliary gas flow rate and 0.8 $Lmin^{-1}$ of a nebulizer flow rate. Argon was used to purge the optics and to form the plasma.

3 Result and Discussion

Voltammetric Response of Cu(II). Prior to membrane deposition, electrochemical behavior of Cu(II) was studied using CV on SPCE. CV scans for 1 to 50 ppm of Cu(II) solutions at a constant scan rate of 100 mVs^{-1} are shown in Fig. 3. The electrochemical oxidation of Cu(0) to Cu(II) by CV can be observed at even very low Cu(II) concentrations. The cyclic voltammograms show the peak current linearly increases as the concentration Cu(II) increased. This is due to increasing the rate of diffusion of Cu(II) towards the electrode surface. The presence of a large amount of electroactive species at higher concentration cause a gradual increase of peak current with the increase of Cu(II) concentration. The system follows a diffusion controlled system [20]. A linear graph was obtained in the concentration range from 1 to 50 ppm (slope = 5754, R^2 = 0.9923).

DOI: 10.1051/matecconf/20165904005

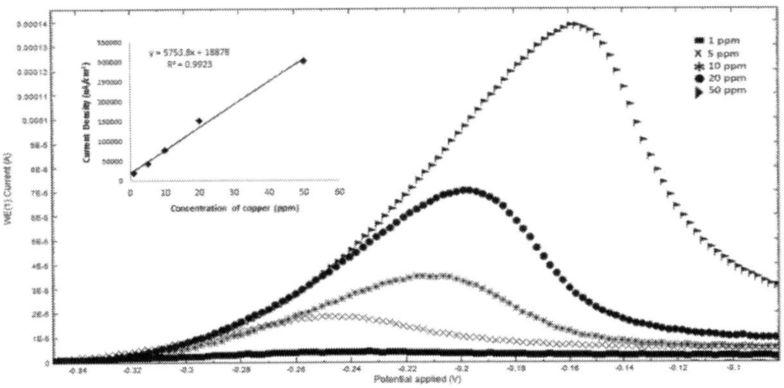

Figure 3. Cyclic voltammograms of 1, 5, 10, 20 and 50 ppm of Cu(II). Insets show the calibration curve of Cu(II) concentration

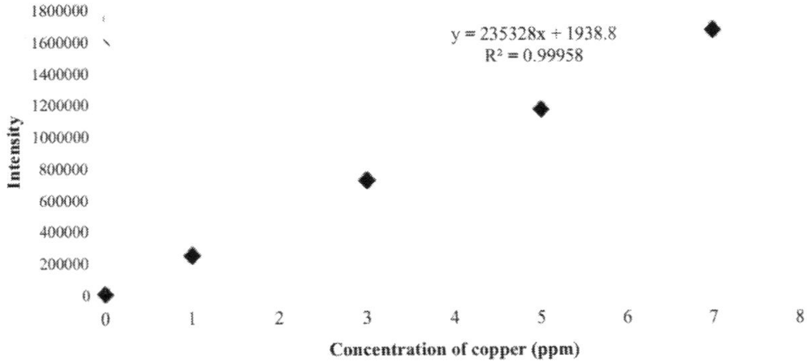

Figure 4. Calibration curve for ICP-OES for 1-10 ppm Cu(II)

ICP-OES of Cu(II). At ICP-OES analysis, a reasonable linearity was obtained upon introducing an increasing concentration of Cu(II) as shown in Fig. 4. A calibration curve established for identification of Cu(II) in GSR. A linearity response was obtained in the concentration range 1 to 7 ppm (slope=235328, R^2=0.9996). The reproducibility of the gold coated SPCE was also evaluated and it was found to be good with relative standard deviation (RSD) of 6.1 % (n=3).

Comparison of CV and ICP-OES. Both methods show the lower sensitivity for identification of Cu(II). However, in terms of practicality CV shows more convenient method. CV analysis on SPCE is a rapid and reliable tool in determination of Cu(II) in gunshot residue. These advantages of SPCE enable the quantification of Cu(II) in GSR on the spot at the crime scene.

Comparison of the performance of SPCE and gold coated SPCE. The performances of Cu(II) on both SPCE and gold coated SPCE were observed. It shows the gold coated SPCE exhibits excellent reproducibility rather than SPCE in aspects of current produced in cyclic voltammograms. Modification of electrodes using nanoparticles enhance electrochemical performance due to their large effective surface area, effective catalysis and fast mass transport [21-22]

DOI: 10.1051/matecconf/20165904005

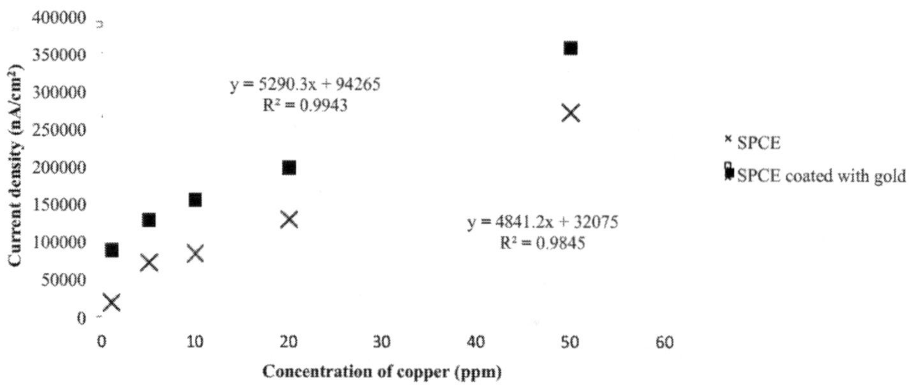

Figure 5. Comparison performances of SPCE and gold coated SPCE for the concentration of 1 to 50 ppm of Cu(II)

GSR Analysis. The copper content in GSR was determined using both cyclic voltammetry and ICP-OES. The Student's t-test was applied to the results obtained by CV and those found using ICP-OES, and comparison is shown in Table 1. The paired Student's t-test indicated that there was no significant difference between the results obtained from both methods at the 94% confidence level. Therefore, it can be concluded that CV is reliable and can be successfully applied to the copper determination in GSR analysis. Based on the analysis of results, it can be seen that the amount of copper present in the GSR from the Rifle Stryer AUG-A1 higher compared to both Rifle Colt M16-A1 and Glock.

Table 1. Results obtained from CV and ICP-OES methods for analysis of GSR samples

Gun	Ammunition	Voltammetric [ppm Cu^{2+}]	ICP-OES [ppm Cu^{2+}]	$t_{calculated}$
Rifle Colt M16-A1	5.56	0.707±0.01	0.748±0.002	0.04
Rifle Stryer AUG-A1	5.56	1.349±0.02	1.432±0.005	0.03
Glock	9	0.680±0.01	0.724±0.005	0.01

4 Conclusion

Cyclic voltammetry analysis on SPCE is the perfect combination for the quantification of copper in GSR for a forensic analyst in the fastest way at the crime scene. The performance of Cu(II) on gold coated SPCE shows excellent reproducibility due to their large effective surface area, effective catalysis and fast mass transport.

Acknowledgement

Financial and technical support from The Ministry of Education through the Fundamental Research Grant Scheme (FRGS) and Universiti Teknologi MARA are highly acknowledged.

References

1. K. Štůlová, T. Vaculovič and V. Kanický: Mendel Net. (2013) p. 940–944
2. M. López-López, J. J. Delgado and C. García-Ruiz: Forensic Sci. Int. Vol. 231, No. 1–5 (2013) p. 1–5
3. Z. Brozek-Mucha and J. Was-gubala: Current Microscopy Contributions to Advances in Science and Tech.(2012) p. 1480–1491
4. M. O. Salles, M. Bertotti and T. R. L. C. Paixão: Sensors Actuators B Chem. Vol. 166–167, p. 848–852
5. M. Morelato, A. Beavis, A. Ogle, P. Doble, P. Kirkbride and C. Roux: Forensic Sci. Int. Vol. 217, No. 1–3 (2012) p. 101–6
6. J. E. S. Sarkis, O. N. Neto, S. Viebig and S. F. Durrant: Forensic Sci. Int. Vol. 172, No. 1 (2007) p. 63–6
7. E. Turillazzi, G. P. Di Peri, A. Nieddu, S. Bello, F. Monaci, M. Neri, C. Pomara, R. Rabozzi, I. Riezzo and V. Fineschi: Forensic Sci. Int. Vol. 231, No. 1–3 (2013) p. 142–9
8. C. A. Woolever, D. E. Starkey and H. D. Dewald: Forensic Sci. Int. Vol. 102, No. 1 (1999) p. 45–50
9. M. O. Salles, J. Naozuka and M. Bertotti: Microchem. J. Vol. 101 (2012) p. 49–53
10. C. Weyermann, V. Belaud, F. Riva and F. S. Romolo: Forensic Sci. Int. Vol. 186, No. 1–3 (2009) p. 29–35
11. J. Bueno, V. Sikirzhytski and I. K. Lednev: Analytical Chem. Vol. 84, No. 10 (2012) p. 4334-9
12. G. Zadora and Z. Brożek-Mucha: Mater. Chem. Phys. Vol. 81, BN. 2–3 (2003) p. 345–348
13. S. Steffen, M. Otto, L. Niewoehner, M. Barth, Z. Brożek-Mucha, J. Biegstraaten and R. Horváth: Spectrochim. Acta Part B At. Spectrosc. Vol. 62, No. 9 (2007) p. 1028–1036

14. B. Cardinetti, C. Ciampini, C. D'Onofrio, G. Orlando, L. Gravina, F. Ferrari, D. Di Tullio and L. Torresi: Forensic Sci. Int. Vol. 143, No. 1 (2004) p. 27–46
15. A. Martiny, A. P. C. Campos, M. S. Sader and A. L. Pinto: Forensic Sci. Int. Vol. 177, No. 1 (2008) p. e9–17
16. C. Love, E. Gilchrist, N. Smith and L. Barron: Forensic Sci. Int. Vol. 231, No. 1–3 (2013) p. 150–6
17. Š. Komorsky-Lovrić and B. Nigović: J. Electroanal. Chem. Vol. 593, No. 1–2 (2006) p. 125–130
18. Sebojka Komorsky-Lovric and I. Novak: Electrochimica Acta. Vol. 98 (2013) p. 153–156
19. A. Chatterjee, R. Wiltshire, K. B. Holt, R. G. Compton, J. S. Foord and F. Marken: Diam. Relat. Mater. Vol. 11, No. 3–6 (2002) p. 646–650
20. F. Haque, M. S. Rahman, E. Ahmed, P. K. Bakshi and A. A. Shaikh: Vol. 61, No. 2 (2013) p. 161–166
21. H. Wan, Q. Sun, H. Li, F. Sun, N. Hu and P. Wang: Sensors Actuators B Chem. Vol. 209 (2015) p. 336–342
22. V. Sosa, C. Barceló, N. Serrano, C. Ariño, J. M. Díaz-cruz and M. Esteban: Analytica Chimica Acta. Vol. 855 (2015) p. 34–40

MATEC Web of Conferences 59, 05001 (2016) DOI: 10.1051/matecconf/20165905001
ICFST 2016

Simulation and Characteristics Analysis on Vibration and Sound Radiation Response for a Small Underwater Robot

Zhi Zhou

Yichang Testing Technology Research Institute, Yichang City 443003, Hubei Province, China

Abstract. An underwater robot is one of the important ocean equipment, and especially its stealth performance has influenced on the vitality in naval warfare. Structure radiation noise as the main source of underwater robot noise, so analysis on vibration and noise radiation is a topic of great concern. The way is used widely that based on fluid-solid coupling modal analysis combined with simulation on vibration and noise-radiation response in order to evaluate stealth performance of underwater structure. In the paper, firstly via finite element method and boundary element method, the modal frequencies and vibration modes of small underwater robot are calculated. Then the surface vibration displacements of underwater robot and the sound pressures of acoustic field under different frequency horizontal exciting force are obtained and analyzed. Lastly, through the analysis of the structural vibration and acoustic performance, the control strategies for structure bending vibration, acoustic radiation and structural acoustic design are proposed.

1 Introduction

With the development of application of underwater robot, it has been extended to underwater assault weapons step by step, so more and more high demands on stealth performance are necessary. However, the vibration and noise prediction and control technology not only affect greatly the stealth performance, but also influence the underwater robot vitality. Thus, the control technology of vibration and radiated noise for underwater robot is researched by navies in the world to decrease the likelihood of detection by sonar, and it is becoming an important evaluation index of underwater robot performance. The kinds of noises for underwater robots are divided into three categories [1-4]: structural noise, propeller noise and flow noise. The structural noise as above mentioned, is the underwater radiation noise by hull shell from the structure vibration caused by power equipment, and it is a major noise source. Especially at a slow speed and at a large depth, structure radiation noise is the main components of the total radiation noise level, and which accounts for about 70%.

For most research on noise prediction method of underwater robot, the finite element method (FEM) and boundary element method (BEM) [5-7] are widely used. The finite element method is put forward in the 1950s, has been quickly applied to the analysis and calculation of acoustic radiation problems. The FEM is used as a solution method for inner sound field of limited region and different shape of the structure boundary. But like finite difference method, it needs to be subdivided in the entire analysis domain, and interpolated with variables, leading to too many calculating degrees of freedom. Meanwhile, for underwater robot, as a case of exterior acoustic radiation problem in infinite domain, the finite element method is difficult to determine the split edge resulting in large calculation errors. In order to solve the above problem, the boundary element method is presented in the 1960s, a kind of half-analytical and half-numerical method which is solved by analytical equations within the domain and by the finite element method for the boundaries of the domain. So the numerical calculation degrees of freedom and too much original information are reduced, and it is suitable for solving infinite area of the sound field.

In the paper, the method combined the finite element method and boundary element method is used so as to analyze vibration and sound radiation response for a small underwater robot. Firstly, the structure vibration including flow-solid coupled modal is obtained by using the finite element method. Then the vibration response on the surface of the shell is set as a sound calculation input condition of the boundary of the analytical field. Lastly, the acoustical problem is solved by using the boundary element method for the infinite domain.

2 The basic principle of FEM and BEM

For underwater robot, the structure is dispersed into finite elements. Under the outer load to underwater robot, the overall dynamics control equation in air as followed:

$$[M_s]\{\ddot{u}\}+[C_s]\{\dot{u}\}+[K_s]\{u\}=\{f_s\} \qquad (1)$$

Where, $[M_s]$, $[C_s]$ and $[K_s]$ are the mass, damping and stiffness matrix of the structure respectively. $\{\ddot{u}\}$, $\{\dot{u}\}$ and $\{u\}$ are the displacement, velocity and acceleration matrix of nodes of the finite element model. $\{f_s\}$ is the excitation force vector.

For the vibration of underwater robot in the sea, the interaction between seawater and outer surface of robot shell is produced physically, namely the fluid- structure coupling effect. The vibration of the shell produces sound pressure, and the pressure induces an opposite force that is exerted on the structural surface. The sound pressure is satisfied with the wave equation as below.

$$\nabla^2 p = \frac{1}{c^2}\frac{\partial^2 p}{\partial t^2} \tag{2}$$

Where, c is the sound velocity. For the calculation process, the fluid domain is divided into numbers of finite elements. The sound pressures of field points, the displacements and corresponding order derivatives of time of mass points can be interpolated by values of the difference nodes. Then sound pressure mathematic variations are divided out. Considering damping effect, the finite element equation is deduced in the fluid acoustic field as followed.

$$[M_f]\{\ddot{p}\}+[C_f]\{\dot{p}\}+[K_f]\{p\}+\rho_f[R]\{\ddot{u}\}=0 \tag{3}$$

Where, $[M_f]$, $[C_f]$ and $[K_f]$ stand for fluid mass, damping and stiffness matrix respectively. $[R]$ represent the coupling mass matrix. The ρ_f is the fluid density. $\{p\}$ is the node pressure matrix. With combining Eq.1 and Eq.3 simultaneous, the whole bending vibration controlling equation of underwater robot under the action from fluid-structure interaction can be deduced as below.

$$\begin{Bmatrix} [M_s] & [0] \\ \rho_f[R] & [M_f] \end{Bmatrix}\begin{Bmatrix} \{\ddot{u}\} \\ \{\ddot{p}\} \end{Bmatrix}+\begin{Bmatrix} [C_s] & [0] \\ [0] & [C_f] \end{Bmatrix}\begin{Bmatrix} \{\dot{u}\} \\ \{\dot{p}\} \end{Bmatrix}+ \begin{Bmatrix} [K_s] & -[R]^T \\ [0] & [K_f] \end{Bmatrix}\begin{Bmatrix} \{u\} \\ \{p\} \end{Bmatrix}=\begin{Bmatrix} \{f_s\} \\ \{0\} \end{Bmatrix} \tag{4}$$

The boundary element is a new numerical method for solving mathematical physics equation. This method is to change differential equations of the physics problem into a boundary integral equation. Then boundary of the region is divided into many finite elements, namely discrete boundary integral equation, so the equations only including the nodes' unknown variables on the boundary are obtained. In final, the numerical solution can be realized.

Under single frequency excitation force load, the Helmholtz integral solutions of acoustic wave equation [7] are calculated based on the equation:

$$p(r_p)=\alpha(r_p)\iint_S [p(r_p)\frac{\partial G(r_p,r_q)}{\partial n} - G(r_p,r_q)\frac{\partial p(r_q)}{\partial n}dSr_q \tag{5}$$

Where r_p is the location of observation point. S is the surface of the sound source. $\alpha(r_p)$ is related to the region that contains the r_p point, and it is also related to the case that if the surface S is smooth and closed or not. $G(r_p,r_q)$ is Green function.

$$G(r_p,r_q)=\frac{e^{-ikR(r_p,r_q)}}{4\pi R(r_p,r_q)} \tag{6}$$

Where $R(r_p,r_q)=|r_p-r_q|$.

3 Simulation and characteristics analysis on vibration response for underwater robot

The underwater robot researched in the paper, is a small robot, which can be used in underwater exploration, underwater investigation, obstacle avoidance and rescue mission. The outside view is illustrated in Fig.1.As only the radial excitation from the propeller is studied in this work, the vibration and radiated noise of the shell caused by axial vibrations of the propeller is not considered. What's more, to make the problem even simpler, the rudder and propeller are neglected in simulation. Its geometric parameters including shell and ribs are shown in Fig.2, Fig.3 and Tab.1. The material of underwater robot is aluminium alloy and its parameters: modulus of elasticity E=82,000 MPa, Poisson's ratio μ=0.3, density of solid ρ_s=2,750 kg/m^3. The density of fluent ρ_f = 1,000 kg/m^3, and sound velocity c=1,500 m/s.

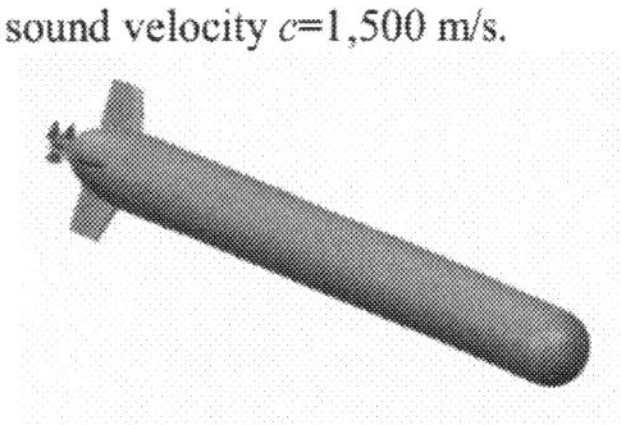

Figure 1. Outside view of underwater robot

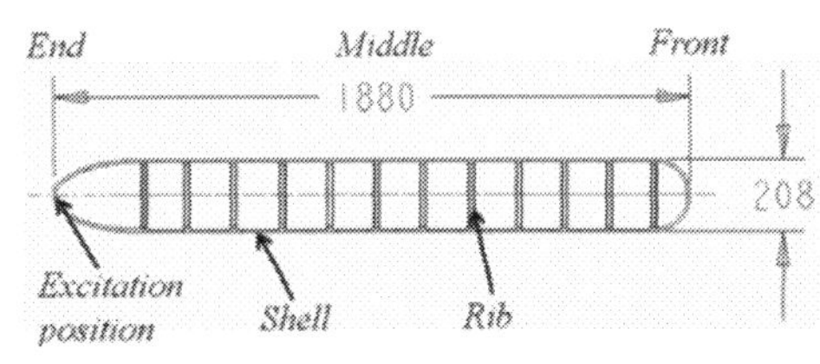

Figure 2. Profile section of underwater robot

DOI: 10.1051/matecconf/20165905001

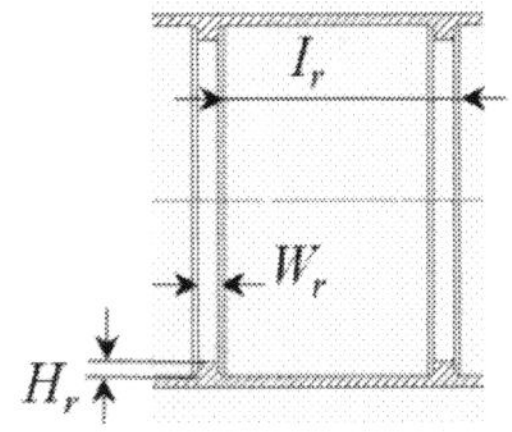

Figure 3. The geometry of ribs

Table 1. Geometric parameters of profile section

Robot shell		
Outer diameter/mm	Thickness/mm	Length/mm
208	6	1,880
Ribs		
H_r/mm	W_r/mm	I_r/mm
8	12	140

It is important to note that the underwater structure modal calculation method is different from the usual structural modal calculation method in the air, and it must be based on the fluid-structure coupling theory for underwater robots. Firstly the geometric model is simplified and meshed into finite element model. Then considered underwater fluid effect, the analysis model is established with a fluid-structure coupling surface. The constraints of the modal analysis are free for six degrees of freedoms. Through the modal analysis and calculation of fluid-solid coupling, the coupling modal frequency and air modal frequency results within frequency range 1Hz-1000Hz as indicated in Tab.2.

Table 2. Modal results of underwater robots

Air modal			Coupling modal		
Order	Frequency/Hz	Modal behavior	Order	Frequency /Hz	Modal behavior
1	349.889	2-node bending /X	1	160.684	2-node bending /X
2	353.724	2-node bending /Y	2	160.69	2-node bending /Y
3	711.084	1-order radial collapse /X	3	279.545	1-order radial collapse /X
4	714.442	1-order radial collapse /Y	4	280.642	1-order radial collapse /Y
5	768.551	2-order radial collapse /X	5	307.134	2-order radial collapse /X
6	771.863	2- order radial collapse /Y	6	308.226	2- order radial collapse /Y
7	879.314	3-node bending /Y	7	381.055	3- order radial collapse /X
8	885.232	3-node bending /X	8	381.81	3- order radial collapse /Y
9	918.091	Globe twist	9	405.143	3-node bending /Y
10	927.198	3- order radial collapse /X	10	405.151	3-node bending /X
11	928.593	3- order radial collapse /Y	11	910.613	Globe twist

From the results, with the reason of axial symmetry of geometric model, many bending vibration modes of the modal results are similar in the horizontal direction (X) and vertical direction (Y). From the Tab.2, we can see that the stiffness of bending modes and radial collapse modes is decreased with the reason of water. Meanwhile, it can be found that the sequence of the coupling modes is different from air modal. Fig.4 and Fig.5 illustrate respectively the 2-node and 3-node horizontal bending vibration mode. Via the bending mode behavior of underwater robot, it is similar to beam.

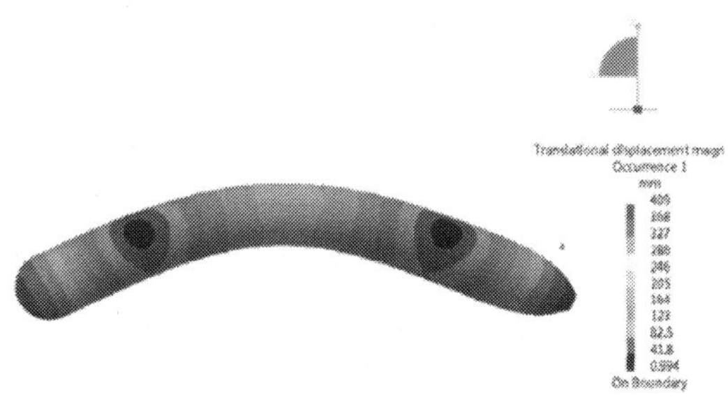

Figure 4. The 2-node bending mode

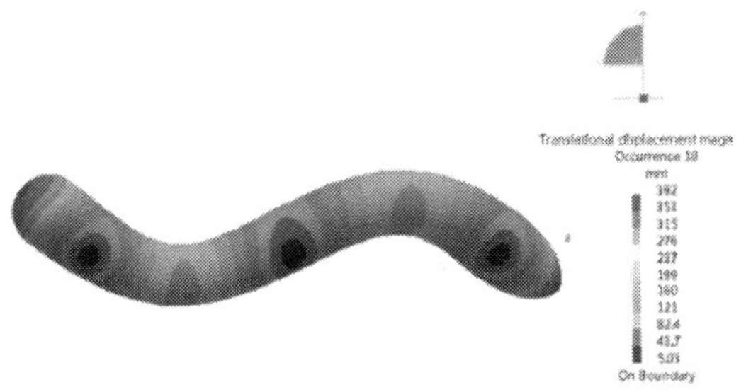

Figure 5. The 3-node bending mode

On the basis of modal analysis, harmonic response is calculated by the modal superposition method. Harmonic response load input conditions: the amplitude of excitation force 40 *N*, frequency range 1Hz-1100Hz, horizontal excitation direction, excitation position at the junction of motor shaft and shell stern section, analysis step 10 Hz. Through calculation and analysis, it is concluded that the horizontal displacement response results of the underwater robot bow section (at front), midship section (at middle) and stern section (at end) as shown in Fig. 6.

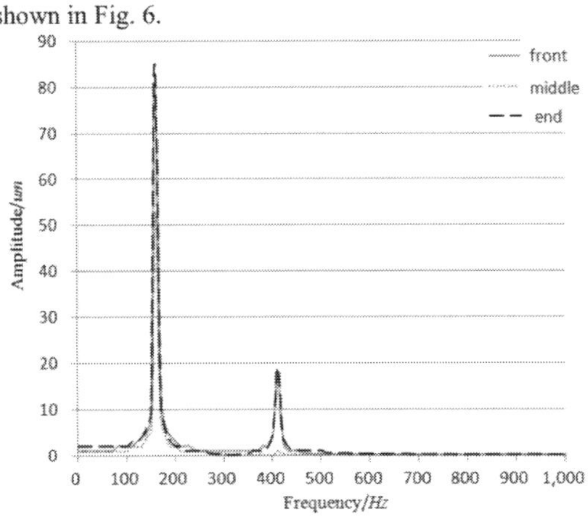

Figure 6. Horizontal displacement response underwater robot

The results indicate that: (1) The horizontal displacement response at 161 Hz is maximum, and the next is at 405 Hz. (2) For 161 Hz and 405 Hz force, the horizontal displacement response at stern section is largest, and the next is at fore section. (3) The transverse displacement response is significantly greater than the longitudinal displacement response under the effect of transverse vibration force. (4) With the increase of frequency, due to the effect of damping, the response is reduced significantly, and represents low frequency characteristics.

The surface displacement nephograms under 161 Hz and 405 Hz exciting force are shown in Fig.7 and Fig.8. It is concluded that the simulation calculation is authentic via making a comparison between the robot displacement harmonic response results and fluid-structure coupling modal results.

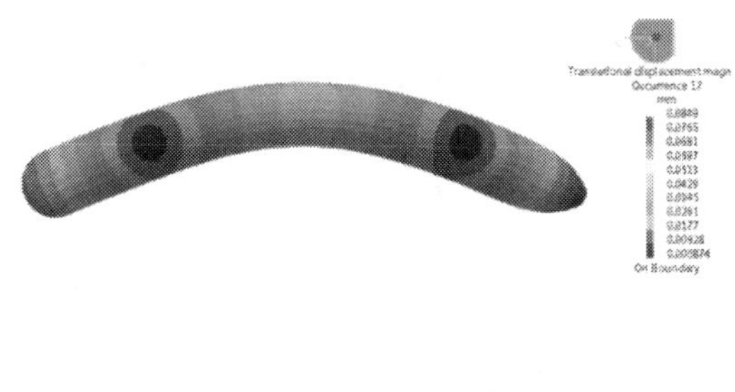

Figure 7. The displacement under 161Hz force

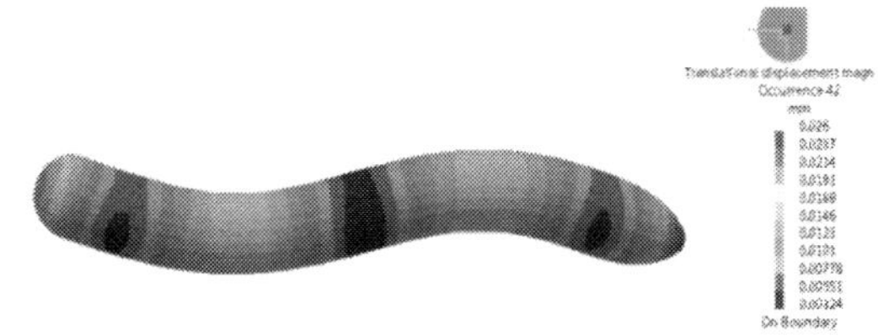

Figure 8. The displacement under 405Hz force

4 Simulation and characteristics analysis on sound radiation response for the small underwater robot

Based on the boundary element theory method, the underwater robot infinite outer acoustic results are obtained by the input displacement load of the above the surface vibration response. In order to reflect the acoustic characteristics of the underwater robot, the plane sets of field test points are created on a vertical plane P1 and a horizontal plane P2 (See Fig.9). The reference sound pressure is 2×10^{-5} *Pa*. The sound pressure response results on plane P1 and P2 under 161 Hz and 405 Hz exciting force are shown in Fig.10-Fig.13, and the sound power level of field points is illustrated in Fig. 14.

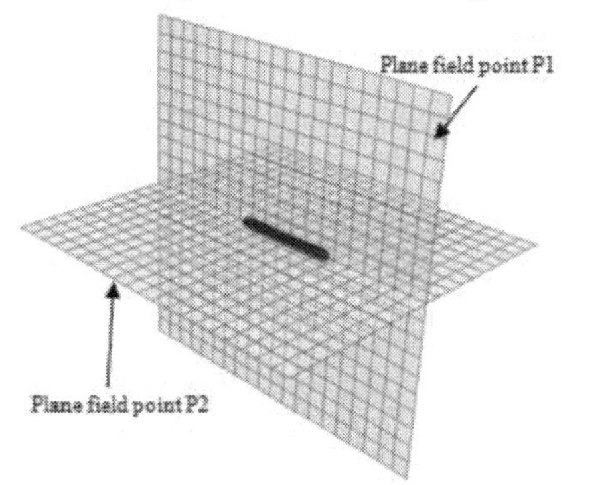

Figure 9. The field points on planes P1 and P2

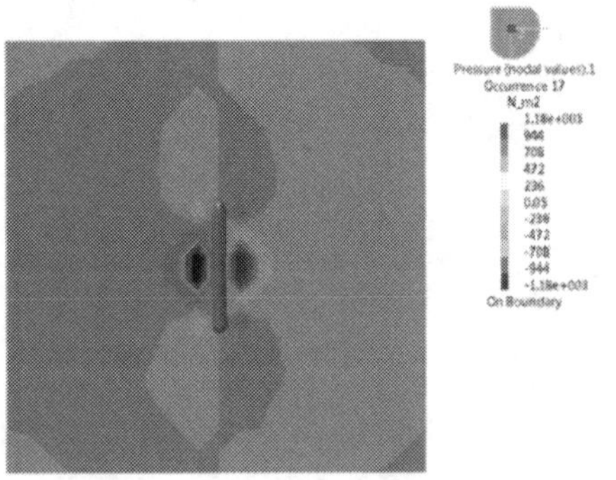

Figure 10. Pressure response on P2 at 161 Hz

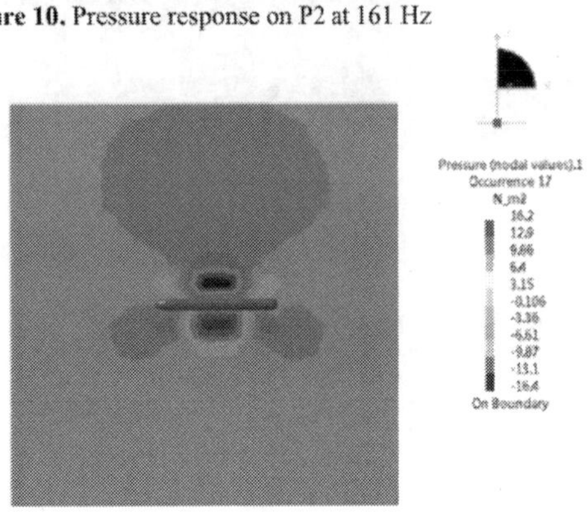

Figure 11. Pressure response on P1 at 161 Hz

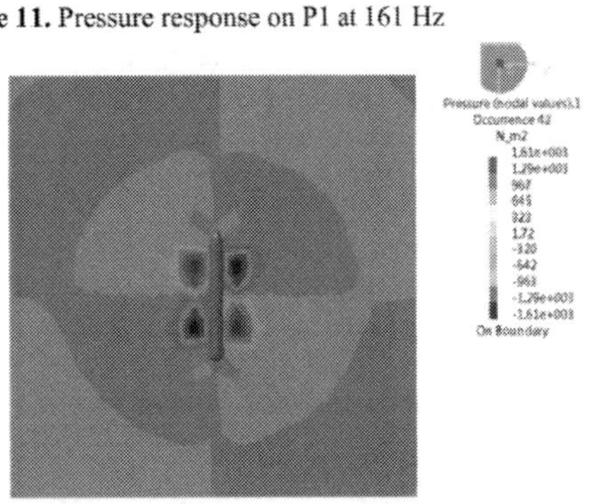

Figure 12. Pressure response on P2 at 405Hz

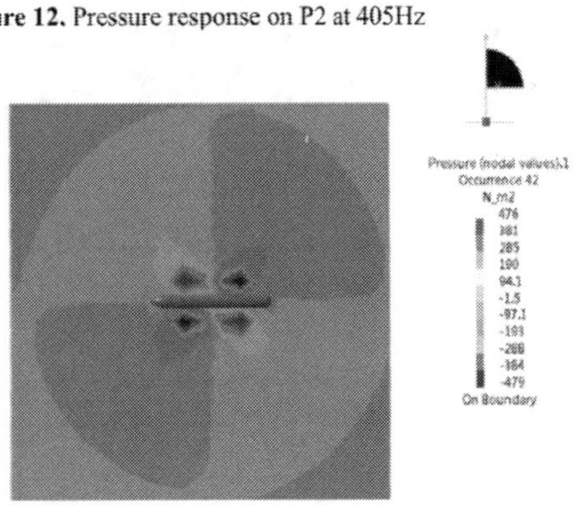

Figure 13. Pressure response on P1 at 405Hz

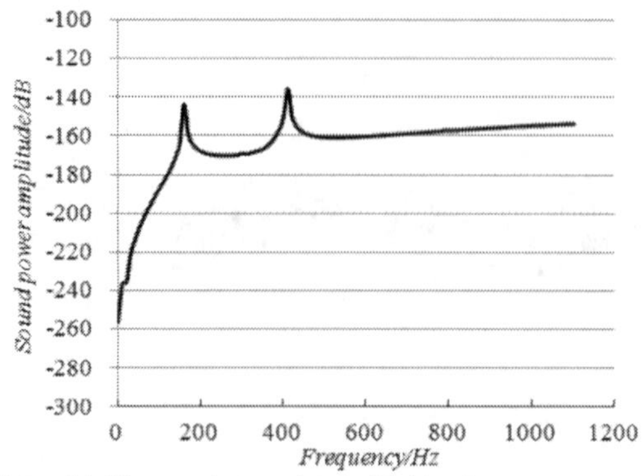

Figure 14. The sound power level of field points

The results indicate that: (1) the sound pressure at the middle side direction of the robot under 161 Hz transverse exciting force is largest. (2) Under 405 Hz transverse exciting force, the sound pressure is large at the peak positions of 3-node vibration mode. (3)The sound power level value is largest at frequency 405Hz; (4) The 3-node vibration mode at 405 Hz has larger amount of contribution to acoustic radiation than the 2-node vibration mode at 161 Hz.

5 Conclusions

By means of combining FEM and BEM, the structure of the underwater robot fluid-structure coupling modal frequencies and vibration modes are obtained. It is found that the sequence and stiffness of modes in water are different from the modes in air. The 2-node bending vibration displacement response is greater than the 3-node vibration displacement response, but the contribution to the sound radiation of the 3-node bending is high. Thus, in order to increase the stealth performance of underwater robot, the 2-node and 3-node vibration modes must to be controlled. What's more, the researched method and results in the paper are important to avoid the coupling resonance frequencies of the robot shell for the dynamic design, the selection of navigational conditions and key equipment, such as propulsion motor, rudder motor and propeller.

References

1. S. Merz, R. Kinns,N.Kessissoglou, Journal of Sound and Vibration, **325**, 266-286(2009).
2. S. Merz, N. Kessissoglou, R.Kinns, *Proceedings of the 14th International Congress on Sound and Vibration* (2007).
3. J. Pan, N. Farag, T. Lin, R. Juniper, *Proceedings of Innovation in Acoustics and Vibration Annual Conference of the Australian Acoustical Society* (2002).
4. O. Rath Spivack, R. Kinns, N. Peake, Journal of Marine Science Technology, **9**, 109-116(2004).
5. Wu G X, R Eatock Taylor, Ocean Engineering,**30**, 387-400(2003).

6. Young D L, Liu Y H, Eldho T I, Engineering Analysis with Boundary Elements, **24**, 307-316 (2000).
7. Chen Z S, Hofstetter G, Mang G A, Methods Appl. Mech. Engrg. , **152**, 147-155(1998).

Stress Analysis of Switchover Module with Al6061-T6 and Al7075-T6 for Hybrid Locomotion Mobile Robot

Khalid Hasnan , Qadir Bakhsh , Bhagwan Das , Aftab Ahmed and Aqeel Ahmed

Universiti Tun Hussein Onn Malaysia, Parit Raja-86400, Batu Pahat, Johor, Malaysia

Abstract. This study presents the design and stress analysis of mobile robot switchover module (track tensioner), which helps the robot to work in hybrid locomotion mechanism. The track tensioner is a loading and unloading module that helps the locomotion system transformability while robot moves either on wheel or track mechanism. The selection of suitable material for its design, the model is therefore analysed under static loading. It was analysed under static load of 200 N on two selected materials Aluminium alloys Al6061-T6 and Al7075-T6, to determine its mechanical behaviour of models. The results concluded that the material Aluminium alloy Al7075-T6 would be more suitable material for robot design as compared to Aluminium alloy Al6061-T6.

1 Introduction

In the design of a mechanical machine element or any system, the designer should cover all the specific needs, which generally involve in the moving component that cause specific kind of motion. The component dimensions/shape, type of material, allowable load/force/pressure and type of work application. Under the stress and failure analysis of any mechanical member, it is necessary to select a proper kind of material for design and all acceptable dimension of model [1-2].

Robot is an autonomous device that performs various tasks in difference applications with preinstalled programs or with external controller by the operator guidance [3]. They are usually performed all dirty, difficult and dangerous tasks as an alternative of humans [4-5]. Mobile robots are generally used to perform various tasks from one point to another point on the ground in different applications. Therefore, the effective locomotion mechanism will give the high flexibility and manoeuvrability to mobile robot to complete all harsh and complex tasks [6-7]. Mobile robots are mostly work in such harsh environments, where the entrance of any human to perform task is not safe or unaffordable [8]. The hybrid locomotion mobile robot designed by Bakhsh [5] was used in this study.

It consists two locomotion mechanisms; wheel and track system as shown in Figure 1. The loading and unloading of wheel to track or vice versa is done with the help of track tensioner unit. In this paper the track tensioner unit is designed and its mechanical behaviour was also analysed under static external load of 200 N. The stress analysis of track tensioner unit was done for two selected materials aluminium alloy Al6061-T6 and Al7075-T6.

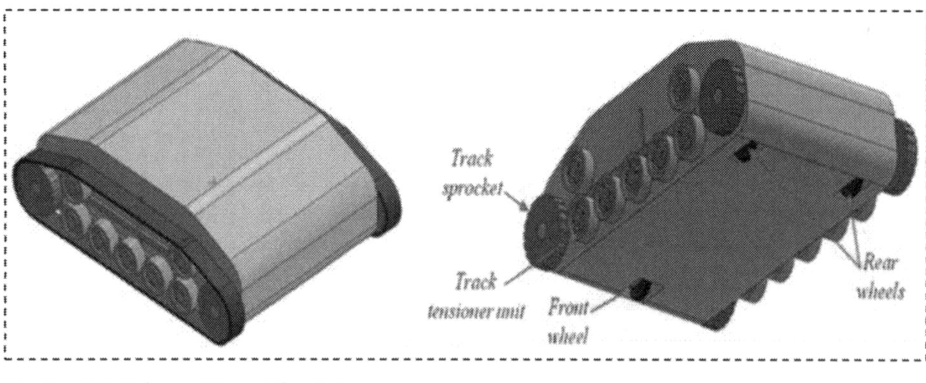

Figure 1. Wheel and Track locomotion mobile robot

2 System design

The Track Tensioner Unit is the most important part and it helps the mobile robot, to move in multi locomotion systems by its interchangeable locomotion phenomena. At smooth and plain path it moves on wheel mechanism, because at low energy consumption it gets high velocity and high manoeuvrability. But if sudden the working environment or path changed to rough and unstructured then robot change its locomotion in track mechanism. Because of this transformation of locomotion system, TTU mostly subjected to external load and stress in terms of weight of robot.

Therefore, it is necessary to analyse the mechanical behaviour of this unit under specific load conditions. So that it would be able to sustain the external applied load up to certain limit or level. This TTU is operated by means of pinion gear that is in mating with rack. The pinion gear is driven by a servo motor, which transmits the power or torque to the rack and pinion unit.

The TTU is designed in two sections or regions, both sections contains of different number of guider drum as its child components. In lower portion of TTU it contains five guider drums mounting, which produced the tension and suspension of track belt. The upper section of TTU contains two guider drums on each corner, those drums used to maintain the length of track belt during robot move on wheel locomotion system. Figure 2 shows the detail of design and dimensions for track tensioner unit and its child components, all values are given in millimeter (mm).

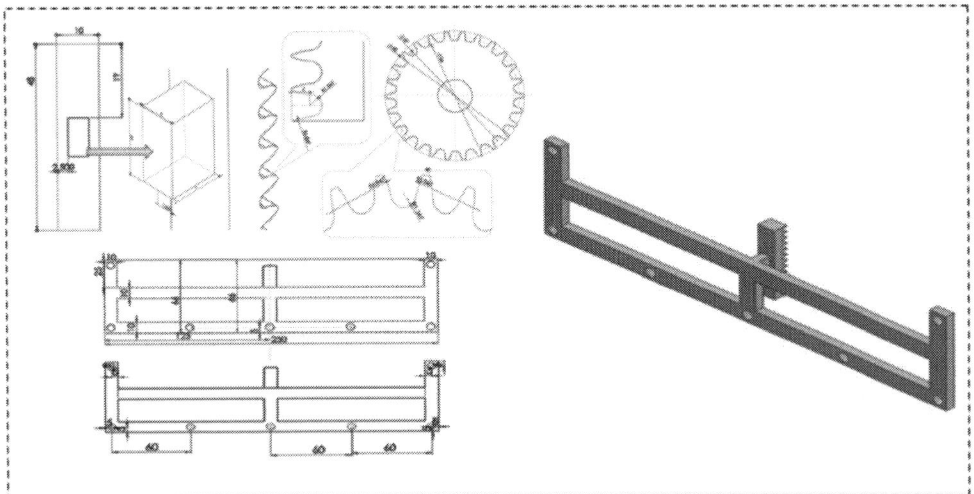

Figure 2. Track tensioner module

3 Stress analysis

Stress analysis is a very important factor for the designer and manufacturer to know the reliability of their design or product [1]. Prior to failure, ductile materials undergo significant plastic deformation, which has a major influence on damage evolution [9]. Mechanical behaviour analysis covers the model failure caused due to the application of external load or force. Therefore, it is necessary to analyse load carrying members for stress and failure due to direct loads, shear, tensional and bending.

In the principle of statics, it is to analyse the type of forces employed on each load carrying model or members. Aluminium alloys are used commonly in most of mobile robot design because of its outstanding characteristics such as; high strength, light weight, corrosion resistance and high thermal conductivity [1]. Aluminium and its alloys are being used successfully in a wide range of applications, from packaging to aerospace industries. Due to their good mechanical properties and low densities, these alloys have an edge over other conventional structural materials [10].

Aluminium alloy Al6061-T6 particular for ground transport systems, when relatively high strength, good corrosion resistance and high toughness are required in conjunction with good formability and weldability, aluminium alloys with Magnesium (Mg) and Silicon (Si) as alloying elements are used, the chemical composition of Al6061-T6 is given in Table 1 [11].

Table 1. Chemical composition of Al6061-T6 [11].

	%Si	%Fe	%Cu	%Mn	%Mg	%Zn	%Cr	%Al
Al6061-T6	0.4-0.8	0.7 max	0.15-0.4	0.15 max	0.8-1.2	0.25 max	0.04-0.35	Balance

Al7075 series are often used in transport applications, including Aircraft fittings, gears and shafts, fuse parts, meter shafts and gears, missile parts, regulating valve parts, worm gears, keys, aircraft, aerospace and defence applications; bike frames, all-terrain vehicle (ATV) sprockets, due to their high strength-to-density ratio [12], the chemical composition of this metal is illustrated in Table 2.

Table 2. Chemical composition of Al7075-T6 [12].

	%Si	%Ti	%Fe	%Cu	%Mn	%Mg	%Zn	%Cr	%Al
Al7075-T6	0.4	Max 0.2	Max 0.5	0.12-2	Max 0.3	2.1-2.9	5.1-6.1	0.18-0.28	87.1-91.4

4 Experimental setup

Firstly, the model is imported into the SOLIDWORKS simulation tool, where the material for designing is selected. The basic information related to the model was given by the tool automatically, Al6061-T6 its mass is 8.843x10-2 kg, volume is 3.275x10-5 m3 and weight is 86.67x10-2 N and Al7075-T6 its mass is 9.203x10-2 kg, volume is 3.275x10-5 m3 and weight is 90.197x10-2 N.

The model was tested for its failure analysis under static load condition, the load 200N applied on the rack part of TTU by means of pinion gear. The constraint details are given in Figure 3, where the point of force or load application is defined at the rack part. The downward load applied on the rack, because it requires lifting the whole body of mobile robot, while it starts moving on track mechanism. At the points where the guider drums are coupled with the TTU are supposed to be a fixed pint. In the simulation the load is applied at the top surface of rack, as the equal intensity of forces applied on the each teeth of rack.

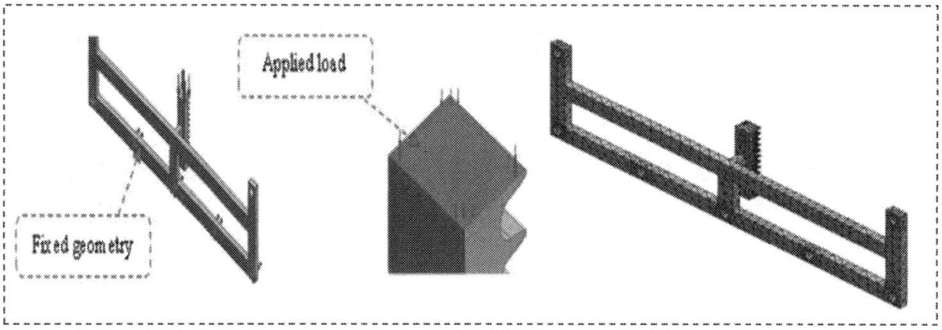

Figure 3. Track tensioner module force constraint and solid meshing

5 Results and discussion

The model undergo for its deformation by von mises stress at 200N with different deformation scale values such as 87.9314 for Al6061-T6 and 91.8219 for Al7075-T6. The simulation result gives the detail of minimum and maximum von mises stress value under the yield strength 275 MN/m2 (Al6061-T6) and 505 MN/m2 (Al7075-T6).

5.1 For Al6061-T6

The minimum value of static nodal stress is 7.838x10-2 kN/m2 at 13959 nodes and maximum value is 56.945x103 kN/m2 at 2008 nodes. The test result for resultant displacement at minimum and maximum elongation in model dimensions under the applied load. The minimum value of displacement is 0 mm at 538 nodes and the maximum value 31.646x10-2 mm at 3143 nodes. The equivalent strain results from the external load; the minimum and maximum values are 1.16279x10-9 at element 3224 and 5.49578x10-4 at element 7667 respectively. The factor of safety from its minimum to maximum range is 4.82918 at node 2008 and 3.40835x105 at node 13959 respectively. The stress-strain curve is shown in Figure 4.

5.2 For Al7075-T6

The minimum value of static nodal stress is 13.327x10-2 kN/m2 at 3237 nodes and maximum value is 60.141x103 kN/m2 at 2036 nodes. The simulation result gives the detail of minimum and maximum elongation in model. The minimum value of displacement is 0 mm at 540 nodes and the maximum value 30.328x10-2 mm at 3171 nodes. The equivalent strain results from the external load, the minimum and maximum values are 6.61454x10-10 at element 6493 and 4.81364x10-4 at element 8051 respectively. The factor of safety from its minimum to maximum range is 8.39693 at node 2036 and 3.78917x106 at node 3237 respectively. The stress-strain curve is shown in Figure 5.

DOI: 10.1051/matecconf/20165905002

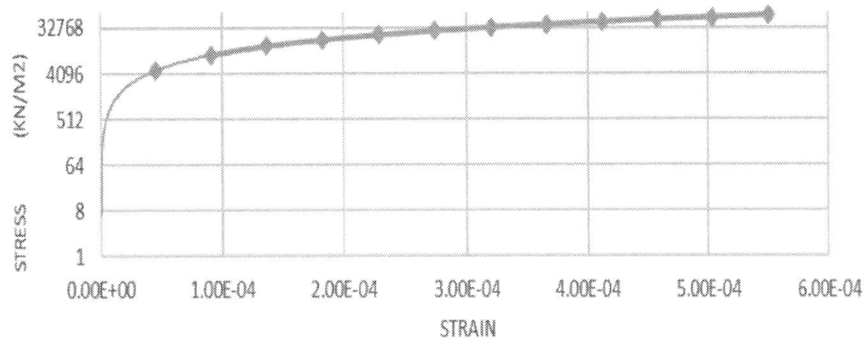

Figure 4. Stress-Strain curve for Aluminium Al6061-T6

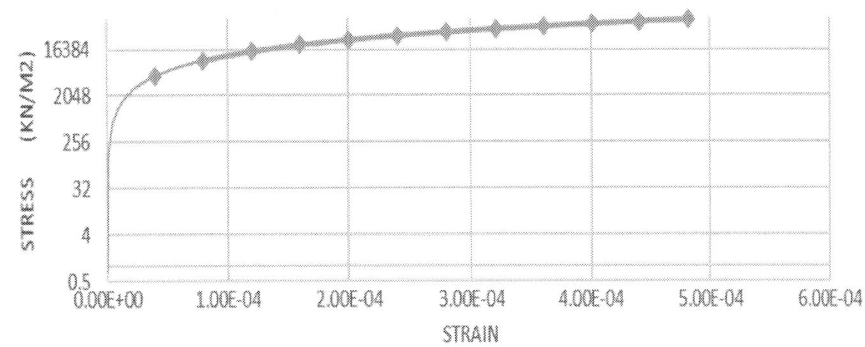

Figure 5. Stress-Strain curve for Aluminium Al7075-T6

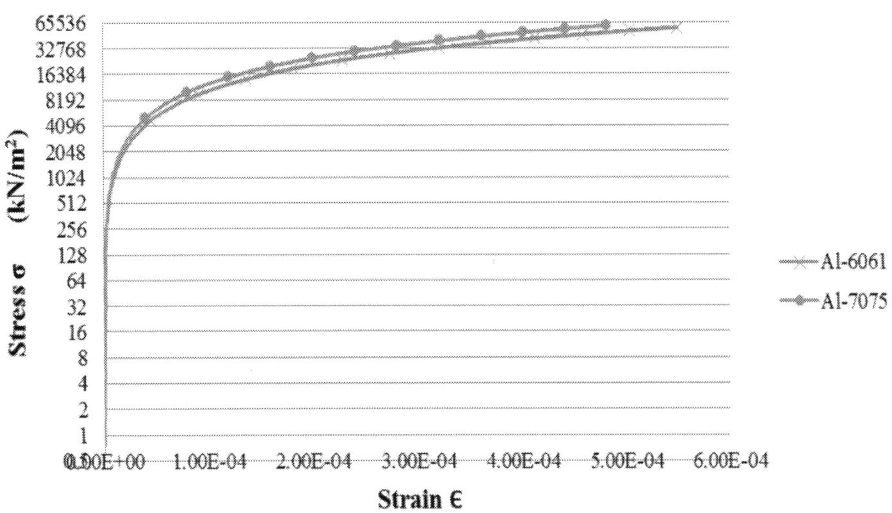

Figure 6. Comparative result

6 Conclusion

For the selection of a suitable material for hybrid mobile robot design, the Aluminium alloys Al6061-T6 and Al7075-T6 were investigated for their stress analysis. For the stability of these materials, the selected parts of mobile robot were then analysed under same load conditions. In this method of FEA, the parts were subjected to payload of 200N by defining the constraint according to their function. The results obtained from simulation were then plotted in the stress-strain curves for both materials, individually as well as comparatively. The results in Figure 6 shows that Al6061-T6 deform more as compared to Al7075-T6, because the stress induced due to the external load caused high elongation and deformation in the Al6061-T6 material. Therefore, the study has concluded that Al7075-T6 material is more suitable for this type of mobile robot.

7 Acknowledgment

The authors would like to thanks the University Tun Hussein Onn Malaysia (UTHM) for financially supporting of this research.

References

1. R. G. Budynas and J. K. Niskett, Shigley's *Mechanical Engineering Design,* 8th ed. (McGraw Hill companies, (2008).
2. Q. Bakhsh, K. Hasnan and A. Ahmed, *Comparative study between wheeled and tracked mobility system for Mobile robot,* J. Applied Mechanics and Materials **393,** 538-543 (2013).
3. J. Defever, B. Francis and T. Geerinck. *Mobile Robot with shared Antonomy.* Vrije Universiteit Brussel, Belguim, MSc Thesis (2004).
4. K. Goris, *Autonomous Mobile Robot Mechanical Design.* Vrije Universitiet Brussel, Belgium, PhD thesis (2005).
5. K. Hasnan, Q. Bakhsh and A. Ahmed, *New Hybrid Locomotion System Design.* Australian Journal of Basic & Applied Sciences, **8(4):** 247-252, (2014).
6. K. Hasnan, Q. Bakhsh, B. Das, A. Ahmed and S. Ayub, *A Novel Hybrid Locomotion Mechanism for Small Mobile Robot,* Indian Journal of Science and Technology, vol, **8(17)** 1- 5, (2015).
7. B. Aisham, Q. Bakhsh, K. Hasnan and A. Ahmed, *Design and Control Architecture of a Small Mobile Robot with Hybrid Locomotion System* Applied Mechanics and Materials, **598**, 619-622, (2014).
8. N. Enayati and F. Najafi, *Design and manufacturing of a tele-operative rescue robot with a novel track arrangement,* Industrial Robot: An International Journal, **38(5),** 476–485, (2011).
9. R. Vignjevic, N. Djordjevic, J. Campbell, and V. Panov, *Modelling of dynamic damage and failure in aluminium alloys,* Int. J. Impact Eng., **49**, 61–76 (2012).
10. B. Mustapha, H. Abdelhamid, and B. Mohamed, *Heat treatment effect on fatigue crack growth,*" International Conference on Integrity, Reliability and Failure, Porto/Portugal, 20-24 July (2009).
11. A. N. Abood, A. H. Saleh and Z. W. Abdullah, *Effect of Heat Treatment on Strain life of Aluminum Alloy AA 6061,*" J. of Materials Science Research, **2(2),** 51-59 (2013).
12. ASM International. Metals Handbook, *Properties and Selection: Nonferrous Alloys and Special-Purpose Materials,* Tenth edition, ASM International, (1990).

Study of an Innovative Indoor Robotic Navigation Approach Based on Beacons and PSD

Wang Zhenxing

Computer Department, Shanghai Second Polytechnic University, China

Abstract. In this paper, innovative indoor navigation methods have been proposed to meet the challenges in robotic navigation systems. The general positioning methods for robotic navigation include vision-based approaches, WIFI beacons, infrared beacons, ultrasonic beacons, etc. However, the common problem with these methods is their inaccuracy. Especially, improving the precision of robotic positioning mechanisms is the key to indoor navigation systems. This paper proposes an approach that combines the external rotating beacon with an internal rotation of position sensitive devices (PSD) which are installed on the robot. While two infrared beams from an external beacon source are equally projected to both sides of the PSD, the robot's position can be calculated precisely. The high performance and accurate results can be achieved by optimizing the rotation aligning time, dividing the working area, and compensating errors with information fusion. In comparison with other generic approaches, this proposed innovative approach requires less computing resources and is easier to implement due to its much lower complexity for the computing algorithms.

1 Introduction

Service robots are widely used in various situations. Route planning and selection will have a great impact on software performance and then the usability of the robots. In order to achieve the functional requirements of service robots, many navigation technologies have been researched. Indoor navigation approaches can be classified into map-based navigation and map-less navigation methods. Their core technology is based upon location identification. Many existing research efforts are based on artificial landmarks such as stripes and coloured objects, but the drawback of these approaches is that they heavily rely on specific tasks which are impractical in many occasions.

In essence, the location perception is essential for a robot to self-navigate in a room autonomously. The perceptive information can be collected from internal status, external signals or both. The internal perception information can be obtained from gyroscope, accelerator, odometer, etc. [1]. The gyroscope can be used to infer the heading direction, but the disadvantage is the accumulation of errors which will eventually mislead the robot. The random drift, temperature variation, circuit noise, different kinds of errors will make the accurate hard to achieve [2]. For MPU6050, the result is that even the robot is in a static state, the best performance of the gyro is shifted 5 degrees every ten minutes with a full-scale range of ±2000DPS. Moreover, while the robot is in moving the shift degree will be more serious, the consequence is that the gyroscope should be corrected periodically. To solve this problem, the digital compass was selected, although this will not accumulate the error thanks to the earth magnetic, nevertheless it is vulnerable to magnetic field interference, such as iron, mental furniture [3]. As the robot has many motors and other execution components, their magnetism will have strong inference to the earth magnetism and will interfere the result of digital compass completely.

On the other hand, the odometer can be used to calculate the moving distance [4]. Some researches on indoor navigation have assumed that the robot wheels will never slide and no error exists. Unfortunately, these preconditions will only exist in the virtual world. In reality, odometer can accumulate error quickly [5]. In most cases the errors come from the difference in wheel radius, gear gap, and meanwhile the floor material also has a dominant influence.

The limitation of the internal sensor mentioned above can be solved by providing the periodically accumulation error correction thanks to exterior navigation signal. The external navigation facilities can be realized through sonar, laser range finder, beacons, virtual wall, vision analysis, RFID, etc. The precision of the measurement is the fundamental and central issue with the performance of the algorithm. According to the results of many research works, if the number of external navigation facilities was increased, the precision will also be improved. Nevertheless, the negative effect to production cost and easy installation will be an inevitable problem if the robot's target market is home applications.

To address the problem mentioned above, this paper proposed a method that uses only one beacon to afford precision indoor navigation. The main idea is to deploy two devices: beacon and receiver, where the beacon shoots two infrared beams alternatively to a one-dimension position sensitive device (1D PSD) that mounted on the receiver. When a narrow beam illuminates the photosensitive strip of the PSD, two photocurrents, each inversely proportional to the distance of the spot from the far ends of the strip, are produced because the PSD outputs the centroid position of the spot. When two spots are on PSD simultaneously, the output position will be the centroid of two spots. Instead of these two beams be generated simultaneously, they should be generated alternatively to make two independent spots exist indifferent time. Then the position of two spots can further be processed by the method proposed in this paper.

2 Related research

2.1 PSD approach

The PSDis a precise distance detection sensor it can detect very small positional changes of the light spot on the active area,in contrast, CCD resolve distances limited by their pixel size and can lose microscopic spot information. Position Sensing Devices are divided into two families: the segmented PSDs and the lateral-effect PSD's. Lateral PSDs are continuous single element planar diffused photodiodes with no gaps or dead areas. These types of PSDs provide a direct readout of a light spot displacement across the entire sensing area. The PSD and its active area char, the general methodusetriangulation measurement system to determine the distance of an objectis illustrated in Fig. 1.

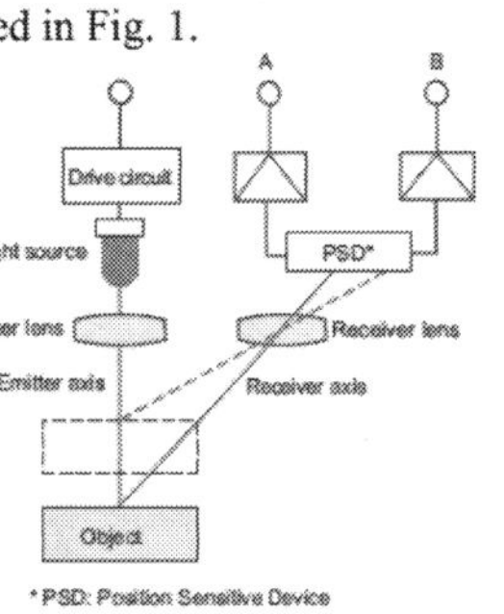

Figure 1. General application about PSD.

The position conversation formula can be written as:

$$x = \frac{Lx \cdot (Ix2 - Ix1)}{2 \cdot (Ix2 + Ix1)} \quad (1)$$

where Lx is the length of active area, Ix1 and Ix2 are currents for both electrodes respectively, x is the distance between light spot position and central of the active area.

The principle of PSD measurement is that the light from the light source is condensed by the lens and directed onto theobject. Light reflected from the object is condensed onto a one-dimensional position sensing device (PSD) by the receiving lens. If the position of the object (the distance to the measuring device) changes, the image formation positions on the PSD will differ and the balance of the two PSD outputs will change. If the two outputs are A and B, calculate A/(A + B) and use appropriate values for the span coefficient "k" and the offset "C" as shown below.

$$\text{Displacements} = \frac{A}{A+B} \cdot k + c \quad (2)$$

3 The proposed PSD approach

To overcome the problem mentioned above as well asreduce the interference from the environment, we modulate the infrared light and beam the PSD directly. As overheat on the spot of the PSD will damage the device,the power of the infrared light should be properly controlled. Fig. 2. is the basic structure of our proposed indoor locating system,it consists of IR sender A and receiver B. On sending side A, around the circumference two laser diodes were installed,their angle is θ,Transmitter A shoots out a concentrated (by the lens) IR-light beam, this beam arrives B straightly and the light is focused on the PSD sensor.

The partially enlarged view of the transmitteris illustrated in right side of Fig. 2. Two laser diodes mounted alongside the circle of A, their angle is θ, they emit two laser beams that flash alternatively and they are perpendicular to the tangent line of the circle. A 1D PSD is mounted on the receiver side B, when the beam is projected onto the surface of PSD, we can calculate the distance of two light spots. Assume a virtual line between the center of two beams, if the virtual line is projected at the center of PSD, then it will perpendicular to the surface of the PSD which means two beam spots lies equally distance d to the center of PSD,then we can easily calculate the distance $L=d/\sin(\theta/2)$. The θ should be selected carefully to satisfy the PSD size as well as the maximum and minimum detection length of L.

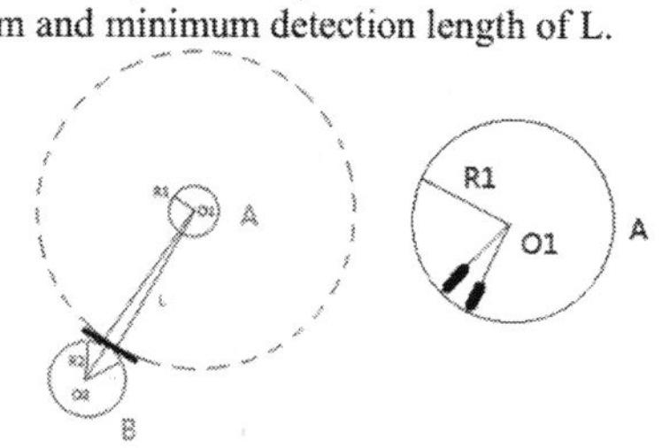

Figure 2. Infrared beams, beacon and robot.

Laser distance and azimuth detection based on PSD. The size of our selected PSD device is 1x15mm and the resolution is 0.1um, the response time is 0.8us. The PSD outputs track the motion of the 'centroid of power density' to an extremely high resolution and ultra-high linearity. The optical components used along with the sensor will add distortion, which can be implemented into the look-up table and thus minimized. To avoid interference from stray light the modulated light is used. The optical suit of A is rotated by a gear step-motor, the stepping angle of the step-motor is 5.625° and the gear reduction ratio is 64, so each step the angle rotation is 0.08789°. An optimal method is to utilize step-motor control function,if the subdivision is 16 then the step can be divided by 16, hence the angle change will be 0.0055°. However, the actual output of the subdivision is influenced by electronic and mechanic parameter of the step-motor. For each step the distance between the intersection of two light beams of A and the dash circle will approximately be d=L*sin(0.08789). Here d<PW where PW is the width of PSD. As the width of the selected PSD is 1.5cm, the maximal L will be 9.77m.

The idea that measuring the distance and azimuth is: the sending side A and receiving side B were mutually rotating in an ad hoc way. A has a starting angle which is detected by a photo interrupter, meanwhile, both step-motors of A and B keep rotating until two beams from A are projected on either side of the PSD of B and the two spots are equally distance to the center of PSD. The rotation angle of A is then recorded for azimuth. At the same time because line O1O2 perpendicular with PSD surface, the distance L can be get very easy.

4 Angular speed optimization

To make the system work properly the rotation angle of A and B should be dynamically adjusted to work in an opposed mode (through beam). As B is also installed on a step-motor,so it can rotate accordingly. Both A and B have a wireless module to make them communicate.

The adjusting formula for opposed mode is:

$$[(\omega 1-\omega 2)\cdot t+(\psi 1-\psi 2)+\pi]\mathrm{mod}(2\pi)=0 \tag{3}$$

where ω1 and ω2 are angle velocity of B and A, ψ1 and ψ2 are initial angle of B and A referenced by the same coordination system.The formula can be optimized, to improving processing time we use the following formula.

$$(\omega 1-\omega 2)\cdot t+(\psi 1-\psi 2)=n\cdot\pi,\ n=1,3,5,\ldots \tag{4}$$

We need to decide the relationship between ω1-ω2 and adjusting time t and find out the optimal value for ω1 and ω2. The formula informs that ω1-ω2 and t are reciprocal function.The negative value of ω1-ω2 will lead to meaningless negative time,so we select the positive quadrant.

We can find from Fig.3 if ω1-ω2→∞,t→0, if ω1-ω2→0,t→∞. While near the intersection point of the reciprocal functioncurve and its tangent line is about -45° (the best tradeoff between t and ω1-ω2 is t=ω1-ω2. It can calculate as:

$$t=\omega 1-\omega 2=\sqrt{n\cdot\pi-(\psi 1-\psi 2)},\ \text{n is integer} \tag{5}$$

Rules for the minimum adjust time t are:

If ψ1>ψ2 and ψ1-ψ2<π
then n=1, means the rotation angle needed will less than PI
If ψ1>ψ2 and ψ1-ψ2>π
then n=2,means the rotation angle needed will less than 2PI
If ψ1=ψ2
then n=0
If ψ1<ψ2 and -2π<ψ1-ψ2<-π
then n= -1,means the rotation angle will less than 3PI
If ψ1<ψ2 and -π<ψ1-ψ2<0
then n= 0

When n was decided, ω1-ω2 was also determined. Let ω2 be a fixed angular speed,then ω1 is calculated by the above formula. By adjusting the speed of step-motor to ω1 that installed on the robot side B,the time for opposed adjusting for A and B will be minimal. While beacon A is installed in a fixed position ψ2, it is easy to determine the X axis of the global coordinate system as it will never change. On the contrary, B is hard to capture the X axis of the global coordinate system when the robot is in moving. To enable the ψ1 can be derived, a gyro can be added to determine the X axis of the global coordinate system. Since the gyro will accumulate error, the initial angle of gyro needs to be revised periodically.

From the above analysis, we know that the worst case for ω1-ω2 is$\sqrt{3\cdot\pi}$, ω1=ω2+3.07 rad/s. This equation can be used when the start angle ψ1 of B is undetermined. Consider the factor of inertia, the step-motor will out of step if it stops suddenly. Therefore, the speed of step-motor needs to be decreased before the opposing mode is reached. The next is the algorithm for location process.

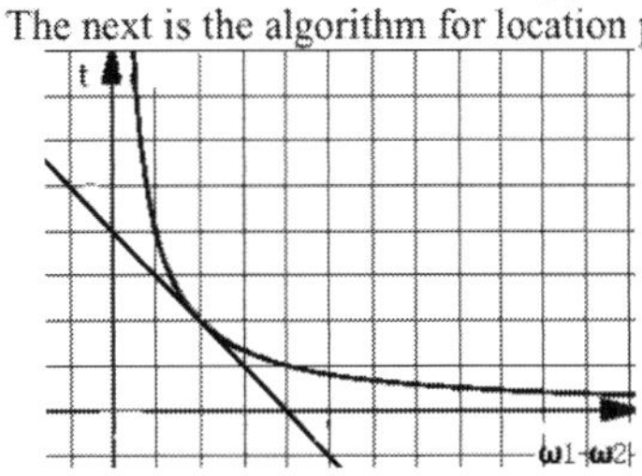

Figure 3. Angular difference with adjust time for opposed mode.

5 Implementation

To fulfill the requirements mentioned above, thesystem described in Fig. 2 can be implemented in two separate devices: a beacon and a receiver. Fig. 4 illustrate the mechanical structure of the beacon. The left part is an internal symmetric shaped infrared component *114*, two infrared diodeswere inserted into the cylinder-shaped hole,*114a* is one of the holes for accommodating diode while *114b*is a very tiny long hole for the beam formalization, hence the structure can avoid scattering problem of two beams. The power of the diode should be carefully adjusted because the PSD saturation limits the spot brightness). The infrared component *114* that installed on the edge of a plate *111* is illustrated in right side of Fig. 4. Plate *111* is rotated by a step-motor *110*. As step-motor will lose step in certain conditions, therefore periodic correction is necessary. Component *113* is a slot optical coupler, each time when *112* passes through the optical coupler will output a pulse. The pulse

DOI: 10.1051/matecconf/20165905003

can be used by processor to correct the accumulated errorof thestep-motor.

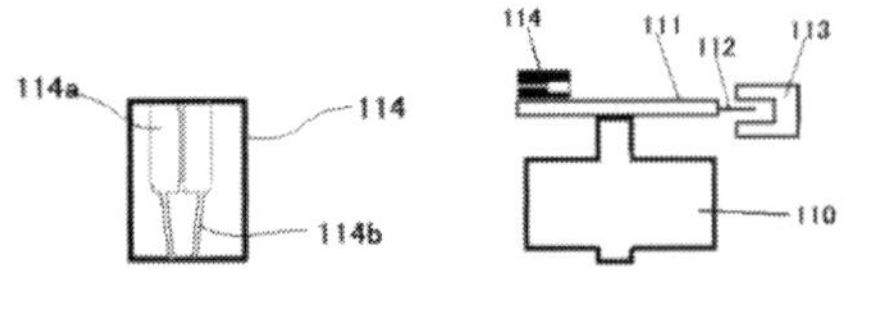

Figure 4. Mechanic structure of beacon.

Fig. 5(a) is adetailedimplementation of the receiver device B in Fig. 2. *120* is aPSD component that installed on a plate *111*, an optical filter 122 which used to reduce the background light level to avoid saturation effects of the PSD is mounted ahead, a lens 121 installed behind the filter, the spot diameter should be designed to about 200um,these are shown in Fig. 5(b). The height of 111 must be same as sender in Fig. 4. When the two beams send out from the sender and projected on the surface of PSD, the software will control the mutual tasks of the sender and receiver to make the two light spots equally projected on either side of PSD center, the algorithm will calculate the exact angle and distance of the receiver side. The beam is modulated and send alternatively,so the analog processing unit on the receiver side need to demodulate and then calculate the spot position and after filter procedure the final result will further processed by the position calculation algorithm.

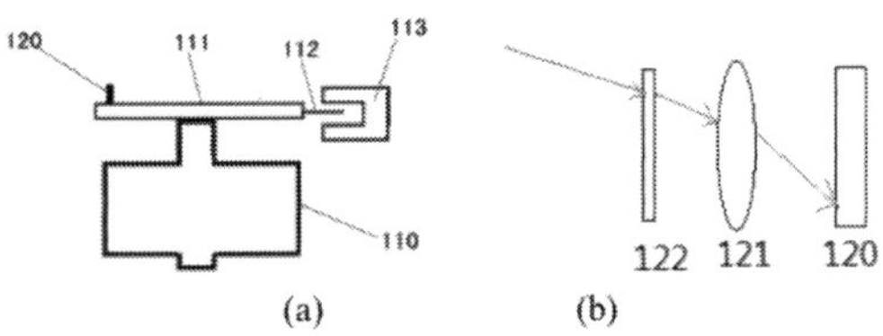

Figure 5. Mechanic structure of receiver.

An important issue for successful implementation is error analysis. Many aspects will influence the performance of this method. On this scenario, errors are sourcing from the beacon and receiver. The errors in the receiver side include PSD error and error of operational amplifier. PSD error will be influenced by voltage drift due to ambient temperature, stress of the sensor, light intensity, nonlinear, electron noise, etc.meanwhile, position resolution is limited by the signal to noise ration of the system. Compared with the receiver error, the beacon error is much more orders of magnitude. The reason is that the infrared diode is mounted on of the step-motor, the mechanic gap between the reduction gears will influence the spot greatly.

$$d = L \cdot \sin\frac{\theta}{2} \approx L \cdot \frac{\theta}{2} \tag{6}$$

According to Fig. 2 a tiny deviation $\Delta\theta$ will generate$\Delta d = L \cdot \frac{\Delta\theta}{2}$, so $\Delta\theta$ is amplified by L. Because the deviation of the step-motor $\Delta\theta$ is inevitable, even the step-motor will generate this error, such as step resolution, it means the above method couldn't work in reality. To solve this problem, this method needs to be revised.

Assume the step-motor's step angle is σ, reduction gear ratio is k, the steps of around circle will be:

$$N = \frac{2 \cdot \pi \cdot k}{\sigma} \tag{7}$$

The circle will be divide by N and the circumference is divided into arc segments, which the segment length will be

$$\Delta s = \frac{2 \cdot \pi \cdot L}{K} \tag{8}$$

when k=64, σ=5.625°, L=10 meter, each arc length will be 0.15m. However, this is only in the ideal case. Fig. 6 is a sectional view of the detail description of the dotted line circle in Fig. 2, it illustrates how the gear clearance of the beacon's step-motor affect the system. In the idea condition, beams from beacon is draw in solid line and their intersection with circumference is in the middle of gray area. For example, position c0 and c1 are the intersection of the beams with a virtual circle in ideal case.Assume the error of gear clearance is Δd, then the shift will +Δd or - Δd, which is the left or right edge of the gray sector. Their intersection with circumference will also change. Such as c0-,c0+,c1-,c1+. Because the characteristic of step-motor, this error will not accumulate.In account of the minimum and maximum shift of the step-motor,the small rectangle and large rectangle is draw on Fig. 6. The arc slices on the circumference can be classified into two categories: transition arc A0 and stop arc A1. The former is expressed with the arc bounded with small rectangle while the latter is bounded with outer rectangle. We define the white section on the disk as a transition area and the gray section on the disk as stop area where the receiver side may enter. According to the mechanic characteristic of step-motor, in arc A0 the light spot movement will much faster than arc A1. So the PSD mounted on the receiver will sense quicker light spot movement in A0 than A1. Assume the state S0 be corresponded with arc A0 and S1 with arc A1.

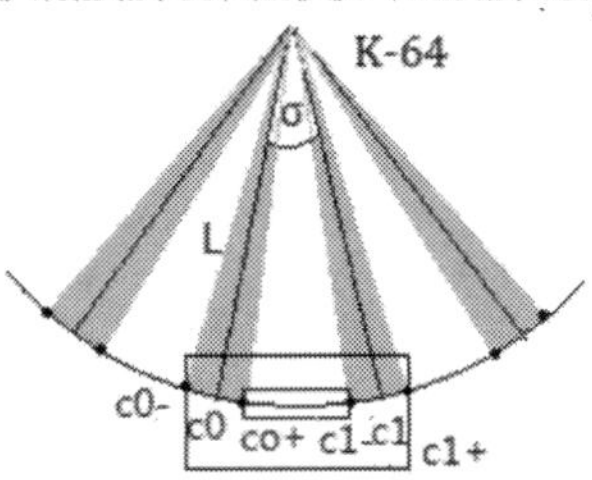

Figure 6. The error influence of step-motor.

5.1 Linear speed determination based on response speed

Assume the angular speed of the beacon side step-motor is $\omega 2$, the linear speed from c0+ to c1- is,$v = \omega_2 \cdot L$, where v must consistent with the response characteristic of PSD. The response speed of PSD is the time required for the generated carriers to be extracted as current by an external circuit. This is an important parameter when detecting a spot light traveling over the active surface at high speeds or using pulse-modulated light for subtracting the background light. Fig. 7 is the diagram of rise time vs. voltage, although increase the reverse voltage can reduce the rise time, but the minimum rise time is still needed to be satisfied. If we select 650nm infrared diode as a beam sender, when the inverse voltage is set in 5v, the rise time is approximately 1us. We can set the reverse voltage to 20v, and assume 2us be the worst case, according to the different infrared transmit diode adopted in beacon design.

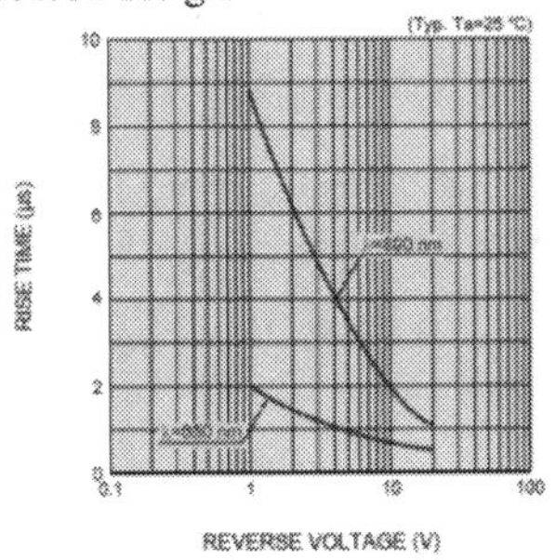

Figure 7. Rise time vs. reverse voltage.

We can therefore compute the linear speed v by:

$$v = \frac{dl}{dt} \tag{9}$$

where dl is the differential of length.

We can omit the imperceptible speed variant of step-motor when it's running at certain speed, the beacon's rotation can be regarded as constant speed, and therefore, the angular speed can be expressed as:

$$\omega 2 = \frac{Lx}{t \cdot L} \tag{10}$$

where t is determined by rise time, Lx is the length of PSD, L the distance from the beacon to the receiver.

For example, if given ω2=3.14 rad/s, L=10m, and Lx=2cm, then t will be 637us, which means the time that the light spot traveling over the active area of PSD will definitely satisfy its response speed.

5.2 Beacon direction tracing

When the system is initialized, the step-motor on the receiver side will rotated and eventually the surface of PSD will almost perpendicular with two light beams.After that the beacon will keep rotating, meanwhile the receiver which is mounted on the robot were moving in a certain pattern according to different application. The receiver also needs a gyroscope to trace the direction of the beacon, and meanwhile a periodic accumulation error elimination function will be called to remove accumulation error. By this method, the PSD on the receiver can be targeted by the beam easily. The following step is searching the beam, because in most cases the beam will only traveling through the PSD very quickly, so when the receiver received signal, it should inform the bean stop rotating and then move in the direction of the tangent line to find the beam again.

5.3 Navigation algorithm

The navigation algorithm is implemented both in sender and receiver, their parameters are mutual transferred by WIFI. A and B will work collaboratively to go through with the algorithm.

1 Device A and B rotate clockwise, angle speed of A equals with angle speed B plus 3.07

2 A detect a beam from B, slow down, inform B by wireless communication

3 After receive a notation from B, A slowdownand sending two beams alternatively

4 receiver moving along tangential line in Fig. 6 to find the beam again, and then slowly adjust the motor speed before the two beam spots projected equally onto either side of PSD center,stop rotating until v=v1=v2, where v1 or v2 is the voltage output for the left spot side or right spot side.

5 Error compensation. According to v, compensate for the error of the operational amplifier and the PSD displacement, get the A/D result v' which is the output voltage of PSD. Find the lookup table and get the distance d according to v'.

6 When spots on either side of PSD are equal, calculate the distance and direction using formula$L = \frac{d}{\sin\frac{\theta}{2}}$, the direction β can be gained by counting the steps from X axis of the globe coordination system.

7 B sending the parameter (L, β) to robot A

8 A stores the location parameter and revise its local position data.

9 Local to globe coordination system transform. In the globe coordination system, position update for A is calculated as: P(x,y)=GP(A)*Rot(β)*L(X,Y),

where P denotes globe coordinate system, GP denotes the globe position value x,y of A; Rot denotes a rotation transform from local angle β to globe angle; L denotes the coordination value for obstacle in the local coordination system of A, that is X,Y. When no obstacle exists, L(X,Y)=1.

10 Goto step 1

6 Application scenario

The algorithm can be applied in cleaning robot. Fig.8 is our prototype: the clean robot. The device described in Fig. 4 is realized in a quadrate box and a step-motor will be installed below it. The main PCB board consists an ARM processor STM32 and a FPGA chip. The main software is processed in arm processor while some signal processing function is implemented in FPGA which can lessen the workload impact on real time processing of the main processor. There is a large hole on top of the robot, a step-motor is installed there and the quadrate box is fixed on the axis of step-motor,see Fig. 8.

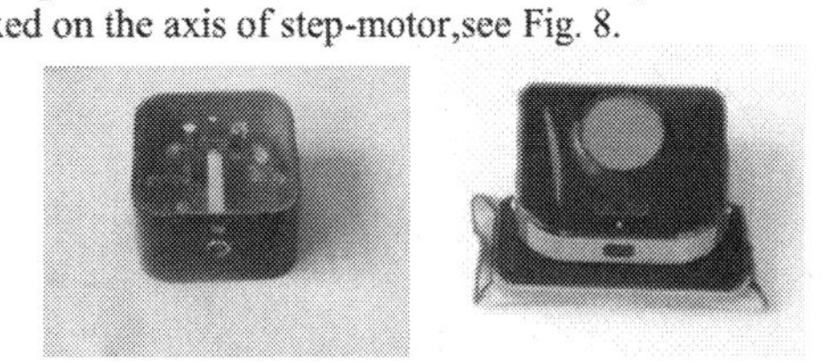

Figure 8. Prototype of clean robot.

 DOI: 10.1051/matecconf/20165905003

7 Conclusion

By adopting a step-motor driven PSD on the robot and an external infrared rotation beacon, and combining the algorithm described in this paper, the current robot distance and azimuth to the beacon can be calculated periodically. Therefore, the internal accumulation error can be eliminated. The pros are: it only needs one external beacon, and meanwhile the precision can be obtained without adopting expensive components. The con is: step-motorisvulnerableand will reduce system liability.

References

1. B. Barshan, H. F. Durrant-Whyte, Durrant-Whyte, "Inertial navigation systems for mobile robots," IEEE Transactions on Robotics and Automation,pp. 328-342, vol.11, 1995.
2. Aggarwal P, Syed Z, Niu X, et al. "A standard testing and calibration procedure for low cost MEMS inertial sensors and units," Journal of navigation, 61(02): 323-336. 2008.
3. Borenstein, Johann, et al. "Mobile robot positioning-sensors and techniques,"naval command control and ocean surveillance center rdt and e div san diego ca, 1997.
4. L. Romero, E. Morales, E. Sucar, "An exploration and navigation approach for indoor mobile robots considering sensor's perceptual limitations," Proceedings of International Conference on Robotics and Automation, pp. 3092-3097, vol. 3, 2001.
5. Yi, S. Y., & Choi, B. W. "Autonomous navigation of indoor mobile robots using a global ultrasonic system," Robotica, pp369-374. 22(04), 2004.

 DOI: 10.1051/matecconf/20165905004

ANALYSIS of Control Force Grasping for a Multifunctional Five Fingered Robot to Pick-up Various of Components

W.Widhiada[1], N.S.Kumara[2] and T.G.T. Nindhia[1]

[1]*University of Udayana, Mechanical Engineering Department, Campus Bukit Jimbaran Bali, Indonesia*
[2]*University of Udayana, Electrical Engineering Department, Campus Bukit Jimbaran Bali, Indonesia*

Abstract. Multi-fingered robot gripper has become popular in the major research topics as grasping an object in robotic systems. The author considers a matter of style-based control model for a multi-fingered robot hand grasping an object with a known geometric characteristics. This paper introduces design process and analysis of contact force the five fingered gripper suitable to handle several of objects. The author applied Simulink/SimMechanics, Support package Arduino and Inventor software packages to facilatate and integrated the design of contact force gripper systems. The advance of PID control is used to control dynamics motions of the five fingered gripper systems. The multifunction finger's gripper is developed to handle the various components. Contact force between fingertips and object surface is computed using the Hooke law concept. The analysis of experiment result shows the optimum of contact forces are achieved to hold the object. The spring and damper algorithm is used to compute the interaction of force between fingertips and object surface.

1 Introduction

During the last three decades, robotic hands have broadly been developed with the aim of copying the human hand in terms of dexterity and adaptive capabilities to function as either a manipulator or as a prosthetic device. Most robotic grippers have been specifically designed only to grasp on certain traditional forms of object. In this study the author has researched into the design and analysis of contact force a multi-functional fingered gripper.

The application of robotic gripper systems has become accepted practice for a variety of tasks including the automation of assembly. However, in the field of automatic assembly there are still several areas to be conquered, one such area is the assembly of relatively small mechanical elements which exhibit non rigid characteristics. For instance we human beings are blessed with tremendous dexterity and the human hand/brain can handle various material characteristics from very stiff materials to extremely compliant materials.

Robots gripper is designed to adapt with an environment in a human manner will often require the hands and such as robot grippers are referred to as humanoid. There are some research papers into human hand behaviour development [1], [2] and [3].

A robot gripper is the physical realisation of an electromechanical system to perform physical handling tasks automatically. A robot griper is an essential element of the robotic system and it is designed to suit industrial application to typically grasp, carry, manipulate and assemble the components. Proper gripper design can simplify the overall assembly, improving overall reliability system, and reduce implementation costs, and grippers have been widely used for automated manufacturing, assembly and packaging, etc.

Robot grippers are generally confronted with a situation in which a few contact points of each finger take place at the same time and must be preserved so that the captured object does not slip from the grasp. To this end, a number of methodologies to reach stable control and dexterous manipulation tasks have been suggested, and a number of related studies published [4], [5] and [6]. Including all these authors proposed a geometric approach to decide displacement and force control of this multi-loop mechanism.

The driver for this research study arose from the difficulty of gripper design process to be an actual griper robot which is used to assembly and manipulate some objects. In this research the authors use both a simulation and an experimental method. An integrated both MATLAB/Simulink and Autodesk-inventor are used to investigate the response of multifunctional of the five fingered robot gripper. This paper has focused upon investigating how to develop the modelling and controlling the contact force of five fingered gripper. The design and a prototype robot gripper five fingers are built in this research. The motion of dynamics robot gripper are controlled by an advance PID control to achieve the best performance include quickly response and high accuracy respectively.

The aim of this research is to develop the multifunction needs for the five fingered gripper to pick up and manipulate some objects. The approach of the spring damper algorithm model is used to compute the contact force between the gripper finger and an object.

This is novel in that the creation of a single gripper that can be precisely controlled the contact force to grasp a range of engineering components of widely different handling characteristics has not, to the author's knowledge, been previously studied. The technical challenge of the work is such that the research will focus on the application of simulation techniques and real prototype.

2 Methodology

2.1 Research description

The development and methods in robotics research have been always affected by the devices used. This is especially true when one considers the profound impact of recent technologies on robotics, especially the development of software modelling environments which have become indispensable when designing the complex systems like three fingered robot gripper.

Grasping tasks are confronted with multi faced at the varied geometrical and physical properties of work pieces. So, before work pieces are grasped that one need to look at the characteristics which define the work pieces type and the way of gripper is presumed to pick it up. The characteristics of workpiece which are of excellent importance for the grasping process.

This paper introduces the demonstration of gripper motion to pick-up several objects. Matlab/Simulink software package is applied to support the computation contact force gripper for each finger and understanding of the various dynamics systems and in particular how both the SimMechanics and support package Arduino toolbox are apllied to interface seamlessly with ordinary Simulink block diagrams to enable the mechanical elements and its associated control system.

The physical model of the five finger robot gripper is built using SimMechanics software and the application SimMechanics has also been verified with identical outputs when compared with Simulink simulations. These studies indicate that each finger of the gripper can be controlled using PID formulation. The author believes an interesting alternative for the gripper, besides its assembly task, is that by changing the trajectory plan and the force control demands the gripper can possibly be used to provide physiotherapy type motions.This paper also demonstrates the power of integrating the software environments of Autodesk-Inventor and SimMechanics to validate the concept of designed force control of a dexterous gripper.

SimMechanics is a toolbox especially development to design the robotic system with a suite of tools to specify bodies and their mass properties, their possible motions, kinematic constraints and coordinates system and to initiate and measure body motions.

In this paper the contact force results of the dynamics of a robotic gripper to pick-up several objects are presented. The design of five fingered robot gripper is drawn in Autodesk-Inventor software as follow the dimension of human finger. The dwaing of gripper design is transfered from phisical model in 3D to xml file into some block diagrams in Simulink/Simmecanics is shown in Figure 1. SimMechanics tool facilates a reserve to deriving equations and performing them with base blocks. The mechanical system should be break down into the building blocks into Simmechanics model.

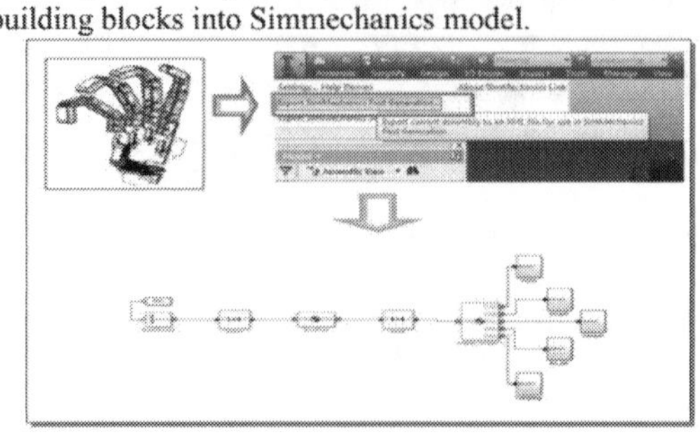

Figure 1. The process of design robot gripper.

The authors have designed the prototype of robot gripper to handle several objects. The contact force between finger tips and the surface of object are computed using the algorithm of force contact. In natural human grasping, the contact fingertip automatically adapts with the material strength. However, the hand struggles to pick up relatively small plane structures. The author has developed a mass spring damper model to determine the contact force response between the gripper finger and grasped object.

2.2 Mathematics model for DC servo motor

The robot gripper has used six DC servo motors. The mathematics model for DC servo motor is presented below.

$$\frac{\theta(s)}{V_a(s)} = \frac{K_{tn}}{s[s^2 J_{eff} L + (L f_{eff} + R J_{eff})s + R f_{eff} + K_{tn} K_b]} \tag{1}$$

where,

R= armature resistor, ohm
L= armature inductance, Henry
Ktn= constant of dc motor
Kb= constant of back voltage
Va(s)= applied armature voltage, volt
$\theta(s)$= feedback voltage, volt
J_{eff}= Moment inertia motor, kg-m^2
f_{eff}=Coefficient viscous motor and load, N-/rad/s

The eq. 1 is the transfer function of DC servo motor which is used to drive the motor in each finger link. Figure 2 shows the mathematics model of DC Motor is applied in block Simulink/Simmechanics.

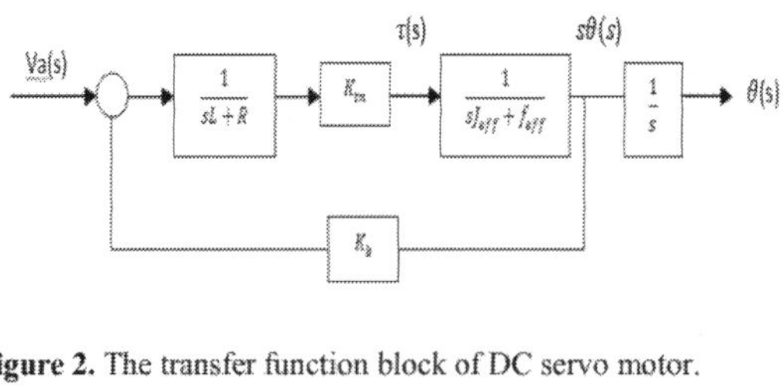

Figure 2. The transfer function block of DC servo motor.

2.3 Trajectory planning

The input references are generated for the trajectory planning to the motion control system, which ensures that the five fingered gripper robot executes the planned trajectories.

Position and orientation of the object are also very important parameters where the position of the an object before the pick operation already sets limits to force induction options as the points where forces can be applied may not always be accessible. Grasping with a single contact surface is called adhesive grasp and this is the classic operating principle for suction grippers. On the contrary, form fit and force fit gripping are basic principles of grasping with more than one contact surface and the combinations of form fit and force fit gripping are frequently used. As described previously, the hand is great to any technical systems in terms of grasping flexibility as it naturally combines numerous grasping types.

2.4 Gripper functionality

Biagiotti proposed the basic concepts for human prosthetic which are the degree of anthropormorphism, the level of dexterity and controllability [7]. Anthropomorphism refers to the capability of prosthesis to the mimic the human hand, party or totally, considering some aspects such as shape, size, consistency, and other more general include colour, temperature, and so on. Dexterity is related to prosthesis functionality and not to shape or aesthetic factors. Generally speaking, with the term "dexterity" he intends the capability of the end-effector, operated by a suitable robotic system, to autonomously perform tasks with a certain level of complexity. An exhaustive review of scientific work done so far about robotic hands dexterity, with a complete list can be found in [8]. The dexterity attribute of prosthetic hands can be divided into two main performance indicates which are grasping and internal manipulation respectively. Grasping is the capability of holding objects inside the end-effectors volume under a given hand configuration. Internal manipulation is the controlled motion of the grasped object inside the hand workspace, with the hand constraining the object with a configuration variable during the manipulation [9].

The controllability is the responds to trend of designing prosthesis that easier to control by user, more than the design of more complex hands, having more degrees of freedom [10]. The controllability can be considered from several points of view such as from the point of view of patients and from the point of view of the controller.

Furthermore, another type of prosthetic hands develops to provide higher dexterity using anthropomorphical principles and improve the grasping function. The better performances of prosthetic hand are achieved by increasing the number of degrees of freedom and provide feedback signals such as position, velocity and grasping forces. Many papers are introduced in the past to mimic the extraordinary flexibility motion and use with a robotic gripper, which come close to the properties of a human hand, but clearly are lacking essential requirements for industrial use, such as robustness and cost [11].

In light of the review into published grippers and the experimental studies into how the range of the components should be grasped, the author believes that a minimum set of three actuated fingers are necessary to achieve the desired grasping functions. Philosophically the gripper is envisaged to mimic the thumb, index, middle, fore finger and a little finger. The gripper finger grasps a variety of tasks such as rigid bodies, and non-rigid bodies. However to cover all the gripping functions the fingers will require to additional functions. Thus the experimental of five fingered robot gripper has proposed to provide the desired grip to enable a variety of components to be correctly handled.

2.5 Robust control strategy

One of the fundamental requirements to grasp the object is the capability to handle interaction between gripper finger and the object. An interaction control method is needed to adapt the forces exerted on the object to avoid damage to both the object and the gripper fingers. Its aim is to control the dynamic behaviour of the gripper finger when contacting the object by controlling the properties of the contact, namely controlling the stiffness and the damping of the interaction.

The force control of robot hand is needed for ensuring that the hand firmly grasps and that corresponding finger in the presence of external forces applied to the object. These forces may require high values of contact force at the fingertip to get the stability response. To solve these problems Widhiada in 2015 published Robust Control For The Motion Five Fingered Robot Gripper adopted the relationship between the interaction force and position trajectory [12]. A considerable body of research has been undertaken on robotic hand system with robot force control and planning related to the entire device and the significant developments are as follows, position control is used when the robot finger must be moved with or without load along a prescribed trajectory through the robot finger workspace. The robust control system of robot fingers is simply a collection of joint controllers each of which is dedicated to a single joint and drives it individually. The reference signals for these controllers are supplied by a joint trajectory generator determining the desired joint trajectory from the desired trajectory of the gripper.

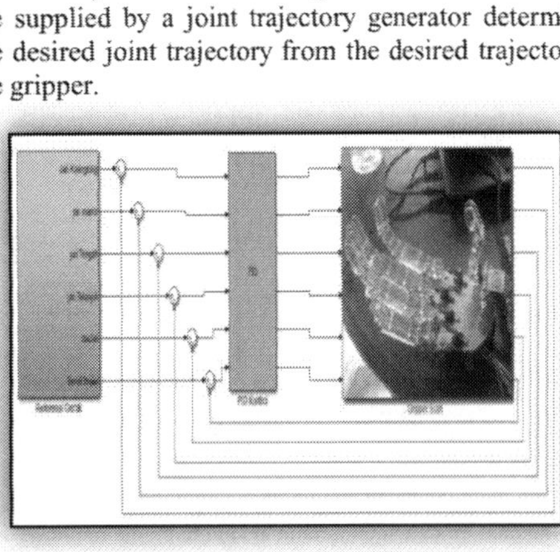

Figure 3. Robot gripper control system.

The advance PID control has been used to control the dinamics motion of each of the fingers surface. By tuning suitable values of the three constant gains in the PID controller algorithm such as gain of Proportional (Kp), Integral (Ki) and Differential gain (Kd) value are determined by auto tuning to achieve a quick response to get the steady state small eror signal. Parameter Kd works on the change of the error signal, so it can reduce dramatically overshoot effects to reach steady state response. Figure 3 shows the basic concept of Force control the Five fingered gripper using PID control.

3. Result and Discussion

Block diagram for computer interface is made in MATLAB/Simulink integrated with support package Arduino Uno. The advance PID control with auto tune is used to control the motions and force of five fingered gripper.

The value of force is used to drive the joint of servo motor can be compuetd from the angular position of joint servo motor. Figure 4 shows a link of gripper finger wich is installed at shaft servo motor.

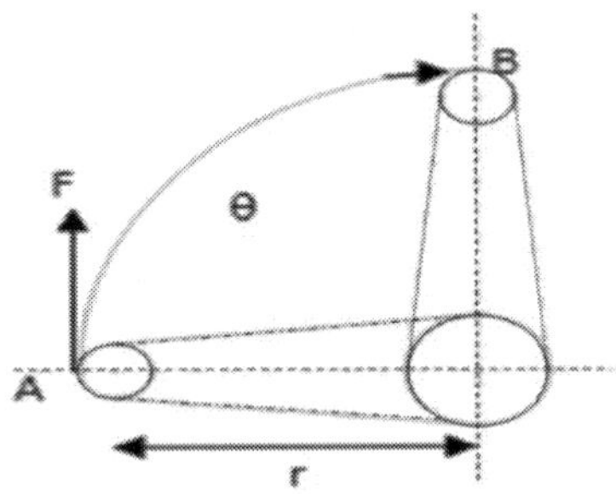

Figure 4. The lever of DC servo motor.

The force can be calculated when the tip of the motor lever moves from point A to B with the angle θ and the length of the lever of r. The first step is to calculate the force (F) which makes the angle of displacement in the form of rad. Calculation results of the force on each lever the motor is shown in the Figure 5 .

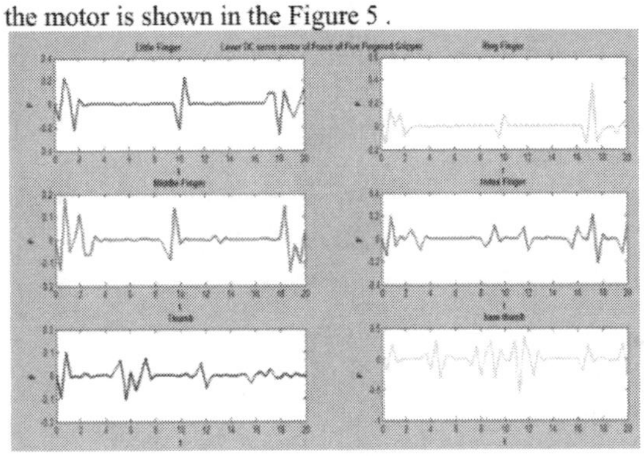

Figure 5. Contact force of servo motor.

The force (F) is a positive value means of the motor lever moves from position 0° to position the angle formed by reference. Conversely, if a force (F) is negative, then the motor lever moves towards 0°.

The contact force of the gripper fingers can be calculated. Spring mounted on the tip of the finger will press the surface of the object. By using Hooke law the contact force between fingertips and an object surface can be computed by the equation 2.

$$F = -k \,.\, \Delta x \tag{2}$$

Figures 6-8 show the gripper grasp an object The contact force can be computed using a mass sprng damper model algorithm. In this test, the gripper will be instructed to hold some objects. The first objects to be held is a bottle. Dynamics motion gripper and the response of each lever of the dc servo motor on the gripper are shown in the graphs as follows.

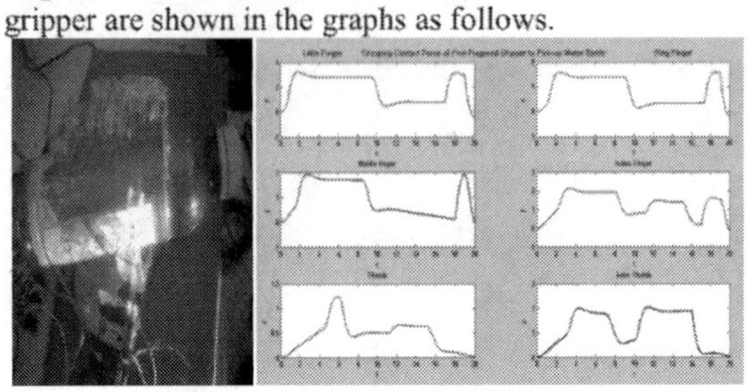

Figure 6. Bootle grasping contact force response finger.

Figure 6 shows the gripper grasp a plastic bottle. Automatically the given contact force for the gripping the bottle is proportional to the weight of the bottle gripped. The maximum force occurs on the little finger is 3.22N, 2.32N ring finger, middle finger 2.82N, 3.21N index finger, thumb 1.23N and 2.01N joint on the big toe. The average force acting on each finger when gripping namely: the little finger of 1.52N, the ring finger of 1,24N, 1.43N middle finger, index finger 1.54N, and 0.63N thumb. Because the bottle tilted position when held resulted in one of the fingers have greater force than the other fingers.

In the second test, the gripper is tested for holding a disk or a small disk with a diameter of 900mm in 20 seconds. The same condition when the gripper is used to grip a disk. The response shown in the graph shows that there are several fingers do not move reaches the reference angle. To force (F) which occurs when holding a disk shown in the figure 7.

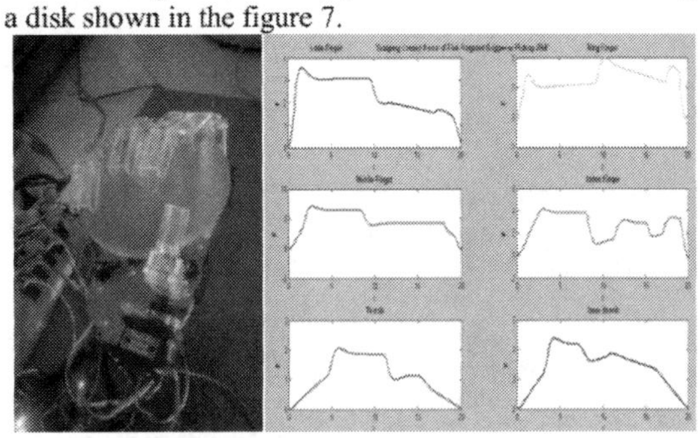

Figure 7. A disk grasping contact force response finger.

The magnitude of force given is proportional to the weight of disk gripped. The maximum force occurs on the little finger is 4.41N, 2.25N ring finger, middle finger 2.91N, 2.75N index finger, thumb 2.05N and 2.86N joint on the big toe. The average force acting on each finger when gripping namely: the little finger of 2.21N, the ring finger of 1.13N, 1.82N middle finger, index finger 1.43N, and 1.92N thumb. Large diameter disk makes moving finger at a small angle so that the force (F) which works also getting smaller.

Figure 8 shows the gripper is holding a pingpong ball. In the third test, a ping-pong ball will be used as an object to be grasped by a gripper in 20 seconds.

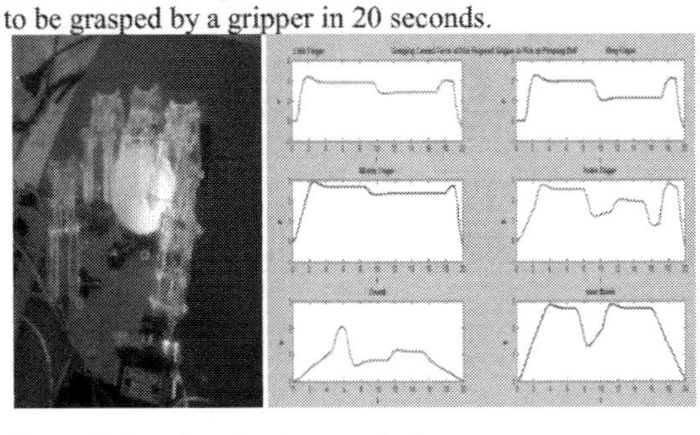

Figure 8. Grasping of a ping pong ball.

The magnitude of the force exerted gripper for holding a disk is proportional to the resistance of the object so that the disk is not damaged. The maximum force occurs on the little finger is 3.54N, 1.96N ring finger, middle finger 7.07N, 4.11N index finger, thumb 2.04N, and joints of the thumb 4.77N. The average force acting on each finger when gripping namely: the little finger of 1.45N, the ring finger of 1.98N, 3.50N middle finger, index finger 2.05N, and 2.40N thumb. At the time of holding a ping pong ball, only three fingers are touching the surface of the ball is shown by the average of the force exerted is greater than the force of two fingers.

4. Summary

The dexterous design five finger gripper functionality is developed based on the philosophy of hand assembly. The creation of a single gripper can be precisely controlled to grasp a range variety of components. The developed of the advance PID controller to control the dynamics motion finger. This PID control provides the best performance for grasping of each component such as giving the smallest error, small overshoot, fast settling time and a stable system. The contact force is applied to determine the interaction forces between the fingertips and grasped object. The Newtonian law, using a spring damper is modelled for force reaction. The contact response force for each finger to holding object is different each other. The input references are provided only for grasping the object. When the finger touched the surface of object, the actual output response is smaller than the input reference. The contact force occurs when the finger is touching the surface of object by force displacement the lever of motor.

References

1. Fite, K.B., Withrow, T.J., Xiangrongshen, Wait, K.W. Mitchell, J.E. And Goldfarb, M., A Gas Actuated Anthropomorphic Prosthesis For Transhumeralamputees, *IEEE Transactions On Robotics*, **Vol. 24**, No. 1, 2008.
2. Kathryn, J., De Laurentis And Mavroidis,C., Mechanical Design Of A Shape Memory Alloy Actuated Prosthetic Hand, *Ios Press, Technology And Health Care*, Pp.91-106, 2002.
3. Massa, B. S. Et Al., Design and Development of an Underactuated Prosthetic Hand, *In Robotics and Automation, Proceedings Icra/Ieee International Conference On*, **Vol. 4**, Pp. 3374-3379, 2002.
4. Roberts, R.K., Paul And Hillberg ,The Effect of Wrist Force Sensor Stiffness on the Control of Robot Manipulators, *IEEE International Conference on Robotics*, pp.269-274, 1985.
5. An, C.H. And Hollerbach, J.M.,Kinematic Stability Issue in Force Control of Manipulators, *IEEE Int .Conf .onRobotics and Automation*, **Vol.4**, pp.897 – 903, 1987.
6. Zhang, H., Kinematic Stability of Robot Manipulators under Force Control" *IEEE International Conference on Robotics andAutomation*, pp.80 – 85, 1989.
7. Biagotti, L., Melchiorri C. And Vassura G. , Control Of A Robotic Gripper For Grasping Objects In No-Gravity Conditions" *Proc. IEEE Int. Conf. Robotics And Automation*, Seoul, Korea,**Vol.2**, Pp. 1427–1432, 2001.
8. Bicchi, A. And Kumar, V, Hands For Dexterous Manipulation And Robust Grasping, *IEEE Transactions On Robotics And Automation*,**Vol.16**, No.6, Pp.652-662, 2000.
9. Cutkosky, M.R. And P.K.Wright, On Grasp Choice, Grasp Models, And The Design Of Hands For Manufacturing Tasks, *IEEE Transactions On Robotics And Automation*, **Vol. 5**, No.3. Pp. 269, 1989.
10. Abboudi, R.L., A Biomimetic Controller For A Multifinger Prosthesis, *IEEE Transactions On Rehabilitation Engineering*,**Vol.7**, No. 2,Pp.121-129, 1999.
11. Lotti, F. And Vassura,G., A Novel Approach To Mechanical Design Of Articulated Fingers For Robots Hands, *Proceedings Of The Ieee/Rsj Intl. Conference On Intelligent Robots And Systems*, Switzerland, Pp.1687-1692, 2002.
12. W. Widhiada, T.G.T., Nindhia And N. Budiarsa, Robust Control For The Motion Five Fingered Robot Gripper, *International Journal Of Mechanical Engineering And Robotics Research*,**Vol.4**, No.3,Pp.226-232, 2015.

MATEC Web of Conferences **59**, 05005 (2016) DOI: 10.1051/matecconf/20165905005
ICFST 2016

Improve Static Performance Identification Of Inductive Proximity Sensor For a Mobile Robot

Nguyen Hoai and Nguyen Le Chau Thanh

Department of Mechanical Engineering, College of Engneering, College of Technology.The University of Danang, Da Nang, Viet Nam

Abstract. Nowadays, inductive proximity sensor and mobile robot are one of the most interesting topics attracting scientists. With enhancement techniques from using sensor for controlling mobile robots which can be developed more in the future. Additionally, measuring of static performance of inductive proximity sensor is also essential because the testing and navigation systems perform more exactly by result information from the experiment by assaying the output signal. In this paper, the hand-made measurement system is used to verify again the static performance of proximity sensor with ferrous, non-ferrous and alloy metal tape. Simultaneously, improve static properties of typical tape metal through the relationship between the output signal and the sensing distance of sensor. According to theoretical basis, the relationship of the static parameters can be determined such as: thickness of metal, sensing distance, dimension of metal, and so on. The testing results have shown the output signal of proximity sensor in the model and theory is not similar. The final data will accommodate some information about the relationship of the thickness of metal tape and the sensing distance of proximity sensor, the different affects of position sensor and optimum operation range of mobile robot.

1 Introduction

The inductive proximity sensor is a common sensor to search a metal guideline for a mobile robot [1]. It uses to detect metallic objects without touching them. The fundamentals of the non-contacting inductive proximity sensor (IPS) and achieved status in year 2002 of the miniaturization of IPS resulting from the high integration level and utilization of modern assembly technologies were presented in [2].

IPS operates on the principle of the inductance change due to change in magnetic field. It detects the presence and absence of the ferrous and nonferrous metal objects without physically touching the object itself. If a metallic object moves into the field, the resulting induced currents add load to the oscillator, stopping oscillation [3]. In fact, when inductive proximity sensor analyzes the position of metal objects needed to be detected by a decrease in the quality of the oscillator circuit's electromagnetic [4], some unknown objects arise under the main tape which is put above the obstruction sheet. This is one of many reasons making noise in the sensor signal [5]. The relationship.

between the sensing distance of proximity sensor and the thickness of the metal is shown in Figure 1.

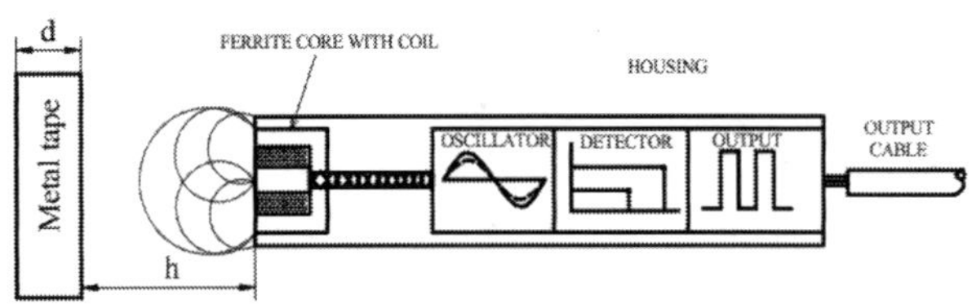

Figure 1. The inductive proximity sensor configuration

Although many kinds of different proximity sensor being developed but its specifications have yet to meet the capabilities of mobile robot control. Therefore, to measure the static performance of an inductive proximity sensor is the key technology to improve the performance of a mobile robot [6]. In this study, static performance testing is the determination of the sensing distance (h) and object thickness (d), continuously by moving the sensor from the edge of beginning to end of metal and go out of it. Finally, the results are analyzed and collated to enhance the continued trend for the control of a mobile robot by using the inductive proximity sensor and also using demands of the industrial process.

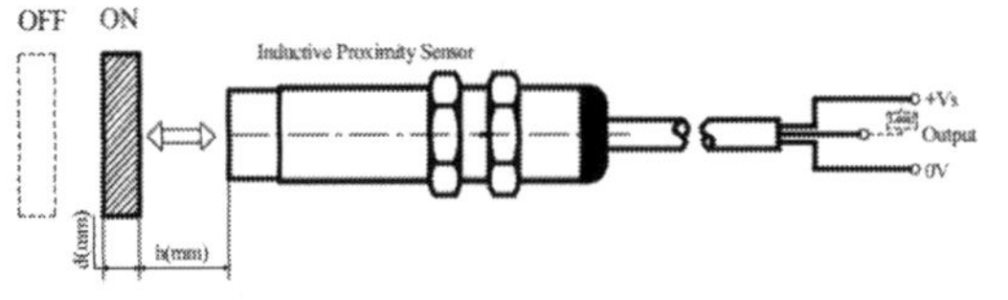

DOI: 10.1051/matecconf/20165905005

Figure 2. Measurement of sensing distance (h) and thickness of the metal tape (d)

2 Testing set-up of measurement system

A testing set-up measurement system is shown in Figure 3. Inductive proximity sensor is held again rotation, but allowed to move up and down. To measure sensing distance, using height gauge is fixed to the sensor. Each sensor has three wire, in which blue wire is connected to the negative power supply, brown wire is connected to the positive power supply through resistor (R=10Ω), and the black wire is the signal wire. The power supply is DC power. To observe and compare the output signal of the sensor, connect the negative wire (blue wire) with the oscilloscope. When an inductive proximity sensor moves into the sensing area and sensor activities, the output sign turns ON. Otherwise, an inductive proximity sensor moves out of the sensor sensing area, the output sign turns OFF. To determine static performance of sensor, many kind of metallic objects, such as: steel, aluminum, copper are chosen.

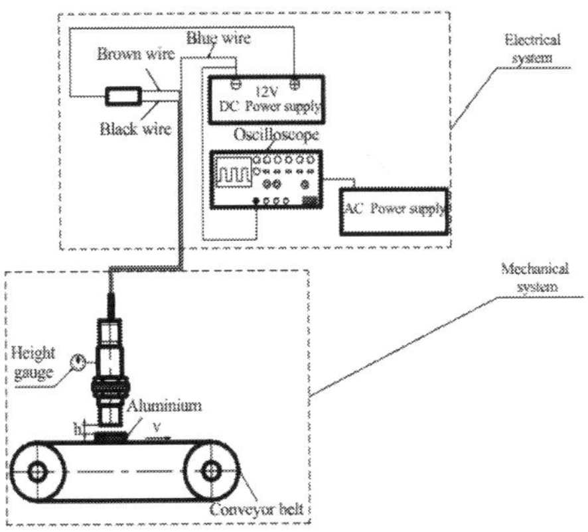

Figure 3. Diagram of measurement system with digital proximity sensor

From result determining the sensing distance of aluminum, keeping the constant sensing distance and move the sensor along the length of the aluminum tape, using measurement equipment (as a Figure 4). Corresponding to the each a value, the shape of the signal (Vavg) will change.

Here: s1 is the smallest distance, in which, the signal shape began to change

s2 is the largest distance, in which, the signal shape changes more than any initial position and the amplitude virtually unchanged over most of the length of the aluminum tape.

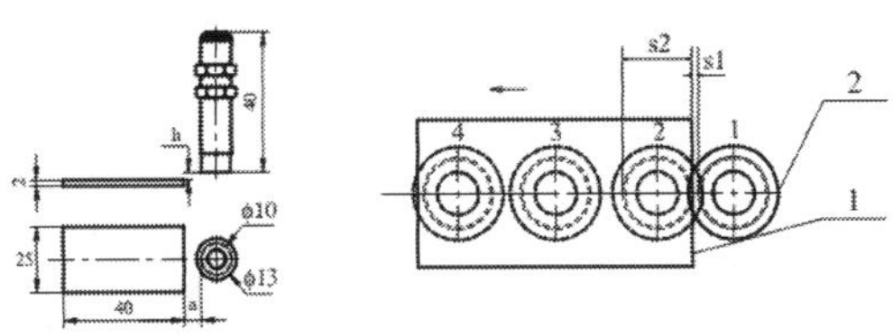

Figure 4. A testing set-up of measurement system for improvement static performance

3 Result and discussion

The curves of the sensing distance (h) depend on materials of ferrous, non-ferrous and alloy. 301 stainless steel is ferrous with excellent corrosion resistance, so the shape of the graph is different to copper and aluminum. With this result, the identification of sensor in the range between 0.06 mm to 0.18 mm thickness is easier than in the range between 0.24 mm to 0.5 mm thickness. When using sensor to measure detecting distance of aluminum and copper, the shape of curves is similar.

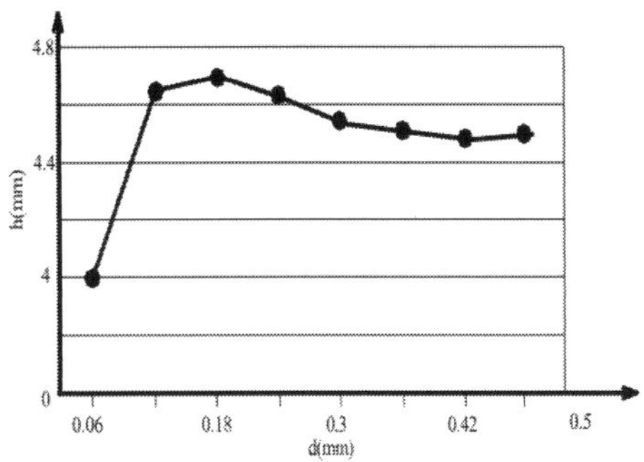

(a)301 Stainless steel

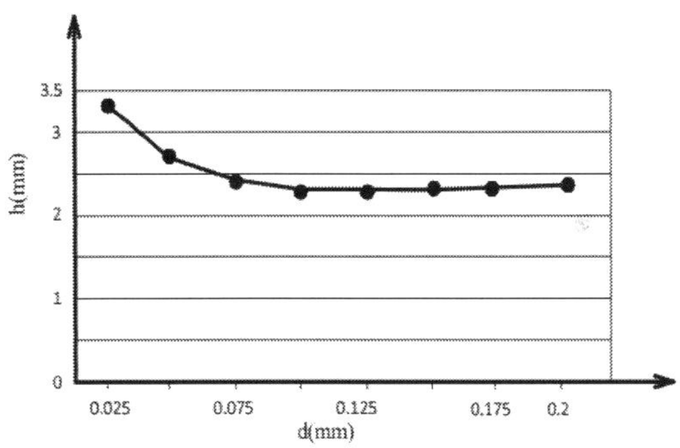

(b)Copper

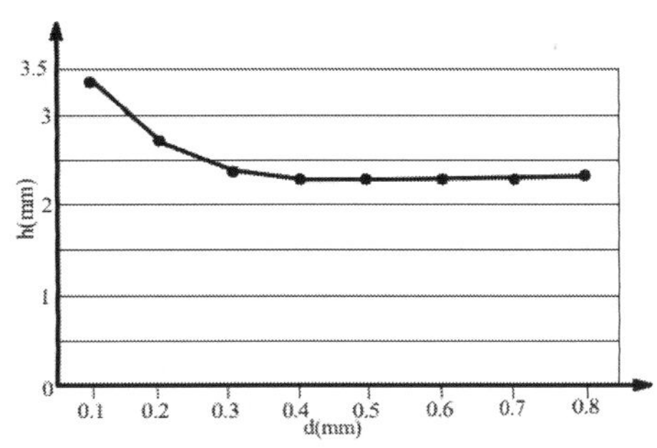

(c)Aluminium

DOI: 10.1051/matecconf/20165905005

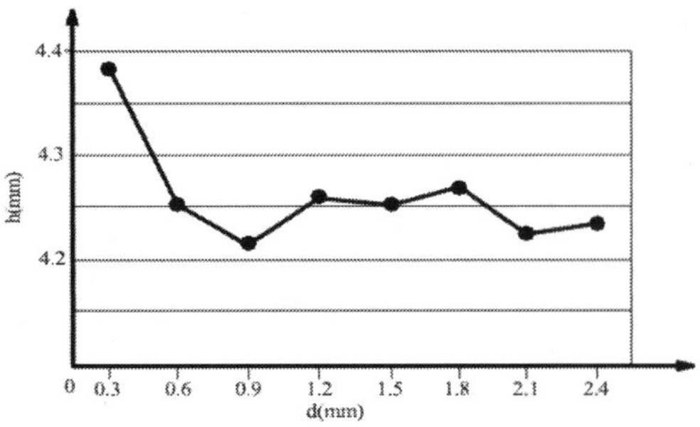

(d) Alloy

Figure 5. Relationship of sensing distance and thickness of objects

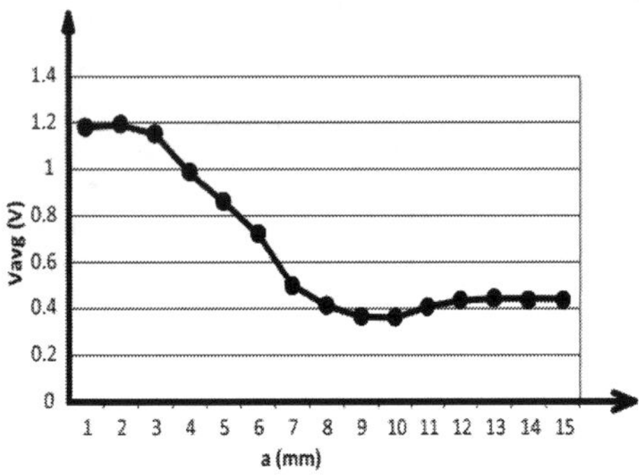

(a) h=0.5 mm

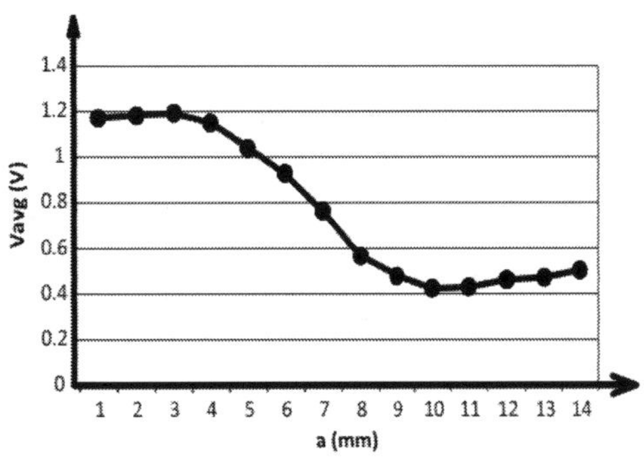

(b) h=0.7 mm

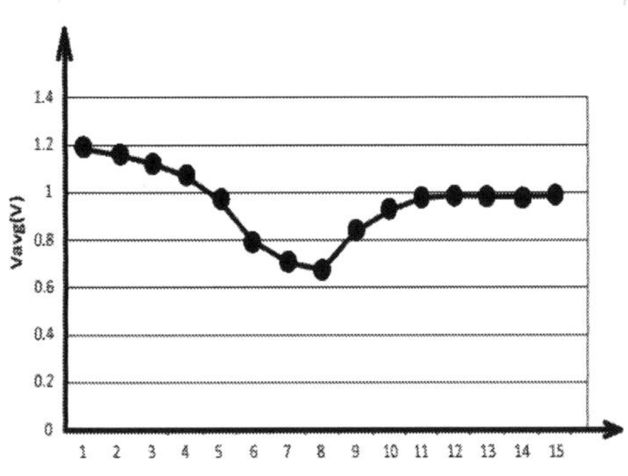

(c) h=1.5 mm

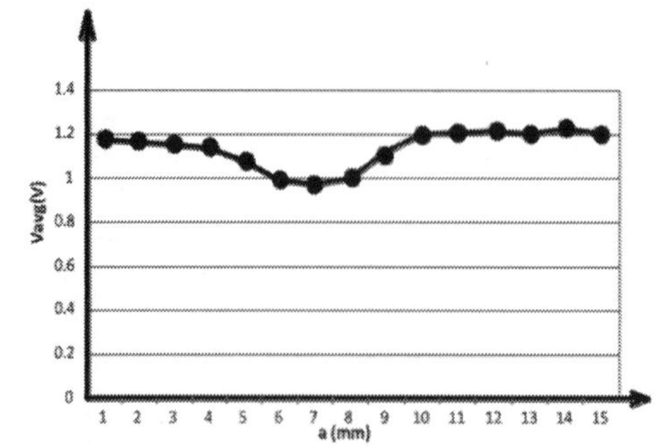

(d) h= 2.5 mm

Figure 6. Relationship of parameters: h,s, V_{avg}

This can be explained by the same attributes (non-ferrous). In these cases, the identification of sensors is easy and quickly from 0.025 mm to 0.125 mm thickness copper and from 0.1 mm to 0.4 mm thickness aluminum. Continuously, with alloy tape, the shape of graph is similar with copper and aluminum's measurement. This is explained by the alloy component contains non-ferrous metals (Cu-Al alloy). So, it is difficult to determine the optimal identification of the proximity sensor for alloy.

In the Figure 6.a-d, with h =2.5 mm (as shown in Fig.6d), the value of the constant average voltage can't be seen, even when moving the sensor into the aluminum tape. Specifically, V_{avg} will increase and exceed the initial value at a= 10 mm (1.2V).

The sensing distance with range from 0.5 mm to 0.7 mm, the received signal is consistent with theoretical knowledge (as shown in Fig.6a, b). When changing a distance from 9 mm to 15 mm, V_{avg} hardly changed. This is the basis for a control mobile robot. In addition, with the smaller sensing distance, the faster initial signal change. With h=1.5 mm,V_{avg} changes from 1.17 V to 1 V (a =1-4 mm), while h= 0.5 mm, Vavg change from 1.17 V to 0.39 V (a =1-4 mm).

4 Conclusion

In summary, for the static performance of proximity sensor, the signals of metal tape that the signals of ferrous, non-ferrous and alloy metal tape have giant differences, as indicated from the experimental results. when the material is ferrous metal, there exists a period of initial thickness of the material, in which the sensing distance is proportional to it. Otherwise, when the material is non-ferrous metal, there exists a period of initial thickness of the material, in which the sensing distance is inversely proportional to it. Additionally, when measuring in the same materials, sensing distance doesn't depend on the initial thickness of that material. This is the key to improve the performance of a mobile robot.

Besides, distance of above 0.7 mm is not only the biggest distance for mobile robot to control more accurate in the process but only more sensitive than other distance.

Reference

 DOI: 10.1051/matecconf/20165905005

1. T.T.Tsung and H.Nguyen, "Measurement of static performance of inductive proximity switch for a mobile robot", Applied Mechanics and Materials, vol.404, pp.502-507, 2013.
2. M.Jagiella and S.Fericeam, "Miniaturized inductive sensors for industrial applications", in Sensors, Proceedings of IEEE, pp.771-778, 2002.
3. R.C. Luo, "Sensor technologies and micro-sensor issues for mechatronics systems",
4. Mechatronics, IEEE/ASME Transactions on, vol.1, pp.39-49, 1996.
5. J.Fraden, Handbook of modern sensors: physics, designs, and applications: Springer, 2004.
6. T.T.Tsung, T.T.K.Vy, and N.Hoai, "The Disturbance of Inductive Proximity Sensor for Mobile Robot", Advanced Materials Research, pp.909-912, 2015.
7. T.T.Sung, T.T.K.Vy, and N.Hoai, "Testing and Improvement of Static Performance of Proximity Sensor for Mobile Robot", Key Engineering Materials, vol.656-657, pp. 719-724,2014.

MATEC Web of Conferences **59**, 06001 (2016) DOI: 10.1051/matecconf/20165906001
ICFST 2016

Quantitative Evaluation of Defect Based on Ultrasonic Guided Wave and CHMM

Le Chen , Yuemin Wang , Haiquan Geng , Wenli Deng1 and Wei Ye
Power Engineering College, Naval University of Engineering, 430033 Wuhan, China

Abstract. The axial length of pipe defects is not linear with the reflection coefficient, which is difficult to identify the axial length of the defect by the reflection coefficient method. Continuous Hidden Markov Model (CHMM) is proposed to accurately classify the axial length of defects, achieving the objective of preliminary quantitative evaluation. Firstly, wavelet packet decomposition method is used to extract the characteristic information of the guided wave signal, and Kernel Sliced Inverse Regression (KSIR) method is used to reduce the dimension of feature set. Then, a variety of CHMM models are trained for classification. Finally, the trained models are used to identify the artificial corrosion defects on the outer surface of the pipe. The results show that the CHMM model has better robustness and can accurately identify the axial defects.

1 Introduction

Guided wave technology is a kind of efficient detection technology, which has a great advantage in the detection of long distance pipeline, and it has been widely used in engineering. The defect size is generally judged according to the amplitude of the defect echo in practical test. We can obtain more structural information when further processing of the guided wave signal, and it's helpful to judge the healthy state of the structure more accurately.

Many scholars have conducted a lot of research on the quantitative analysis of the defects based on the guided wave signals. Demma [1] put forward a method of calculating the reflection coefficient of the notch by studying the SH0 guide wave on plate, and pointed out that the reflection coefficients and defect length are not monotonically increasing relation. Peter [2] proposed a matching pursuit method to decompose the front and back signal of the defect, and estimated the defect size by comparing the differences of the decomposed signal.

Hidden Markov model (HMM) is applied in speech signal recognition in the early, and introduced into the fault diagnosis field [3]. The HMM and fault diagnosis are the same in essence, that is to say, the true state is "hidden", which can be detected only through observation vector. Therefore, the application of HMM in the field of fault diagnosis is feasible.

The raw data of pipeline axial defects is obtained by the longitudinal guided wave technique, and the characteristics are extracted using wavelet packet decomposition. The KSIR method is proposed to reduce the dimension of feature data. Finally, the CHMM classified model is applied to the recognition of the defect axial length, achieving the preliminary quantification.

2 Theory Fundament

2.1 Theory of CHMM

In contrast to the conventional HMM, the observed value of CHMM (Continuous CHMM) is continuous, which is generally assumed to be generated by Gauss PDF (probability density function). In practical applications, the assumption of Gauss PDF is often unable to meet the requirements, so the linear combination of several Gauss PDFs is used to simulate the observation sequence. The combination of a sufficient number of Gauss PDFs can approximate any PDF, and the mean and covariance matrix of each Gauss PDF can be obtained by the statistics of a large number of observation samples.

HMM can be expressed as $\lambda = (\pi, A, B)$, where π for the initial probability matrix, A for the state of transtion-probablity matrix, B for the probability matrix of the observation. In CHMM, B is a set of probability density functions, $B = \{b_{jo}\}$:

$$b_{jo} = \sum_{k=1}^{M} c_{jk} N(\mathrm{O}, \mu_{jk}, U_{jk}) = \sum_{k=1}^{M} \frac{c_{jk}}{(2\pi)^{D/2} \mid U_{jk} \mid^{1/2}} \exp\left[-\frac{(o-\mu_{jk})(o-\mu_{jk})^T}{2U_{jk}} \right] \quad (1)$$

The significance of each parameter in the formula (1) is as follows:

M: the number of Gauss probability density functions used to fit the distribution.

c_{jk}: the weight coefficients of the k-th Gauss probability density function in state S_j.

N: Gauss probability density function.

O: observation sequence.

μ_{jk}: the mean vector of the k-th Gauss probability density function in state S_j.

U_{jk}: the covariance matrix of the k-th Gauss PDF in state S_j.

Therefore, CHMM can be represented as $\lambda = (\pi, A, C, \mu, U)$. The training algorithm of CHMM can refer to [4, 5]

2.2 KSIR - dimension reduction method

Sliced Inverse Regression (SIR) is an effective method to reduce the dimension of data, which is proposed by Li [6]. There is a non-linear relationship between the physical characteristics, which limits the application of SIR. Kernel Sliced Inverse Regression (KSIR) [7] can adapt to the nonlinear relationship and obtain higher classification accuracy with less feature vectors. The basic concept of KSIR is to map the samples to higher dimensional space by a nonlinear transformation, and then to calculate the inverse of the slices. That is:

$$\Phi:\ x_i \rightarrow \phi(x_i) \tag{2}$$

Where $\phi(x_i)$ belongs to a higher dimensional space than R^p.

The model $y = g(X\beta_1, X\beta_2, \cdots, X\beta_M, \varepsilon)$ proposed by Li is changed to

$$y = g(\Phi(X)\beta_1, \Phi(X)\beta_2, \cdots, \Phi(X)\beta_M, \varepsilon) \tag{3}$$

$(\beta_1, \beta_2, \cdots, \beta_M)$ is called the effective dimension reduction space, and each vector β_m is effective dimension reduction direction. The calculation procedure of KSIR is as follows:

1. Standardize $\Phi(X)$ to meet

$$\sum_{i=1}^{n} \Phi(x_i) = 0 \tag{4}$$

The covariance matrix

$$\Sigma_{xx} = \frac{1}{n}\sum_{i=1}^{n} \Phi(x_i)\Phi(x_i)^T \tag{5}$$

2. Divide $\Phi(X)$ into H slices according to the value of y, and the sample number of the h-th slice is n_h.

3. Calculate the sample mean of each slice, and the weighted covariance matrix.

$$\bar{\Phi}_h = \frac{1}{n_h}\sum_{y(h)\in h} \Phi_h \tag{6}$$

$$\Sigma = \frac{1}{n}\sum_{h=1}^{H} n_h(\bar{\Phi}_h)\cdot(\bar{\Phi}_h)^T \tag{7}$$

4. Calculate the eigenvalues $\lambda_1 > \lambda_2 > \cdots > \lambda_p$ and the corresponding eigenvectors $\beta_1, \beta_2, \cdots, \beta_p$ of the covariance matrix Σ.

5. Project the $\bar{X}$ along the first M eigenvectors, and the results are M-dimension feature component. The selection of M is based on the cumulative contribution rate of the characteristic value. In practical application, the general selection contribution rate is 80%-90% or higher.

2.3 The identification process of pipe axial defect length based on the KSIR-CHMM

To identify the axial length of pipeline defect based on CHMM, the guided wave signals are collected and the feature is extracted, and KSIR algorithm is used to reduce the dimensions of feature data for the application of CHMM. Then, the CHMM classifier is trained, which will be used to classify accurately as different axial length. The flow chart of the identification process based on the KSIR-CHMM is shown in Figure 1.

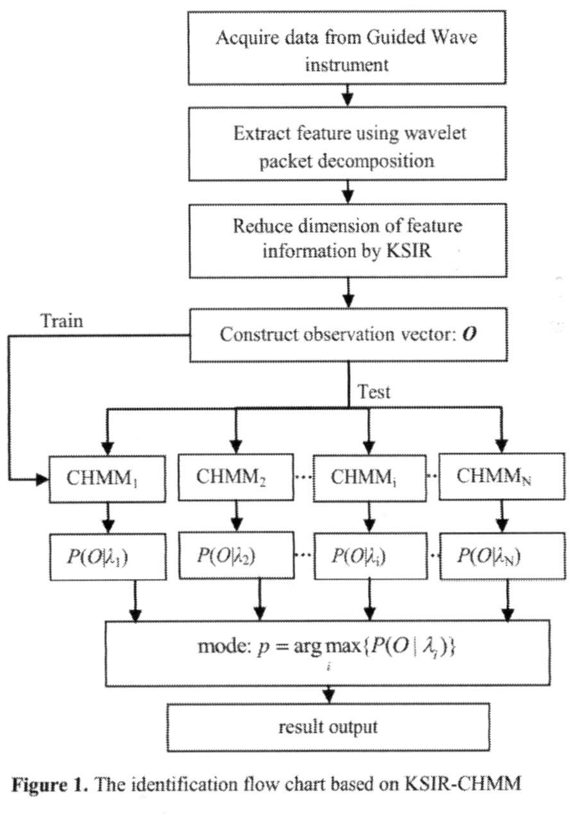

Figure 1. The identification flow chart based on KSIR-CHMM

3 Experimental Verification

3.1 Experimental Setup.

The magnetostrictive guided wave detection system is used to detect the steel pipe with an artificial corrosion on the outside surface, as shown in Figure 2, and a series of guided wave signals are recorded with the increase of defect length. The specifications of the test pipeline are shown in Table 1. The excitation signal frequency is 53 KHz, and the sampling frequency is 5 MHz. The signal in the dashed box as shown in Figure 3 is used for testing. Table 2 shows the samples of axial length for classification training.

Defect Axial length

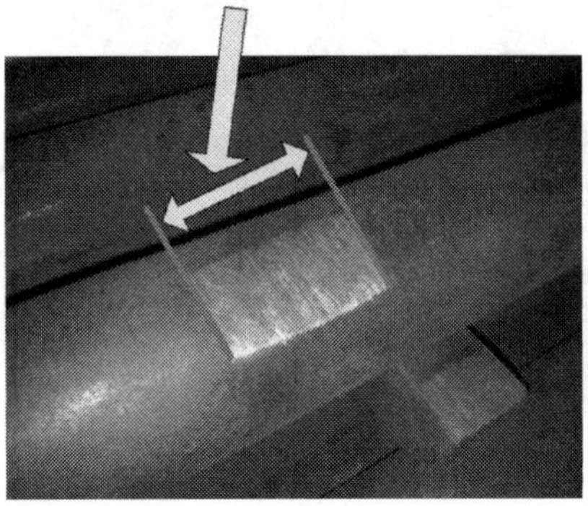

Figure 2. Photo of the artificial corrosion

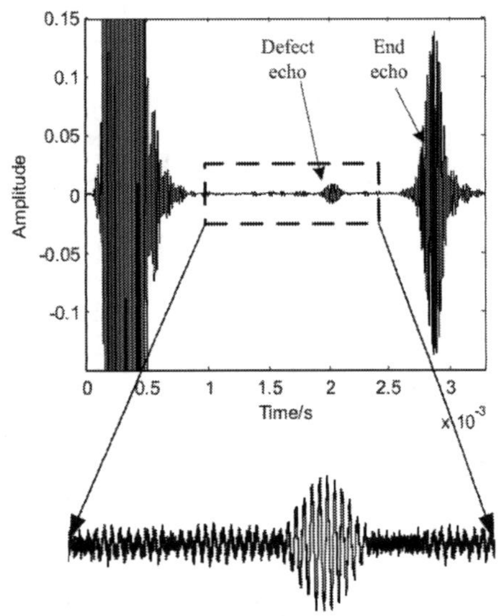

Figure 3. Testing sample signal

Table 1. Pipeline specifications to be tested

Item	Specifications /mm
Length of pipe	6500
Outside diameter	57
Thickness	3.5
Defect position	4333
Defect depth	2.25

Table 2. Samples of axial length for classification training

Damage level	**Defect axial length /mm**
CLASS 1	4.5 6.6
CLASS 2	9.8 12.5
CLASS 3	19.5 22.8
CLASS 4	25.5 29.2
CLASS 5	35.5 39.8

3.2 Wavelet Packet Decomposition

The signal is decomposed by three-layer wavelet packet, and reconstructed in 8 frequency bands. Five parameters for each band are extracted, including root mean square value, variance, skewness, kurtosis and the energy of the frequency band. Finally, a 40-dimension feature vector is constructed.

3.3 Dimension Reduction by KSIR

The features set are normalized before reduced. The cumulative contribution rate of 85% is chosen to determine the reduced dimension of feature. In the end, the 40-dimension feature vector is reduced to 4-dimension.

3.4 CHMM Model Training

Figure 4 shows the training process of different CHMM classifier, and we can see that each classifier achieves convergence in less than 20 steps.

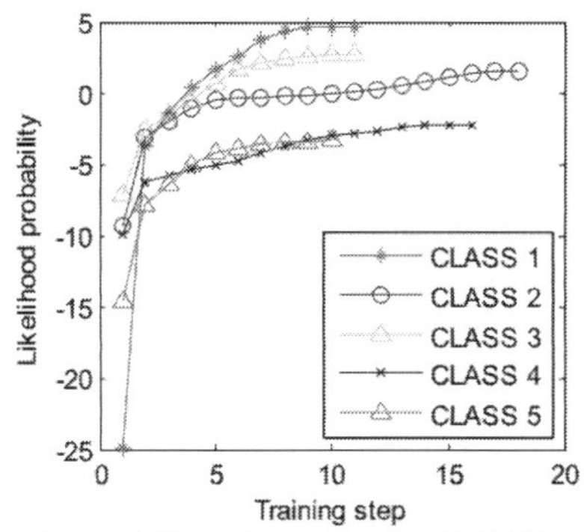

Figure 4. The training process of CHMM

3.5 Application of CHMM Model

Five samples with different axial length of the defect (5mm, 10.9mm, 21.1mm, 27.5mm and 37.3mm, corresponding to CLASS 1 to CLASS 5 respectively) are used to verify the model. The identification results used CHMM are shown in Figure 5. For example, Figure 5(a) is the result of 5mm-length defect by trained CHMM, and the maximum value of likelihood probability corresponds to CLASS 1, which means the sample belongs to CLASS 1. The results show that the CHMM models can accurately identify the axial defects, and achieve the objective of preliminary quantitative evaluation.

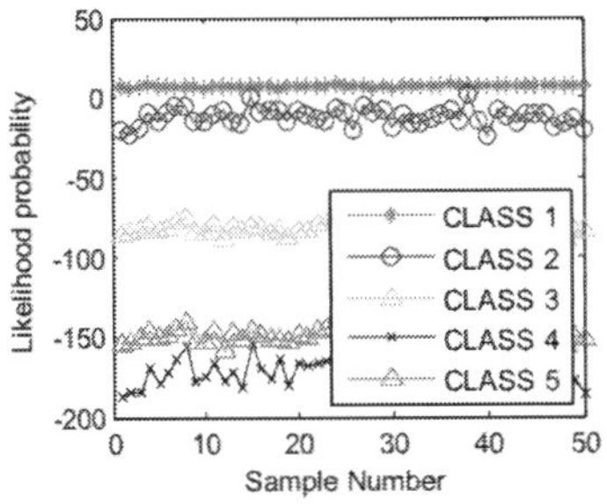

(a) 5mm

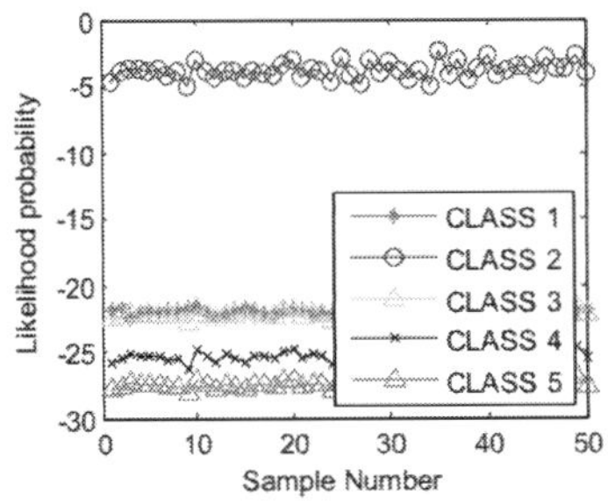

(b) 10.9mm

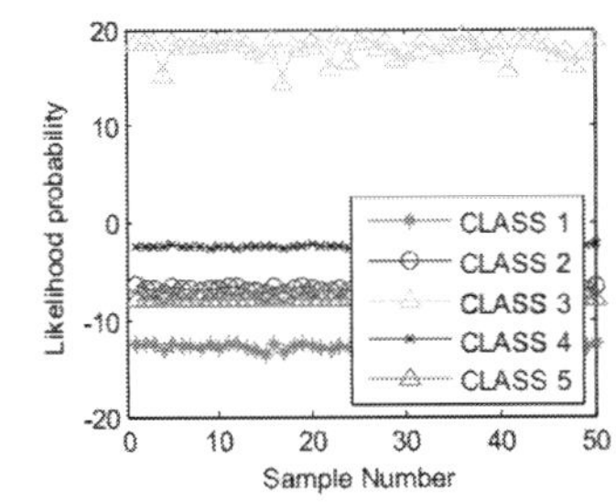

(c) 21.1mm

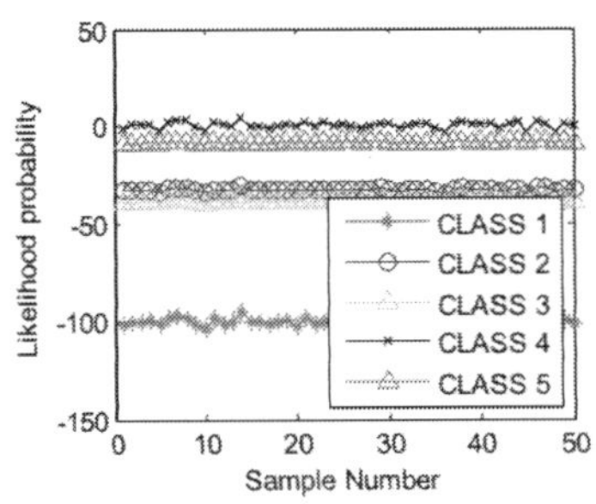

(d) 27.5mm

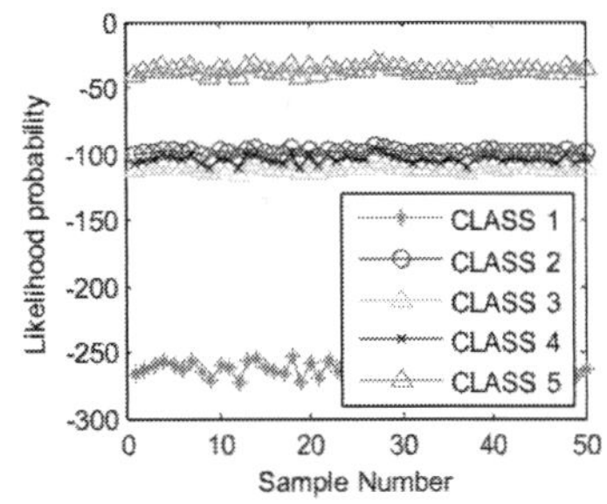

(e) 37.3mm

4 Summary

The axial length of pipe defects is not linear with the reflection coefficient, and the length of the defect cannot be judged directly by the reflection coefficient. Wavelet packet decomposition is applied to extract features, then, KSIR is used to reduce the feature dimension, and finally the CHMM model is used for classification. The following conclusions can be obtained.

1. KSIR can be used to reduce the dimension of feature matrix, which will be helpful to reduce the computation time and improve the accuracy in classification process.

2. The CHMM model can effectively distinguish the different axial length of defects, and the state is given in the form of likelihood probability. The experimental results show that the method has good practicability in pipeline health assessment.

References

1. A. Demma, P. Cawley and M. Lowe, J.Acoust. Soc. Am., **113** (2003).
2. P. W. Tse and X. Wang, NDT & E Int., **54** (2013).
3. L. Y. Zhao, L. Wang and R. Q. Yan, Entropy, **17** (2015).
4. L. R. Rabiner, B. H. Juang, S. E. Levinson, and M. M. Sondhi, At & T Technical Journal, **64** (1985).
5. J. M. Lee, S. Kim, Y. Hwang, and C. Song, J. Sound. Vib., **276** (2004).
6. K. C. Li, J. Am. Stat. Assoc., **86** (1991).
7. H. M. Wu, J. Comput. Graph. Stat., **17** (2008).

DOI: 10.1051/matecconf/20165906002

Bevel Gearbox Fault Diagnosis using Vibration Measurements

Dennis Hartono[1], Dunant Halim[1], Achmad Widodo[2]and GethinWyn Roberts[3]

[1]*Department of Mechanical, Materials and Manufacturing Engineering, The University of Nottingham Ningbo China, China 314100*
[2]*Department of Mechanical Engineering, Diponegoro University, Semarang, Indonesia 50275*
[3]*Department of Civil Engineering,The University of Nottingham Ningbo China, China 315100*

Abstract. The use of vibration measurementanalysis has been proven to be effective for gearbox fault diagnosis. However, the complexity of vibration signals observed from a gearbox makes it difficult to accurately detectfaults in the gearbox. This work is based on a comparative studyof several time-frequency signal processing methods that can be used to extract information from transient vibration signals containing useful diagnostic information. Experiments were performed on a bevel gearbox test rig using vibration measurements obtained from accelerometers. Initially, thediscrete wavelet transform was implementedfor vibration signal analysis to extract the frequency content of signal from the relevant frequency region. Several time-frequency signal processing methods werethen incorporated to extract the fault features of vibration signals and their diagnostic performances were compared. It was shown thatthe Short Time Fourier Transform (STFT) could not offer a good time resolution to detect the periodicity of the faulty gear tooth due the difficulty in choosing an appropriate window length to capture the impulse signal. The Continuous Wavelet Transform (CWT), on the other hand, was suitable to detection of vibration transients generated by localized fault from a gearbox due to its multi-scale property. However, both methods still require a thorough visual inspection. In contrast, it was shown from the experiments that the diagnostic method using the Cepstrumanalysis could provide a direct indication of the faulty tooth without the need of a thorough visual inspection as required by CWT and STFT.

1 Introduction

In order to accurately diagnose possible faults in a gearbox, theextraction of features from the impulsive signal associated with the gearbox vibration is crucial. Alarge number ofsignal processing methodshave beenproposedto identify the impulsive features from vibration signals, particularly using time-frequency analysis such as the Wigner-Ville distribution,Cepstrum,Fast Fourier Transform (FFT), and the amplitude-phase demodulation.

In most cases, vibration signals from a faulty gearbox will consist of a series of impulses that repeat at a particular defect frequency [1].An impulsesignal is a non-stationary signal that cannot be captured by conventional steady-state spectrum analysis, such as FFT analysis, because of the main assumption of FFT analysis that the underlying signal must be stationary.

The Short Time Fourier Transform (STFT) [2-4], however, can overcome the aforementioned problem. However, the main problem with the method concerns about the constant resolution for all frequencies because it uses the same window for the analysis of the entire signal. On the other hand, the CWT has a multi-scale property alower scale wavelet hasa high time resolution and a lower frequency resolution, but a higher scale wavelet has a high frequency resolution and a low time resolution[5]. Compared to STFT, CWT has the flexibility in adjusting its time-frequency resolutionwhich means thatit is more suitable to analyzethe impulsive signals associated with a faulty gearbox.

On the other hand, Cepstrum analysis has been widely used for signal periodicity detection in the frequency spectrum, especially for speech analysis that contains harmonics in its voice frequencybandwidth.Cepstrum can be considered as the frequency spectrum of a frequency spectrum.In gearbox vibration signals, harmonics in the form of sidebands equally spaced at an integer multiple of carrier frequencies or gearmesh frequencies. The defects that introduced as the periodic patterns in the vibration signals can be considered as a system with multiple echoes [6]. Therefore,Cepstrum can be used to identify the periodicity observed from the harmonics associated with certain gear faults.

The objective of this study is to investigate theeffectiveness of severaltime, frequency or joint time-frequency signal processing methods to identify the fault location in a gearbox.The need of each method for a thorough visual inspection for the fault location identification was investigated. Once the vibration measurements were obtained from the gearbox, FFT was carried out to evaluate the frequency contents of the signal.Discrete Wavelet Transformwas thenused to extract the useful frequency bandwidth to be analysed further.Finally, the performance ofseveral signal processing techniques were compared for gearbox fault diagnosis.

2 Theoretical Background

2.1 Discrete Wavelet Transform

The Discrete Wavelet Transform (DWT) is an alternative method of the Continuous Wavelet Transform,which is discretized in terms of scale and translations. It is widely used in signal compression and filtering [7, 8]. Equations 1-4 providethe basic analysis of DWT decomposition scheme

$x(t) = \hat{x}_0(t) + \sum_{n=-\infty}^{\infty}\sum_{m=-\infty}^{0} T[m,n]\psi_{m,n}(t)$(1)

$\hat{x}_0(t) = \sum_{n=-\infty}^{\infty} S[0,n]\emptyset(t-n)$ (2)

$\psi_{m,n}(t) = a_0^{-\frac{m}{2}}\psi(a_0^{-m}t - nb_0)$(3)

$\emptyset_{m,n}(t) = {}^{1}\!/_{2^{\frac{m}{2}}}\,\emptyset(2^{-m}t - n)$

(4)where $\psi_{m,n}(t)$ and $\emptyset_{m,n}(t)$ are the respective scaled wavelet and scaling functions, a_0=2 and b_0=1. *T[m,n]* is a typical wavelet coefficient.*S[0,n]* is the aggregate of all details for scales greater than or equal to 1 (coarser scale). In practice, DWT is normally implemented using a series of filter banks. Interested reader can refer to [9] for further details.

2.2 Short Time Fourier Transform (STFT)

The STFT is considered as the most straight forward time-frequency signal processing methods. It is based on slicing the waveform of interest into several segments and then performing the usual FFT analysis on each of these segments. The usual FFT is applied within each segment. The idea is that the underlying signal is assumed to be stationary within each segment.It is formulated as [10]:

$X(t,f) = \int_{-\infty}^{\infty} x(\tau)w(t-\tau)e^{-j\pi f\tau}\, d\tau$ (5)

where $w(t-\tau)$is the window function and τ is the variable that slides the window across the underlying waveform.

2.3Continuous Wavelet Transform (CWT)

The Continuous Wavelet transform is a linear transformation decomposing the underlying signal $x(t)$ into elementary functions called wavelets. Wavelets are local functions in time that can be described as[11]:

$W(a,b) = \frac{1}{\sqrt{a}}\int_{-\infty}^{\infty} x(t)h^*\left(\frac{t-b}{a}\right)dt$(6)

where the basis function for the wavelettransformisgiven in term of translation parameter*b*anddilation parameter*a*with the followingmotherwaveletfunction:

$h_{a,b}(t) = \frac{1}{\sqrt{a}}h^*(\frac{t-b}{a})$ (7)

2.4 Cepstrum

Cepstrum can be regarded as the spectrum of a spectrum. It can detect the periodicities in the frequency domain. There are several forms of Cepstrum analysis, yet most of them can be considered as a spectrum of a logarithmic spectrum as formulated below[12]:

$C_p(\tau) = F^{-1}(log[S_{xx}(f)])$ (8)

whereτisquefrency (s),S_{xx}isusual FFTspectrum and$F^{-1}$is the Inverse Fourier Transformmethod.

3Experimental Apparatus

Experimentson asingle-stage bevel gearbox were carried out on atest-rig from the Machinery Fault Simulator-Spectra Quest located at Diponegoro University as shown in Figure 1.

Figure 1. The gearbox test rig.

During the experiment, a three-phase AC motor was used as the driver of the system and amagnetic brake with the third scale was incorporated as a load. Three accelerometers sensor were used to capture the vibration signals from the test rig,although only the axial measurementwas used in the analysis. Tooth breakage gear fault was introduced in this work and compared with the healthy gear. The broken gear usedin the experiment was depicted at Figure 2.

Figure 2. The bevelgear with a faulty tooth.

For the experiment,5.12 kHz sampling rate was used to sample the vibration signals to gather eight seconds of data. A tachometer sensor from Lattice Instruments was also used to provide the phase information from the gear, which then will be used to pre-process the vibrations signal by TSA method to get rid of the asynchronous component from the faulty gear under consideration.

4Results and Discussions

The raw vibration signal (in terms of axial acceleration output from the accelerometer) of a faulty gear in axial direction is presented in Figure 3. It can be seen that it is not straightforward to extract the fault information by merely observing the time domain behaviour of the raw vibration signal. There are several observed spikes that occurred and may provide useful information for detecting the periodicity of impulsive signal from the fault gear. However, the spikes may also be contributed by the measurement noise obtained during the data acquisition process.

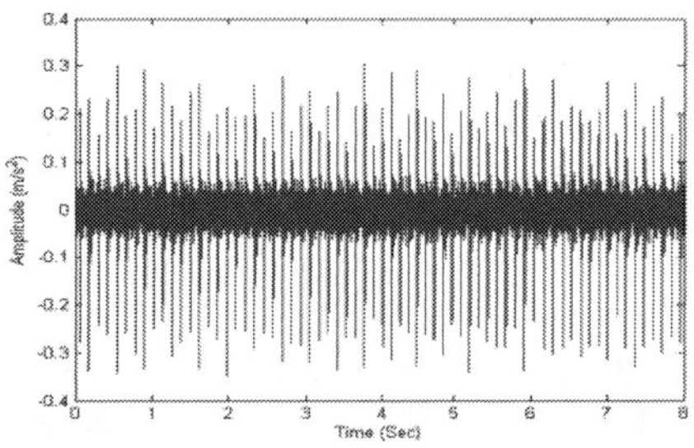

Figure 3. Vibration signal from a faulty gearbox.

Another method to analyze the behaviour of the signal from the faulty gear is by utilizing the spectral analysis of FFT as presented in Figure 4. As mentioned in introduction, gearbox signals are characterized bysidebands around gearmesh frequency and its harmonics. Sidebands can thus provide the useful information that indicates the faulty tooth existence [13].

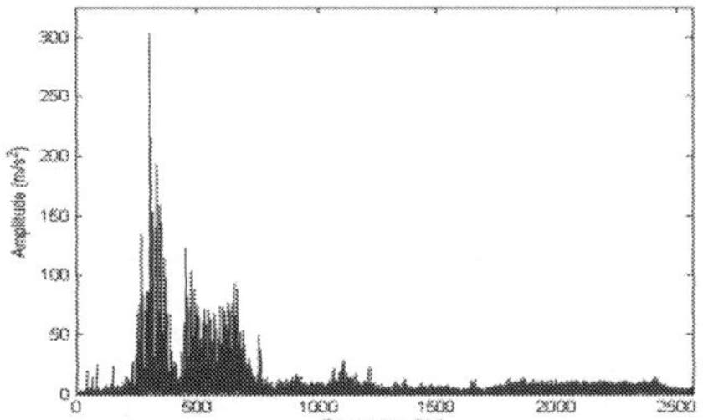

Figure4.FFT of vibration signal from a faulty gearbox.

FFT can show sidebands patterns,offering diagnostic information that is harder to visualize from time domain. However, it does not contain information about how frequency contents evolve over time and it is not suitable to analysenon-stationary signal, such as impulse signal typically observed in this situation. Moreover, it is usually impossible to observe the sidebands around gearmesh harmonics becausesomesidebandscanbecovered by the noisethatcan havehigher amplitudes compared to those of the sidebands.

For further analysis, DWT is utilized to extract useful information from the signal from the faulty gear. From FFT plot in Figure 4, it can be concludedthat the important frequency bandwidth that needs to be analyzedis approximately in the range of 0-700 Hz due to strong amplitude of sidebands around third and fourth gearmesh frequencies.This can be achieved by decomposing the vibration signals by 3 level DWT so it can provides signals in frequency range from 0-640 Hz. Due to the fact that the experiment was carried out under steady-state condition, only 0.4s segmented part of the de-noised axial vibration signal was chosen to be furtheranalyzed by the time-frequency analysis.

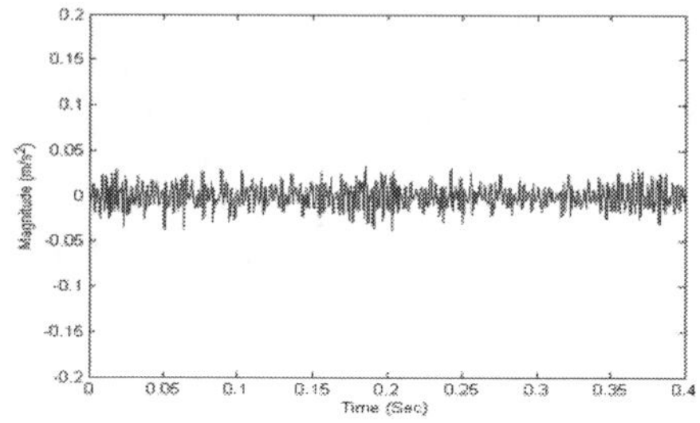

Figure 5.DWT of vibration signal from a healthy gearbox.

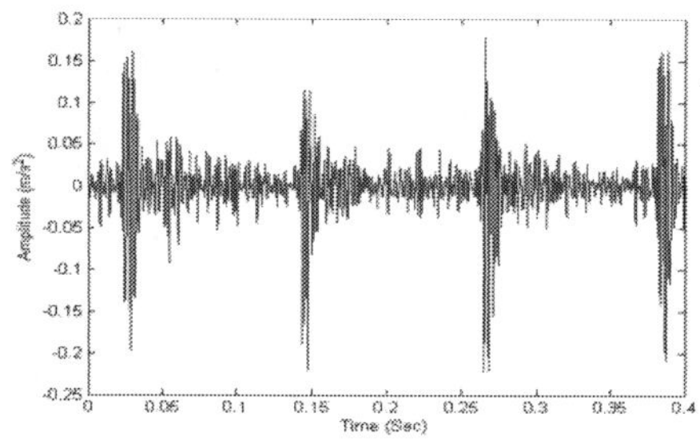

Figure 6.DWT of vibration signal from a faulty gearbox.

Figures5 and 6show the signals after DWT decomposition for healthy and faulty gearboxes, respectively. It is observed that the healthy gearbox shows a regular pattern of vibration amplitude compared to the broken gearbox.The plot also showsthe impulsive patterns that correspond to the impact generated by faulty tooth when it meshed with the other gear. Another way to extract the periodicity of signal is by analysing the time domain signal in the joint time-frequency representation. Such a representation isnot only able to provide the time representation of the signals but also able to provide information about how the frequency contents of the underlying signal evolve over time, which cannot be achieved by FFT analysis.To obtain such a representation, another method for the time-frequency analysis will be needed as presented in this work.

DOI: 10.1051/matecconf/20165906002

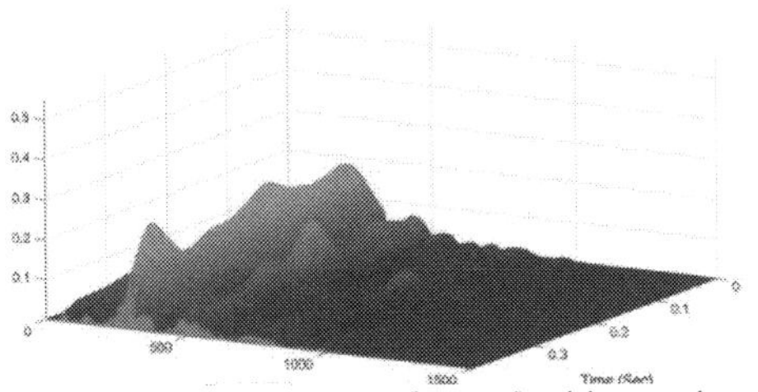

Figure 7.STFT of vibration signal from a healthy gearbox.

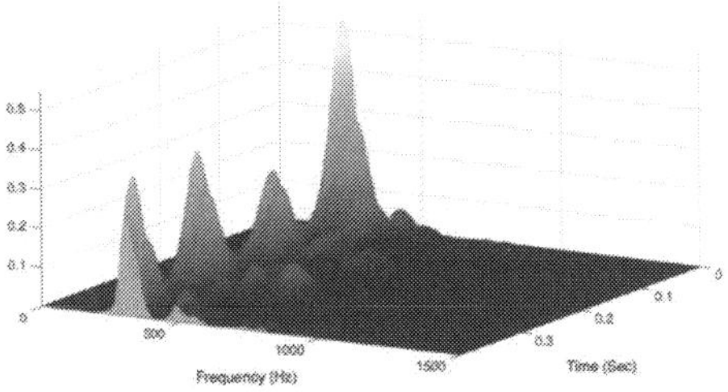

Figure 8.STFT ofvibration signalfrom a faulty gearbox.

Initially,STFT was incorporated to capture the impulse signal characteristics from the vibration signal. STFT is considered as the simplest method for time-frequency analysis to evaluate how the frequency components of signals evolve over time. STFT plot of healthy and broken gearboxes are presented in Figures7 and 8, respectively. As observed from the figures,signalsassociated with healthy and broken gearboxes can be differentiated based on their patterns. However, the constant time-frequency window used in the analysis usually makes STFTnot suitablefor analyzing complex vibrations signals that require the multi scale signal analysis[5]. In other words, once the length of the window is chosen, the entire analysis will be based onfixed time and frequency bandwidths. Figure 8 showed clearly that the poor time resolution of STFT in presenting the precise timing of the impulsive event. This timing is quite important to determine the time period of the faulty gear that can be used to identify which gear contains the fault[7].

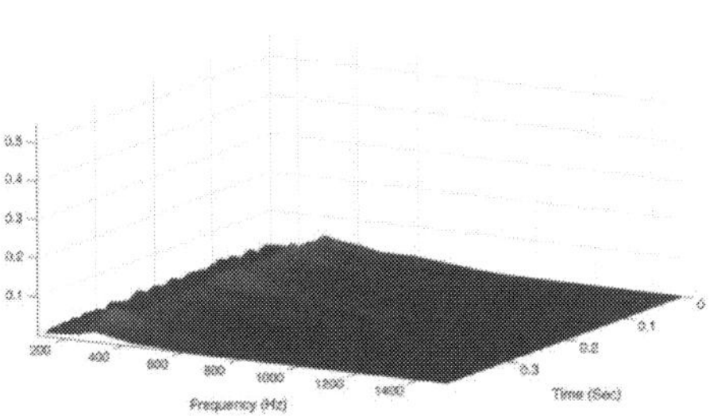

Figure 9.CWTof vibration signalfrom a healthy gearbox.

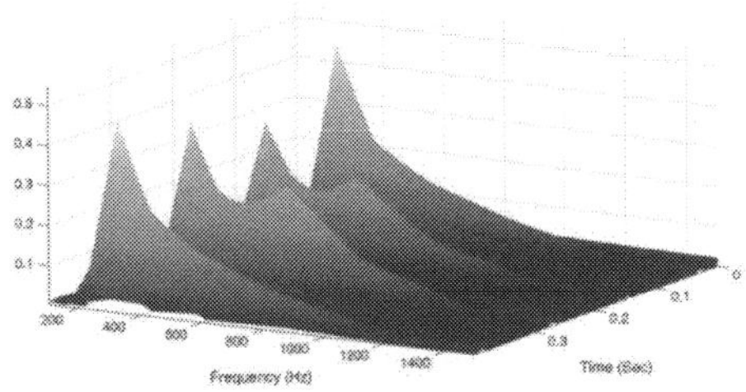

Figure 10.CWTof vibration signal from a faulty gearbox.

The CWT as one of the most prominent tools in detecting transient signals offers a better compromise for time-frequency signal analysis. It has a fine localization of frequency content in the low frequency region and a fine localization of time contentat the high frequency region, which makes it as an excellent tool to diagnose gearbox fault[5]. As the impulsive feature of vibration signal exhibits a short time duration but a wide frequency bandwidth, CWT is one of the best candidates to be used to detect the pattern of vibration signal from the faulty gear. Figures9 and 10represented the scalogram (the magnitude of CWT coefficients) plots of vibration signal from the healthy and faulty gearboxes.Figure 10 clearly revealed the transient signal characteristics and the period of the impulse signal that reflected the rotation frequency of the faulty toothusing the joint time-frequency representation. The signal periodwas approximately 0.1195 second between two successive impulses. These two figures showed that CWT was not only able to distinguish the pattern between the healthy and broken gears but also offered a better time resolution to detect the periodicity between two impulses, compared to STFT.Since the periodicity was inversely related to the speed of the faulty gear, so the method was also able to determine which gear was generating the impulse due to faulty tooth.

Although CWT overcamethe STFT limitations,it still requireda thorough inspection to the plot of the scalogramto provide information about thefaulty gears, which was not a straight forward process. Based on this consideration, the Cepstrum method was investigated since it could offer a better solution to detect the signal from the faulty gears. This can be doneby detecting the sidebands pattern from the signal from the faulty gears around the gearmesh frequency that presents in FFT, sincecepstrum reveals the periodicity of spectrum [12]. Since gearbox vibration typically contains harmonics in itsspectrum(e.ggearmesh frequency, sidebands) which makes Cepstrum a good candidate to detect such harmonics for gear fault diagnosis. Figure 11 showed the Cepstrum of vibration signal from the faulty gearbox.The quefrency of the signal from the faulty gear was observed to be 0.11947 s.This indicated that there were some sidebands around the gearmesh frequency spaced at 8.37 Hz (i.e. 1/0.11947), which corresponded to the rotational frequency of the faulty gear.It was shown that Cepstrum method could eliminate the need for a manual visual inspection for determining the periodic pattern of the faulty tooth signal, in contrast to the CWT method.

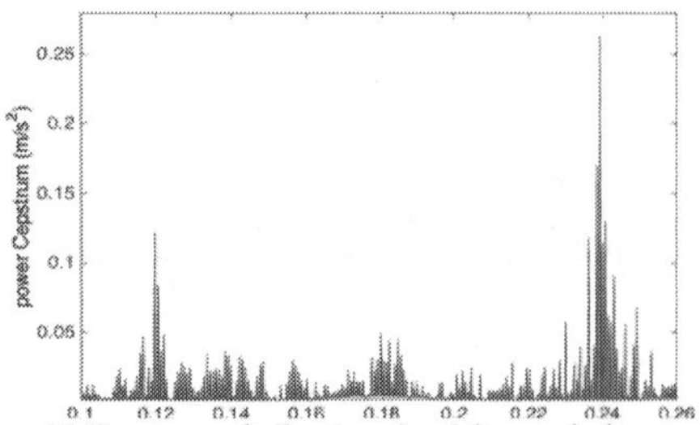

Figure 11. Cepstrum of vibration signal from a faulty gearbox.

Conclusions

A comparative studyof several gear fault diagnostic methods using vibration measurements has been done. The aim of the study is to detect the existence of a faulty gear with a broken tooth from a bevel gearbox test-rig.Several time-frequency signal processing methods were compared: FFT, STFT, CWT and Cepstrum methods.Initially, DWT was implemented to extract only the useful frequency content of the raw vibration signal that containedthe fault features. It was further observed thatFFT could reveal the sideband patterns that contained information about which particular gear contained a faulty tooth. However, such informationwas difficult to identify due to the need for thorough visual inspections of such sidebands that might be corrupted by the measurement noise. In this case, STFT offereda better solution by providing the simultaneous time-frequency analysis to detect the impulse signals generated by the faulty tooth. Itsusefulness for fault diagnosis, however,was still restricted by theduration-bandwidth principle that compromisedthe time and frequency resolutions. Appropriate window function and length should thus be taken care to be able to extract the gear fault features. In this work, STFT gavea poor time resolution, which is insufficient to indicate the accurate timing of fault features generated by gearbox. On the other hand, CWT has been shown to be a better candidate to analyze thetransient signals since it was able to offer a better solution compared to STFT. However,a thorough visual inspection was still required. Finally, the experiments showed the advantage of the Cepstrum analysis in providing the gear fault information without the need of a thorough visual inspection.

Acknowledgment

The authors gratefully acknowledge the support provided by Ningbo Science and Technology Bureau-International Cooperation Programme(Project: 2012D10029), China.

References

1. R.B. Randall,*Vibration-based Condition Monitoring: Industrial, Automotive, and Aerospace Applications* (2011)
2. G. Chen, J. Chen, G. Dong, H. Jang, Appl. Acst**87**, 131-141 (2015)
3. G. He, K. Ding, W. Li. X. Jiao, Ren. Engy**87**, 364-375 (2016)
4. O. D. Mohammed, M. Rantatalo, Mech. Sys. Sig. Proc**66-67**. 612-624 (2016)
5. L. Jedlinski, J. Jonak, Appl. Sft. Cmpt**30**, 636-641 (2015)
6. S. Jolivet, S. Mezghani, M. E. Mansori, B. Jourdain, J. Mnfct. Sys **37**, 467-471 (2015)
7. S. Jin, J. S. Kim, S. K. Lee, J. Mech. Sci. Tech **29**, 3165-3173 (2015)
8. V. Skrickij, M. Bogdevicius, R. Juvenicius, Appl. Acst**106**. 51-62 (2016)
9. S. Mallat, *A Wavelet Tour of Signal Processing, The SparseWay*(2009)
10. N. H. Chandra, A. S. Sekhar, Mech. Sys. Sig. Proc**72-73**, 105-133 (2016)
11. W. He, Q. Miao, M. Azarian, M. Pecht, Mech. Sys. Sig. Proc**64-65**, 149-161 (2015)
12. B. Liang, S.D. Iwnicki, Y. Zhao, Mech. Sys. Sig. Proc 39, 342-360 (2013)
13. G. Dalpiaz, A. Rivola, R. Rubini, Mech. Sys. Sig. Proc**14**, 387-412 (2000)

Vision based error detection for 3D printing processes

Felix Baumann and Dieter Roller

University of Stuttgart, IRIS, D-70569 Stuttgart, Germany

Abstract. 3D printers became more popular in the last decade, partly because of the expiration of key patents and the supply of affordable machines. The origin is located in rapid prototyping. With Additive Manufacturing (AM) it is possible to create physical objects from 3D model data by layer wise addition of material. Besides professional use for prototyping and low volume manufacturing they are becoming widespread amongst end users starting with the so called Maker Movement. The most prevalent type of consumer grade 3D printers is Fused Deposition Modelling (FDM, also Fused Filament Fabrication FFF). This work focuses on FDM machinery because of their widespread occurrence and large number of open problems like precision and failure. These 3D printers can fail to print objects at a statistical rate depending on the manufacturer and model of the printer. Failures can occur due to misalignment of the print-bed, the print-head, slippage of the motors, warping of the printed material, lack of adhesion or other reasons. The goal of this research is to provide an environment in which these failures can be detected automatically. Direct supervision is inhibited by the recommended placement of FDM printers in separate rooms away from the user due to ventilation issues. The inability to oversee the printing process leads to late or omitted detection of failures. Rejects effect material waste and wasted time thus lowering the utilization of printing resources. Our approach consists of a camera based error detection mechanism that provides a web based interface for remote supervision and early failure detection. Early failure detection can lead to reduced time spent on broken prints, less material wasted and in some cases salvaged objects.

1 Introduction

Contemporary consumer grade 3D printers are often cube shaped with a size in the range of 50 cm edge length. This is due to the mechanics surrounding the build area that is on average 15x15x15 cm in size for consumer grade printers. With a weight of about 10 kg they could be placed on the desk of the user but are recommended to be placed away from the user due to potentially harmful fumes emitting during the printing process.

Different technologies for 3D printing or additive manufacturing exist[1], ranging from metal-powder or ceramic[2] based laser sintering over laminated object modelling, resin based stereolitography to thermoplastics based fused deposition modelling (FDM). The most common class for consumer grade 3D printers is FDM.

FDM printers are cheap to manufacture as they are mechanically simple and do not require complex and expensive components. Their makeup consists of 3 axes that are controlled by 3 stepper motors. These axis and motors are the main components, beside the print-head and print-bed and are affixed in a constellation that enables 3 degrees of freedom along these axes. This method offers reduced granularity of the objects compared to stereolitography. Applications are in rapid prototyping [3].

As it is recommended to utilize the printers in rooms different than the workplace of the user direct supervision is inhibited. Printing errors [4],[5] occur for the following reasons:

- Misalignment of the print-bed
- Misalignment of the nozzle
- Clogging of the nozzle
- Depletion of printing material or disrupted material flow
- Lack or loss of adhesion to the print-bed
- Vibration or shock (from the printer or another source)

As a goal misprints should be reduced because they consume material and occupy the printing device for long periods of time without producing a useful object. Printing times on FDM machines can range up to 20 hours for sufficiently complex and large objects.

In order to alleviate the supervision problem users and manufacturers place video cameras (web cams) in or at the printer to facilitate remote supervision capabilities. Other error detection approaches facilitate laser scanners [6], [7], currents [8] or thermography[9]. This can help to receive information on potential problems and to assess the printing progress remotely but requires the user to watch the video stream either constantly or in intervals in order to get information on problems and errors. The goal is to support the user in detecting printing errors by utilizing machine vision [10]. The benefits are reduced material waste, reduced occupancy of printing resources due to early detection of print errors and the prevention of completion of the printing process for flawed or broken objects and reduced time effort for the end user due to the alleviation of the need for constant supervision.

For our approach we utilize consumer grade video cameras mounted before the 3D printer (Makerbot Replicator 2X [11]) and video analysis software. This work describes the design decisions and the underlying foundations for this system.

This work is structured as follows: In this section the problem and motivate our approach is described by displaying the consequences of the problem. In section [**2 Error Classes**] we describe error classes present in this area and explain which we address with our approach. In the following section [**3 Optical Error Detection**] we describe our approach to visual error detection in FDM. Section [**4 Results**] displays the results from an experiment conducted to validate this approach. In section [5 **Discussion**] we discuss our approach from the software side and conclude with an outlook in [**6 Conclusion and Future Work**].

2 Error classes

In order to provide successful error detection it is important to define the errors to be encountered beforehand. We will lay out the errors that can arise and which classes we can detect with our approach. We define five error classes of which we can detect class one, two and three with our system. Classes four and five are not detected due to complexity of their detection. We propose these error classes as no systematical research on error classification in FDM is present.

2.1 Detachment

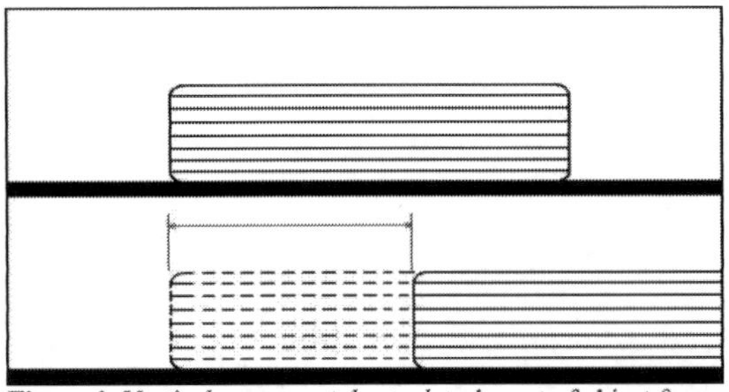

Figure 1. Vertical movement due to detachment of object from printing-bed.

The most prevalent error and the easiest to detect is the detachment of the object from the print-bed. The object must not detach for the object to be printed successfully. In figure 1 this error is presented in two sketches. The striped rectangle with rounded corners represents the object to be printed and is attached to the print-bed (represented by the thick black bar below the object). In a print without errors this object is firmly attached to the print-bed and does not move in any direction relative to the print-bed. The printer in our approach has a print-bed that has one degree of freedom in the Z-direction (height) and a print-head with two degrees of freedom in X and Y direction (left/right and forward/backward). Other configurations of printers might need an adjustment because there the print-bed can move in two dimensions. In our sketch the detachment is visible as the object has moved to the right and the original position is marked with dashed lines. The displacement between the original and the actual position is called the displacement error. In [12] this error type is described as a result of material warping. This error type/class can occur

• When the temperature of the print-bed fluctuates and the object cools non-uniformly

• When the print-bed is not adjusted and calibrated and therefore the distance between the nozzle and the print-bed varies

• Due to vibration or shock that detaches the object

2.2 Missing material flow

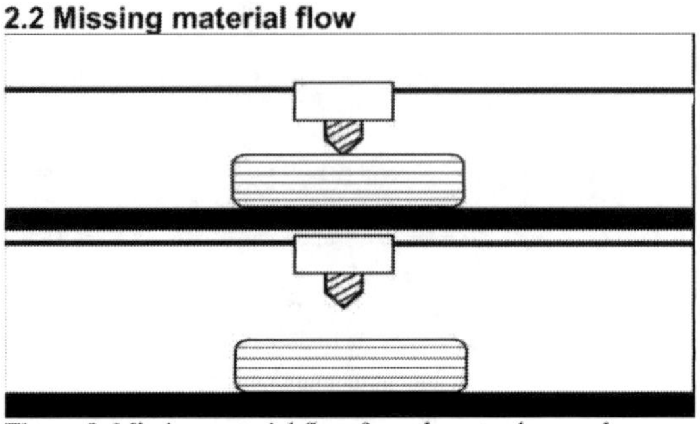

Figure 2. Missing material flow from the extruder nozzle.

With this error type the thermoplastic does not flow through the nozzle of the print head. The print head moves along its predetermined tool-path without extruding the filament. The height of the object stays constant and the object moves along the print-beds Z dimension downwards. The object is not complete and will not be completed. If the material flow is interrupted only temporarily the object will be printed defective as one or more layers of material are missing. If the material flow is interrupted only briefly the object might be printed completely with only minor defects as layers can compensate for missing lower layer parts to a small degree. In [13] this error is detected using acoustic emissions from the printer. In figure 2 this error type is depicted. The object is displayed as a striped rectangle with rounded corners, the print-bed is shown as the thick black line below the printing object and the print-head or extruder is displayed as the white rectangle with a striped pentagon below. The upper part of the sketch shows the case that the object is printed normally without errors and the extruder is adjacent to the object. The lower part shows the case where the material flow has stopped, the extruder does not touch the object anymore and moves along its tool-path. In the lower case an abort signal can be sent to the printer as this error is not recoverable.

This class of error can occur due to

• Clogged Nozzle/Extruder

• Air-bubbles in the extruder

• Jammed material (filament string is broken or extruder motor cannot grip the filament for transportation into the nozzle)

• Depleted material

2.3 Deformed object

As a further error class deformed objects that are not printed according to their CAD template is described. These errors can occur at areas with an overhang or at bridges as described in [14]. Overhangs and bridges are sections of the object where parts of the lower layer are intentionally missing. The printer extrudes material into the air spanning a path between two points with lower layer support (in case of a bridge) or over the edge of a lower layer supporting material (for overhangs). For bridging the printer tries to compensate by moving at higher speeds but sagging does usually occur for longer bridges as the material is drawn downwards due to gravity. When printing a filigree or unstable structure or when the printing parameters are chosen unsuitable regarding the supporting structures the object can topple over (With or without partial detachment from the print-bed) or collapse. Following this the object is either pushed aside and around by the extruder if detached completely from the print-bed or the extruder cannot position itself at the correct position in order to complete the object due to missing layers and adhesion points. The object or parts of the objects can in this case become stuck to the extruder and be partially re-molten. The extruder keeps extruding material that forms a clot around the nozzle requiring extensive cleaning and or maintenance of the extruder. This error class has features from the error classes of missing material flow [2.2 Missing Material Flow] and loss of adhesion from the print-bed [2.1 Detachment] and therefore can be detected utilizing methods designed for them.

2.4 Surface errors

This error class is easily detectable for human users but very hard in a machine vision system. This class encompasses errors that manifest themselves in unclean, broken or in other ways deviating surfaces from the model. FDM printing has by design a problem with smooth surfaces and very sharp structures. In [15] the reasons and occurrences of this error class are discussed. Smooth surfaces are not possible as the material is deposited in rows by the extruder, either overlapping partially or with a gap. Sharp structures are not possible due to the thickness of the deposited material that can be regarded as string like.

These errors stem from a mismatch between the chosen parameters for slicing and printing and the objects geometry. Printing fine structures with high speeds is error prone.

2.5 Deviation from model

This error is hard to detect automatically as the object is printed without any of the previously mentioned errors but differs from the model (i.e. template) e.g. in size or structure. In [16] this error class is discussed as appropriate test objects are discussed and proposed. The object does not exhibit obvious flaws and is printed successfully. This error class can occur when the user is using unsuitable parameters for slicing and printing. User actions like erroneous scaling or rotation can also lead to these errors. Errors due to shrinkage or warping when cooled can also lead to deviations from the model. This error class is hard to detect automatically as no obvious defects occur. In order to detect these errors the object needs to be compared to the model. There is a problem in acquiring a 3D dimensional representation of the object while printing as rotational 3D Laser scanners are prohibitive due to the common design of 3D printers. Inferring 3D geometry through multi camera systems is possible but expensive due to the multitude of angles that need to be analysed for a complete 3D geometry suitable for comparison with the model. As models can be fine detailed a camera system with a high resolution is necessary, varying lighting situation and various coloured materials need to be taken into account.

3 Optical error detection

In order to detect errors from the error classes defined in [2.1 Detachment] to [2.3 Deformed Object] this work utilizes an auxiliary thresholding algorithm. With this algorithm we segment the digital image into binary images with a clear distinction between the object and the background. Our algorithm segments an image from a Playstation EyeCam camera with an OmniVision chip of the resolution of 640 pixels by 480 pixels into the object that is represented in white in the following images. We have chosen this specific camera as it provides a fixed focus and is broadly available at low cost. High cost system can be inhibitive for the utilization by users of such systems. The frame rate is set to 25 frames per second with the camera capable of providing up to 120 frames per second in reduced resolution.

3.1 Camera position calibration

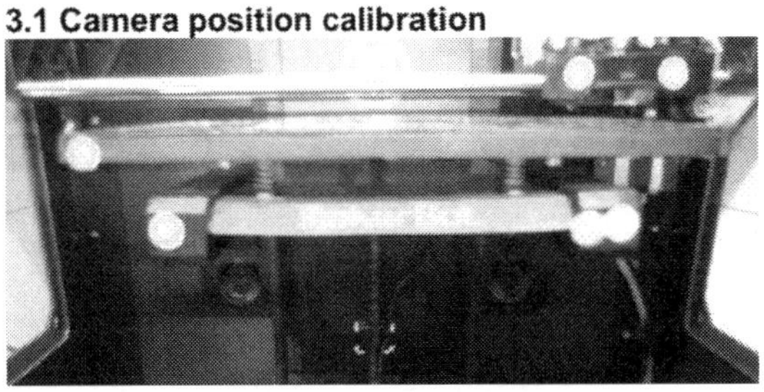

Figure 3. Visual marker positions.

Our system requires the geometry of the 3D printer and the relation to the position of the camera system to be known. We utilize optical markers (see fig. 3) to determine specific parts of the 3D printer and derive the layout from them. Circular markers are utilized to determine a horizontal ground line representing the print-bed. The print-head is identified with a pair of circular markers horizontally aligned and the printing nozzle identified by the distance between the circular markers halved. With this information the video frame can be cropped to the area of interest in order to save computational effort.

3.2 Thresholding

To segment the image into object (foreground) and background we utilize a thresholding algorithm with a threshold determined by the colour of the filament. We question the user for a marking of the object that is printed in the video frame. The user selects the colour with a mouse click from the video frame. We utilize a HSV (Hue-Saturation-Value) colour model for the selection as the native RGB (Red-Green-Blue) colour model is unsuitable for the stable detection of the object due to lighting changes and shadows on the object. Our system also supports a thresholding colour selection with sliders for a HSV combination in case the material stays constant for consecutive prints.

3.3 Utilized Software

We use OpenCV[17] in Version 3.0 and the Python API as this software provides recent algorithms for image detection, manipulation and recognition. We use the packages GUI Features, Core Operations, Imaging Processing and Video Analysis.

3.4 Algorithm

Our Algorithm is composed of the following steps:
1. Selection of thresholding Value via In-Picture Colour selection or HSV Combination selection via sliders
2. Conversion of original RGB video frame to HSV
3. Detection of visual markers for image cropping i.e. reduction to the area of interest (Pre-Processing [3.4.1 Pre-Processing])
4. Thresholding of image to binary image with the selected threshold value
5. Blob detection and marking of candidate object
6. GOTO Step 2

Step 1 is performed by the user's interaction and ensures the appropriate thresholding value for further image analysis. We convert the original 640x480 pixel RGB image into HSV for easier selection of areas with the same colour, this alleviates problems due to changing lighting situations or shadows occurring on the object. Our system utilizes two different methods to detect movement of the object, differential [see 3.4.2 Differential Images] imaging and blob detection [see 3.4.3 Blob Detection]. The threshold parameters are user selected for every printed object and therefore not subjective to algorithmic selection.

3.4.1 Pre-processing

For pre-processing the image is examined and the area of interest is determined by analysing a set of visual markers. We investigate deriving the positions of the print-bed and print head with an algorithm for line and edge detection but find this to be too noisy due to colour grading of the parts to be detected and the large number of parallel lines in the design of our printer.

Visual markers (see fig. 4) are used to indicate the significant parts of the print-head and print-bed which are identified by our algorithm. We utilize CV HoughCircles[18] to detect circular shapes in the image and reduce the number of possible false-positives by averaging the circle position over a 100 frame period and discard circles that are not in a defined dimensional relationship with each other[19]. With the visual markers identified and the image cropped to the area of interest the algorithm determines the print-bed upper surface by finding associated parallel lines (CV CannyEdge Method) in a defined distance to a virtual line between the visual markers for the print bed. The nozzle tips are inferred from a known dimensional relationship between the visual markers and the nozzle tip.

The pre-processing is composed of the following steps:
- Detect Visual Markers
- Determine Area of Interest and crop to it
- Detect print-bed Upper Surface
- Detect print-head Nozzle Tip

In Figure 4 the relationship between the markers is shown with two false positive circle detections (Circle at right-hand bottom side, and circle at the top of the print object which is limited with a yellow line).

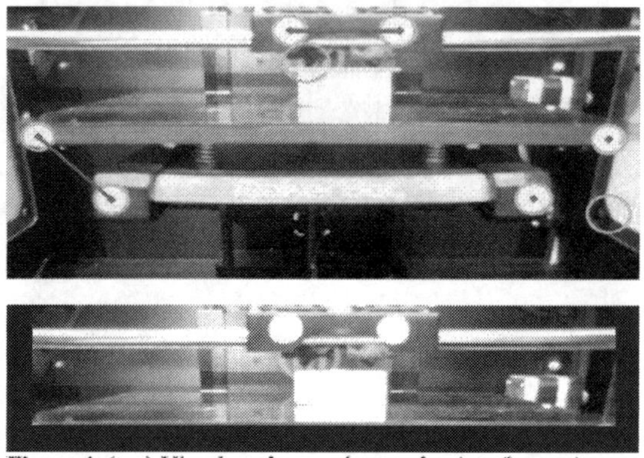

Figure 4. (top) Visual markers and control points (bottom) cropped area of interest.

3.4.2 Differential images

Subtraction of two consecutive images (i.e. video frames) yield a differential image. The differential image f_{diff} displays the temporal differences between two frames f_1 and f_2: $f_{diff}(x, y) = f_1(x, y) - f_2(x, y)$. More than two consecutive frames can be used to calculate differential images leading to movement detectable over longer periods of time.

Our approach calculates f_{diff} from 3 consecutive images (f_{t-1}, f_t, f_{t+1} where t indicates a time stamp). We utilize 25 frames per second so this differential image spans a period of t=3/25 seconds. Stationary objects (i.e. background) are removed from the image frame. Only moving objects and picture errors remain as the content of the image.

On an error free print the object does not change its horizontal location but will only grow vertically. A summation of pixels indicating horizontal change is an indicator for rapid horizontal movement i.e. detachment of the object from the print-bed. We integrate 3 consecutive frames for detection for a short response time to detachment detection. This approach is error prone as pixel errors from the camera can occur and also the area of the print-head needs to be cropped out precisely. Unstable objects that bend while printing are also prohibitive for this approach.

DOI: 10.1051/matecconf/20165906003

3.4.3 Blob detection

Areas with constant features (i.e. same colour, see fig. 5) in a digital image are called blobs. Our algorithm searches for blobs (see fig. 6) with the colour selected by the user as described in [3.2 Thresholding]. Multiple blobs can be detected in the image and the largest blob is selected as the candidate for the object. The algorithm tries to repair missing links between connected blobs in order to completely describe the object. After the candidate object is identified our algorithm describes the object with an enclosing bounding box. The centre of the bounding box is tracked in consecutive video frames for movement detection. As the object grows during the printing process, the bounding box and the centre of the bounding box is updated consequently. A small tolerance for object and centre detection is implemented as the object blob is not detectable completely congruent over time. If our algorithm detects a deviation of the centre of the detected object blob larger than the hard coded threshold this is indicative for a sudden movement of the object. If the object is detected to be not moving in the following frames only a warning to the user is issued. This can be due to pixel errors or mis-detection of the object block. Consecutive larger changes of the detected centre will trigger an error to the user or send the abort signal to the printer as per configuration.

Figure 5. Extracted blob from blue object on printing-bed.

Figure 6. Blue object on printing-bed.

3.4.4 Algorithm for missing material flow detection

For the detection of missing material flow from the extruder the algorithm for the blob detection [see 3.4.3 Blob Detection] is used and extended (see fig. 7). The upper border of the object bounding box is used as a reference line. This object reference line is compared to the reference line for the print-head lower boundary (i.e. nozzle tip, see fig. 8) calculated by the image pre-processor [3.4.1 Pre-Processing].

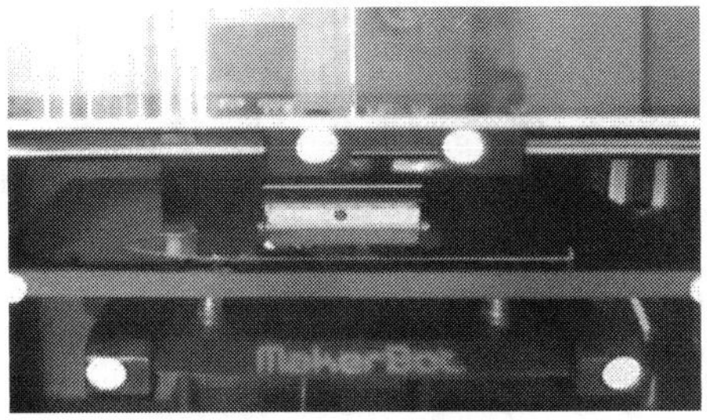

Figure 7. Missing filament detection with nozzle reference line and object bounding box.

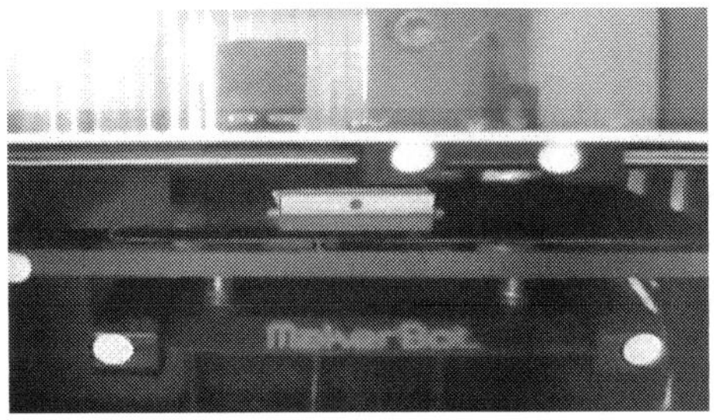

Figure 8. Blue object marked with bounding box.

Our test system has a height fixed print-head and is only moving in 2 dimensions (X and Y). We position the camera frontal to the print-head so that changes in the Y direction do not alter the apparent height in the video frame. When the camera is position at an angle to the print-head, the apparent height can change and this needs to be tracked utilizing the visual markers on the print-head. In the error free case the objects upper reference line and the print-head reference line are congruent. If the material flow stops the object moves down along the Z Axis with the print-bed and the print-head continues its operation. The object upper reference line and the print-head reference line diverge. We compare these reference lines in consecutive frames and issue a warning when the difference between them grows between a defined threshold $Thresh_{Flow1}$. In case the difference continues to grow over a second Threshold $Thresh_{Flow2}$ we issue an error to the user and abort the print as per configuration.

4 Results

For validation of our approach we devise a test with two different test objects. Test object A (see fig. 9) is a block of 39x20x5 mm printed on a stand of 1x25x10 mm.

It is printed in order to detect missing material flow when supply of filament is cut during print. Test object B (see fig. 10) is a triangular block for detection of deformed object failures and detachment. The standing area is chosen to be very small so that the object can topple over during print. Both test objects are printed 10 times each and the detection of the failures is observed with the interface. Test run A with test object A is executed with 5 objects printed without cutting the filament flow in order to ensure that no false positive detection occurs and with 5 objects printed with the filament flow cut off mid-print in order to ensure positive recognition of the failure detection.

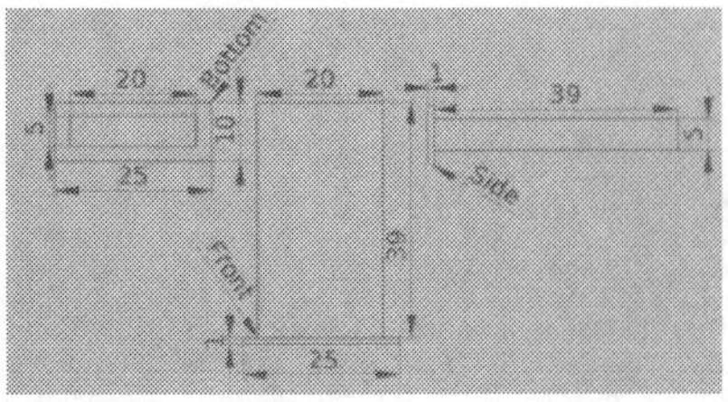

Figure 9. Test Object A.

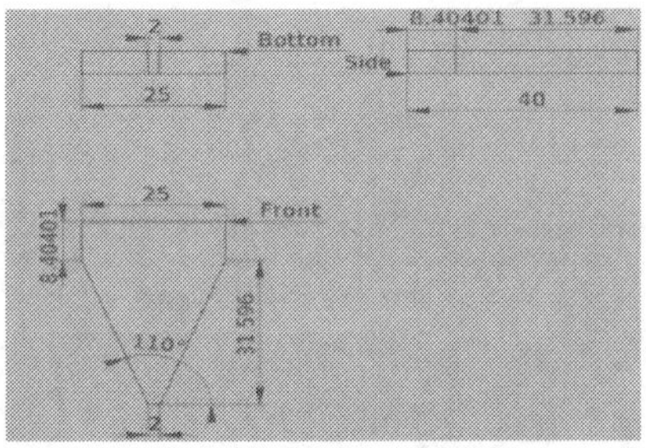

Figure 10. Test Object B.

Test run B with test object B is executed 10 times undisturbed with the object toppling over randomly 6 times. In test run A the undisturbed prints yield a number of warnings during the print due to temporary detection problems and one mis-detected print failure. This results in a false detection rate of one out of five (20% false positive for material flow detection). Test objects with cut-off filament are detected earliest with an update to the displayed frame (ca. 5 seconds) and one print not detected as failure to do mis-detection of the extruder position. This results in four out of five objects correctly classified as failed (80% detection rate for material flow failures). For test run B six objects (see fig. 11) out of ten toppled over. Four objects are correctly classified as failed resulting in 60% detection rate for object detachment. The remaining prints again are classified with a number of warnings and two objects are temporarily marked as failed. The temporary failures detected are corrected within a time span of 30 to 90 seconds. Lighting changes can be identified as a source of increased detection errors. For detachment detection the detection rate is 60% in our experiment. The algorithm implemented is reasonably stable but is to be improved for resilience against lighting changes, marker mis-detection.

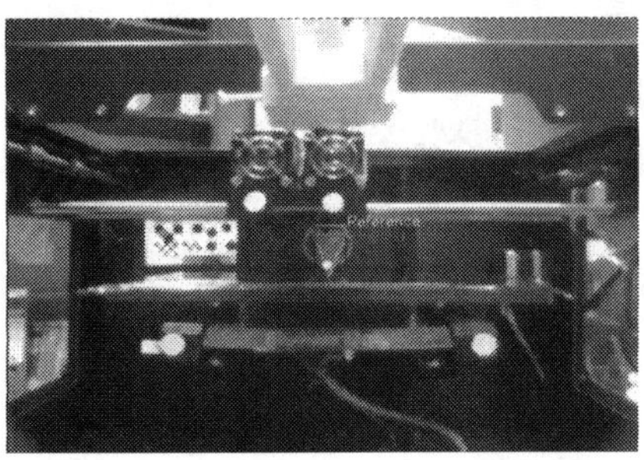

Figure 11. Test object b during print with offset extruder detection line.

5 Discussion

One obstacle in stable in stable error detection in our approach is pixel errors from the camera that can be compensated for by software interpolation and smoothing of the original video frame or to some extend with higher quality video cameras. Smoothing out the image feed consumes additional computational resources. Our system is capable of analysing a video stream with a resolution of 640x480 pixel and a framerate of 25 frames per second with an CPU utilization rate of about 20 percent on an Intel Pentium M 1.5 GHz computer. For display in the web interface the Portable Network Graphics (PNG) format is chosen for compatibility reasons. When this interface is optimized for a specific browser more efficient display methods for example WEBM can be implemented, this will increase the refresh rate of the displayed stream in the web interface which is approximately five seconds with the current setup. The second obstacle is change in lighting for the object. This can be alleviated by a fixed lighting system for the 3d printer. This fixed lighting system is set to be placed at multiple angles in front of by the side of the 3D printer in order to avoid the casting of shadows. One difficulty we have encountered with additional lighting is that reflections can destabilize our algorithm as some surfaces of the 3D printer are highly reflective. Increasing the video resolution and framerate can yield better images suitable for detection of errors currently not detectable as surface problems. This will also increase the computational effort. For our test system an increase of resolution and framerate was found to be not necessary as the error classes we want to detect are detected with a high degree of certainty. Our approach is not suited for flat objects because of the camera position. It is also not suitable for materials that are of similar colour as the 3D printer as the difference in colour is needed for the thresholding and identification of the object. Strategies on how to handle detected failures (e.g. issue abort command to printer) and when to handle detected failures (in order to prevent action on false positives) are to be devised and are dependent upon the scenario the printer is operated in. Those strategies can be implemented in our prototype as external scripts. With adaptions our prototype can be used in scenarios for cloud based manufacturing where remote supervision is to be provided to the users.

6 Conclusion and future work

With this work we propose an algorithmic implementation of error detection based on video camera data for FDM 3D printers. From the five classes of proposed error classification we are able to detect three. We show the validity of our approach by an experiment that shows the algorithm is capable of a 60 to 80 percent detection rate for failure detection with a false positive detection rate of 60 to 80 percent. Improvements described in [5 Discussion] are expected to bring the rate of false positives down and the detection rate up. For future work the implementation of additional error detection algorithms and stabilization and improvement of implemented algorithms is advised. In following prototypes the interface and algorithms are to be implemented in an integrated interface for cloud based manufacturing. For future research we suggest implementing and extending our basic approach into multi camera systems and stereo vision camera systems. This will improve the object detection capability as there is better understanding for the scene geometry and therefore for object detection in the depth dimension. Furthermore we suggest looking into IR cameras for this setup as this can reduce glaring effects and reflection with appropriate lighting.

Acknowledgement

We would like to thank Daniel Nägele, Lan Jiang and Robert Gänzle for their work in this project.

References

1. F. Klocke and W. König, Fertigungsverfahren 3: Abtragen, Generieren und Lasermaterialbearbeitung, Springer, ISBN 9783540489542 (2007)
2. T. Fang, M. A. Jafari, S. C. Danforth, A. Safari, Signature analysis and defect detection in layered manufacturing of ceramic sensors and actuators, *Machine Vision and Applications*, **Vol. 15 Iss. 2**, DOI: 10.1007/s00138-002-0074-1 (2013)
3. T. Pham and S. S. A. Dimov, Rapid Manufacturing: The Technologies and Applications of Rapid Prototyping and Rapid Tooling, Springer, ISBN 9781852333607 (2001)
4. C. Bayley, L. Bochmann, C. Hurlbut, M. Helu, R. Transchel and D. Dornfeld, Understanding Error Generation in Fused Deposition Modeling, *SPRING TOPICAL MEETING* **Vol. 57** pp. 98-103 (2014)
5. R. Anitha, S. Arunachalam and P. Radhakrishnan, Critical parameters influencing the quality of prototypes in fused deposition modelling, *Journal of Materials Processing Technology* **Vol. 118** pp. 385-388 (2001)
6. T. Craeghs, F. Bechmann, S. Berumen, J.-P. Kruth, Feedback control of Layerwise Laser Melting using optical sensors, *Physics Procedia* **Vol. 5 Part B**, pp. 505–514, DOI: 10.1016/j.phpro.2010.08.078 (2010)
7. J. J. Ballyns, D. L. Cohen, E. Malone, S. A. Maher, H. G. Potter, T. Wright, H. Lipson and L. J. Bonassar, An Optical Method for Evaluation of Geometric Fidelity for Anatomically Shaped Tissue-Engineered Constructs, *TISSUE ENGINEERING: Part C* **Vol. 16 No. 4** pp. 693-703, DOI: 10.1089/ten.tec.2009.0441. (2010)
8. C. Kim, D. Espalin, A. Cuaron, M. A. Perez, E. MacDonald and R. B. Wicker, A Study to Detect a Material Deposition Status in Fused Deposition Modeling Technology, *IEEE Int. Conf. On Adv. Int. Mechatronics AIM*, pp. 779-783 (2015)
9. H. Krauss, T. Zeugner and M. F. Zaeh, Layerwise Monitoring of the Selective Laser Melting Process by Thermography, *Physics Procedia*, **Vol. 56**, pp. 64-71, DOI: 10.1016/j.phpro.2014.08.097 (2014)
10. S. Kleszczynski, J. zur Jacobsmühlen, J. T. Sehrt and G. Witt, ERROR DETECTION IN LASER BEAM MELTING SYSTEMS BY HIGH RESOLUTION IMAGING, *Proceedings of the Solid Freeform Fabrication Symposium*, URL: http://www.lfb.rwth-aachen.de/files/publications/2012/JAC12a.pdf (2012)
11. B. Pettis, A. K. France and J. Shergill, Getting Started with MakerBot, Published by O'Reilly, ISBN 9781449338657 (2013)
12. T.-M. Wang, J.-T. Xi and Y. Jin, A model research for prototype warp deformation in the FDM process, *The International Journal of Advanced Manufacturing Technology*, **Vol. 33, Iss. 11**, pp. 1087-1096 (2007), DOI: 10.1007/s00170-006-0556-9
13. H. Wu, Y. Wang and Z. Yu, In situ monitoring of FDM machine condition via acoustic emission, *International Journal of Advanced Manufacturing Technology*, pp. 1-13 (2015), DOI: 10.1007/s00170-015-7809-4
14. N. Volpato, J. A. Foggiatto, D. C. Schwarz, The influence of support base on FDM accuracy in Z, *Rapid Prototyping Journal*, **Vol. 20 Iss 3** pp. 182-191, (2014), DOI: 10.1108/RPJ-12-2012-0116
15. P. Chennakesava and Y. S. Narayan, Fused Deposition Modeling - Insights, *International Conference on Advances in Design and Manufacturing (ICAD&M'14)*, ISBN 978-93-84743-12-3, (2014)
16. S. P. Moylan, J. A. Slotwinski, A. L. Cooke, K. K. Jurrens and M. A. Donmez, Proposal for a standardized test artifact for additive manufacturing machines and processes, *Proceedings of the 23rd International Solid Free Form Symposium - An Additive Manufacturing Conference*, pp. 902-920, (2012)
17. G. Bradski and A. Kaehler, Learning OpenCV: Computer Vision with the OpenCV Library, O'Reilly Media, ISBN 9780596554040 (2008)
18. H. K. Yuen, J. Princen, J. Illingworth and J. Kittler, Comparative study of Hough transform methods for circle finding, *Image and Vision Computing* **Vol. 8** pp. 71-77 (1990)
19. B. Jähne, Digitale Bildverarbeitung, Springer, ISBN 9783540249993 (2005)

MATEC Web of Conferences **59**, 06004 (2016) DOI: 10.1051/matecconf/20165906004
ICFST 2016

Investigating the Role of Wind Turbine Pitch using CFD

Dorit Sobotta[1], Robert Howell [2] and Lou Jing [3]

[1]*German Aerospace Centre, Rotorcraft, Institute of Flight Systems, Lilienthalplatz 7, 38108 Brunswick, Germany*
[2]*Department of Mechanical Engineering, The University of Sheffield, Western Bank, Sheffield, S10 2TN, UK*
[3]*A*STAR, Institute of High Performance Computing, Fusionopolis, 1 Fusionopolis Way, #16-16 Connixes, 138632 Singapore*

Abstract. Horizontal axis wind turbines are an attractive renewable energy source due to their very low carbon emission during their life cycle. In this study the effect of pitching the rotor blades of the NREL Phase VI rotor has been investigated in detail using CFD in order to allow a detailed torque analysis. Initial investigations have shown that at low rotational speeds the inboard section of the blade is responsible for the majority of the power generation. As the rotational speed increases the power producing section is shifted towards the blade tip. These trends are less pronounced when the blade is pitched which allows the blade to generate significantly more power at low rotational speeds. The improved low speed performance however comes at the cost of a significantly lower maximum power output. These findings are particularly relevant for turbines operating in unfavourable wind environments and for small scale turbines which solely rely on their aerodynamic torque for starting.

1 Introduction

Wind turbines have experienced a significant increase in their number over the last years as they have a very low carbon footprint of only 4.64 g of CO_2 equivalent per kWh over their life cycle, most of which occur during their manufacturing and construction [1]. A mechanism common to all of today's large scale rotors is their ability to adjust their blade pitch based on their current wind environment in order to maximise their power output. Several studies, such as those of the WKA-60 rotor [2], have been conducted to investigate the effect of blade pitch. These studies typically only address their torque and power characteristics but not the underlying physical phenomenon, although fully understanding relevant characteristics could aid in the design process of blades that perform well over a wide range of different wind environments. This could not only help designing more efficient large scale rotors but also more efficient small scale rotors as they often operate in unfavourable wind environments and solely rely on their aerodynamic torque for rotor starting.

It is the aim of this study to investigate these short-comings by analysing the effect of blade pitch which modifies the blade surface inclination with respect to the rotational plane. This in turn influences the pressure-to-torque translation and hence the rotor's power output. The modified surface inclination also causes a change in power output by provoking a different pressure field around the blade. This is also reflected in a different near blade flow field. In order to quantify these effects, a Computational Fluid Dynamics (CFD) analysis has been conducted using the geometry of the Phase VI rotor from the National Renewable Energy Laboratory (NREL) as a reference rotor [3]. Due to the availability of extensive experimental data, the two-bladed upwind rotor with a radius of 5.029 m shown in Figure 1 also served for validating the CFD model.

Figure 1. Experimental NREL Phase VI Turbine in NASA's Ames Research Centre [3].

2 Computational Methods

In order to investigate the flow around the turbine using CFD, suitable computational methods have to be established first. This section is concerned with the evaluation of suitable turbulence models and meshes which arise due to the closure of the Reynolds Averaged Navier Stokes (RANS) equations and the spatial discretisation of the flow field respectively. In this study the CFD solver Ansys Fluent versions 12 [4] and 14 along with the meshing software Pointwise Gridgen version 15 [5] has been used.

2.1. Turbulence Model Selection

For computational efficiency only one and two equation turbulence models have been considered for this study. The one equation Spalart-Allmaras (S-A) model and selected models from the two equation k-ε and k-ω turbulence model families have been validated against experimental aerofoil data from the NACA 65_4-421 [6]. For the simulation a high resolution 2D O-mesh has been used which extends for 20 chord lengths radially outwards. The aerofoil has 320 nodes on its surface which are clustered towards its leading and trailing edge.

The first cell height is such that it satisfies the corresponding turbulence models near wall requirement and extends with a growth rate of 1.06. Table 1 summarises the simulated error of the aerodynamic coefficient. Based on the findings the k-ω SST model has been chosen for all further investigations due to its comparatively little overall error of the lift, drag and moment coefficient c_L, c_D and c_M respectively. This model has often been used for turbines [7].

Table 1. Percentage Error of Investigated Turbulence Models.

	S-A	k-ε		k-ω	
		Stand.	RNG	Stand.	SST
c_L	-41	-8	-5	-14	-22
c_D	+131	+82	+41	+57	+20
c_M	-25	+30	+33	+21	+6

2.2 Mesh Selection

After the identification of a suitable turbulence model, a suitable turbine mesh that allows fast, yet accurate computation has been established. A variety of different mesh types including fully structured, hybrid and un-structured have been used in literature. For the current study a hybrid mesh has been adopted which consists of a fully structured high resolution boundary layer mesh, an unstructured transition region and a structured far-field block as outlined in Figure 2. The graph also outlines the domain extent and the boundary conditions.

Table 2. Cell Count x10^6 of Mesh Depicted in Figure 2.

Mesh	Total	Boundary Layer
Very Coarse	4.2	0.6
Coarse	5.2	1.3
Medium	6.2	2.6
Fine	12.6	6.9

The outlined mesh has been used with meshes of 4.2 to 12.6 million cells as displayed in Table 2. While the rotational speed of the two bladed rotor was kept constant at 72 rpm, while the wind speed was adjusted to match the blade tip to wind speed ratio λ, shown in Figure 3. The resulting distribution of the pressure coefficient c_{Pr} indicates that the very coarse mesh shows significant mesh dependence at the inboard and outboard blade sections where the flow around the blade is partially stalled at a λ of 2.5. At lower and higher λ all grids yield mesh independent results which agree reasonably well with experimental data. Consequently the 5.2 million cell mesh has been used for the analysis in this work.

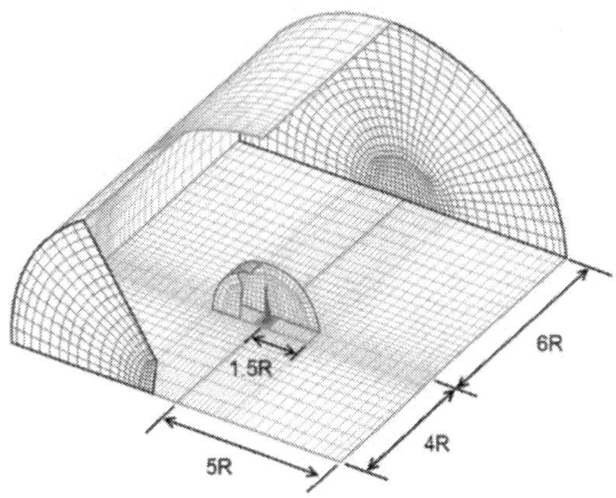

Figure 2. Mesh Topology, Boundary Conditions and Domain Extent.

3 The Source of Rotor Power

In order to gain a better understanding of the source of the power produced by a Horizontal Axis Wind Turbine (HAWT), an analysis of its local torque generation characteristics has been conducted in this section. Figure 4 shows the NREL Phase VI turbine's power and torque coefficients, c_P and c_T, as a function of λ from a stationary rotor. For this and all following studies the wind speed was kept constant at 6 m/s while the rotor speed has been adjusted to simulate 0.5 λ increments.

From Figure 4 it can be seen that the rotor reaches its maximum c_P of 0.29 at a λ of 5.6, while its maximum torque has already been reached at λ = 4.5. In order to further increase its power production beyond this λ, the decline of its torque production is thus compensated by a higher rotational speed. This already indicates a high sensitivity of the torque the rotor produces towards its highest attainable c_P. A succeeding study investigates the source of the torque along the blade radius to identify critical blade regions at different λ. This provides the base to identify the source of differences in rotor power when pitching a blade.

To gain an overview of the radial torque of NREL Phase VI rotor, the spanwise torque distribution has been computed and normalised with the total torque of the respective λ. The resulting graph is depicted in Figure 5. The normalised torque contour has been overlaid with the local Reynolds number, Re, and the local geometric angle

 DOI: 10.1051/matecconf/20165906004

of attack, α, which is the angle between the local chord line and the resulting velocity vector from the wind speed and rotational velocity. The lines therefore give an indication of the expected local aerodynamic performance; however, their accuracy decreases with increasing λ as the induced velocities have not been accounted for. Furthermore, the stripy appearance of the torque contour is due to the underlying mathematical method and does not have any physical meaning.

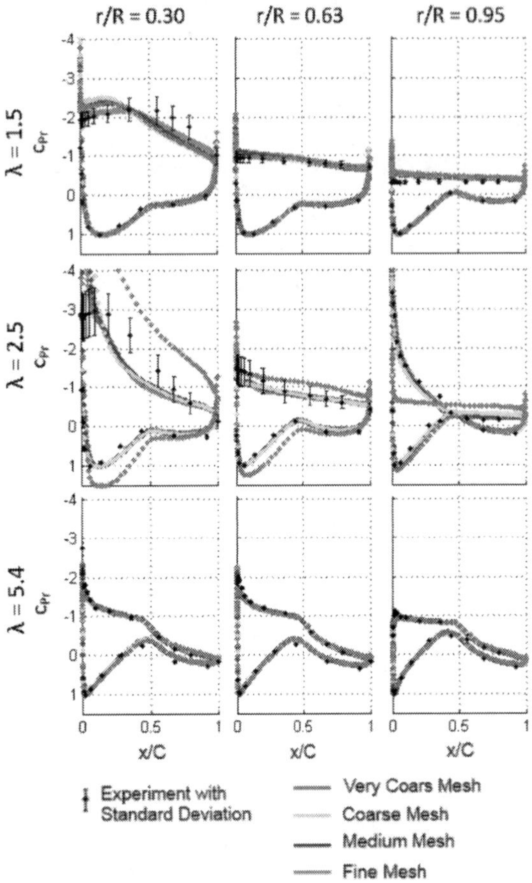

Figure 3. c_{Pr} Plots of the Mesh Independence Study of the NREL Phase VI Rotor.

It can be seen that at low λ the torque production is predominantly determined by the large blade chord near the blade's root. When Re is nearly uniform along the blade span at $\lambda \approx 2$ the torque production peaks at the blade root. The angle of attack only plays a secondary role. As λ increases, the angle of attack at the blade tip drops below that of the blade root. When α further drops below approximately 22°, the torque production at the blade tip begins to overtake that at the blade root and Re plays a comparatively insignificant role. Based on the current analysis three distinct λ sections have been identified:

- $0.0 \leq \lambda < 2.0$: performance dominated by Re
- $2.0 \leq \lambda < 4.3$: performance dominated by α
- $4.3 \leq \lambda \leq 5.5$: decreasing c_T with increasing λ

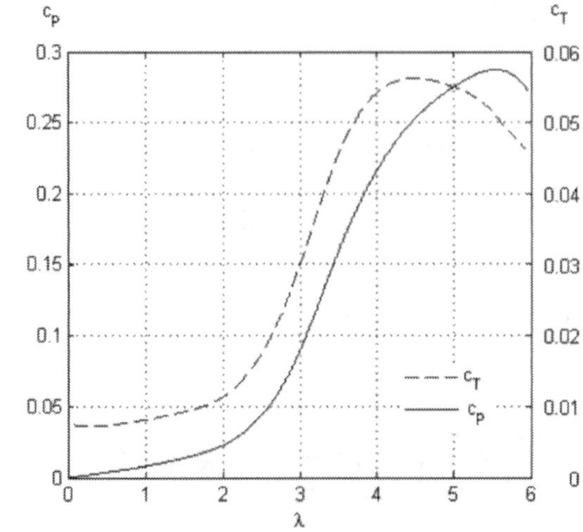

Figure 4. Power and Torque Curve, NREL Phase VI Rotor.

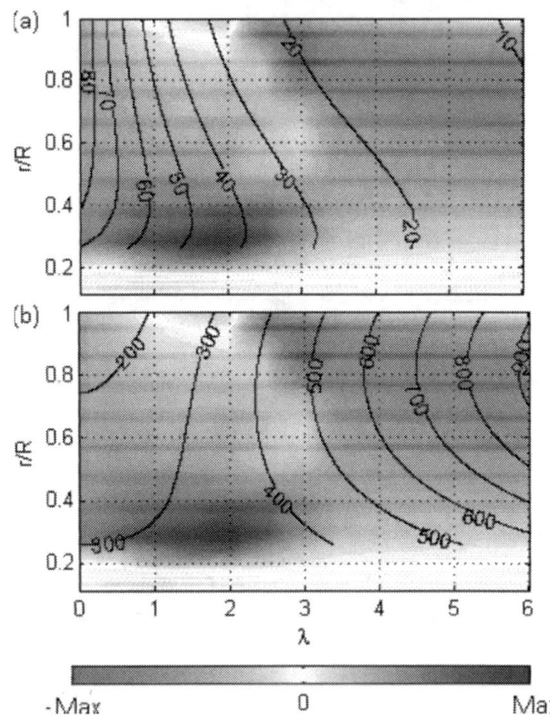

Figure 5. Normalised Radial Torque Distribution of Phase VI Rotor at 0° Pitch vs. λ, Superimposed with Geometrical (a) α and (b) Re x 10^3.

When analysing wind turbines, static pressure contours are often presented. These, however, do not allow any direct conclusions about the overall blade performance to be made. In this study the importance of the surface orientation and the blade pitch on translating a given pressure into torque acting in the rotational plane has been highlighted. A given pressure may act to produce useful torque or try to slow the blade down depending on where it acts, for instance near the leading edge or trailing edge of the blade. For demonstration purposes the analysis illustrated in Figure 6 has been conducted using a stationary blade as this produces almost exclusively positive and negative static pressure on the pressure and suction surface respectively. Despite each blade surface experiencing no static pressure reversal, the resulting torque distribution shows a change in sign on each surface. This is caused by the orientation of the local blade surface orientation with respect to the rotational plane. The alignment of these two lines is referred to as the zero-inclination-line. When a given static pressure passes this line a change in the direction of the torque produced is thus induced.

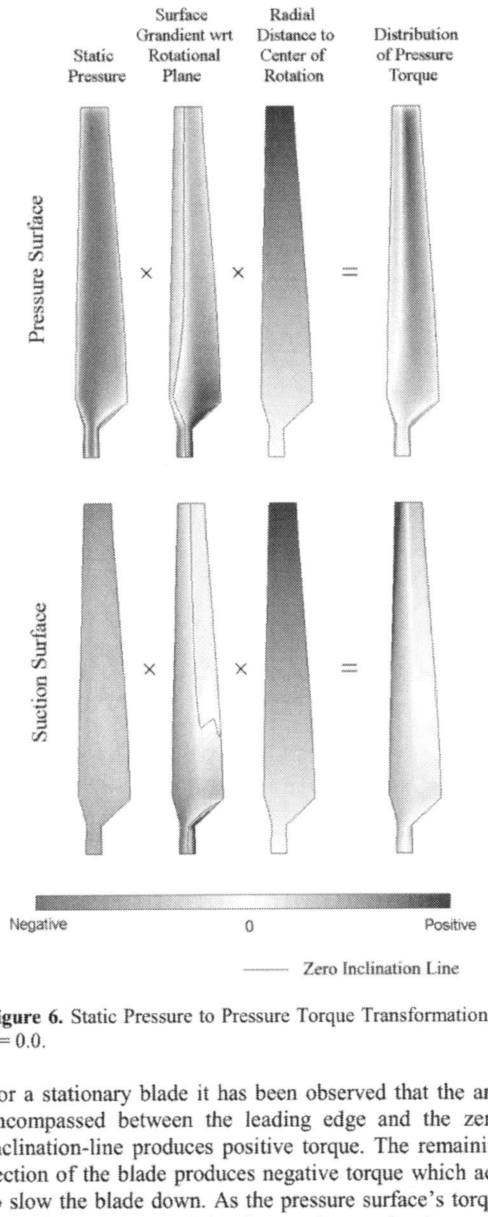

Figure 6. Static Pressure to Pressure Torque Transformation at $\lambda = 0.0$.

For a stationary blade it has been observed that the area encompassed between the leading edge and the zero-inclination-line produces positive torque. The remaining section of the blade produces negative torque which acts to slow the blade down. As the pressure surface's torque favouring area outweighs the suction surface's torque retarding area, 63% of the total torque is produced by the pressure when the blade is stationary.

4 Effect of Blade Pitch

After having established relevant criteria influencing the performance of the NREL Phase VI rotor, the blade pitch has been analysed by pitching the NREL Phase VI rotor by 10° into the wind, thus reducing the angle of attack. This relatively large pitch has been chosen to clearly demonstrate the effect of pitching in 2 sections.

4.1. Blade Shape

The first section of the analysis investigates the change of the pressure-to-torque translation properties of the Phase VI rotor when applying pitch. Figure 7 shows the surface orientation of the pitched blade which may be compared to Figure 6. For an easier comparison the line of where the surface orientation of the unpitched rotor is aligned with the rotational plane is also included.

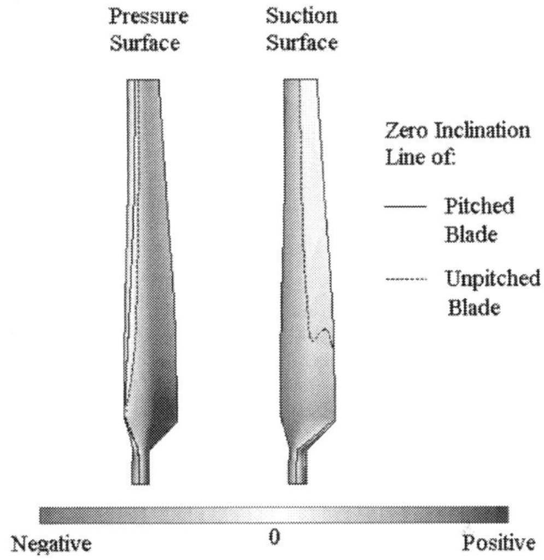

Figure 7. Differences in Blade Surface Orientation of Pitched and Un-Pitched Blade.

Figure 7 indicates that the zero-inclination-line on the suction surface is completely shifted towards the leading edge. Consequently, when the suction surface experiences a negative static pressure, more torque is generated due to a more favourable pressure-to-torque translation. The pressure surface also experiences an increase in its torque production for a given positive pressure distribution as the zero pressure surface line has also been shifted towards the leading edge.

4.2. Torque Distribution

The combined effect of the modified blade surface orientation on each blade surface of the rotor torque is illustrated in Figure 8. As expected, the pitched rotor reaches a lower maximum power at a lower λ than the reference turbine. Its torque generation at low λ, however, is superior to that of the unpitched turbine. When the blade is stationary the pitched rotor produces 1.8 times the torque of the unpitched blade. Here the suction surface particularly benefits from the better pressure-to-

torque translation shown in Figure 7 as it produces 2.2 times the c_T of the reference blade.

The rapid increase of c_T with increasing rotational speed from $\lambda = 2.3$ onwards for the Phase VI rotor is now shifted towards 1.4 which allows the turbine to perform well in the low λ regime. The following torque peak which occurs at $\lambda = 2.8$ is reduced by 22% when pitching the rotor by 10°. Thereafter, the pitched rotor reaches its optimum performance of only $c_T = 0.127$ at $\lambda = 3.2$.

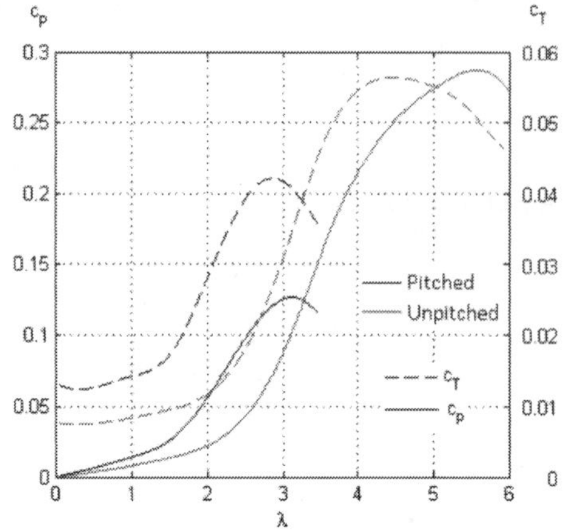

Figure 8. Power and Torque Curve of the Pitched NREL Phase VI Rotor.

A more detailed analysis of the radial dependence of the torque is shown in Figure 9. In contrast to the unpitched blade, the pitched blade shows a very weak Re dependence at low λ, thus also allowing the outboard section to produce useful torque. The relatively evenly distributed radial torque causes the following increase in the torque production at the blade tip to be much less pronounced than for the unpitched blade. It has also been noted that the pitched blade experiences its c_T increase only when α drops below approximately 20°. As this occurs at significantly lower λ, the pitched turbine reaches its optimum rotational speed sooner than the unpitched rotor and thus delivers a lower power output.

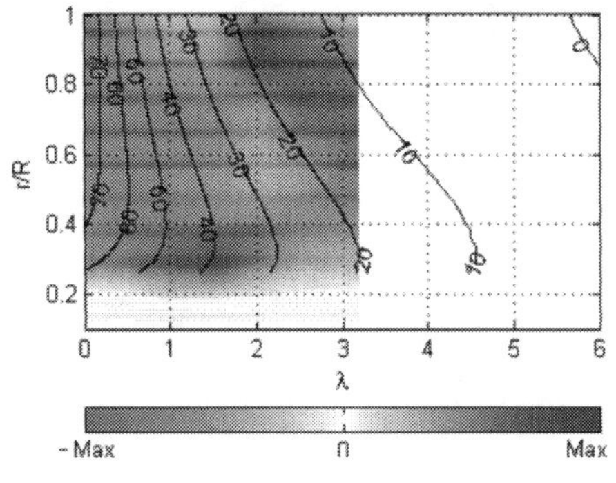

Figure 9. Normalised Radial Torque Distribution of Phase VI Rotor at 10° Pitch vs. λ, Superimposed with Geometrical α.

The observed trends make the pitch of a blade of a HAWT a suitable mechanism for adjusting to low or unsteady wind environments. Particularly smaller rotors which tend to be placed where they are needed rather than where the wind is best may benefit from pitched blades. When they accelerate from rest they solely rely on their aerodynamic torque as they are usually not fitted with a motor to assist turbine starting for economic reasons.

5 Conclusions

It has been identified that at very low λ the rotor power is dominated by the local Reynolds number and thus its local chord length. This effect has been observed to diminish when the blade is pitched. As λ increases above 2 the resulting torque rapidly increases as α drops below 22°. For the unpitched blade the increase in torque continues until $\lambda = 4.3$. Thereafter the gain in the power output is solely due to an increase in rotational speed.

Differences in the torque produced when pitching a blade have been attributed to different pressure-to-torque translation characteristics which are determined by the blade's local surface orientation. The effect of the surface orientation, however, is twofold as it also affects the static pressure distribution around the blade. Pitching a blade causes it to reach its optimum local angle of attack sooner and thus prohibits the blade from reaching high rotational speeds. This thereby reduces a pitched blade's maximum power output. Before reaching its maximum power however, the pitched blade consistently delivers a higher power output than the unpitched blade which makes it suitable for low wind speed environments.

References

1. N.N., "Carbon Footprint of Electricity Generation", Parliamentary Office of Science and Technology, Number 268, London, UK, 2006.
2. E. Hau, *Windkraftanlagen - Grundlagen, Technik, Einsatz, Wirtschaftlichkeit*, 4th edition: Springer, Berlin Heidelberg, Germany, 2008.
3. M.M. Hand, *et al.*, "Unsteady Aerodynamics Experiment Phase VI: Wind Tunnel Test Configurations and Available Data Campaigns", NREL/TP-500-29955, NREL, Colorado, 2001.
4. N.N., *User's Guide*, Ansys Fluent 12.0, Canonsburg, Pennsylvania, 2009.
5. N.N., *Version 15 User Manual*, Pointwise Gridgen, Forth Worth, Texas, 2006.
6. P. Devinant, *et al.*, "Experimental study of wind-turbine airfoil aerodynamics in high turbulence", *Journal of Wind Engineering and Industrial Aerodynamics*, vol. **90**, pp. 689-707, 2002.
7. Y.M. Park, *et al.*, "Numerical simulation of wind turbine scale effects by using CFD", in: *45th AIAA Aerospace Sciences Meeting and Exhibit*, Reno, Nevada, 2007, pp. 2547-2556.

Development of fast scattering model of complex shape target for seminatural tests of onboard proximity radars in real time mode

Andrei K. Likhoedenko[1], Victor B. Suchkov[2]

[1]*Higher School of Economics, Department of applied mathematics, 101000, ul. Myasnitskaya, 20, Moscow, Russia*

[2]*Bauman Moscow State Technical University, Department of autonomous information and control systems, 105005, ul. Baumanskaya 2-ya, 5/1, Moscow, Russia*

Abstract. Problems of creation of models of real time of complex shape targets on the basis of use of their polygonal models are considered. Formulas for radar cross section of multipoint model of target and power of input signal of onboard radar are described. Technique of semi-natural tests of onboard radar detector on the base of multipoint model of target is proposed. Results of digital simulation of input signals of the onboard radar detector of the target from the aerodynamic target on the basis of their multipoint models are given.

1 Introduction

For simulation and efficiency estimation of onboard proximity radar detectors of targets semi-natural tests are widespread amongst researchers and institutes [1-2]. Semi-natural tests of proximity inboard radars can be implemented on the base of hardware stand with channel of target signal imitation. In case of semi-natural tests for onboard radar detectors imitation of target signal is carried out by calculation of scattering electromagnetic fields from complex shape targets in near zone of radar-location [3]. Such type of calculation could be developed on the base of exact electrodynamic methods such as method of moments (MoM), Multilevel Fast Multipole Method (MLFMM).

But in case of electrically large size target it would be possible to apply Physical Optics (PO) [3] with Physical Theory of Diffraction (PTD) on the base of polygonal model of target [4]. But the polygonal model is practically unsuitable for fast estimation of the operation of onboard radar during its tests at semi-natural stand because of huge computer resources of calculation of targets with large number of polygonal elements even with use of parallel algorithms [5]. So it is necessary to develop fast scattering model of target for operative semi-natural tests and efficiency estimation.

2 Technique of creation of fast scattering model of the target

For creation of fast scattering model of the target we can use its polygonal model. As one of advanced CAD system for creation of complex shape target we can use PTC Creo Parametric. It is possible to create target with high level of complexity of target surface with taking into account of edges in Creo Parametric. Some examples of polygonal models of targets created in Creo Parametric are shown at Figure 1 and Figure 2.

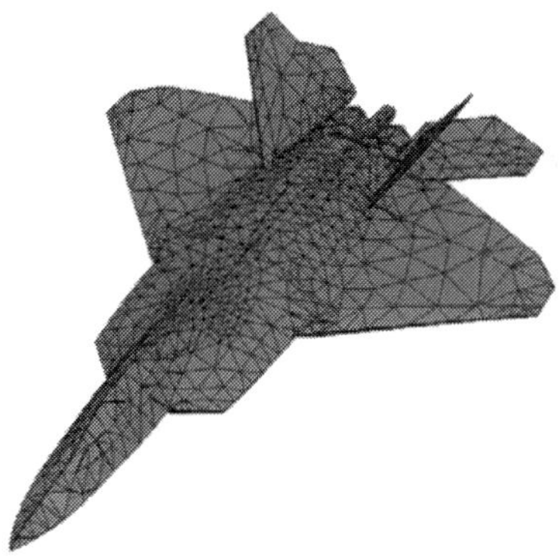

Figure 1. Polygonal model of fighter F-22

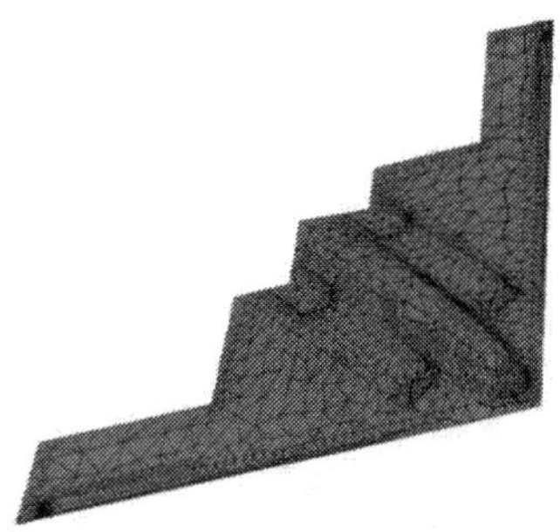

Figure 2. Polygonal model of bomber B-2

Resulting intensity of the scattered electromagnetic field from complex shape target on the base of its polygonal model:

$$\dot{E}_{scat\,\Sigma} = \sum_{m=0}^{N_\Sigma-1} \dot{E}_{sm} = \sqrt{\frac{P_i D_i Z_0}{2\pi}} \sum_{m=0}^{N_\Sigma-1} F_{im}\dot{\rho}_{sm} \frac{\exp(-\mathrm{i}\,kR_{im})}{R_{im}R_{sm}}, \quad (1)$$

where P_i is the power of the signal radiated by the transmitting system, W; D_i is the directive gain of the transmitting antenna in the direction of the maximum radiation; $Z0=120\pi$ is the wave impedance of air, Ω; $R_{i(s)m}$ is the distance between the phase center of the antenna and reflector, m; $F_{i(s)}$ is the value of the directivity function of the transmitting antenna in the direction toward the reflector; $\dot{\rho}^s_{im}$ is the complex reflection coefficient of the reflector, m.

For creation of fast scattering model of the target we can use multipoint target which can be formed on the base of calculation of scattering fields from local parts of polygonal model. Thus polygonal model of the target is divided in different cubes. Each cube limits specified part of target surface (Figure 3).

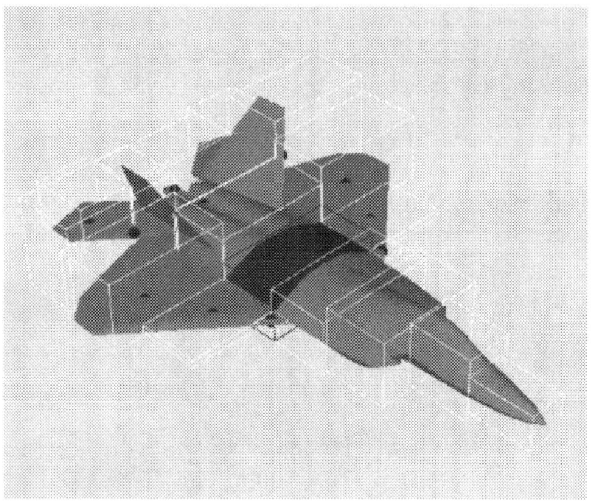

Figure 3. Formation of multipoint model of fighter F-22 on the base of local cubes

Description of elementary reflectors of polygonal model is based with use of local points $P_n = \{X_n \;\; Y_n \;\; Z_n\}$ and triangles $T_k = \{P_{1k} \;\; P_{2k} \;\; P_{3k} \;\; N_k\}$ (Figure 1). For formation of the multipoint model array of the local cubes Ψ_k of specified sizes have to be created. Each cube limits specified part of target. Furthermore, we can form array of triangles of polygonal model T_{kj} which specified cube includes.

As a result, for each m-th cube there will correspond the "brilliant" point. Thus multipoint model of the target consists of array of "brilliant" points each of them will correspond surface equivalent of complex target:

$$P_m = \{X_m \;\; Y_m \;\; Z_m\}, \; m = 1..N_m,$$

where m – number of brilliant point, $N_m = N_{cub}$ - number of brilliant points (cubes).

For each cube of target (brilliant point) diagram of radar cross section (RCS) have to be calculated in the local coordinate system of brilliant point X_m, Y_m, Z_m (Figure 4). Calculation of RCS could be carried out in azimuth and angles of vision range $\alpha_0 = 0, \alpha_{max} = 360^0$, $\beta_0 = 0, \beta_{max} = 360^0$, in the local coordinate system of brilliant point for specified distance to radar detector. RCS of target have to be calculated by PO and PTD methods on the base of formula (1) for scattering field of target on the base of its polygonal model. RCS of the local part of the target could be described on the base of known formula:

$$\sigma^m(R,\alpha,\beta) = 4\pi R^2 \frac{\left|\dot{E}_s^{(m)}(R,\alpha,\beta)\right|^2}{\left|\dot{E}_i^{(m)}(R,\alpha,\beta)\right|^2}. \quad (2)$$

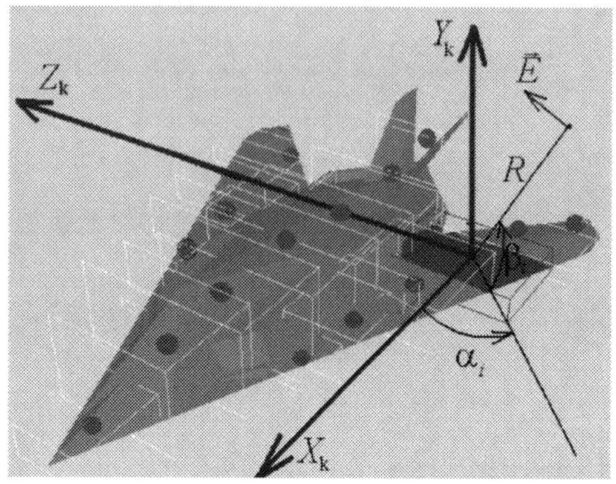

Figure 4. Formation of local coordinate system of local cube

As a result, for each brilliant point of the target array of RCS in local coordinate system $\sigma_{i,j} = \sigma(\alpha_i, \beta_j)$ will be formed. Multipoint model of fighter F-22 with surface diagram of RCS one of point on the distance R=5 m from radar detector is shown at Figure 5.

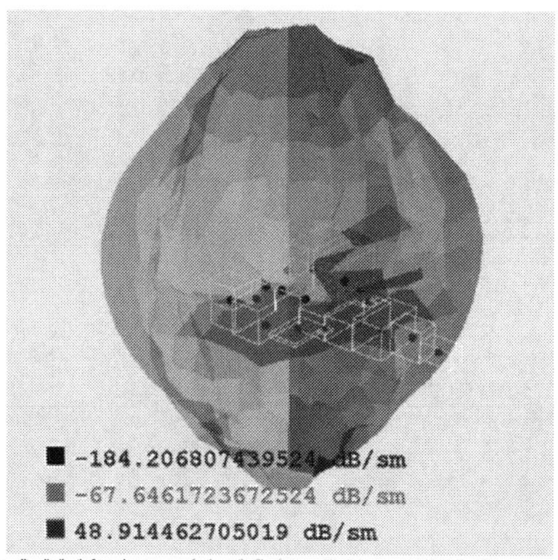

Figure 5. Multipoint model of fighter F-22 and RCS diagram one of its point

Each RCS array have to be saved in text file for specified brilliant point. So number of text files with RCS diagrams of brilliant points are the source data for calculation input signal of onboard radar detector at semi-natural tests.

 DOI: 10.1051/matecconf/20165907001

3 Technique of semi-natural tests of onboard radar detector on the base of multipoint target

For calculation of input signals of onboard radar detector from target we have to describe the scheme of mutual position of onboard radar and the target (Figure 6).

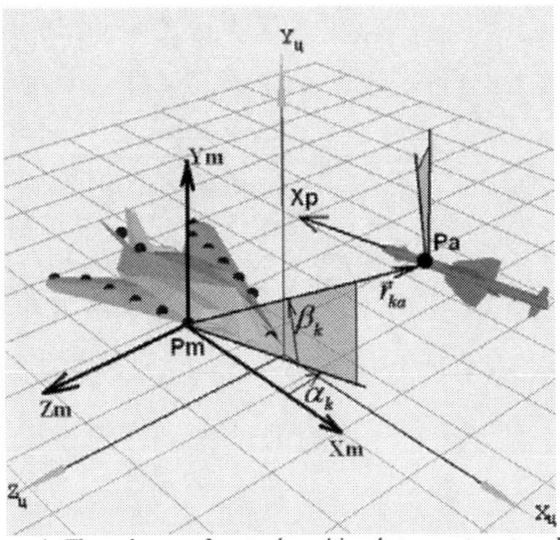

Figure 6. The scheme of mutual position between target and onboard radar

Trajectory of target in coordinate system of onboard radar is direct line from start point of target $P_{t0} = \{x_{t0}, y_{t0}, z_{t0}\}$ to last point of target $P_{tk} = \{x_{tk}, y_{tk}, z_{tk}\}$. For each position of the target we can calculate the power of input signal of target with use of arrays of RCS of each brilliant point. So power of the input signal from target for specified position of target on trajectory of its movement could be obtained by formula:

$$P_{refl.} = \frac{P_i D_i D_s \lambda^2}{(4\pi)^2} \left| \sum_{k=1}^{N_m} \frac{F_{ik} F_{sk} \dot{\rho}_m}{R_m^2} \right|^2, \tag{3}$$

$$\dot{\rho}_m = \sqrt{\frac{\sigma_m}{4\pi}} \exp\left(-i2\frac{2\pi}{\lambda} R_m\right), \tag{4}$$

where P_{refl} is the power of the signal reflected from target, W; D_s is the directive gain of the receiving antenna in the direction of the maximum radiation; R_m is the distance between the phase center of the antenna and brilliant point, m; $F_{i(s)}$ is the value of the directivity function of the transmitting (receiving) antenna in the direction toward the brilliant point; $\dot{\rho}_m$ is the complex reflection coefficient of the brilliant point, m; σ_m is RCS of the brilliant point of target from multipoint model, m^2.

As an example the dependence of power of input signal of onboard radar detector from fighter F-22 at the trajectory of its movement is shown on Figure 7. Dependence of input power was calculated for typical onboard radar detector which operates at frequency f=10 GHz. Moreover, the scheme of mutual position between fighter F-22 and onboard radar with antenna pattern is also shown at Figure 7.

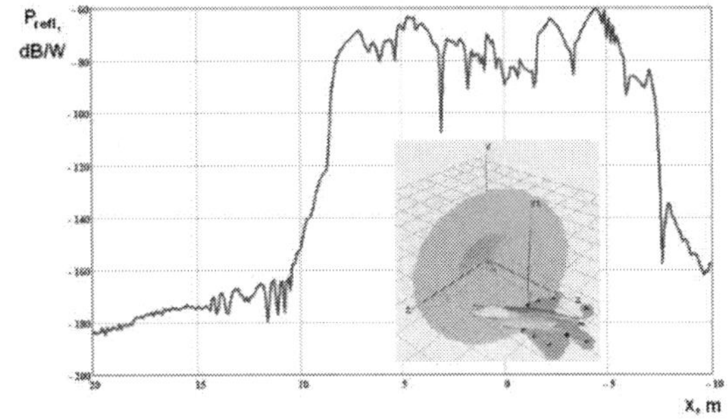

Figure 7. Dependence of power from fighter F-22 from coordinate x

Calculation of power of reflected signal from fighter F-22 on the base of its multipoint model (Figure 7) has taken not more than 1 sec. So multipoint model of complex shape target (like at the Figure 5) could be considered as a fast scattering model of target. Developed multipoint model of the target could be applied in semi-natural tests of onboard proximity radar detector.

For each point of target's multipoint model the following parameters could be defined:

- power of reflected signal from specified m-th brilliant point of target:

$$P_{refl.(m)} = \frac{P_i D_i D_s \lambda^2}{(4\pi)^2} \left| \frac{F_{ik} F_{sk} \dot{\rho}_m}{R_m^2} \right|^2; \tag{5}$$

- time delay of reflected signal from m-th brilliant point:

$$\tau_m = 2R_m / c; \tag{6}$$

where R_m is distance between onboard radar detector and m-th point.

So for each brilliant point of the target attenuation by voltage of generating signal from transmitter of onboard radar detector could be defined as following:

$$\gamma_m = \sqrt{\frac{P_{refl(m)}}{P_i}}, \tag{7}$$

where P_i - amplitude of power from output of generator, W.

On the base of formulas (5-7) input signal of onboard radar detector from target on the base its multipoint model could be defined as:

$$U_{inp}(t) = \sum_{k=1}^{n} u_{gen}(t - \tau_m) \cdot \gamma_m, \tag{8}$$

where n – number of points of target; $u_{gen}(t)$ - voltage from output of generator of radar, V.

With use of definition of input signal of radar from complex target (8) on the base of multipoint model of target the functional scheme of stand for semi-natural tests of onboard radar is shown at Figure 8. So the stand for semi-natural tests have to consist of onboard radar detector, imitator of the target reflected signal and computer with special software for control of operation of the radar detector.

Imitator of target reflected signal is developed for simulation of real physical signal from target in real time mode in laboratory by creation of attenuation and time delay of generator signal. Operation of imitator is realized with use of control device and special software. Thus imitator consists of the following elements:

- operating dynamic time delay line;
- electrically operating attenuator in real time mode.

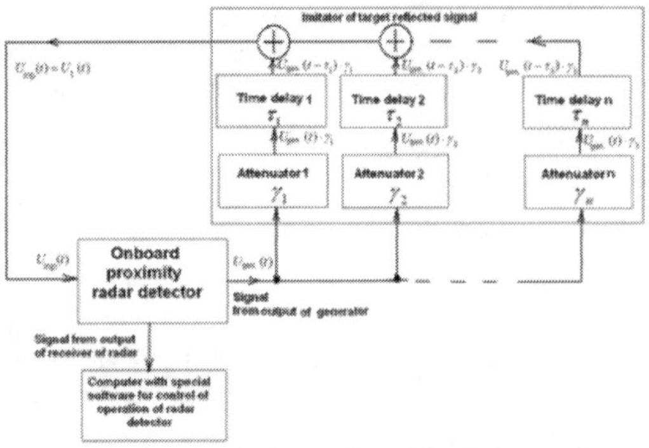

Figure 8. The functional scheme of stand for semi-natural tests of onboard proximity radar detector

The functional scheme of stand (Figure 8) could be applied for operative estimation of efficiency of onboard radar detector in real time mode for widespread mutual positions between target and onboard radar detector.

4 Conclusion

In this paper we have shown that high speed of calculation of input signals of onboard radar detector from complex shape targets could be achieved by use of multipoint model of target. To investigate availability of creation of fast scattering model, the general formulas of RCS of each specified local part of target corresponding to brilliant point of fast scattering model are developed. The results of calculation of power from fighter F-22 on the base of its multipoint model show that process of calculation has taken not more than 1 sec. So multipoint model of complex shape target could be considered as a fast scattering model of target with very high speed of calculation. From the research results, it is seen that the values of power of input signal of onboard radar could be applied for establishing of attenuations in the functional scheme of the stand of semi-natural tests.

References

1. Yin Jun, Deng Jia-hao, Wang Wei, *3 ICCEA Proceedings* (Beijing, 2004)
2. Xiangjun Li, Jing Li, Tie Li, *10 ISAPE Proceedings* (Xian, 2012)
3. A. B. Borzov, V. B. Suchkov, B. I. Shakhtarin, Yu. A. Sidorkina, J. of Comm. Tech. and Electr., **59**, 13 (2014)
4. V. V. Akhiyarov, A. B. Borzov, V. B. Suchkov, B. I. Shakhtarin Yu. A. Sidorkina, J. of Comm. Tech. and Electr., **60**, 8 (2015)
5. V. B. Suchkov, M. I. Baev, A. K. Likhoedenko, *5 IWCSE Proceedings* (Moscow, 2015)

Direction Tracking of Multiple Moving Targets Using Quantum Particle Swarm Optimization

Hongyuan Gao , Jia Li and Yanan Du

College of Information and Communication Engineering, Harbin Engineering University, Harbin, China

Abstract. Based on weighted signal covariance (WSC) matrix and maximum likelihood (ML) estimation, a direction-of-arrival (DOA) estimation method of multiple moving targets is designed and named as WSC-ML in the presence of impulse noise. In order to overcome the shortcoming of the multidimensional search cost of maximum likelihood estimation, a novel continuous quantum particle swarm optimization (QPSO) is proposed for this continuous optimization problem. And a tracking method of multiple moving targets in impulsive noise environment is proposed and named as QPSO-WSC-ML. Later, we make use of rank-one updating to update the weighted signal covariance matrix of WSC-ML. Simulation results illustrate the proposed QPSO-WSC-ML method is efficient and robust for the direction tracking of multiple moving targets in the presence of impulse noise.

1 Introduction

In array signal processing domain, the estimation of DOA has been a top research subject in the past several decades, and the technology of DOA estimation is of great interest for many sensor systems such as radar, sonar, electronic surveillance, seismic exploration and so on [1]. DOA estimation makes use of the signal received by the sensor array to estimate the direction of the target, hence, the algorithm of estimation directly affects its performance. Numerous improved methods have been proposed after conventional methods, such as the maximum likelihood (ML) estimation and approximate maximum likelihood estimation [2-3]. In the spatial spectrum estimation, the method is simple and has excellent performances in both asymptotic and threshold domains. Nevertheless, the process of implementation is complex, which involves a multidimensional nonlinear optimization problem with an increased computational burden. With the rapid development of intelligent computing technology, scholars began to estimate the directions of arrival by all kinds of intelligent algorithms, which have undoubtedly been powerful tools to overcome the drawback of the method [4-6].

As conventional DOA estimation methods assumed that the noise environment is Gaussian noise, however, in the practice, many random signals and noises are not Gaussian distribution, such as lightning atmospheric noises, the underwater acoustic signals, the line instantaneous peak voice signals of the communications as well as a lot of man-made noises. Above noises have a notable peak value characteristic, and may be described by different characteristics of the SαS process in which second order moment and high order moment do not exist, so some low order moments are designed to solve this problem [7-8]. At present, dynamic tracking is studied in the Gaussian noise environment, but dynamic tracking of the impulse noise environment is still a theory difficulty.

Therefore, in order to solve dynamic tracking problem of the impulse noise environment, a WSC-ML method is proposed, then the QPSO algorithm is designed to solve tracking equation of the WSC-ML method under impulsive noise environment. The proposed tracking method of multiple moving target is called as QPSO-WSC-ML. This method can avoid the decompositions of the fractional lower order moments matrix which should be repeated in the method based on subspace tracking, and performs better than other methods based on subspace tracking in the aspect of tracking precision.

2 Tracking model of multiple moving targets

Consider a uniform linear array consisting of N identical omni-directional antennas, M narrowband sources with wavelength λ impinge on the array from $\{\theta_1,\theta_2,...,\theta_M\}$. Assume that the distance between neighborhood antennas is $\lambda/2$. The kth snapshot sample data which is received by linear array is defined as

$$\boldsymbol{x}(k)=\boldsymbol{A}(\boldsymbol{\theta})\boldsymbol{s}(k)+\boldsymbol{n}(k) \quad (1)$$

where $\boldsymbol{x}(k)$ is the $N\times 1$ vector of the signals received by the array sensors, $\boldsymbol{s}(k)$ denotes the $M\times 1$ signal vector, $\boldsymbol{A}(\theta)=[\boldsymbol{a}(\theta_1),\boldsymbol{a}(\theta_2),...,\boldsymbol{a}(\theta_M)]$ represents steering matrix, $\boldsymbol{n}(k)$ denotes the $N\times 1$ complex SαS distribution noise vector. The mth steering vector toward direction θ_m is written as $\boldsymbol{a}_m = \left[1, \mathrm{e}^{-\mathrm{j}2\pi d_2 \sin(\theta_m)/\lambda}, ..., \mathrm{e}^{-\mathrm{j}2\pi d_N \sin(\theta_m)/\lambda}\right]^{\mathrm{T}}$ $(m = 1,2,...,M)$, where d_i $(i = 2,..., N)$ represents the distance between the ith sensor and the first sensor.

Since second order moment and high order moment do not exist in impulse noise environment, we define weighted received signal vector as $\bar{\boldsymbol{x}}(k) = [\bar{x}_1(k), \bar{x}_2(k), ..., \bar{x}_N(k)]^{\mathrm{T}}$, $\bar{\boldsymbol{x}}(k) = \boldsymbol{x}(k) / (\max_{1 \le i \le N} |x_i(k)|^{\sigma})$, where σ represents a constant. The weighted signal covariance (WSC) matrix of the first snapshot sampling data can be estimated by the following equation.

$$\boldsymbol{C}(1) = \bar{\boldsymbol{x}}(1)\bar{\boldsymbol{x}}^{\mathrm{H}}(1) \quad (2)$$

For the $(k+1)$th snapshot sample data, the weighted signal covariance matrix is updated as follows.

$$\boldsymbol{C}(k+1) = \mu \boldsymbol{C}(k) + (1-\mu)\bar{\boldsymbol{C}}(k+1) \quad (3)$$

where $\bar{\boldsymbol{C}}(k+1) = \bar{\boldsymbol{x}}(k+1)\bar{\boldsymbol{x}}^{\mathrm{H}}(k+1)$ is the covariance matrix of the $(k+1)$th weighted snapshot sample data, and $\boldsymbol{C}(k)$ is weighted signal covariance matrix of the kth received data, μ denotes the updating factor.

The tracking problem is to estimate the angle $\theta_m (m = 1,2,..., M)$ of the sources from k samples ("snapshots") of the received signals. The maximum likelihood estimation of the sources named WSC-ML is described as

$$\bar{\boldsymbol{\theta}} = \arg\max_{\boldsymbol{\theta}} [\mathrm{trace}(\boldsymbol{P}_{\boldsymbol{A}(\theta)}\boldsymbol{C}(k))] , \quad (4)$$

where trace( ) represents the trace function of the matrix, $\boldsymbol{P}_{\boldsymbol{A}(\theta)} = \boldsymbol{A}(\boldsymbol{\theta})(\boldsymbol{A}^{\mathrm{H}}(\boldsymbol{\theta})\boldsymbol{A}(\boldsymbol{\theta}))^{-1}\boldsymbol{A}^{\mathrm{H}}(\boldsymbol{\theta})$ represents the projection matrix of steering matrix $\boldsymbol{A}(\boldsymbol{\theta})$, $\boldsymbol{C}(k)$ is weighted signal covariance matrix of the kth sampling data.

The DOA estimation of the proposed WSC-ML could be considered as a classical continuous optimization algorithm, hence, we use the proposed continuous QPSO algorithm as the optimization method to search for the global optimal solution of WSC-ML and find the directions of moving targets.

3 Direction tracking of multiple moving targets based on QPSO

3.1 Quantum particle swarm optimization

Quantum particle swarm optimization is a novel intelligent optimization algorithm inspired by social and quantum behavior metaphor of particles. In the proposed QPSO algorithm, a quantum position can be used to produce the corresponding particle's position which represents a latent solution. The quantum position of the qth particle is defined as

$$\mathbf{y}_q = [\mathbf{y}_{q1}\ \mathbf{y}_{q2} \cdots \mathbf{y}_{qM}] = \begin{bmatrix} \alpha_{q1}\ \alpha_{q2} \cdots \alpha_{qM} \\ \beta_{q1}\ \beta_{q2} \cdots \beta_{qM} \end{bmatrix}, \quad (5)$$

where α_{qm} and β_{qm} represent a pair of quantum bits of the mth $(m = 1,2,...,M)$ quantum bit position of the qth quantum position. Additionally, $|\alpha_{qm}|^2 + |\beta_{qm}|^2 = 1$, $0 \le \alpha_{qm} \le 1$ and $0 \le \beta_{qm} \le 1$.

There are Q particles in the space of M dimensions in the quantum particle swarm. $\mathbf{y}_q = [\mathbf{y}_{q1}, \mathbf{y}_{q2}, ..., \mathbf{y}_{qM}]$ $(q = 1,2,..., Q)$ denotes the quantum position of the qth particle. $\mathbf{v}_q = [v_{q1}, v_{q2}, ..., v_{qM}]$, $\mathbf{p}_q = [p_{q1}, p_{q2}, ..., p_{qM}]$ and $\mathbf{p}_g = [p_{g1}, p_{g2}, ..., p_{gM}]$ denote the velocity of the qth $(q = 1,2,..., Q)$ particle, the best quantum position (e.g. the local optimal quantum position) of the qth particle and the best quantum position (e.g. the global optimal quantum position) of the whole quantum particle swarm, respectively. Therefore, the velocity of the qth particle is updated as

$$v_{qm}^{t+1} = w v_{qm}^{t} + c_1 r_1 (p_{qm}^{t} - \beta_{qm}^{t}) + c_2 r_2 (p_{gm}^{t} - \beta_{qm}^{t}) , \quad (6)$$

where c_1 and c_2 are the acceleration coefficients that control the maximum step length of flying to both the local optimal quantum position and the whole optimal quantum position, r_1 and r_2 represent the uniform random numbers between 0 and 1. Additionally, w denotes the weighted coefficient between 0.4 and 0.9, which is expressed as

$$w = w_{\max} - t \times (w_{\max} - w_{\min}) / t_{\max} , \quad (7)$$

where $w_{\max}$, $w_{\min}$, t and $t_{\max}$ denote the maximum weighted coefficient, the minimum weighted coefficient, the current number of iterations and the maximum number of iterations, respectively.

If the particle's velocity v_{qm}^{t+1} exceeds the boundary of upper bound and lower bound, v_{qm}^{t+1} will be subject to

$$v_{qm}^{t+1} = \begin{cases} v_{m,\max}, v_{qm}^{t+1} > v_{m,\max} \\ -v_{m,\max}, v_{qm}^{t+1} < -v_{m,\max} \end{cases} , \quad (8)$$

where $v_{m,\max}$ represents the maximum boundary of the mth dimension of velocity.

The evolutionary process of quantum position is mainly completed through quantum rotation gate [9-10]. In the algorithm, the mth dimension of quantum position of the qth particle y_{qm} is updated as

$$\begin{aligned} \boldsymbol{y}_{qm}^{t+1} &= \mathrm{abs}[\boldsymbol{U}(v_{qm}^{t+1})\ \boldsymbol{y}_{qm}^{t}] \\ &= \mathrm{abs}\left(\begin{bmatrix} \cos(v_{qm}^{t+1}) & -\sin(v_{qm}^{t+1}) \\ \sin(v_{qm}^{t+1}) & \cos(v_{qm}^{t+1}) \end{bmatrix} \boldsymbol{y}_{qm}^{t}\right) \end{aligned} \quad (9)$$

where abs( ) represents the absolute value function which restrains quantum bit in the range of [0,1], and v_{qm} denotes the mth dimension of velocity of the qth particle. Additionally, t represents the number of iterations of QPSO.

3.2 Direction tracking of multiple moving targets based on QPSO

In QPSO, the initial searching bound of the first snapshot is limited to defined domain of direction. As for the other snapshots, the searching space will change with the past searching value and the related parameter when the signal sources move. And the searching bound is derived as

$$\begin{cases} z_m^{\min}(k+1) = z_m^c(k+1) - \varepsilon^k \left|z_m^{\min}(k) - \bar{z}_m(k)\right| - r \\ z_m^{\max}(k+1) = z_m^c(k+1) + \varepsilon^k \left|z_m^{\max}(k) - \bar{z}_m(k)\right| + r \end{cases} \quad (10)$$

where ε denotes the convergence factor which affects the rate of convergence of the searching space, r represents the searching radius of searching space, $z_m^{\min}(k)$ represents the mth-dimension of searching lower bound of the kth snapshot within searching space, $z_m^{\max}(k)$ represents the mth-dimension of searching upper bound of the kth snapshot within searching space, $\bar{z}_m(k)$ represents the mth-dimension of estimated value of the kth snapshot and $z_m^c(k+1)$ represents the central value of the (k+1)th snapshot within searching space. Moreover, $z_m^c(k+1)$ is written as

$$z_m^c(k+1) = \delta z_m^c(k) + (1-\delta)\bar{z}_m(k) \quad (11)$$

where δ denotes the genetic factor.

Suppose that $\mathbf{z}_q = [z_{q1}, z_{q2}, \ldots, z_{qM}]$ represents the qth particle's position which is the corresponding mapping state of quantum position $\bar{\mathbf{y}}_q = [\beta_{q1}, \beta_{q2}, \ldots, \beta_{qM}]$, where $0 \le \beta_{qm} \le 1$ $(m = 1, 2, \ldots, M)$. The mth dimension of position of the qth particle is obtained by $z_{qm} = z_m^{\min} + \beta_{qm}(z_m^{\max} - z_m^{\min})$, $z_{qm} \in [z_m^{\min}, z_m^{\max}]$, where $z_m^{\min}$ and $z_m^{\max}$ represent the lower bound and upper bound of the mth dimension of variable, respectively.

The initial population of QPSO is created randomly within the quantum domain and the fitness function is used to evaluate the status of each particle. For the proposed tracking method, the fitness of the qth particle's quantum position is calculated as follows.

$$f(\mathbf{z}_q) = \mathrm{trace}[\boldsymbol{P}_{A(\mathbf{z}_q)}\boldsymbol{C}(k)] \quad (12)$$

According to introduction and analysis above, multiple targets tracking method of QPSO-WSC-ML is described as follows.

Step 1: According to the first snapshot sample data, initialize weighted signal covariance matrix $\boldsymbol{C}(1)$, where $\boldsymbol{C}(1) = \bar{\boldsymbol{x}}(1)\bar{\boldsymbol{x}}^{\mathrm{H}}(1)$.

Step 2: Initialize searching boundary with definition domain of variants, and initialize quantum position and velocity of quantum particle swarm by random method.

Step 3: Compute the fitness of the quantum position of each particle by (12), and initialize the local optimal quantum position and the global optimal quantum position.

Step 4: Number of iterations for direction finding of the kth snapshot is set as $t_{\max} = H \cdot \mathrm{round}[z_1^{\max}(k) - z_1^{\min}(k)]$, where H is set as constant and round( ) represents rounding function, and update the velocity and the quantum position of each particle.

Step 5: According to the corresponding position of the quantum position, compute corresponding fitness of the quantum position.

Step 6: Update the local optimal quantum position of each particle and the global optimal quantum position of the whole quantum particle swarm.

Step 7: If the end condition is not met (the end condition is set as maximum number of iterations $t_{\max}$), then $t = t+1$, and back to Step 4; else output the global optimal position and go to next step.

Step 8: Update weighted signal covariance matrix by $\boldsymbol{C}(k+1) = \mu\boldsymbol{C}(k) + (1-\mu)\bar{\boldsymbol{C}}(k+1)$ and the searching bound is updated by (10).

Step 9: If $k \ge k_{\max}$, where $k_{\max}$ represents the maximum number of sampling, end the loop; else $k = k+1$ and go to Step 2.

4 Simulation results and evaluation

In this section, we will give the simulation of the proposed QPSO-WSC-ML algorithms in impulsive noise background via computer simulations. Consider that direction tracking system is equipped with uniform linear arrays with half wavelength space, and the number of antenna is set as 8. The maximum number of sampling $k_{\max}$ is set as 500, and the impulsive noise is assumed to follow the standard SαS distribution with the characteristic exponent α. Generalized Signal-to-Noise Ratio (GSNR) is computed by $GSNR = 10\lg[E(|s(k)|^2)/\gamma]$, where $E(|s(k)|^2)$ represents the signals' average power and γ represents the dispersion of impulse noise.

In the following, in order to illustrate the superiority of the proposed QPSO-WSC-ML, we compare QPSO-WSC-ML with PSO-FLOM-ML which was proposed in [11]. For QPSO, some parameters are set as follows,

DOI: 10.1051/matecconf/20165907002

$\varepsilon = 0.995$, $\mu = 0.95$, $r = 3$, $\delta = 0.8$, $H = 6$, $\sigma = 0.7$, $w_{max} = 0.9$, $w_{min} = 0.4$, $v_{m,max} = 0.2$, $c_1 = c_2 = 1$. For PSO and QPSO, the size of population is set as 20.

Two independent signal sources coming from $\theta_1(k) = [-12 - 0.01k]^{\circ}$ and $\theta_2(k) = [28 + 0.01k]^{\circ}$ are considered. For two independent signal sources, Fig. 1 and Fig. 2 depict the direction tracking performance of the two tracking algorithms with $\alpha = 1.85$ and GSNR=4dB. Moreover, Fig. 1 and Fig. 2 show that QPSO-WSC-ML is more robust than PSO-FLOM-ML.

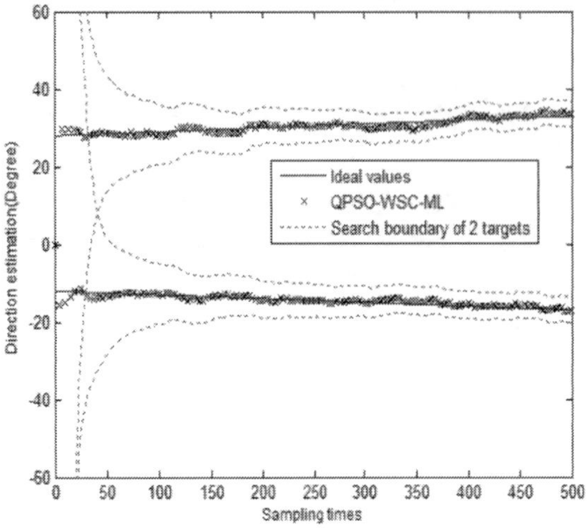

Figure 1. Tacking of two independent signal sources based on QPSO-WSC-ML

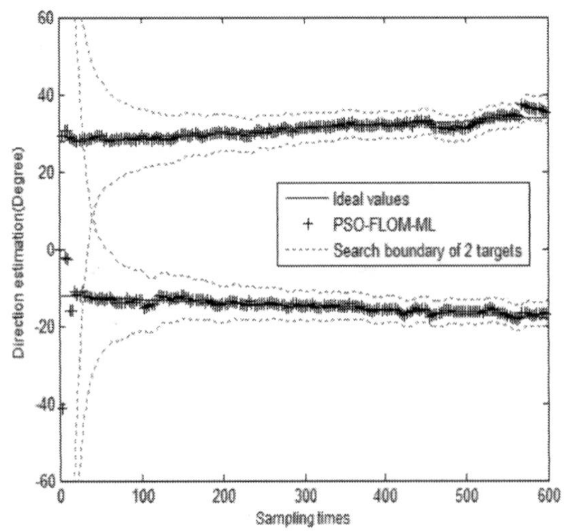

Figure 2. Tacking of two independent signal sources based on PSO-FLOM-ML

Similarly, three independent signal sources coming from $\theta_1(k) = [-30 + 5\sin(2\pi k/500)]^{\circ}$, $\theta_2(k) = [13 + 5\sin(2\pi k/500)]^{\circ}$ and $\theta_3(k) = [48 + 5\sin(2\pi k/500)]^{\circ}$ are considered. Fig. 3 and Fig. 4 demonstrate the direction tracking performance of the two tracking algorithms for three independent signal sources with $\alpha = 1.5$ and GSNR=10dB. It can be seen that the tracking performance of QPSO-WSC-ML is better than that of PSO-FLOM-ML.

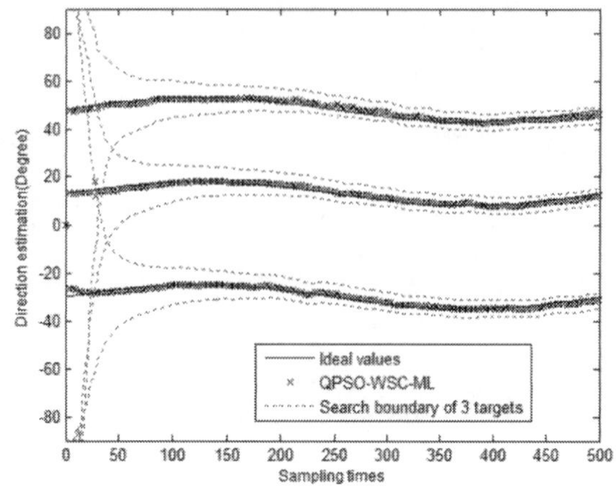

Figure 3. Tacking of three independent signal sources based on QPSO-WSC-ML

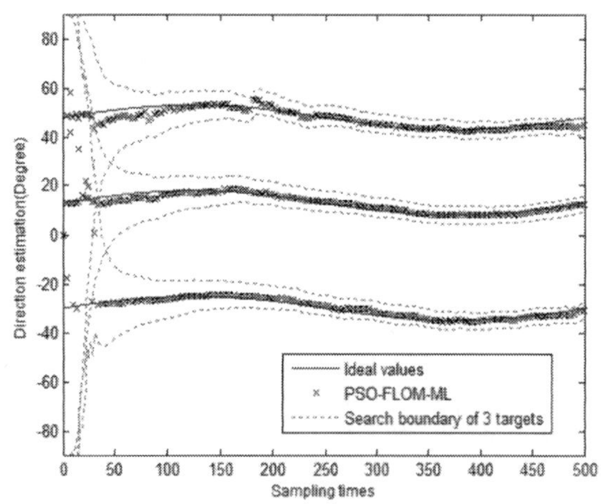

Figure 4. Tacking of three independent signal sources based on PSO-FLOM-ML

At last, Fig. 5 and Fig. 6 demonstrate the direction tracking performance of the two tracking algorithms for three coherent signal sources with $\alpha = 1.5$ and GSNR=10dB. Compared with PSO-FLOM-ML, QPSO-WSC-ML has more robust tracking performance. From Fig. 3-6, we can also see that it is easy to track independent signal sources with the same simulation condition.

DOI: 10.1051/matecconf/20165907002

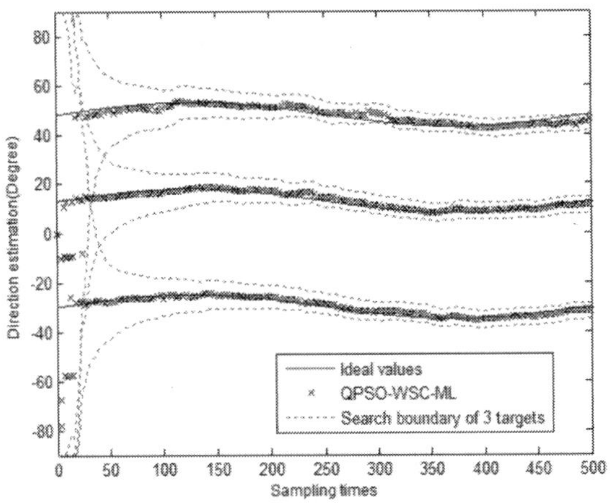

Figure 5. Tacking of three coherent signal sources based on QPSO-WSC-ML

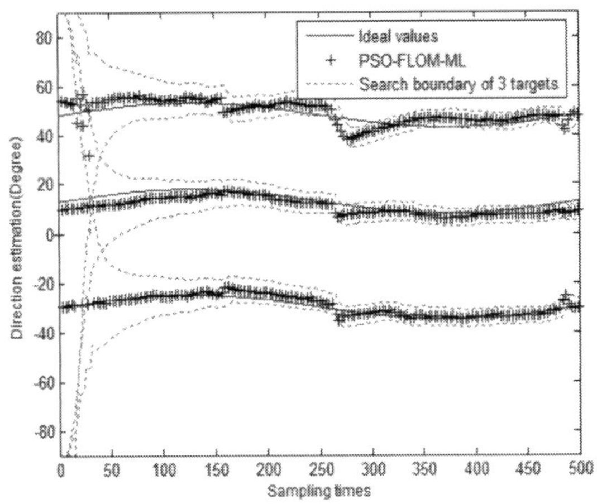

Figure 6. Tacking of three coherent signal sources based on PSO-FLOM-ML

5 Conclusion

Based on the WSC-ML and QPSO, we have proposed a novel method which is named as QPSO-WSC-ML for tracking multiple moving targets in the impulse noise environment. By QPSO for the continuous optimization problem of the direction tracking, we can reduce the computational cost of WSC-ML and improve the searching accuracy of previous PSO algorithm. Our future work is to use the proposed QPSO-WSC-ML method to MIMO radar, and we will try to convert the method into the impulsive noise environment.

References

1. H. Krim, and M. Viberg, *IEEE Signal Processing Magazine*, **13**, 67-94 (1996)
2. X. Mestre, P. Vallet, P. Loubaton, *Signal Processing Conference (EUSIPCO), 2013 Proceedings of the 21st European IEEE*, 1-5 (2013)
3. J. Y. Lee, R. E. Hudson, K. Yao, IEEE Systems Journal, **8**, 131-141 (2014)
4. Z. Y. He, Z. Q. Zhao, K. Yang, J. Ouyang, *Computational Problem-Solving (ICCP), 2011 International Conference on IEEE*, 405-408 (2011)
5. Y. L. Sit, M. Agatonovic, T. Zwick, *Radar Conference (EuRAD), 2012 9th European*, 298-301 (2012)
6. W. Shi, J. Huang, Y. S. Hou, Journal of Systems Engineering and Electronics, **23**, 173-178 (2012)
7. K. V. S. Hari, V. Lalitha, Proc. IEEE International Conference on Acoustics, **7882**, 2580-2583 (2011)
8. J. B. Long, D. F. Zha, H. B. Wang, *International Conference on Mechatronic Sciences, Electric Engineering and Computer (MEC)*, 3504-3508 (2013)
9. H. Y. Gao, C. Q. Xu, International Journal of Computing Science & Mathematics, **4**, 321-331 (2013)
10. H. Y. Gao, C. W. Li, Acta Phys. Sin., **63**, 128802-130 (2014)
11. D. Y. Zhao, M. Diao, L. L. Yang, C. Chen, Journal of Shandong University, **40**, 133-138 (2010)

MATEC Web of Conferences **59**, 07003 (2016) DOI: 10.1051/matecconf/20165907003
ICFST 2016

Research on Cross-eye Jamming Technique based on Phase Diversity Processing

Yu Daobin[1], Wang Hongyan[2] and Wu Yanhong[1]

[1]*The Equipment Academy, The Optoelectronic Equipment Department && Science and Technology on Complex Electronic System Simulation Laboratory, Beijing, China*
[2]*The Equipment Academy, The Information Equipment Department && Science and Technology on Complex Electronic System Simulation Laboratory, Beijing, China*

Abstract. The cross-eye jamming technique is a common method for monopulse tracking radar, which can cause a displacement towards the wrong direction and result in an angle deception. This paper proposes a modification of the traditional cross-eye jamming system, and the phase diversity processing of the received signal is introduced, and its precise expression of the offset is derived. The phase difference between the two branches of the signal processing is controllable, which is in the vicinity of 180 ° phase shift, reducing the coordination requirements of the two branches. Simulation with different parameters is conducted so as to prove the conclusion, and the flexible control of the jamming effort is realized by this new cross-eye jamming technique.

1 Introduction

The monopulse tracking radar has strong ability against the traditional interference technology which poses a great challenge to the jamming part, and relevant research becomes urgent. An important way to interfere with the general monopulse tracking radar is the application of spatially separated sources of interference. The phase characteristic of spatially separated jammer may be coherent or irrelevant. Non-coherent sources of interference between phases mean a random relationship, and it's easy to be eliminated or tracked. Coherent interferences have determined or synchronous relationship between phases of multiple jammers and its apparent power may be located outside the centre of the constant angle, and it has less restrictions. Therefore, the application of coherent interference is an effective means for interference monopulse tracking radar [1-3].

Cross-eye interference is a typical coherent interference technology, domestic and foreign scholars have done some extensive researches. The model's analysis under cross-eye interference environment, the angle's control of aiming axis and system simulation for monopulse tracking radar are key issues, which are the cross-eye interference theoretical basis for studying and improving cross-eye interference structure [4-6]. Currently, the available methods for improving cross-eye such as by increasing the number of interference sources and divided branches are proposed. Although it can increase the performance of the system, but to a certain extent, also increase the complexity of interference control and system's construction [7, 8].

This paper presents a new method for cross-eye interference based on phase diversity processing, which can result in a flexible and controllable interference effects. On the basis of traditional cross-eye interference principle, the procedure of phase diversity processing and its structure are given. The obtained interference effect is derived from a mathematical model, and the conclusions are proved through related simulation finally.

2 Analysis of Cross-eye Jamming

Cross-eye interference takes advantage of the limited resolution of the angle tracking radar system, enabling optimum parameter's design for maximum directional error. So the stronger interference guides, the more that the radar pointing to the wrong target. Cross-eye interference is based on the front waveform's distortion effect. When the radar's tracking direction and the target's moving direction is in a certain relationship, a large tracking error will generate which causing a stronger displacement from the aiming point, creating plenty of false targets with different angles information. The false point is moving to a new location in the range of the target of interest, and it has nothing to do with the distance, but can achieve a large offset distance. Advantages of cross-eye jamming technology are ensured when the two sources of coherent radiation interference signal's phase are matched to 180 °, but the angle that the radar signal reaches the jammer won't affects. In this case, the radar can not distinguish echoes and interfering signals, only to see a false target decade meters' far away from the real target and track it.

Traditional cross-eye jamming technology under all circumstances requires that the two interfering signals are transmitted to maintain the correct phase and power relationship. On one hand, the amplitude-frequency characteristics of the two channels are of high consistency requirements, and are particularly sensitive to the interference pattern of aircraft movement, allowing little yaw angle. A slight deviation will reduce interference effect greatly. On the other hand, when meeting the optimum conditions for interference, the two sources of interference power will offset mostly, resulting in the consequence that less interference will enter the radar receiver. Therefore, cross-eye interference technology based on phase diversity processing is proposed. Its principle is that the interference of one branch is given a plurality of diversity in the vicinity of 180 ° phase shift, so the signal on the two paths will have phase differences, which can produce different offset distance in a certain range over a period of time. It is similar to the multipath effects, overcoming the disadvantages above and producing controlled angle deception jamming for monopulse tracking radar.

3 Phase diversity processing and its jamming effects' analysis

3.1 New cross-eye interference system

Cross-eye interference technology for monopulse tracking radar takes advantage of two independent forwarding antennas, and each antenna has a transmitting and receiving antenna connected by transmission lines to an amplifier generating the interference signal. Above all, the interference system has further improvement that the phase diversity processing module is brought in, which should be set on a branch circuit for phase diversity signal processing. By adding this module, the amplitude ratio and phase difference between two coherent sources of interference can be designed in any combination, resulting in different distances offset. By matching the magnitude and phase timely, interference can be achieved and produce more false angles information. Improved cross-eye interference system is as shown in Fig. 1:

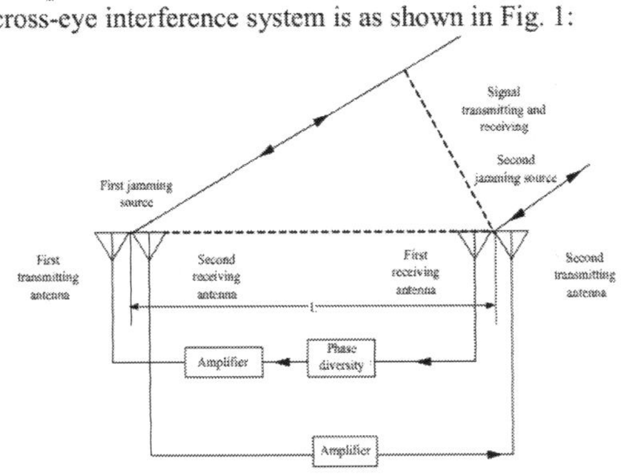

Figure 1. Schematic of Improved cross-eye interference system

The cross-eye interference structure shown above functions as follows:

(1) The two independent interference sources obtain a monopulse tracking radar signals emitted by radar respectively;

(2) In the first branch, the obtain interference signal is amplified by power amplifier and transmitted back to the monopulse tracking radar;

(3)In the second radar, the obtained interference signal after phase diversity processing is also amplified by power amplifier and transmitted back as well;

After these interference signals with echo signals all received and processed by the monopulse tracking radar, a plenty of offset distance will generate. Near the target's real position, false targets can be made flexibly controlled, generating different angles decoy jamming effects.

3.2 Phase diversity processing procedure

By using the average logical method, the process can be depicted as follows:

(1) Signal splitter and phase modulation;

After receiving the radar signal, it's divided into p branch ($p \geq 2$) through a rectangular pulse train with equidistant spacing. Radar target echo signal on each branch is delayed to obtain a path of the sub-phase signal waveform from the same signal, and all sub-phase modulated signals from all the branches are set different time delayed so as to obtain phase modulated output in different period.

The original radar signal is expressed as:

$$s(t) = rect(t / T_p) \cdot e^{j2\pi(f_0 t + \mu_r t^2 / 2)} \quad (1)$$

Different signal splitter have different delay characteristics, and the obtained signal of each branch is expressed as:

$$\begin{cases} s_1(t) = rect(\frac{t-\tau_1}{T_p}) \cdot \exp\{j2\pi[f_0(t-\tau_1) + \mu_r(t-\tau_1)^2/2]\} \\ s_2(t) = rect(\frac{t-\tau_2}{T_p}) \cdot \exp\{j2\pi[f_0(t-\tau_2) + \mu_r(t-\tau_2)^2/2]\} \\ \vdots \\ s_p(t) = rect(\frac{t-\tau_p}{T_p}) \cdot \exp\{j2\pi[f_0(t-\tau_p) + \mu_r(t-\tau_p)^2/2]\} \end{cases} \quad (2)$$

After the phase modulation, the obtained signal can be written as follows:

$$\begin{cases} s_1'(t) = s_1(t) \cdot e^{j2\pi\phi_1} = rect(\frac{t_r-\tau_1}{T_p}) \cdot \exp\{j2\pi[f_0(t-\tau_1) + \mu_r(t_r-\tau_1)^2/2 + \phi_1]\} \\ s_2'(t) = s_2(t) \cdot e^{j2\pi\phi_2} = rect(\frac{t_r-\tau_2}{T_p}) \cdot \exp\{j2\pi[f_0(t-\tau_2) + \mu_r(t_r-\tau_2)^2/2 + \phi_2]\} \\ \vdots \\ s_p'(t) = s_p(t) \cdot e^{j2\pi\phi_p} = rect(\frac{t_r-\tau_p}{T_p}) \cdot \exp\{j2\pi[f_0(t-\tau_p) + \mu_r(t_r-\tau_p)^2/2 + \phi_p]\} \end{cases} \quad (3)$$

(2) Signal samples

According to the sub-phase signal stored in the control logic, the signal sampling time is selected after the phase modulation process, which is selected based on pre-stored logic. The n $(2 \leq n \leq p, n、p \in Z^+)$ branch is sampled in real time to obtain phase-modulated sub-pulse signal. This process can be represented by the formula as follows:

$$\begin{cases} s_1''(t) = s_1'(t) \bullet [rect(\frac{t-t_1}{t_1}) * \sum_{i=0}^{N_1-1} \delta(t-\sigma_i)] \\ s_2''(t) = s_2'(t) \bullet [rect(\frac{t-t_2}{t_2}) * \sum_{i=0}^{N_2-1} \delta(t-\sigma_i)] \\ \vdots \\ s_p''(t) = s_p'(t) \bullet [rect(\frac{t-t_p}{t_p}) * \sum_{i=0}^{N_p-1} \delta(t-\sigma_i)] \end{cases} \quad (4)$$

At this point, the radar signal is split into N sections, where $N_1 + N_2 + \cdots + N_p = N$.

(3) Signal synthesis

According to pre-stored control logic, $m(2 \le m \le n,\ m,\ n \in Z^+)$ phase-modulated sub-pulse signals synthesize from n selected sampled signal branches. The obtained interference signal after the sub-sampled radar pulses phase modulation in all signal path combine at last. The interference signal has the same pulse length compared with the original signal, and it's expressed as:

$$J(t) = s_1''(t) + s_2''(t) + \cdots + s_p''(t) \quad (5)$$

Above is the phase diversity handling process. The process of phase diversity interference signals' generation is shown in Fig. 2.

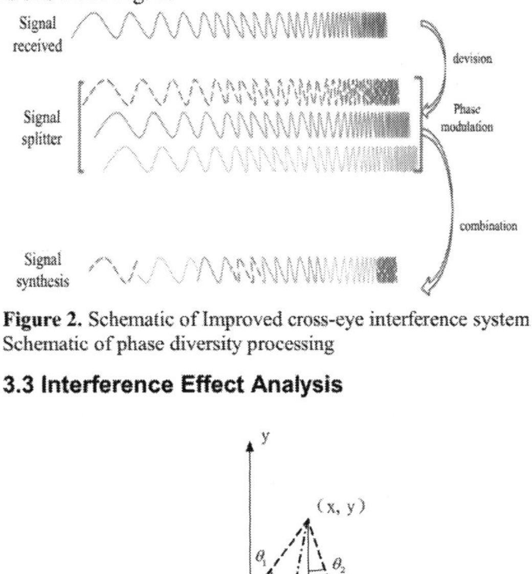

Figure 2. Schematic of Improved cross-eye interference system Schematic of phase diversity processing

3.3 Interference Effect Analysis

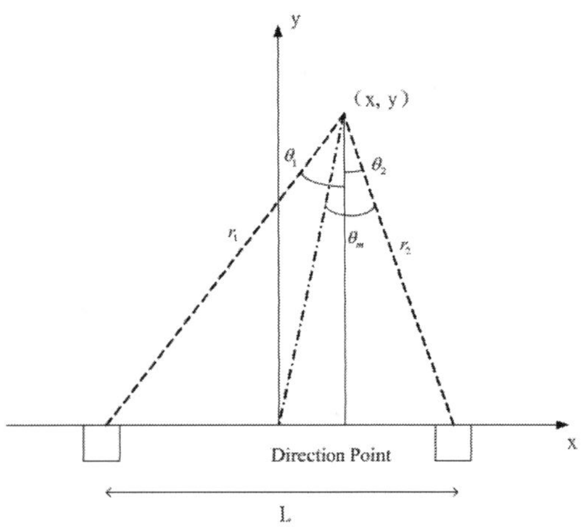

Figure 3. Interference of double coherent sources

Mathematical model of double source interference is shown in Fig. 3. Provided the double sources of interference spacing L, the coherent interference source's measurement is given by the monopulse angle tracker:

$$\Delta_1 = k\theta_1\Sigma_1, \quad \Delta_2 = k\theta_2\Sigma_2 \quad (6)$$

k stands for the coordinate's factor. Thus the indicated θ_i of monopulse tracking angle equals to:

$$\theta_i = \frac{1}{k}\frac{\Delta}{\Sigma} = \frac{\Delta_1 + \Delta_2}{\Sigma_1 + \Sigma_2} = \frac{\theta_1\Sigma_1 + \theta_2\Sigma_2}{\Sigma_1 + \Sigma_2} \quad (7)$$

Set $\frac{\Sigma_2}{\Sigma_1} = ae^{j\Phi}$, the above formula can be further written as:

$$\theta_i = \frac{\theta_1 + \theta_2 ae^{j\Phi}}{1 + ae^{j\Phi}} \quad (8)$$

Both sides multiply $(1 + ae^{-j\Phi}) / (1 + ae^{-j\Phi})$, and apply Euler's formula to obtain the angle's value:

$$\theta_i = \theta_m - \frac{\Delta\theta}{2}\frac{1 - 2ja\sin\Phi - a^2}{1 + 2a\cos\Phi + a^2} \quad (9)$$

The real part of the equation represents the interference source's indication angle by double-coherent monopulse radar:

$$\theta_i = \theta_m - \frac{\Delta\theta}{2}\frac{1 - a^2}{1 + 2a\cos\Phi + a^2} \quad (10)$$

Imaginary part represents the direction of deviation angle, so offset distance is available at this time

$$\Upsilon_{miss} = R \cdot \tan\left[\frac{\Delta\theta}{2}\frac{1 - a^2}{1 + 2a\cos\Phi + a^2}\right] \quad (11)$$

With a small angle approximation $\tan(\cdot)$, considering $\Delta\theta = L\cos\Psi / R$ and thereby obtaining

$$\Upsilon_{miss} = \frac{L \cdot \cos\Psi}{2}\frac{1 - a^2}{1 + 2a\cos\Phi + a^2} \quad (12)$$

Where a stands for the amplitude ratio of two coherent sources of interference, Φ stands for the phase difference between the sampled phase-modulated sub-pulse radar signal to the radar echo signal path during all jamming periods, and Ψ stands for two interference source's connection angle between the vertical midpoint and radar sighting axis.

By phase diversity processing, different phases result in different offset distance's value at different times as follows:

$$\frac{\Sigma_2}{\Sigma_1} = \begin{cases} ae^{j\Phi_1} & 0 < t < t_1 \\ ae^{j\Phi_2} & t_1 < t < t_2 \\ \vdots & \\ ae^{j\Phi_n} & t_{n-1} < t < T \end{cases} \quad (13)$$

The final expression of offset distance obtained:

$$\Upsilon_{miss\,i}(t) = \frac{L \cdot \cos\Psi}{2}\frac{1 - a^2}{1 + 2a\cos[\Phi_i(t)] + a^2}(t_{i-1} < t < t_i) \quad (14)$$

The radar measures different offset distances in different time periods, and the interference effect obtained finally is shown in Fig.4:

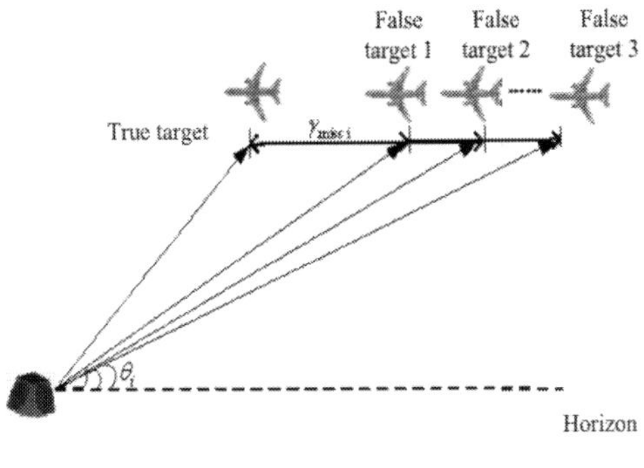

Figure 4. Monopulse tracking radar interference effect

4 Simulation analysis

Interference with airborne cross-eye system is simulated as an example to verify the above conclusion. Figure 5 shows the relationship between offset distance from the real target under different calculated amplitude ration and phase difference combination's variation, where $L=10m$, $\theta_i \approx 4°$. Assuming that the signal applied has achieved the desired signal interference ratio. Traditional cross-eye jamming technical parameters used are: $\Phi=180°$, $a=1$, then obtaining the maximum offset distance.

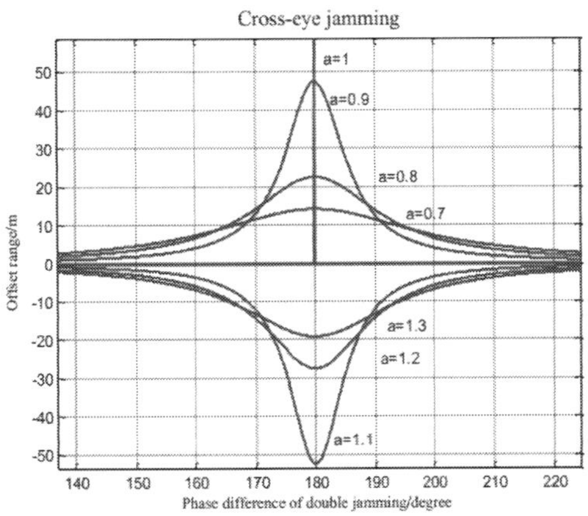

Figure 5. Change the offset range in cross-eye jamming

Through the phase diversity processing technique in a fixed amplitude ratio set, where different number of paths may be provided, five and eight branches applications are selected to be compared sharing average logic, respectively changing in the [160°,200°] and [150°,200°] spaced interval, and the obtained offset range's variation is shown in Figure 6. The phase offset is obtained from the variation in the diversity processing in 3s by the random selection of each branch as shown in Figure 7.

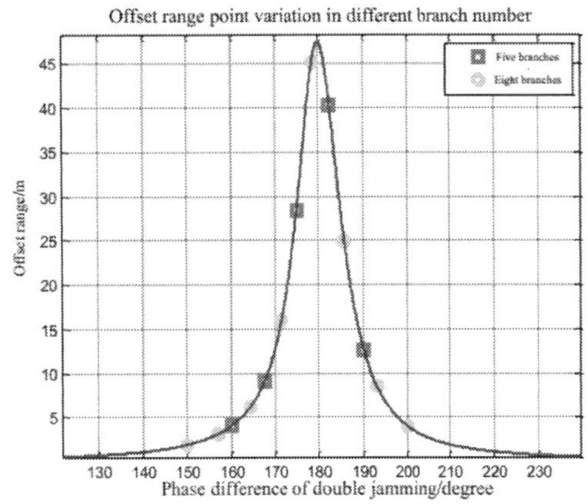

Figure 6. Offset range point variation of different branch number in fixed amplitude ratio

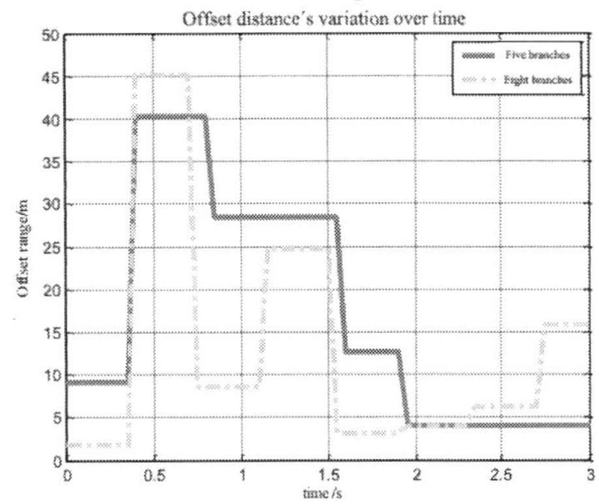

Figure 7. Offset distance's variation over time

In figure 6, we can see that in a certain angle's region, the values are chosen in flexibility, causing different offset distance. The value's number is determined by the branches' number. The more the phase diversity processing branches are, the more values the offset range can be determined, no matter long or short, so as to reach a flexible jamming mode.

In figure 7, we can see that as time goes on, the offset ranging from 3m to 45m. In the five branches case, it reaches 40m during 0.3s to 0.8s, and it's enough to form a false target compared with the range L between the two interference sources. The eight branches case is more competitive than the former one, because it reaches a larger offset, and it shifts more frequently due to the advantages that more branches bring in.

By analysing the simulation results, some conclusions are listed as follows:

(1) The parameter set's selection process has great flexibility, where the number of branches and signal synthesis logic and so on can be flexibly combined to set, producing more false targets in short distance near the real target. In different periods it has different offset distances which are similar to the effect of multipath interference, generating multiple false targets flicker effect;

DOI: 10.1051/matecconf/20165907003

(2) Since the phase diversity processing's parameter set can be treated with a certain regularity, so that the radar can not overcome the effects of angular error by just accumulating through the data, as a result that such interference is superior to the non-coherent multiple interference sources such as glint interference;
(3) The phase difference between the two branches of the signal processing is controllable, which is in the vicinity of 180 ° phase shift to meet the requirements. It reduces the coordination requirements of the two branches, and will be better in practical engineering applications.

5 Summary

This paper presents a new cross-eye jamming technique based on phase diversity processing method through improvements in system architecture, and flexible and controllable interference effects can be achieved. This jamming method using phase diversity signal processing, generating a plenty of false targets with different false angles information, which has an obvious advantages compared with the traditional one. Less demanding on the system's coordination makes it easily realized. How to take advantage of the signal amplitude ratio of the two antennas to reduce the use of cross-eye antenna spacing requirements and get more improvement is one of the main directions of future research

References

1. Schleher D C. Electronic Warfare in the information age [M] America: Artech House, 1999: 256-271.
2. Falk L. Cross-eye jamming of monopulse radar[C] Waveform Diversity and Design Conference. Pisa: IEEE Press, 2007: 209-213.
3. Meyer G J. Using cross-eye techniques to counter radio frequency agile monopulse [D]. Ohio: Air Force Institute of Technology, 1997: 156-168.
4. Cao Fei, Liu Qingyun, Xin Zengxian. Mathematical Modeling of Cross-eye Jamming[J]. Modern Radar, 2013, 35(6): 69-72.
5. Cheng Ning, Jiang Yong. Research on Angle Deception Using Interference of Coherent Dual-Sources to Angle Tracking of Mono-Pulse[J]. Electronic Warfare, 2009, 126(3): 6-9.
6. Liu Yangtao, Hao Zhimei. Research on Jamming Simulation to Angle Tracking System of Monopulse Radar[J]. Aerospace Electronic Warfare, 2012, 28(4):62-64.
7. Harwood N.M., Dawber W.N., King D.J., et al Multiple-element Cross-eye[C]. IET Radar Sonar Navigation, 2007, 1(1): 67–73.
8. Wang Huiping, Zhang Youyi. A New Method to Realize Cross-eye Jamming Technique[J]. Warships Electronic Warfare, 2007, 30(6): 23-25.

MATEC Web of Conferences **59**, 07004 (2016) DOI: 10.1051/matecconf/20165907004
ICFST 2016

Analysis of LFM-waveform Libraries for Cognitive Tracking Maneuvering Targets

Wang Hongyan[1], Yu Daobin[2] and Wu Yanhong[2]

[1]*The Equipment Academy, The Information Equipment Department && Science and Technology on Complex Electronic System Simulation Laboratory, Beijing, China*
[2]*The Equipment Academy, The Optoelectronic Equipment Department && Science and Technology on Complex Electronic System Simulation LaboratoryBeijing, China*

Abstract. Based on the idea of the waveform agility in cognitive radars, the waveform libraries for maneuvering target tracking are discussed. LFM-waveform libraries are designed according to different combinations of chirp parameters and FrFT rotation angles. By applying the interact multiple model (IMM) algorithm in tracking maneuvering targets, transmitted waveform is called real time from the LFM-waveform libraries. The waveforms are selected from the library according to the criterion of maximum mutual information between the current state of knowledge of the model and the measurement. Simulation results show that waveform library containing certain amount LFM-waveforms can improve the performance of cognitive tracking radar.

1 Introduction

Cognitive radar can select transmitter configuration according to characteristics of the targets and environment and can improve the performance of detection and tracking [1]. It is adaptively, intelligently and dynamically.

The cognitive radar is proposed by Simon Haykin in 2006[2]. The components of the cognitive radar are discussed in his correlative papers [3-6]. The tracking filter and transmitter of the cognitive radar are regarded as a closed-loop system with the environment. The transmitted waveform is selected in real time to track targets effectively by taking advantage of all available information. Study on optimal waveform design and scheduling based on signal-to-noise ratio (SNR) maximisation or information theory is preferred for cognitive radar. The optimization is too complex to obtain analytical solution and also have large amount of calculation. Probably the transmitter cannot tune into the optimal waveform in the observation period.

We have taken an information theoretic waveform optimization, which choose optimal waveform in several observation times for global optimization [7-9]. It is the problem of waveform scheduling. Targets tracking based on the waveform library Theory on design waveform library is introduced by some scholars [10-15], which is practical and operational for waveform agility in radars systems. It has a finite library of waveforms in the tracking system. An optimal choice is made based on the certain consideration for next transmitted waveform. The basic target tracking system with waveform library is shown in figure 1.

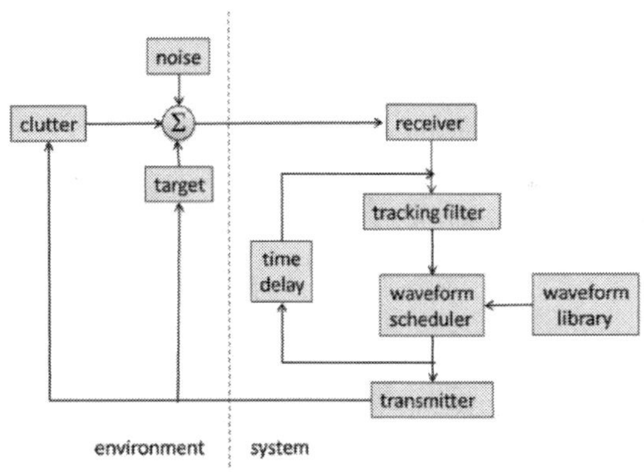

Figure 1. Target tracking system with waveform library

The idea of selecting waveforms simply based on the four different types according to their ambiguity functions was introduced in the paper of A. W. Rihaczek [10]. Work has been done in the area of waveform libraries design and scheduling algorithm by Cochran [11]. The design of waveform libraries was discussed using the information-theoretic measures of effectiveness and the results of simulations comparing one-step ahead and multi-step ahead scheduling with an unscheduled approach are summarized. In the paper of Savage, waveform library was obtained by implementing a fractional Fourier transform (FrFT) for tracking accelerating targets within an interacting multiple model (IMM) framework [12]. A waveform selection approach for tracking the maneuver target described as polynomial prediction model is proposed by Tan. FrFT is also used to rotate the measurement error ellipse to select the optimal waveform [13]. The papers of Suvorova were discussing waveform library design and waveform selection criteria for target tracking by [14-15]. With this criterion the LFM and LFM-rotation waveform libraries need only the maximum and the minimum of chirp rate.

We focus on the design of LFM-waveform library for maneuvering target tracking. The IMM is introduced for the maneuvering targets of multiple possible dynamical models. Simulations are performed to validate the design of library and information-theoretic criteria for adaptive selection of a waveform.

2 Problem Formulation

The baseband signal is $s(t)$ and he transmitted pulse signal, $s_T(t)$, is [13]

$$s_T(t)=\sqrt{2}\,\mathrm{Re}\left\{\sqrt{E_T}s(t)e^{j\omega_c t}\right\} \tag{1}$$

where ω_c is carrier frequency, E_T is emitted energy, $\mathrm{Re}\{\cdot\}$ denotes the real part. The received signal is

$$s_R(t)=\sqrt{2}\,\mathrm{Re}\left\{[\sqrt{E_R}e^{j\phi}s(t-\tau)e^{j\nu t}\quad n(t)]e^{j\omega_c t}\right\} \tag{2}$$

where ϕ is stochastic phase, E_R is received energy, $n(t)$ is white Gaussian noise. τ is the time delay and ν is the Doppler shift. Range and the radial velocity of target can be described by the following equation:

$$r=\frac{c\tau}{2},\ \dot{r}=\frac{c\nu}{2\omega_c} \tag{3}$$

The salient transform of the down conversion of $S_R(t)$ is its ambiguity function [12, 13]:

$$A(\tau,\nu)=\int_{-\infty}^{\infty}s\left(t-\frac{\tau}{2}\right)s^{*}\left(t+\frac{\tau}{2}\right)e^{-j\nu t}dt \tag{4}$$

The ambiguity function is a two-dimensional function of time delay and Doppler shift. The maximum of $A(\tau,\nu)$ is at (τ,ν), which is the measurement of time delay and Doppler shift under ideal conditions. Then range r and velocity $\dot{r}$ can be calculated by equation 3.

The measurement noise covariance matrix $N(\theta_k)$ is waveform dependent:

$$N(\theta_k)=TJ^{-1}T^T \tag{5}$$

where θ_k is the vector of waveform parameters, J is the Fisher information matrix corresponding to the τ and ν, J^{-1} is the Cramer-Rao lower bound (CRLB) of the unbiased estimator of range and Doppler of a target as $A(\tau,\nu)$ at its maximum point.

The interacting multiple model (IMM) Kalman filter is approximate to optimal filter for linear dynamic model with Gaussian noise. It is not for the maneuvering target tracking. The interacting multiple model (IMM) filter is adaptive for the target with deceleration, acceleration or swerve dynamics. clutter and noise are not taken into account in the discus [12].

The IMM has structure as figure 2. It is the weighted summation to estimations of each filter with different tracking model. We assume that changes in target trajectory can be modelled as n dynamic models, that is $m_1, m_2, \ldots, m_n$. The state equation and measurement equation of dynamic model m_i are

$$X_i(k+1)=F_iX_i(k)+G_iu_i(k) \tag{6}$$

$$Z_i(k)=H_iX_i(k)+\omega_i(k) \tag{7}$$

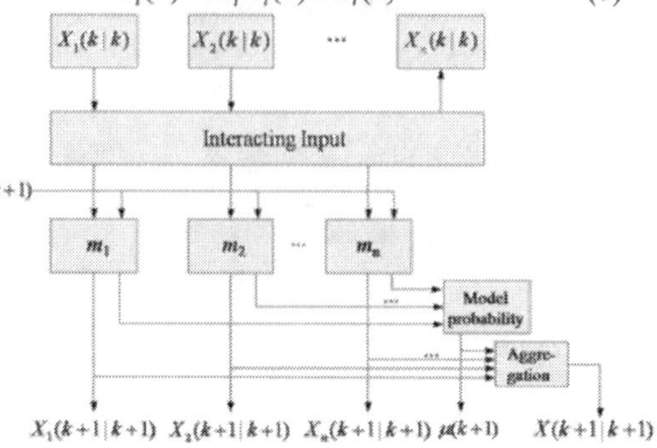

Figure 2. Diagram of IMM

where $i=1,\ldots,n$ is the dynamic model at time . We write $X_i(k)$ for the state of the track and $Z_i(k)$ for the measurement at time k. F_i is the state propagation matrix, H_i is the measurement matrix and G_i is Process noise matrix. Process noise is denoted by u and the measurement noise by ω. These both are zero mean, white, and uncorrelated Gaussian noise sequences with covariance matrix $Q_i(k)$ and $N(\theta)$ respectively.

Let $\mu_i(k-1|k-1)$ denotes the initial probability of dynamic model i and M_k denotes the model set at time k. We assume the transition probabilities of dynamic models as π_{ij}, $i,j\in 1,\ldots,n$. The algorithm of IMM is described as follows.

2.1 Initialization

The predictive model probability is

$$\mu_j(k|k-1)=\sum_{m_j\in M_{k-1}}\pi_{ij}\ {}_i(k\quad 1) \tag{8}$$

The initial conditional weight is

$$\mu_{ij}=\pi_{ij}\ {}_i(k-1)/\ {}_j(k|k-1) \tag{9}$$

The initial state estimate is calculated using

$$X_j^0=\sum_{m_j\in M_{k-1}}X_i(k-1)\mu_{ij} \tag{10}$$

and covariance is

$$P_j^0=\sum_{m_j\in M_k}[P_j(k-1|k-1)\quad (X_j(k\quad 1|k\quad 1)\quad X_j^0)\cdot(X_j(k-1|k-1)-X_j^0)^T]\mu_j(k-1) \tag{11}$$

$X_i(k-1|k-1)$ denotes the output of ith filter at time $k-1$.

2.2 Estimation

Each filters then use X_j^0 and P_j^0 as update of the state estimate and error covariance and calculate according to its model.

2.3 Update model probability

We assume the likelihood function and the probability of model as follows

$$L_j(k)=N[\tilde{Z}_j(k);O,S_j] \tag{12}$$

$$\mu_j(k)=\frac{\mu_j(k\mid k-1)L_j(k)}{\sum_{m_j\in M_k}\mu_i(k-1)L_i(k)} \tag{13}$$

where $\tilde{Z}_j(k)$ is the innovation and S_j is its covariance matrix. $N[\cdot]$ denotes normal distribution.

2.4 Aggregation

Aggregate state estimate and covariance information are given by

$$X(k\mid k)=\sum_{m_j\in M_k}X_j(k)\mu_j(k) \tag{14}$$

$$P(k\mid k)=\sum_{m_j\in M_k}[P_j(k\mid k)+(X_j(k\mid k)-X(k))\cdot(X_j(k\mid k)-X(k))^T]\mu_j(k) \tag{15}$$

The initial input will be the measurement z_0 or the parameters θ_0 of waveform. Our motivation is to select optimal waveform of transmitter within the cognitive loop. Then we select z_0 as the initial input.

3 LFM waveform library

3.1 Waveform library design

The goal of waveform library design is to produce a waveform library that is limited, yet sufficient to provide a choice of high performance waveform.

When the measurement error ellipse is orthogonality with the one of the predicted estimated states, the waveform is the optimal for purpose of target tracking [16]. The measurement covariance matrix is dependent on the waveform. It is a selection of a waveform with certain measurement error. Within IMM framework, waveform library for each model consists of arbitrary transforms of a single waveform.

We describe the linear frequency modulation (LFM) or "chirp" waveform library. In this case the waveform is $s(t)=s_0(t)\exp(jbt^2)$. b is the chirp rate supported by the radar. The waveform library is constructed of the transforms with different rotation angles of LFM waveform. It is conducted by the fractional Fourier transform.

For this library the corresponding measurement covariance matrices N_{θ_i} are given by the equation

$$\begin{cases}N_{\theta_i}=S^{-1}N_{\theta_0}S\\ S=S(\varphi)*S(b)\end{cases} \tag{16}$$

where φ is the rotation angle and $|S|=1$.

$$S(\varphi)=\begin{bmatrix}\cos\varphi & -\sin\varphi\\ \sin\varphi & \cos\varphi\end{bmatrix},S(b)=\begin{bmatrix}1 & 0\\ 2b & 1\end{bmatrix} \tag{17}$$

3.2 Waveform selection criteria

For the adaptive selection, we would like to select the waveform, while the uncertainty of the dynamic model of the target is minimized at each echo. If we take a single step ahead approach, the waveform is the one, which minimizes the expected entropy of the probability distribution of the model at the next epoch. This implies that the mutual information between the model and the measurement needs to be maximized [13].

By the criterion of maximum mutual information between the current state of knowledge of the model and the measurement, we select the waveform from the library satisfying

$$\theta_{opt}=\underset{\theta\in\{\theta_1,\theta_2,\ldots,\theta_n\}}{\arg\min}\left(-\sum_{j=1}^{n}P_j(k\mid k-1)\log\frac{|N_{j,\theta}|}{\sum_{l=1}^{n}P_l(k\mid k-1)\log N_{l,\theta}}\right) \tag{18}$$

where the $P_j(k\mid k-1)$ are the predictive model probabilities.

4 Simulations

The simulations were performed to validate the ideas above. There are multiple maneuvers for the target trajectories in range and Doppler, which are described in figure 3.

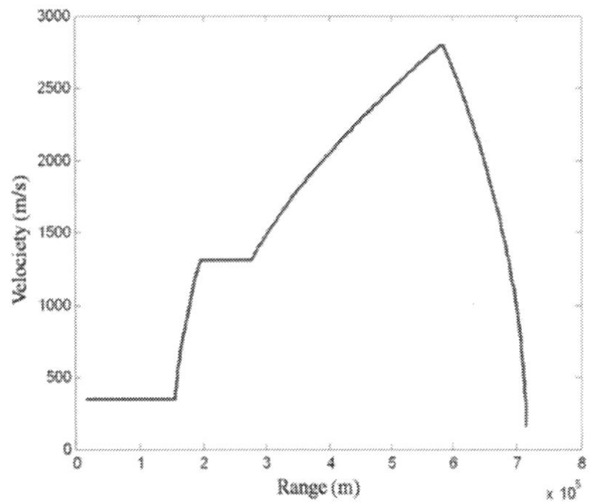

Figure 3. Manoeuvres for the target trajectory

The measurement interval is $\Delta t=2$. The simulation was run with the following parameters:

$$F=\begin{bmatrix}1 & \Delta t\\ 0 & 1\end{bmatrix},H=\begin{bmatrix}1 & 0\\ 0 & 1\end{bmatrix},G=\begin{bmatrix}0\\ 1\end{bmatrix} \tag{19}$$

The accelerated velocity of each dynamic model is $a_i=50,20,0,-20,-50,i=1,2,\ldots,5$. The initial probability of each dynamic model is 0.2. The model transition probability matrix of the Markov chain is

$$\Pi = \begin{bmatrix} 0.6 & 0.1 & 0.1 & 0.1 & 0.1 \\ 0.1 & 0.6 & 0.1 & 0.1 & 0.1 \\ 0.1 & 0.1 & 0.6 & 0.1 & 0.1 \\ 0.1 & 0.1 & 0.1 & 0.6 & 0.1 \\ 0.1 & 0.1 & 0.1 & 0.1 & 0.6 \end{bmatrix} \quad (20)$$

It is assumed that $N(0) = \begin{bmatrix} \sigma_1^2 & 0 \\ 0 & \sigma_2^2 \end{bmatrix}$. The range and range-rate standard deviations are $\sigma_1 = 10$ and $\sigma_2 = 1$ 。

In consider of the bandwidth of the transmitter, the chirp rate is limited. We define the waveform library L_1 containing one waveform ($b_1 = b_{max}$) and L_2 containing two waveforms ($b_1 = b_{max}, b_2 = b_{min}$). Based on the above considerations, we also define the waveform libraries containing six waveforms and ten waveforms.

$$L_6 = \{b_1, (b_1, \phi = \ 0.2\pi), (b_1, \phi \ \ 0.4\pi), b_2, (b_2, \phi = 0.2\pi), (b_2, \phi = 0.4\pi)\}$$

$$L_{10} = \{b_1, (b_1, \phi = \ 0.1\pi), (b_1, \phi \ \ 0.2\pi), (b_1, \phi \ \ -0.3\pi), (b_1, \phi = -0.4\pi), b_2, (b_2, \phi = \ 0.1\pi), (b_2, \phi \ \ 0.2\pi), (b_2, \phi = 0.3\pi), (b_2, \phi = 0.4\pi)\}$$

The Monte Carlo simulations were performed 100 times. The accumulative error of the filter is given by

$$e_s(k) = e_s(k-1) \ \ \frac{1}{M} \sum_{i=1}^{M} [X_i(k) \ \ \hat{X}_i(k/k)] \quad (21)$$

The choice of the optimal waveform from the LFM waveform library and the accumulative estimation errors are simulated with different initial estimation covariance. The different initial estimation covariance is taken into account in simulation.

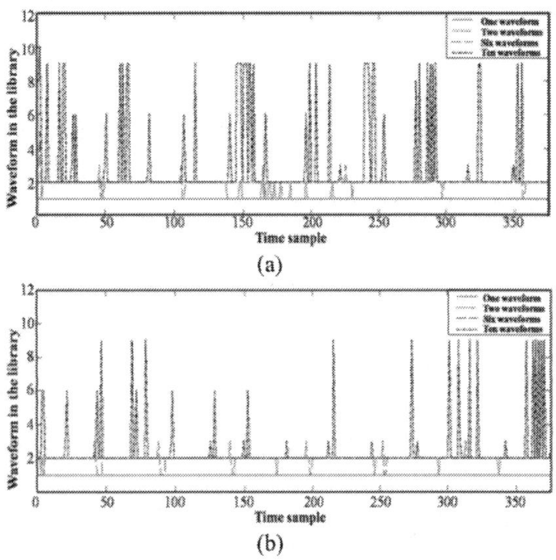

(a)

(b)

Figure 4. Waveform scheduling with different initial estimation covariance

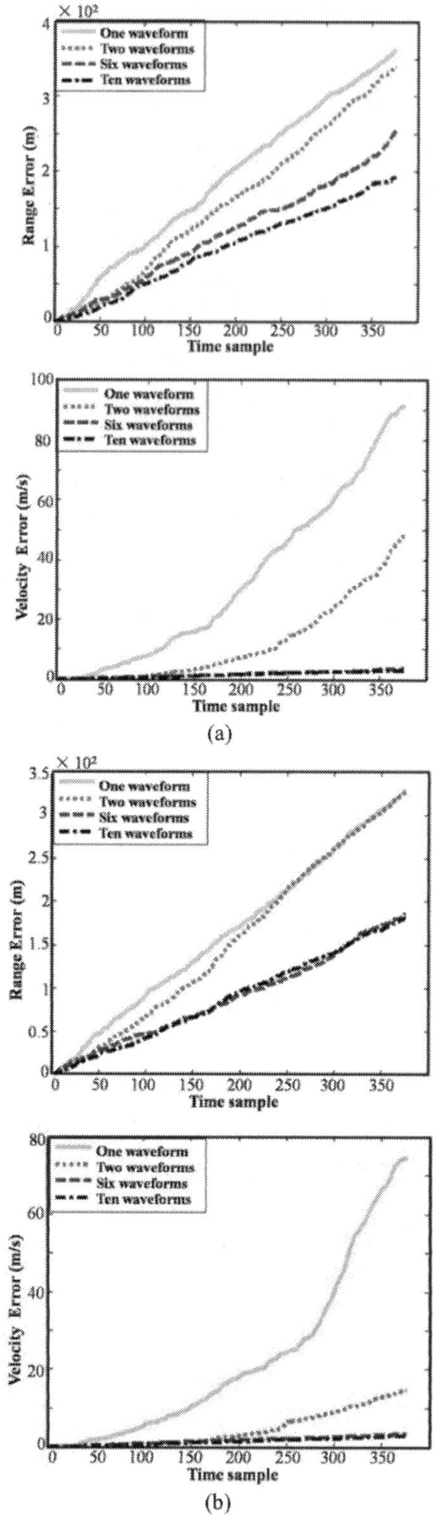

(a)

(b)

Figure 5. Accumulative errors in range and velocity for waveform selection with different initial estimation covariance

The initial estimation covariance for results, shown in Figure 4(a) and Figure 5(a), is larger than the other. The curves in fugre 4 illustrate that, if the initial estimation covariance is larger, the waveform agility is frequently. The waveform scheduling is stably with the estimation covariance decreased. The waveform scheduling is related with the trajectories.

The range errors of results using L_1 and L_2 are similar, while the velocity error is better for library L_2.For the accelerated target in the distance, the choice of the waveform is frequently to maintain the velocity tracking error, but the range tracking error is deteriorated. The waveform library containing more waveforms is not necessarily effective to improve tracking performance.

The tracking performance with waveform library L_6 and L_{10} are improved remarkably. further increasing of waveforms is not validated to error reduction and has huge computation. The design of the waveform library should be the compromise between performance and complexion.

5 Conclusions

The construction method of LFM waveform library for cognitive radar is investigated in this paper. The interact multiple models tracking filter's theory is elaborated. With the waveform selection criteria derived on the basis of information theory, a strategy of waveform library construction through combination of fractional Fourier transformation and chirp rate is analyzed. The simulations shows that construction of waveform library with a number of waveforms can improve the tracking performance.

References

1. Goodman N A, P R Venkata, M A Neifeld. "Adaptive waveform design and sequential hypothesis testing for target recognition with active sensors" [J]. *IEEE Journal of Selected Topics in Signal Processing*, 2007, 1(1), pp. 105-113.
2. Haykin S. "Cognitive radar: a way of the future" [J]. IEEE Signal Processing Magazine, 2006, 23(1):30-40.
3. Haykin S. "Cognitive dynamic systems" [J] *IEEE IC-ASSP*, 2007: 1369-1372.
4. Haykin S, Yanbo Xue and Timothy N. "Davidson O-ptimal Waveform Design for Cognitive Radar" [J], *IEEE Asilomar*, 2008, 3-7.
5. Arasaratnam I, Haykin S. "Cubature Kalman Filters" [J], *IEEE Transaction on automatic control*, 2009, 54(6), 1254-1269.
6. Haykin S, Amin Zia, Arasaratnam I, et al. "Cognitive tracking radar" [J], *IEEE Xplore*, 2010, 1467~1470.
7. Lu J, Hu W, Yu W. "Adaptive beam scheduling algorithm for agile beam radar in multi-target tracking" [C]. Shanghai, 2007.
8. Li Y, Sira S P, Moran B, et al. "Adaptive Sensing of Dynamic Target State in HeavySea Clutter" [C]. 2007.
9. Boers Y Driessen H, Zwaga J. "Adaptive MFR parameter control: fixed against variable probabilities of detection" [J]. *IEEE Proceedings-Radar, Sonar and Navigation*, 2006, 153(1): 2-6.
10. Rihaczek A W. "Radar Waveform Selection-A Simplified Approach" [J]. *IEEE Transactions on Aerospace and Electronic Systems*, 1971, AES-7(6): 1078-1086.
11. Cochran D, Suvorova S, Howard S D, et al. "Waveform Libraries" [J]. *IEEE Signal Processing Magazine*, 2009, 26(1): 12-21.
12. Savage C O, Moran B. "Waveform selection for maneuvering targets within an IMM framework" [J]. *IEEE Transactions on Aerospace and Electronic Systems*, 2007, 43(3): 1205-1214.
13. TAN Jiajia, ZHANG Jianqiu. "New waveform selection approach to tracking maneuver targets"[J], *Systems Engineering and Electronics*. 2011, 33(3): 515-522, 543.
14. Suvorova S, Howard S D, Moran W. "Waveform libraries for radar tracking applications" [J], *Waveform Diversity Conference*, 2004.
15. Suvorova S, Howard S D, Moran W. "Generalized Frequency Modulated Waveform Libraries for Radar Tracking Applications" [J], *IEEE Asilomar*, 2009.
16. Niu R , Willett P. "Tracking considerations waveform for range and range-rate measurement" [J], *IEEE Trans. On Aerospace and Electronic Systems*, 2002, 38(2): 467~487.

Research on Phased Array Radar Resource Management in Searching Mode

Junqing. Wu , Ting. Cheng

School of Electronic Engineer, UESTC, 611731 Chengdu, China

Abstract. Multifunction phased array radar, with rational management of time resource and energy resource, can search and track multiple targets simultaneously. A resource management optimization model for phased array radar in searching mode has been proposed, whose objection is to minimize the total resource consumption under the guarantee of searching performance with controllable parameters: search frame period, dwell time and signal duty circle. Both radial cumulative detection probability and tangential cumulative detection probability are considered in the searching performance. The consideration of tangential cumulative detection probability as one of the constraints is the main contribution here. The solution to this model by applying genetic algorithm suggests that this model is more suitable in actual scene.

1 Introduction

In modern battlefield environment, target characteristics and attack warfare have become increasingly complex, and make the continuous extension of radar functional requirements. The less resource the searching task consumes, the more resource can be used for other tasks, which can obtain a better performance in target detecting and tracking.

The resource management [1] in searching mode is actually the parameter control of the searching tasks. Billiam[2][3] obtained the optimal search frame period when there is no high prior load in radar system, which not only can meet the tracking initiation condition, but also consume the least radar resources. Based on this, Jianbin Lu [4] derived optimal searching model of multifunction phased array radars with limited searching resources and proposed the criterion for the optimum parameter choice of search frame period and detection range in detail. Litao Zhang studied searching parameters of phased array radar from the point of view of concept and system and presented the selection principle and method of a number of searching parameters [5]. Reference [6] discussed the influence on searching performance caused by the dwell time of a searching beam, scan time and false alarm probability, then designed the optimal operation parameters for searching performance. Above research only focus on the situation when target fly to the target directly, only the racial cumulative detection probability is considered. However, because of the limit of the angle range of searching area, target may cross the angle border at a short distance with both radial and tangential velocity. However, the target has both racial and tangential speed; the tangential detection performance should not be omitted.

A resource management algorithm for phased array radar in searching mode is presented, which sets the search frame period, dwell time, signal duty circle as controllable parameters to minimize the total recourse consume under the guarantee of searching performance. Resources include time resource and energy resource. The searching performance can be indicated in tangential cumulative detection probability and radial cumulative detection probability. Here, we consider the velocity of target have both radial and tangential velocity.

This paper is organized as follows. The resource management optimization model in searching mode is formulated Section II. In Section III we utilize Genetic Algorithm to solve this optimization problem. The simulation result is presented in Section IV. Section V is the conclusion.

2 Resource management optimization model in searching

The criterion for the optimum choice of parameter is the value which will give the required searching performance with minimum resource consumption. Considering a phased array radar which is performing the searching task, the controllable parameters include search frame period T_f, dwell time t_B and signal duty circle η. First of all, we formulate the relationship between controllable parameters and resource consumption..

2.1 Objective function

Let η, t_B, T_f be the parameters of searching task in the surveillance space. Search frame period T_f and dwell time t_B reflect the time resource consumption, η indicates the energy resource consumption. Considering the time and energy resource consumption synthetically, the objective function can be calculated as,

$$\min_{(\eta, t_B, T_f)} c_1 \frac{t_B N_B}{T_f} + c_2 \eta \tag{1}$$

where N_B is the number of dwell beams in searching area. The values of c_1, c_2 show the importance of different resources.

2.2 Constraint conditions

Here, the relationship between controllable parameters and searching performance is formulated. That is, we formulate the constraint conditions. In [2], searching performance is defined in terms of the track formation range. It supposed that the velocity of the target only have a radial component. However, in actual scene, target approach also with a tangential velocity component. Therefore, similarly to the concept of track formation range, the track formation angle is defined in this paper. We use these two indices to describe the searching performance.

Firstly, we show the case that target approach to radar only with radial velocity. Donate the largest operational range provided by radar system as R_s. Assume the target is approaching with a radial velocity of v , track formation range is R_t .Thus the radial cumulative detection probability 2 is

$$P_d(R_t) = \frac{1}{\Delta r}\int_{R_s-\Delta r}^{R_s}\left\{1-\prod_{i=0}^{\left\lceil \frac{R_s-R_t}{\Delta r}\right\rceil}\left[1-P_{dc}\cdot P_{di}(r-i\Delta r)\right]\right\}dr \tag{2}$$

where $\lceil . \rceil$ means rounding down to the nearest integer, $\Delta r = vT_f$ is the radial range during T_f . P_{dc} is the probability of successful confirmation, $P_{di}(r-i\Delta r)$ is the single detection probability at range $r-i\Delta r$,which is related to the signal to noise ratio.

$$P_{di} = P_{fa}^{\frac{1}{1+\mathrm{SNR}_i}} \tag{3}$$

where P_{fa} is the probability of false alarm. According to the radar function 7, signal to noise ratio can be written as

$$SNR_i = \frac{P_t \eta G_t G_r \lambda^2 \sigma t_B}{(4\pi)^3 \left(r - ivT_f\right)^4 kT_0 F_n L_{tot}} \tag{4}$$

In order to guarantee the searching performance, the radial cumulative detection probability has to reach a certain threshold P_D .Then

$$P_d(R_t) \geq P_D \tag{5}$$

Secondly, we analyze the case which target approach to radar only with tangential velocity. Donate the angle searching range as $[\theta_1, \theta_2]$ .Assume the target is approaching with a tangential velocity of v_x , track formation angle is θ_t ,

$$\theta_t = (\theta_2 - \theta_1)Q, Q \in (0,1) \tag{6}$$

That is to say, when target fly from angle θ_1 to $\theta_s, \theta_s = \theta_1 + \theta_t$, the tangential cumulative detection probability should reach a certain threshold to guarantee the searching performance. Fig.1 shows a visualization definition of track formation angle. Suppose that the target fly into the searching area at time t_0 , and fly to angle θ_s at time $t_{\max}$.

$$t_{\max} = t_0 + \frac{\theta_t R_x}{v_x} \tag{7}$$

R_x is the radial range when target fly into searching area along tangential direction.

Now calculate the tangential cumulative detection probability during $[t_0, t_{\max}]$. Fig.2 shows the changes of radar beam pointing and target's direction with time. In fig.2, bold line describes the relation of radar beam pointing to time. This is a periodic function of T_f .

To a certain beam, the range of beam pointing is $\left[B_w(i-1)+\theta_1, B_w(i)+\theta_1\right]$ during the time interval $\left[T_f(j-1)+(i-1)t_B, T_f(j-1)+it_B\right]$.The dash dot line represents the target angle position changing with time which can be expressed as

$$\theta_t(t) = \frac{v_x(t-t_0)}{r_x} + \theta_1 \tag{8}$$

The intersections of these two lines represent that the radar illuminates the target. But the dwell time on the target is different in different encounter time. Here, just call the dwell time on target the real dwell time.

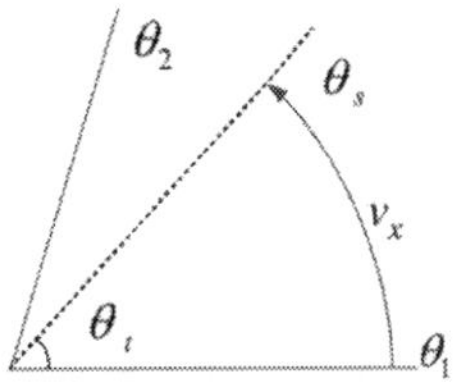

Figure 1. Visualized definition of track formation angle.

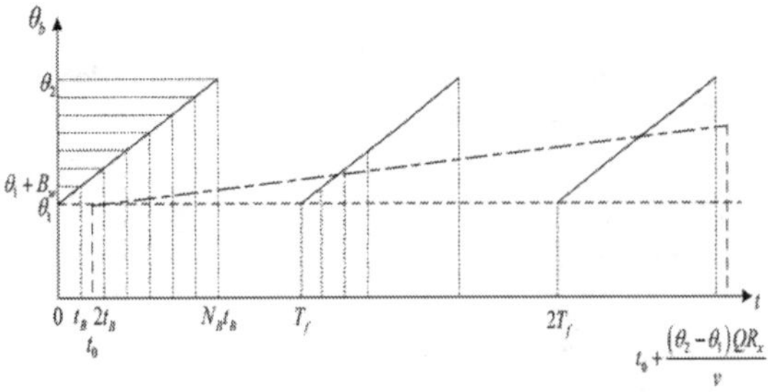

Figure 2. Changes of radar beam pointing and target's direction with time

As the total number of searching beam is I during $[t_0, t_{\max}]$, then we can obtain the tangential cumulative detection probability as,

$$\prod_{n_{count}=1}^{I}\left[1-P_{dc}\cdot P_d\left(t_B'\left(t_0\right),n_{count}\right)\right] \tag{9}$$

Where

$$I = I_1 + I_2 \tag{10}$$

$$I_1 = \left\lceil \frac{\left(t_{\max}-t_0\right)}{T_f} \right\rceil * N_B = \left\lceil \frac{\theta_{\max} r_x}{v_x T_f} \right\rceil * N_B \tag{11}$$

$$I_2 = \min\left[\bmod\left(\frac{\left(t_{\max}-t_0\right)}{T_f}\right) * T_f / t_B, N_B\right] \tag{12}$$

Where t_B' can be calculated as follows.

Supposing at t_0 moment, target fly into the searching area with angle position θ_1 .At this time, the serial number of dwell beam is n_0 .

$$n_0 = \begin{cases} \left\lfloor \dfrac{t_0}{t_B} \right\rfloor, & \dfrac{t_0}{t_B} < N_B \\ 0, & \dfrac{t_0}{t_B} > N_B \end{cases} \tag{13}$$

Then start the beam counter n_{count} , which is started from 0 and finished in $I-1$. So to the n_{count} -th beam, the counter of search frame period n_{T_f} is

$$n_{T_f} = \left\lfloor \frac{n_0 + n_{count}}{N_B} \right\rfloor \tag{14}$$

The radar beam serial number n_{search} ($n_{search} \geq 1$) is

$$n_{search} = \begin{cases} N_B, & n_0 + n_{count} = kN_B, \\ \bmod\left(\dfrac{n_0 + n_{count}}{N_B}\right), & else, \end{cases} \tag{15}$$

where $k = 1, 2, \ldots$ The beam pointing is

$$\left[\theta_{s1}, \theta_{s2}\right] = \left[\theta_1 + \left(n_{search}-1\right)B_w, \theta_1 + n_{search}B_w\right] \tag{16}$$

The beam acting time is

$$\left[t_{s1}, t_{s2}\right] = \left[\left(n_{T_f}-1\right)T_f + \left(n_{search}-1\right)t_B, \left(n_{T_f}-1\right)T_f + n_{search}t_B\right] \tag{17}$$

Assuming in the moment interval $[\tilde{t}_1, \tilde{t}_2]$, target is illuminated by radar, so its beam position is between $[\theta_{s_1}, \theta_{s_2}]$. The $[\tilde{t}_1, \tilde{t}_2]$ can be solved by

$$\theta_{s_1} = \frac{v\left(\tilde{t}_1 - t_0\right)}{r} + \theta_1,\ \theta_{s_2} = \frac{v\left(\tilde{t}_2 - t_0\right)}{r} + \theta_1 \tag{18}$$

If $[t_{s_1}, t_{s_2}]$ and $[\tilde{t}_1, \tilde{t}_2]$ intersect, the real dwell time t_B' is the length of intersection. If the intersection is empty, the target is not illuminated. That is, $t_B' = 0$ and $P_d\left(t_B'\right) = 0$.

However, the calculation above supposes that t_0 is known. In actual scene, it is a random variable, which is assumed to be uniform in general case. Therefore, the statistical tangential cumulative detection probability can be written as

$$P_d(\theta_t) = \frac{1}{T_f}\int_0^{T_f}\left\{1 - \prod_{n_{count}=1}^{I}\left[1-P_{dc}\cdot P_d\left(t_B'\left(t\right),n_{count}\right)\right]\right\}dt \tag{19}$$

Furthermore, we have time constraint,

$$\frac{N_B t_B}{T_f} \leq 1 \tag{20}$$

This means the time use factor should less than 1.

So the resource management optimization model of phased radar in searching mode can be given by,

$$\min_{(\eta,t_B,T_f)} c_1 \frac{t_B N_B}{T_f} + c_2 \eta$$

$$s.t. \begin{cases} P_d(R_t) \ge P_D \\ P_d(\theta_{\max}) \ge P_D \\ \dfrac{N_B t_B}{T_f} \le 1 \end{cases} \quad (21)$$

3 Genetic algorithm solving optimization model

As described in Eq.21 the formulated resource management optimization model is nonlinear. Genetic algorithm [8] is one kind of adaptive random searching algorithms that simulate the evolution process and mechanism in nature to solve the optimization problems, it can automatically gain and simulate the knowledge about search-space, therefore, find the best solution. Based on the genetic algorithm, the resource management algorithm for phased radar is obtained. Fig.3 shows the flow chart of it.

The value of objective function is considered as the adaptive value of genetic algorithm. The smaller the value of objective function is, the greater the adaptive value can be. If there is no individual to meet all the constraints in the initial population, we can continue to crossover and mutate to rebuild a new population until appropriate individual appears. When the evolutional generation has approached its maximal value, the algorithm ends. Otherwise the reorganized population will continue circulating.

4 Simulation

Consider a uniformly linear array with $M = 256$, where the unit gap is the half of the wavelength. Here, constant parameters include wavelength $\lambda = 5.45cm$, Boltzmanns constant $k = 1.38 \times 10^{-23} J/K$, standard temperature $T = 230K$, noise figure $F = 2$, and the effective area duty ratio of antenna $\eta_e = 0.5$. The angle range of the searching section is $[-40^\circ, 40^\circ]$, the largest operational range is $R_s = 200km$. In case of Swelling-I, set the average RCS of target to be $1m^2$, the probability of false alarm to be $P_{fa} = 10^{-6}$, the threshold of $P_D = 0.95$. The radial velocity and tangential velocity are the same as $v_r = v_t = 1.5Ma$. The weighting of different resource are $c1 = 0.5, c2 = 0.5$. Signal duty circle changes in $\eta \in [0, 0.25]$, dwell time changes in $t_B \in [0, 26]ms$ and search frame period is chosen in $T_f \in [0, 20]s$. In genetic algorithm, the number of individuals in population is 1000, the maximal number of

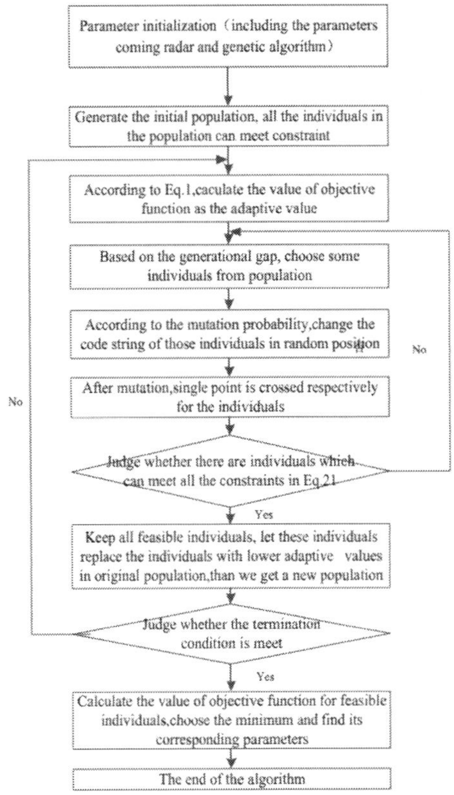

Figure 3. The flow chart of genetic algorithm

generations is 300, the crossover probability is 0.7, and the mutation probability is 0.05, its generational gap is 0.95. Firstly, we only considered the radial cumulative detection probability and time constriction. Fig.4 to Fig.6 are the changing curves of the dwell time, the search frame period and the signal duty circle respectively. When track formation range is small, radar can accumulate many times, in other words, the desired single detection probability is reduced. With the increase of track formation range, in order to meet the requirement of radial cumulative detection probability, the search frame should be smaller. In Fig.4, the dwell time increased before 120km, and then keeps the maximum. Fig.5 shows the changing tendency of search frame period. When track formation range is larger than 120km, search frame period becomes smaller to increase the accumulation times. The signal duty circle increase just as Fig.6 shows.

Then, use these optimal parameters values to calculate the tangential cumulative detection probability. Suppose the Rx=30km,Q=0.3.The result is showing in Fig.7.

As we can see from Fig.7.,the tangential cumulative detection probability is less than 0.95 before 150km, so above parameters allocation cannot guarantee the required searching performance. So we should consider all the constraint conditions, including the radial

DOI: 10.1051/matecconf/20165907005

cumulative detection probability, tangential cumulative detection probability and time limit.

In actual scene, the track formation range and track formation angle is always known. However, the radial range when target fly into searching area along tangential direction is a random variable. So the following results are the optimal parameters in different R_x . Here, we suppose $R_t = 140km$, $Q = 0.3$. Fig.8 to Fig.10 are the changing curves of the dwell time, the search frame period and the signal duty circle respectively. Parameters fluctuated at certain value in a small range. From the result we can see, R_x doesn't have great influences when R_x is larger than $40km$. So we choose dwell time as 26ms, the search frame period as 14.7s, and signal duty circle as 22.3%. However, when R_x is less than 40km, the dwell time is less than 24ms, and the search frame period is less than 13s.So in order to meet the required searching performance, we choose dwell time as 24ms, the search frame period as 8.2s, and signal duty circle as 21.3% when R_x is less than 40km.

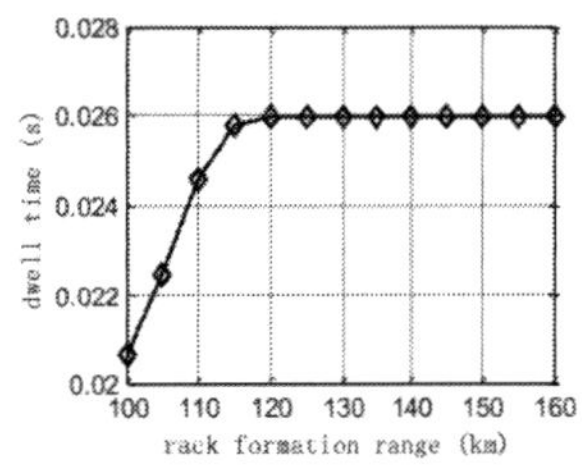

Fig.4. The relationship between dwell time and Rt

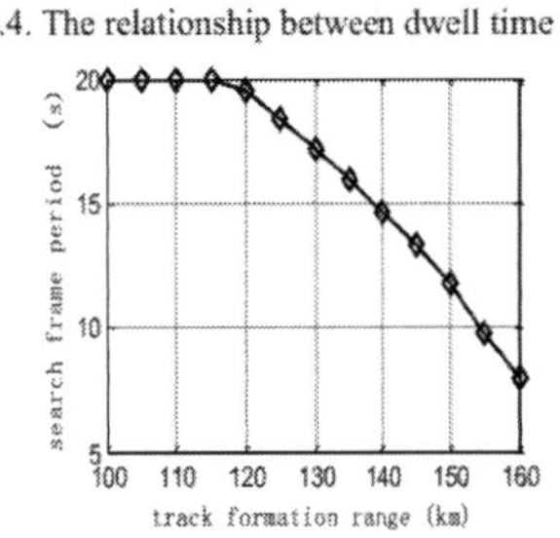

Fig.5. The relationship between search frame period and Rt

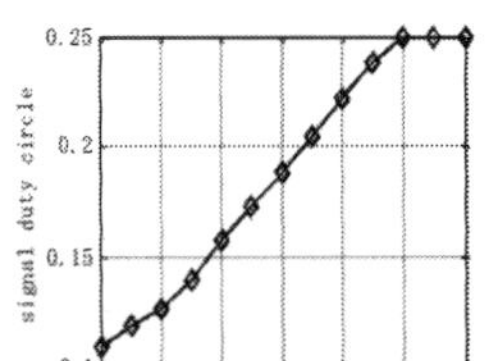

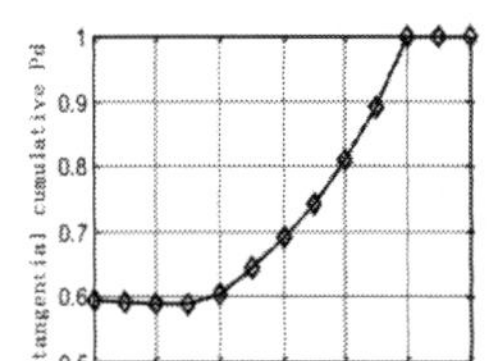

Fig.6. The relationship between signal duty circle and Rt(left)
Fig.7 The relationship between tangential cumulative detection probability and track formation range (right)

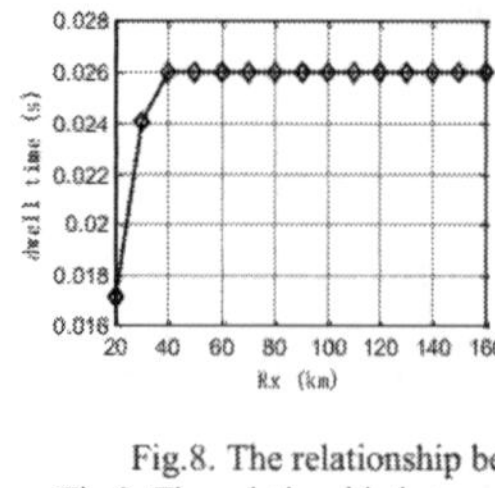

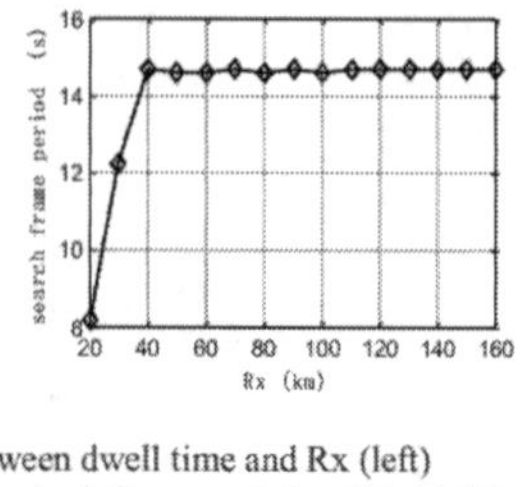

Fig.8. The relationship between dwell time and Rx (left)
Fig.9. The relationship between search frame period and Rx(right)

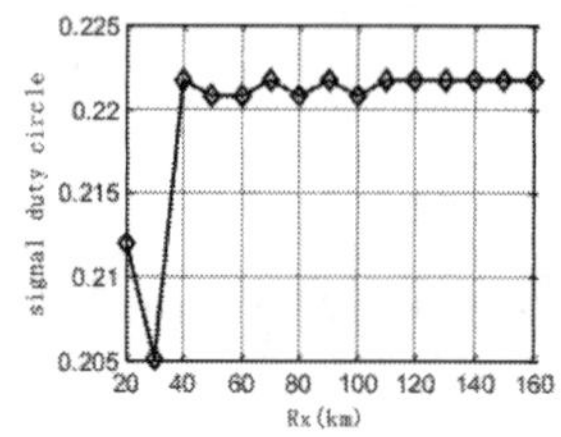

Fig.10. The relationship between signal duty circle and Rx

5 Conclusions

The resource management of phased array radar in searching mode is vital for the multifunction of it. An optimal mode is formulated whose objection is to minimize the total recourse consumption and constraint conditions include tangential cumulative detection probability, radial cumulative detection probability and time limit. The consideration of tangential cumulative detection probability is the main contribution here. The calculation of the tangential cumulative detection probability is given in detail. The constraint considered in this model can guarantee a more reliably searching performance. Simulation results demonstrate the necessity of the tangential constraint and the effectively of the resulting resource management algorithm

References

1. Michael Zatman. "Radar Resource Management for UESA,". IEEE International Conf, 2002, 73-76
2. E. R. Billam, "Parameter optimisation in phased array radar," in Real-Time Management of Adaptive Radar Systems, IEE Colloquium on, London, 1992, pp. 34-37.
3. E. R. Billam, "Design and performance considerations in modern phased array radar," IEE Colloquium on, London, 1982, pp. 15-19.
4. J. B. Lu, W. D. Hu, and W. X. Yu, "Research on optimal search performance of phased array radars with limited resources," Systems Engineering and Electronics, Vol.26, No.10, pp. 1388-1390, May 2004.
5. L. T. Zhang, D. Li, and G. Y. Wang, "A study on search parameters in phased radar," Modern Radar, Vol.30, No.10, pp. 20-25, October 2008.
6. H. R. Zhang, H. W. Yang, and W. X. Yu, "Design of optimal search operation parameters for phased array radar," Acta Armamentarii, vol. 33, no. 9, pp. 1062-1065, Sep.2012.

7. M. Y. Zhang, X. G. Wang, *Radar Systems,* Beijing: Publishing House of Electronics Industry, 4nd, 2013, ch6.
8. Holland J H. *Adaption in Natural and Artificial Systems*. Ann Arbor, USA: University of Michigan Press, 1975.

LFM Radar Convolution Jamming Suppression Based on Oblique Projection in FrFT Domain

Wang Jun, Ping Fulong, Zhao Yuan and Tang Bin

School of Electronic Engineering, UEST of China, Sichuan Chengdu 611731

Abstract. Convolution false-targets jamming against LFM fire-control radars generates range and velocity false targets which are coherent with target echo, which increases the difficulty of jamming detection and suppression and makes the victim radar system lost the track of real target. To combat against this type of jamming, the uncorrelated characteristic between the jamming and echo in FRFT domain is discussed firstly. Thus, an oblique projection operator which is capable of suppress convolution false-targets jamming is built. The simulation results showed the operator have little effect on the echo and proved the effectiveness of this method.

1 Introduction

The function of fire-control radar system is targets tracking through their range and velocity information. This radar system concentrates on one target only when process tracking. Electronic countermeasure (ECM) techniques against fire-control radar systems are enhanced by using digital radio frequency memory (DRFM), which is a challenge for radar systems to combat the impact of DRFM. With the application of DRFM, convolution false targets jamming can be applied as a simple ECM scheme to deceive fire-control radar, therefore making it hard to distinguish the real target from the false one. It also decreases the tracking performance and even makes the system fail to track the real target [1]. Moreover, convolution jamming can also generate multi-targets after CFAR processing, which increases the calculation burden for data processing and blankets the echo from the real target. It is noted that, range-gate pull off (RGPO), if analysed in one PRI, can also be regarded as convolution false range target jamming. RGPO makes fire control radar lost its track from real target and turn to track the false one. Thus, the application of convolution false target jamming against fire-control radars makes the usage of Electronic counter countermeasure (ECCM) with false target suppression performance in the radar systems an exigency.

In order to suppress active false target jamming, Luo [2] proposed the oblique projection based on the uncorrected feature between SMSP jamming and echo signal, by building a projection operator to suppress the jamming. Lu [3] studied pulse diversity to cancel range and velocity false target jamming. This method shows an outstanding property. But it is still a puzzle to cancel jamming when jamming is over-lapped with echo in time or frequency domain.

This paper proposed a jamming suppression method to combat against convolution false target jamming based on the assumption that the radar system has detected the echo through jamming detection. Fortunately, a few researches have focused on LFM active jamming detection like ref. [4]. Based on the previous researches, this paper discusses the uncorrelated characteristic between echoes and jamming based the "thumbtack" feature of LFM signal in FRFT domain [5]. Finally, build an oblique projection operator according to [6] and [7] to suppress the jamming. With the application of the oblique projection, jamming suppression can be processed instantaneously. The paper is organized as follows: Section II models target echo and convolution jamming. Section III proves the uncorrelated property and built an oblique projection operator, the procedure of this method is also shown in this section. Section IV shows the simulation results.

2 Convolution jamming modelling

Consider the setup where the fire-control radar transmits LFM signal for moving target detection. The jammer intercepts and re-transmits modulated jamming to combat fire-control radar system. The modulation can be in time delay, frequency shift and even both.

Herein, $S(t)$ represents the real target echo. In the time domain, $S(t)$ is defined as

$$S(t) = rec\left(\frac{t-\tau}{T}\right)\cdot \exp\left(j\phi_0\right)\cdot \exp\left(j2\pi f_c(t-\tau-\Delta\tau) + jk\pi(t-\tau-\Delta\tau)^2\right) \tag{1}$$

where φ_0 is the initial phase, f_c is the carrier frequency of transmitted LFM signal, $f_d(t)$ denotes the Doppler frequency of real target. $k = B/T_p$ denotes the chirp rate, B and T_p represents the band-width and pulse-width respectively. $\tau = 2R(t)/c$ denotes the delay of echo, while $R(t)$ is the range function of the real target. To simplify the derivation, suppose $\varphi_0 = 0$, and the Doppler frequency $f_d(t)$ is constant for the real target.

The transmitted jamming is

$$J(t) = S(t) * f(t) \tag{2}$$

where $f(t)$ represents the modulation function of the convolution jamming and $*$ denotes the convolution operation.

For range-velocity false target

$$f(t) = \sum_{i=1}^{N_i} \delta(t-\tau_i)\exp\left(j2\pi f_{d_i} t\right) \tag{3}$$

where N_i represents the ith false target, τ_i and f_{d_i} represents the range and velocity of ith false target, respectively. While $\delta(t)$ is the Dirichlet function. Substitute equ.(3) into equ.(2),

$$J(t) = \sum_{i=1}^{N_j} S(t-\tau_i)\exp\left(j2\pi f_{d_i} t\right) \tag{4}$$

the received signal of fire-control radar can be represented as,

$$S_r(t) = S(t) + J(t) + n(t) \tag{5}$$

where $n(t)$ represents the Gaussian white noise , $f(t)$ represents the type of jamming, it can be in range, velocity or the range-velocity convolution jamming.

3 Oblique projection operator modelling

3.1 Basis of Fractional Fourier Transform

Convolution jamming and echo usually over-lap in time and frequency domain. Detection can only indicate the position of echo. But when jamming and echo overlapped in time domain, there will be two target indications in one range gate, and then it is difficult to suppress the jamming simply. Thus, it is necessary to suppress the jamming through signal processing perspective. However, after rotating the "optimal angle" in a fractional domain, the LFM signal will show a "thumbtack" property. According to [5], when rotating the optimal angle α, the definition of $S(t)$ in FrFT domain is given by

$$S^{\alpha}(u) = A_{\alpha}\exp(j\pi u^2\cot\alpha)\cdot T\frac{\sin[\pi(f_0 - u\csc\alpha)T]}{\pi(f_0 - u\csc\alpha)T} \tag{6}$$

$$A_{\alpha} = \frac{\exp[-j\pi\,\mathrm{sgn}(\sin\alpha)/4 + j\alpha/2]}{|\sin\alpha|^{1/2}}, \tag{7}$$

$$f_0 = f_c + f_d$$

Similarly, the corresponding expression of convolution jamming $J(t)$ in FrFT domain can be represented by

$$J^{\alpha}(u) = A_{\alpha}\exp(j\pi u^2\cot\alpha)T\cdot \sum_{i=1}^{N_j}\frac{\sin[\pi(f_0 + \Delta f - u\csc\alpha)T]}{\pi(f_0 + \Delta f - u\csc\alpha)T} \tag{8}$$

where $\Delta f = f_{dj} - k\tau_j$. f_{dj} and τ_j denotes the Doppler frequency and delay of false target, respectively.

Note that in equ.(8),the derivations have the same structure which are the range and velocity false target. This united structure proves that in FrFT domain, the range false target acts the same way as the velocity false target.

According to [8], the definition of fractional order cross-correlation is

$$(S\otimes_{\alpha} J)(v) = e^{j2\pi(v^2/2)\cos\alpha\sin\alpha}\cdot \int S(t)J^*(t - v\cos\alpha)e^{j2\pi tv\sin\alpha}dt \tag{9}$$

where $\otimes_{\alpha}$ denote fractional convolution in rotate angel α.

To simplify the derivation, suppose the delay of echo is zero and the carrier frequency is also set to zero, substitute equ.(1) and equ.(2) into equ.(9).

$$R_{\alpha}(v) = (S\otimes_{\alpha} J)(v) = e^{j\pi v^2(\cos\alpha\sin\alpha - \cos\alpha) + j\pi k\tau^2}\cdot \int_0^T \exp\{j\pi[2v(\cos\alpha - \sin\alpha) - 2k\tau]t\}dt \tag{10}$$

Define

$$\rho_{\alpha} = \frac{R_{\alpha}(0)}{\sqrt{\|S_{\alpha}(v)\|_2\,\|J_{\alpha}(v)\|_2}} \tag{11}$$

as fractional cross correlation coefficient between echo and convolution jamming, $\|\cdot\|_2$ represents 2-Norm of a vector. According to [9], a traditional DRFM jammer will bring an extra time delay greater than 40ns. Assume there is only one convolution range false target, which has the minimized delay τ. Substitute this minimized delay into equ.(11) and it is easy to find out that ρ_{α} is approximates to be zero. Namely, echo and convolution jamming is uncorrelated in FRFT domain. It is observed in (8) that velocity, range-velocity false target act the same in FrFT domain, thus, the fractional cross correlation between convolution jamming and echo is approximately to be zero.

3.2 Oblique Projection Operator

Behrens and Scharf discussed the suppressions of unrespect signal using subspace oblique projection in

[6,7].The projection operator is capable of suppressing the undesired component without information loss.

Define V as a Hilbert space. H is the sub-space of V and $H = H_1 \oplus H_2$, where $\oplus$ denotes the direct sum of sub-space. Herein, H_1 and H_2 is uncorrelated (disjoint) sub-space of H. $S \in \mathbb{C}^{n\times m}$ and $J \in \mathbb{C}^{n\times k}$ are two matrix with full column rank. The range of S is H_1 and the range of J is H_2, they are disjoint.

As defined in [6, 7], the oblique projection operator is

$$\begin{aligned} E_{J|S} &= J(J^H P_S^{\perp} J)^{-1} J^H P_S^{\perp} \\ E_{S|J} &= S(S^H P_J^{\perp} S)^{-1} S^H P_J^{\perp} \end{aligned} \quad (12)$$

where $P^{\perp}$ is orthogonal projector.

The received signal in FrFT domain can be represented as

$$S_r^{\alpha}(u) = S^{\alpha}(u) + J^{\alpha}(u) + FrFT(n(t)) \quad (13)$$

Fire-control radar system usually focus on only one target when process target tracking, based on the theory of oblique projection, the desired echo sub-space is the span of column vector

$$A = span\{S^{\alpha}\} \qquad S^{\alpha} \in C^{n\times 1} \quad (14)$$

Similarly, the undesired jamming sub-space is the span of range and velocity false target jamming rotated α

$$B = span\{J^{\alpha}\} \qquad J^{\alpha} \in C^{n\times 1} \quad (15)$$

According to Section III, S^{α} is uncorrelated with J^{α}. Usually, it is impossible to obtain the orthogonal projector $P^{\perp}$. Thus, according to [10], the statically sub-space oblique projection operator is,

$$E_{A|B} = A(A^H R_B^{+} A)^{-1} A^{*} R_B^{+} \quad (16)$$

where, $R_{s_r} = E[S_r^{\alpha}(u) \times (S_r^{\alpha}(u)^H)]$, $(\cdot)^{+}$ represents The Moore-Penrose Inverse. The formation of operator requires the information of received signal and target component. The range of the oblique projection operator $E_{A|B}$ is $S^{\alpha}(u)$ .And the null space of $E_{A|B}$ is $J^{\alpha}(u)$. Namely

$$E_{AB} \cdot S^{\alpha} = S^{\alpha}, E_{AB} \cdot J^{\alpha} \to 0 \quad (17)$$

According to the orthogonality property between noise space and signal space (including echo and jamming)

$$E_{AB} \cdot FrFT(n(t)) \to 0 \quad (18)$$

Equ(17) and equ(18) show that the estimation of echo in FrFT domain is

$$\hat{S}^{\alpha}(u) = E_{AB} \cdot S_r^{\alpha}(u) \to S^{\alpha}(u) \quad (19)$$

3.3 Procedure of Convolution Jamming Suppression

Step1.Fractional Fourier Transform of received signal according to equ.(6) and equ.(8).

Step2. Target detection and echo extraction

Suppose delay and Doppler frequency of real target is well estimated, the echo reaches its peak at u_0 in the FrFT domain. We can then extract target component in FrFT domain through fractional filtering. The residue is applied to estimate jamming component.

Step3.Formation of oblique projection operator based on equ.(16)and equ.(19).

Step4.Processing inverse FrFT to re-construct echo

Step5.Jamming suppression evaluation

Signal to noise jamming ratio improvement factor I_{SJNR} is applied to evaluate the effect of the oblique projection operator. I_{SJNR} defined as

$$\begin{aligned} I_{sjnr} &= SJNR_{out} - SJNR_{in} \\ &= 10\log\left(\frac{\|\mathbf{A}\theta\|_2^2}{\|y - \mathbf{A}\theta\|_2^2}\right) - 10\log\left(\frac{\|\mathbf{A}\theta\|_2^2}{\|x - \mathbf{A}\theta\|_2^2}\right) \end{aligned} \quad (20)$$

The likelihood factor is applied to evaluate the distortion after jamming suppression.

$$f_l = \frac{S_t(t) \cdot S_e^{*}(t)}{\sqrt{|S_t(t)|^2 \cdot |S_e(t)|^2}} \quad (21)$$

The simulations are carried out in order to verify the effective of the method proposed.

4 Simulations

4.1 Range False Target Suppression

The simulations are carried out in the following scenario, the bandwidth of transmitted LFM signal $B = 5MHz$ and the pulse-width $T_p = 10\mu s$. There exist three false targets 100m, 300m and 750m away from the real target. It is noted that the target 100m away from the real target falls in the same range gate, thus it is impossible to cancel the jamming through range gate. These false targets have the same Doppler frequency as the echo. Fig.1 shows the improvement factor of signal to jamming ratio I_{SJNR} through 100 times Monte Carlo experiment when SJR is $0dB$ and $-3dB$, while SNR ranges from $-6dB$ to $6dB$ with step of $1dB$.

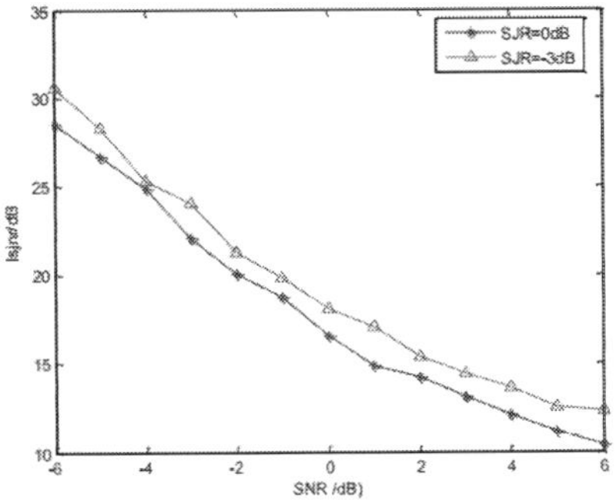

Figure 1. The improvement factor

From the simulation result shown in Fig.1, this method is valid to suppress the Range false targets and improve SJNR. However I_{SJNR} shows a downward trend with SNR which is reasonable because the oblique projection operator can suppress noise simultaneously.

Fig.2 shows the likelihood of target echo indicates the distortion after jamming suppression with the same simulation scenario

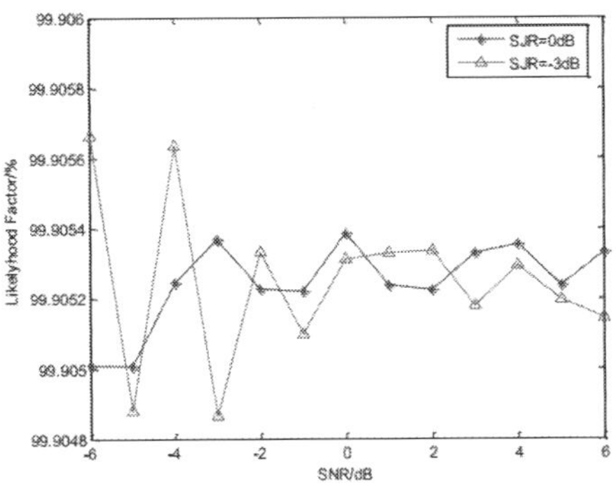

Figure 2. Likelihood factor

Fig.2 shows the echo after jamming suppression does not have too much distortion. The likelihood factor remains approximately 1 which indicates that the projection operator is capable of preserve the echo component through jamming suppression.

4.2 Range-Velocity Joint False Targets Suppression

The simulations are carried out in the following scenario: The bandwidth of Radar and the pulse duration is the same as above, to simplify the analysis, supposing the carrier frequency of the fire-control radar is $1GHz$. There exist three range-velocity joint false targets at 300m, 750m and 2400m away from echo. However, these false targets have the same Doppler frequency equals to $2KHz$.

Fig.3 shows the improvement factor of signal to jamming ratio I_{SJNR} through the 100 times of Monte Carlo experiment under which the SJR is $0dB$ and $-3dB$, while SNR ranges from $-6dB$ to $6dB$ with a step of $1dB$.

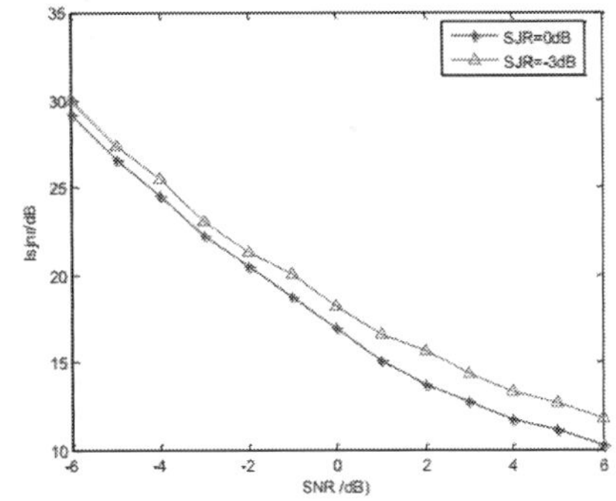

Figure 3. The improvement factor

According to Fig.3, this method is also valid for range-velocity joint jamming suppression. Same as range false target suppression, this improvement factor shows a downward trend.

Simulation in Fig.4 shows the likelihood of target echo.

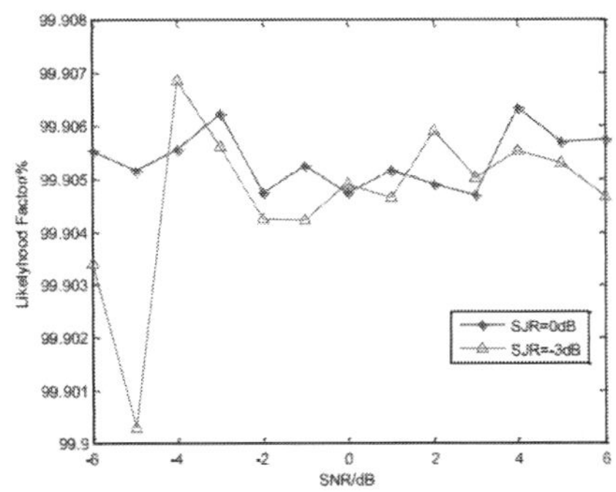

Figure 4. Likelihood factor

Fig.4 shows the signal after jamming suppression which does not have too much distortion.

5 Conclusion

In this paper, the technique based on the oblique projection operators in FRFT domain is proposed. Suppression of convolution jamming and noise can be realized simultaneously. Once a target is detected in the range-gate or velocity-gate of the fire-control radar, the information in FRFT domain can be obtained. The simulation results indicate that the method based on oblique projection in FRFT domain is efficient for false target suppression.

6 Acknowledgment

This work Project supported by the National Defense Pre-Research Foundation of China (NO.9140A07030713DZ02101).

References

1. W. Zhang, X. Zhou, and Y. Tan, "Analysis of the range false target deception jamming effect against fire control radar," Sichuan Ordnance Journal, 28(4), (2007).
2. S. C. Luo, "An algorithm of deception jamming suppression based on blind signal separation," Journal of Electronics & Information Technology, 33(12), 2801-2806, (2011).
3. G. Lu, B. Tang, and G. Gui, "Deception ECM signals cancellation processor with joint time-frequency pulse diversity," IEICE Electronics Express, 8(19), 1608-1613, (2011).
4. R. Chen and Y. M. Wang, "Detection of frequency-shifting jamming to LFM pulse-compression radar based on FRFT," Radar Science & Technology, (2013).
5. S. Peleg and B. Porat, "Linear FM signal parameter estimation from discrete-time observations," IEEE Trans on Aerospace and Electronic Systems, 27(4), 607-616, (1991).
6. R. T. Behrens and L. L. Scharf, "Signal processing applications of oblique projection operators," IEEE Transactions on Signal Processing, 42(6), 1413-1424, (1994).
7. R.T. Behrens and L. L. Scharf, " Corrections to 'signal processing applications of oblique projection operators' [Correspondence]," IEEE Trans. on Signal Processing, 44(5),1300, (1996).
8. X. J. Yang, "Theory and applications of local fractional fourier analysis," Chinese Journal of Mechanical Engineering, 1(4), (2012).
9. D. Curtis Schleher. Electronic warfare in the information age, MA: Artech House, Norwood,Massachusetts, (1999).
10. M. L. Mccloud and L. L. Scharf, "A new subspace identification algorithm for high-resolution DOA estimation," IEEE Transactions on Antennas & Propagation, 50(10), 1382-1390, (2002).

A U-Shaped Slot UWB Antenna with Flexible and Wide Tunable Dual Notch Band

Zhongmin Zhang and Yongchang Sun

College of Information and Communication Engineering, Harbin Engineering University, Harbin, China

Abstract. A coplanar waveguide (CPW) fed ultra-wideband (UWB) antenna with flexible and wide tunable dual band-notched characteristics is proposed in this paper. The dual band-notched function is achieved by using an U-shaped slot inserted into the ellipse radiation patch and by using an elliptic parasitic slit placed near the ground plane. The wide tunable band-notched characteristic is implemented by adjusting the length of U-shaped slot and by adjusting the length of ellipse parasitic slit. The design aims to achieve wide reconfigurable band-notched features on the UWB antenna. The simulated results indicate that the proposed antenna has a wide bandwidth (VSWR under 2) from 2.9GHz to 12.6GHz with fractional bandwidth of 125%, and has a wide tunable notch band center frequency from 4.5GHz to 12.4GHz.

Keywords. Ultra-wideband; slot antenna; notch band; wide tunable; U-shaped

1 Introduction

In recent years, the attention on ultra-wideband (UWB) communication technology has been increasing due to the Federal Communication Commission (FCC) allocated the UWB spectrum from 3.1GHz to 10.6GHz [1]. UWB communication has many advantages, such as wide bandwidth, high data rate wireless transmission, low power consumption, and so on [2]. However, in such a wide UWB frequency spectrum, there are several narrow-band systems like wireless local area network (WLAN) working at 5.15-5.825GHz, ITU 8GHz band operating at 7.725-8.275GHz, WiMAX system operating at 3.3-3.7GHz, HIPERLAN/2 in Europe working at 5.15-5.725GHz, X-band satellite communication links operating at 7.9-8.4GHz [1] [3] [4]. These narrow-band signals may interference with UWB communication system. To avoid the interference from other nearby frequency, UWB communication systems need a notch band function. The solution mainly has the following points. One is add some band-stop filter components [5], but this method increase the complexity of the system. The other one is using the antenna which integrated many band-notched techniques. Such as electromagnetic band gaps (EBG) [6], T-shaped (tree-like) elements [7], band-rejected slot in feed line [8], rectangular slot [9], parasitic slits [10], L-shaped slots [11] and Ring-Resonator slot [12]. However, these antennas mostly aimed at certain frequencies, such as WLAN or WiMAX, and cannot achieve a wide range of adjustments to reconfigure the notch band.

With the development of science and technology, more and more in narrow-band communication systems may join in the band of 3.1GHz to 10.6GHz. In order to avoid these potential interference, reconfigurable become another notable direction of UWB antenna design, which can be used as single band-notched, dual band-notched, or completely as the normal UWB antenna. Most of these antennas used to adjust the corresponding ideal switch, to achieve a wide tunable notch frequency [7], [13]-[15]. But, these antennas have a complex structure, which is hard to adjust or fabricate. One effective way is designing a wide range adjustment antenna with simple structure, so we can adjust the performance of the antenna conveniently.

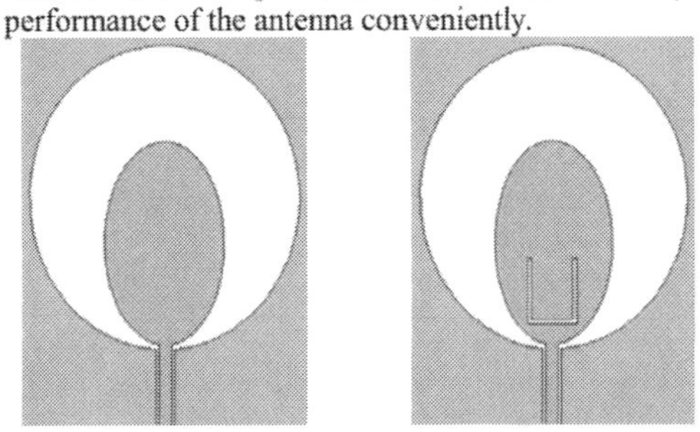

(a) Antenna-1 (b) Antenna-2

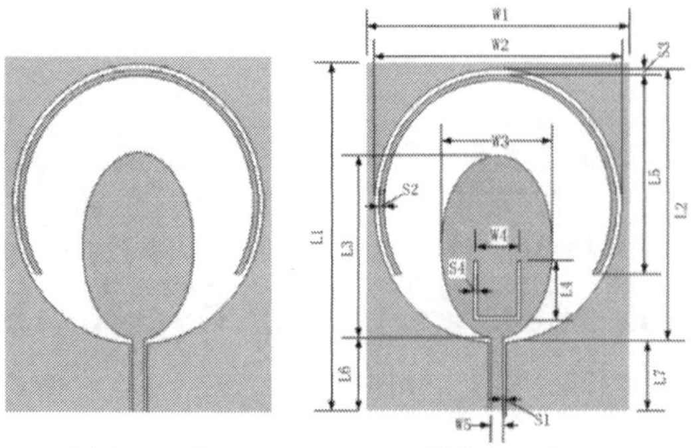

(c) Antenna-3 (d) Antenna-4

Figure 1. Proposed antenna geometry and design evolution

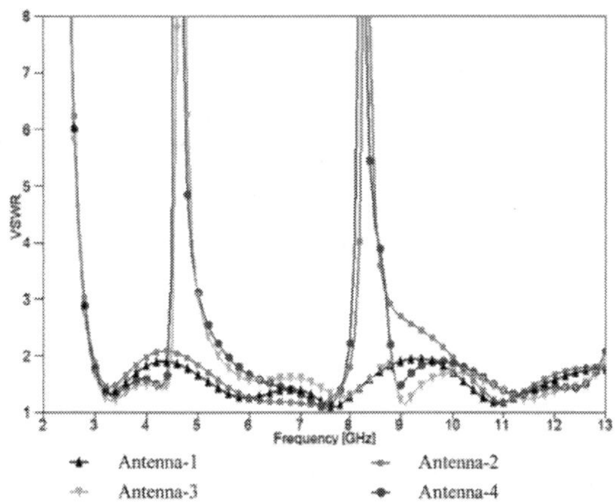

Figure 2. Four antennas band-notched characteristic

In this paper, we propose a simple structure and low-cost coplanar waveguide (CPW) fed ultra-wide band (UWB) antenna with a wide range tunable dual notch band. The dual band-notched function is implemented by inserting an U-shaped slot inside the ellipse radiation patch and by using an elliptic parasitic slit placed beside the ground plane. The center frequencies of these notch bands can be tuned by adjusting the length of the U-shaped slot and the elliptic parasitic slit. It is observed from simulated results that the proposed UWB antenna could operate on 2.9GHz to 12.6GHz with the VSWR less than 2, and the notch bands center frequency could be adjusted from 4.5GHz to 12.4GHz. In addition, the proposed antenna can also be used as a single notch band UWB antenna or normal UWB antenna by removing the U-shaped slot or elliptic parasitic slit, without use any switch, which can achieve a wide stop-band adjusting range. Therefore, the designed antenna has simple structure, wide tunable dual notch band, omnidirectional radiation patterns, small size (30.6mm x 23mm), and easy reconfigurable functions, which makes it very suitable for UWB communication system. The details of the design procedure and the performance of the proposed antenna are described in the following sections.

2 Antenna design

The geometry and design evolution of proposed UWB antenna is displayed at Figure 1. First of all, an UWB basic antenna without U-shaped slot and elliptic parasitic slit is illustrated in Fig. 1(a) and represented as Antenna-1. This antenna contains an elliptic slot etched near the ground plane, an ellipse radiation element, and a 50 Ohm CPW-fed structure. The relative dielectric constant and loss tangent of 1.6mm thickness substrate is 2.65 and 0.002, respectively. The CPW-fed structure contains a transmission line with a gap between the feed-line and the ground plane is S1 = 0.2mm, and the width of transmission line is W5 = 1mm.

To achieve band-notched functions, U-shaped slot is inserted in ellipse radiation patch, as shown in Fig. 1(b) and elliptic parasitic slit is placed beside the ground plane, as shown in Fig. 1(c). These antennas are both single band-notched topologies which are denoted as Antenna-2 and Antenna-3, respectively. In order to implement the dual band-notched feature, U-shaped slot and elliptic parasitic slit are both employed to obtain two band-stop characteristic in the meanwhile. The proposed antenna is shown in Fig. 1(d), and defined as Antenna-4.

By adjusting the length of the U-shaped slot and the elliptic parasitic slit, we can control the center frequency of two notch bands independently. In the design, one notch band which operating at 5GHz WLAN band, is realized by placing the elliptic parasitic slit, we can tune this notch band center frequency by changing the length of L5, and the other stop band to prevent the interference from the X band application which operating at 7.9-8.8GHz, is achieved by inserting the U-shaped slot to the ellipse radiation patch, we can tune this notch band center frequency by modifying the length of L4. The notch band adjustment does not affect each other.

A full-wave simulation solver, High Frequency Structure Simulator (HFSS), is used to model the antenna and give the simulation results. The proposed antenna optimized parameters are shown below: L1 = 30.6mm, W1 = 23mm, L2 = 24mm, W2 = 21.6mm, L3 = 16mm, W3 = 9.6mm, L4 = 5.4mm, W4 = 4mm, L5 = 17.5mm, W5 = 1mm, L6 = 6.4mm, L7 = 6mm, S1 = 0.2mm, S2 = 0.5mm, S3 = 0.5mm, S4 = 0.3mm.

3 Notch band characteristic

We analyze the notch band characteristic of the proposed antenna by HFSS in this section. Fig. 2 shows the standing-wave ratio (VSWR) of four antennas mentioned above. We can find that the band-notched performance of these antennas is an obvious difference. For antenna-1, there is no U-shaped slot and no elliptic parasitic slit, which has a bandwidth of 9.7GHz without any notch band. It can be used as fundamental UWB antenna which covers all UWB band. The antenna-2 is based on the antenna-1, increase a U-shaped slot in the radiation patch which makes it become a single band-notched antenna. The notch band is operating at 7.9-8.8GHz. It is designed for preventing the interference from satellite communication (7.9-8.4GHz) and any other system which works within this band. The antenna-3 is also based on the antenna-1, place an elliptic parasitic slit near the ground plane. This antenna has single notch band features which operating at 4.5-5.8GHz. This band stop function can be used for mitigating the interference from WLAN (5.15-5.825GHz), C band system (part of 4.5-5GHz), and so on. The antenna-4 inherited the merit of the antenna-2 and antenna-3, makes it becomes a dual band-notched UWB antenna.

The simulation results, as shown in Fig. 3, demonstrate the current distribution of each antenna. The Fig. 3(a) and Fig. 3(b) display the current distributions of antenna-2 which operates at 4.8GHz and 8.2GHz, respectively. In this case, the current at 4.8GHz is working normally but the current at 8.2GHz concentrates on the U-shaped slot which consumes a lot of energy. While the Fig. 3(c) and (d) are indicate that the elliptic parasitic slit of antenna-3 consumes a large of energy at 4.8GHz but 8.2GHz works well. Fig. 3(e) and (f) illustrate the current distribution of antenna-4 at 4.8GHz and 8.2GHz. We can find that the current of 4.8GHz and 8.2GHz are both notched by elliptic parasitic slit and U-shaped slot. Compare with the results of the above, we can also be found that the stop-band performance of antenna-4 is almost equal to the superposition of the antenna-2 and antenna-3, in other words, they can work together and not affect each other.

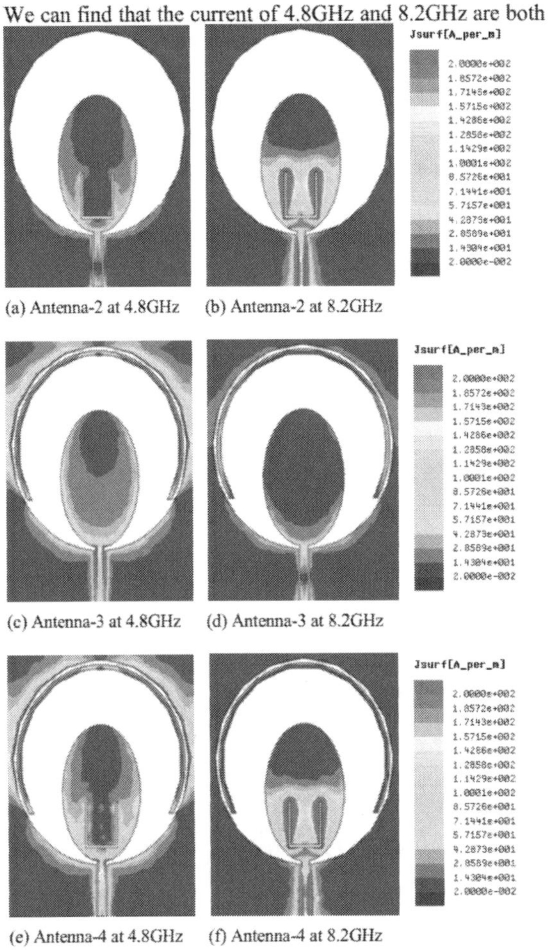

(a) Antenna-2 at 4.8GHz (b) Antenna-2 at 8.2GHz

(c) Antenna-3 at 4.8GHz (d) Antenna-3 at 8.2GHz

(e) Antenna-4 at 4.8GHz (f) Antenna-4 at 8.2GHz

Figure 3. Current distribution on corresponding frequency of each antenna

Combined with Fig. 2 and Fig. 3, we can make a conclusion that the antenna-4 has two notch band both 4.5-5.8GHz and 7.9-8.4GHz, which the U-shaped slot contribute to the higher notch band and the elliptic parasitic silt lead to the lower notch band, and these notch band can work independently. Therefore, the antenna-4 can reduce all of the potential interference mentioned above. Besides, the single notch band UWB antenna can be obtained by removed the U-shaped slot or elliptic parasitic slit from antenna-4 and the procedure of reconfigure antenna-4 as a conventional UWB antenna is the same.

4 Parametric effects

In this section, we will study the influence of various

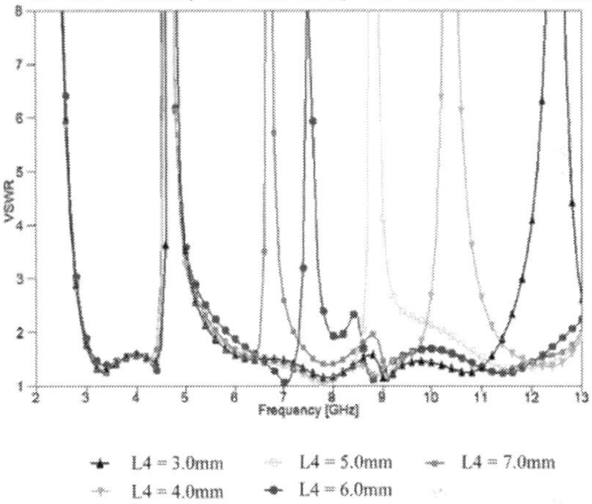

Figure 4. Effects of L4

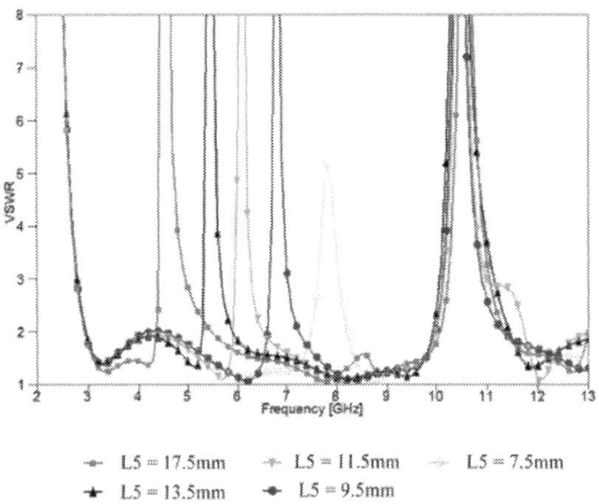

Figure 5. Effects of L5

antenna parameters on the stop-band characteristics. The HFSS is employed to obtain the simulation results and all the simulation target is antenna-4. Fig. 4 shows the effects of varying L4. We can see that the parameter L4 has a great influence on the higher stop band. With the length of the L4 increasing, the center frequency of the higher notch band moves to low frequency and the bandwidth of the notch band is reduced, while the lower notch band hardly changes. The higher notch band can be tuned from 6.6GHz to 12.4GHz, which is a very large tunable range. This is because the varying L4 can largely adjust the total length of the U-shaped slot, which also increases the current path along the U-shaped slot in a wide tunable range. We can adjust it to prevent the potential interference like ITU 8GHz band application, satellite communication system, X band system, and any other narrow band systems which operating from 6.6-12.4GHz. The effects of varying L5 are shown in Fig. 5. With the increasing of L5, the center frequency of higher notch band and the bandwidth of the notch band are both changed slightly, while the center frequency of lower notch band moves to low frequency obviously. Thus, the varying L5 is also a key factor, same as L4, influencing the corresponding notch band center frequency, which can be tuned from 4.5GHz to 8.3GHz. This range covers HIPERLAN/2 (5.15-5.725GHz), WLAN (5.15-5.825GHz) and also covers ITU 8GHz band (7.725-8.275GHz). Fig. 6 shows the effects of varying S2. With the increment of S2, the center frequency is few changed both lower notch band and higher notch band, but the bandwidth of lower notch band is expanded. The effects of varying S3 are demonstrated in Fig. 7. It is indicated that the center frequency of lower notch band moves to high frequency and the bandwidth of lower notch band is reduced and the bandwidth of higher notch band is increased with the increment of S3. The reason is that the increased S3 has influence on the total length of the elliptic parasitic slit, and also affect the coupling between ground plane and elliptic parasitic slit. Besides, adjusting S3 can also alter the coupling between radiation patch and elliptic parasitic slit. Thus, we can find that the bandwidth of higher notch band caused by U-shaped slot is also changed slightly. Fig. 8 describes the effects of S4. With the increment of S4, the center frequency of higher notch band moves to high frequency and the lower notch band is not changed. The effects of varying W4 are displayed in Fig. 9. It can be seen that the center frequency of higher notch band is decreased with the increment of W4. In addition, the bandwidth of higher notch band is also improved because of increasing

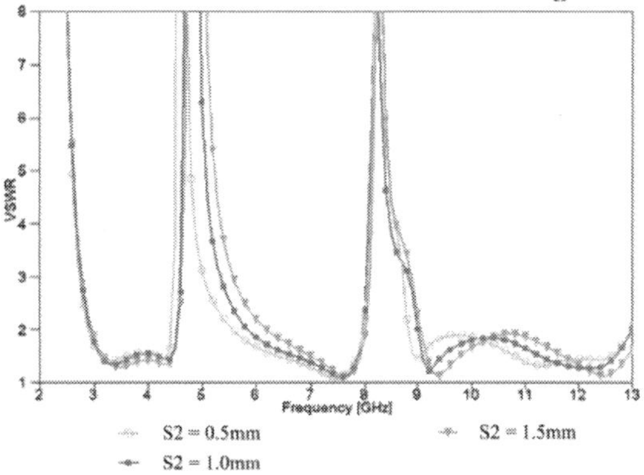

Figure 6. Effects of S2

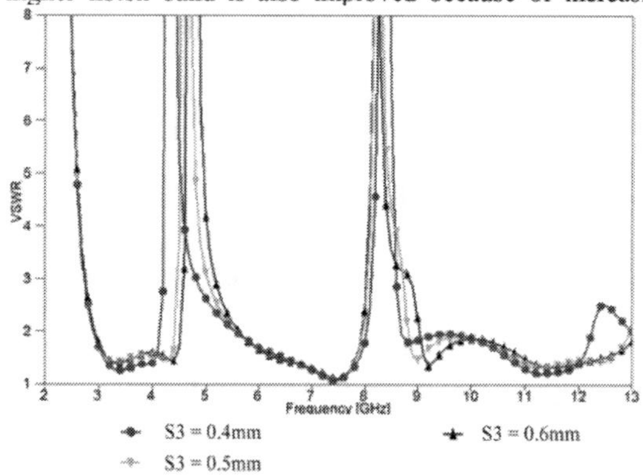

Figure 7. Effects of S3

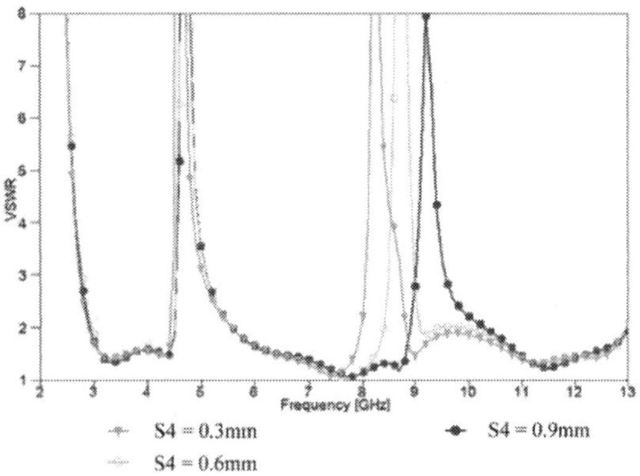

Figure 8. Effects of S4

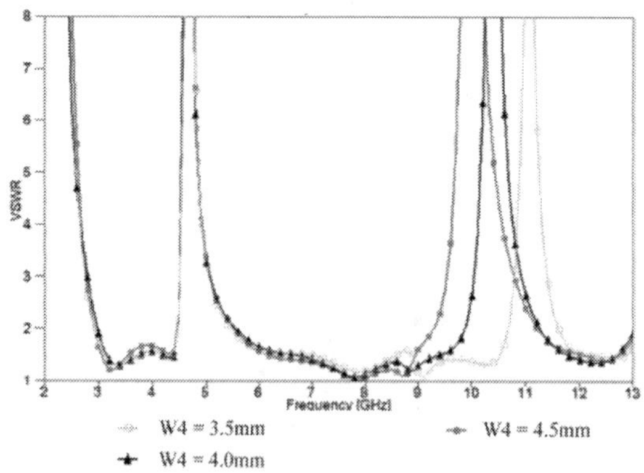

Figure 9. Effects of W4

DOI: 10.1051/matecconf/20165907007

W4 not only change the resonance length of the U-shaped slot but also alter the distributive capacitance and distributive inductance among two parallel branches of the U-shaped slot.

From what has been discussed above, the major parameters that affect the stop band center frequency are L4 and L5, and the major parameters that affect the bandwidth of corresponding notch band are S2 and W4. We do not recommend adjusting S3 and S4 because either the tuned results are not very obviously, can be replaced by other parameters or these parameters can cause some uncertainly effects, which had been displayed in Fig. 7, 12-13GHz.

5 Radiation

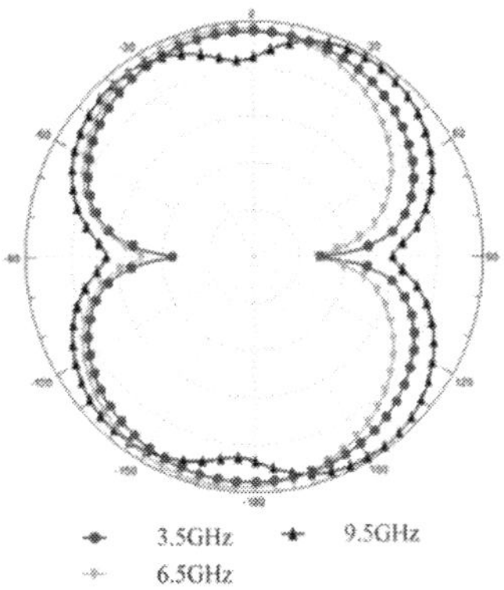

(a) X-Z plane

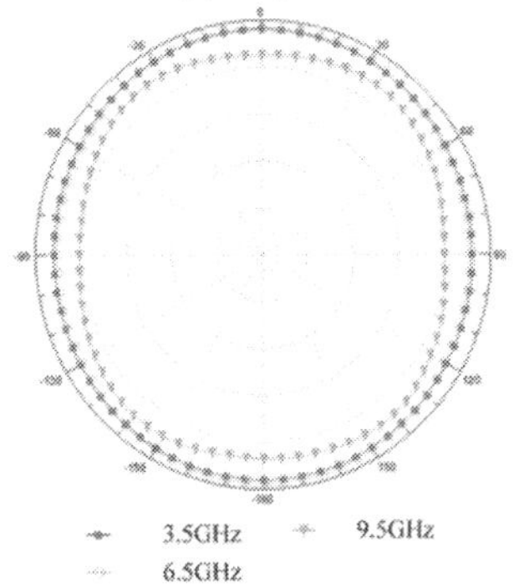

(b) Y-Z plane

Figure 10. Radiation patterns of antenna-4

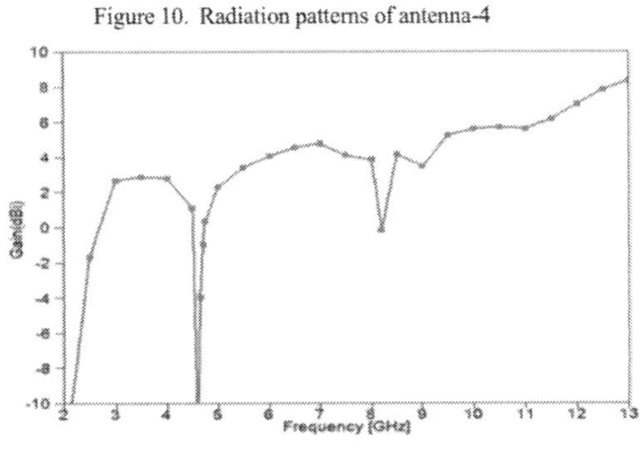

Figure 11. Peak gain of antenna-4

The radiation patterns of the antenna-4 at 3.5GHz, 7.5GHz and 11.5GHz are demonstrated in Fig. 10. We can see that the proposed antenna has a good omnidirectional radiation characteristic in the Y-Z plane and monopole-type radiation characteristics in the X-Z plane. Despite the slight deterioration at higher frequency, which is caused by inserting the U-shaped slot to the radiation patch, the proposed antenna still has good omnidirectional radiation characteristics. Fig. 11 illustrates the peak realized gain with frequency for the designed antenna over UWB spectrum range (2-13GHz). It can be seen that, as discussed above, the peak gains of two notch bands which operating at 4.5GHz and 8.2GHz are dropped, and decrease about -10.2dBi in the lower notch band and about -0.3dBi in the higher notch band. Besides, following the higher notch band, we can see the gain has a little drop, but the gain in this wave trough is still greater than 2dBi. Therefore, the little mismatch is not enough to affect the normal work of the proposed antenna.

6 Conclusion

In this paper, we designed a simple, low cost, and wide tunable dual notch band UWB antenna. The wide adjustable stop bands are realized by U-shaped slot and elliptic parasitic slit, respectively. These notch bands work independently, and we can adjust the notch band features by some parameters. For the center frequency of these notch bands, we can change the parameter L4 and L5 to achieve a wide tunable range. For the bandwidth of these notch bands, we can control the parameter S2 and W4 to obtain a proper bandwidth. In addition, the proposed antenna can also be reconfigurable as a single notch band antenna or a conventional UWB antenna by removing the U-shaped slot or the elliptic parasitic slit. Besides, the proposed antenna has omnidirectional radiation pattern, and the peak gain also indicates the good performance of the proposed antenna.

Acknowledgment

This paper is funded by the International Exchange Program of Harbin Engineering University for Innovation-oriented Talents Cultivation and this paper is also supported by National Nature Science Fund of China (No. 61101141), the authors wish to thank for these supports.

References

1. D. Sarkar, K. V. Srivastava, and K. Saurav, "A Compact Micro strip-Fed Triple Band-Notched UWB Monopole Antenna," *IEEE Antennas and Wireless Propag. Lett.*, vol. 13, pp. 396-399, 2014.
2. Z. H. Tu, W. A. Li, and Q. X. Chu, "Single-Layer Differential CPW-Fed Notch-Band Tapered-Slot UWB Antenna," *IEEE Antennas and Wireless Propag. Lett.*, vol. 13, pp. 1296-1299, 2014.
3. R. Azim, M. T. Islam, and A. T. Mobashsher, "Dual Band-Notch UWB Antenna With Single Tri-Arm

Resonator," *IEEE Antennas and Wireless Propag. Lett.*, vol. 13, pp. 670-673, 2014.

4. C. T. Chuang, T. J. Lin, and S. J. Chung, "A Band-Notched UWB Monopole Antenna With High Notch-Band-Edge Selectivity," *IEEE Trans. Antennas and Propag.*, vol. 60, No. 10, pp. 4492-4499, Oct. 2012.
5. B. Mohammadi, A. Valizade, P. Rezaei, and J. Nourinia, "New design of compact dual band-notch ultra-wideband bandpass filter based on coupled wave canceller inverted T-shaped stubs," *IET Microw. Antennas Propag.*, vol. 9, Iss. 1, pp. 64-72, Aug. 2014.
6. H. A. Majid, M. K. A. Rahim, M. R. Hamid, N. A. Murad, and A. Samsuri, "Reconfigurable Band Notch UWB Antenna using EBG Structure," presented at the IEEE Asia-Pacific Conference on Applied Electromagnetics, Johor, Malaysia, December 8-10, 2014.
7. M. Naser-Moghadasi, R. A. Sadeghzadeh, T. Sedghi, T. Aribi, and B. S. Virdee, "UWB CPW-Fed Fractal Patch Antenna With Band-Notched Function Employing Folded T-shaped Element," *IEEE Antennas and Wireless Propag. Lett.*, vol. 12, pp. 504-507, 2013.
8. F. Zhu, S. Gao, A. TS Ho, R. A. Abd-Alhameed, C. H. See, T. W C Brown, J. Li, G. Wer, and J. Xu, "Multiple Band-Notched UWB Antenna With Band-Rejected Elements Integrated in the Feed Line," *IEEE Trans. Antennas and Propag.*, vol. 61, No. 8, pp. 3952-3960, Aug. 2013.
9. M. Ojaroudi, and N. Ojaroudi, "Ultra-wideband Small Rectangular Slot Antenna With Variable Band-Stop Function," *IEEE Trans. Antennas and Propag.*, vol. 62, No. 1, pp. 490-494, Jan. 2014.
10. R. Azim, M. T. Islam, A. T. Mobashsher, "Design of a Dual Band-Notch UWB Slot Antenna by Means of Simple Parasitic Slits," *IEEE Antennas and Wireless Propag. Lett.*, vol. 12, pp. 1412-1415, 2013.
11. J. W. Wang, J. Y. Pan, X. N. Ma, and Y. Q. Sun, "A Band-Notched UWB Antenna With L-Shaped Slots and Open-Loop Resonator," in *Proc. of 2013 IEEE International Conf. on Applied Superconduc-tivity and Electromagnetic Devices*, Beijing, 2013, pp. 312-315.
12. K. D. Xu, Y. H. Zhang, R. J. Spiegel, Y. Fan, W. T. Joines, and Q. H. Liu, "Design of a Stub-Loaded Ring-Resonator Slot for Antenna Application," *IEEE Trans. Antennas and Propag.*, vol. 63, No. 2, pp. 517-524, Feb. 2015.
13. Y. S. Li, W. X. Li, and Q. B. Ye (October 2013). A CPW-Fed Circular Wide-Slot UWB Antenna with Wide tunable and Flexible Reconfigurable Dual Notch Bands. *The Scientific Word Journal* [Online]. Volume 2013, Article ID: 402914, 10 pages. Available: http://dx.doi.org/10.1155/2013/402914
14. Y. S. Li, W. X. Li, "A Circular Slot Antenna with Wide Tunable and Reconfigurable Frequency Rejection Characteristic Using Capacitance Loaded Split-Ring Resonator for UWB Applications," *Wireless Personal Commu.*, vol. 78(1), pp. 137-149, Mar. 2014.
15. D. Zhao, L. Lan, Y. Han, F. Liang, Q. Zhang, and B. Z. Wang, "Optically Controlled Reconfigurable Band-Notched UWB Antenna for Cognitive Radio Applications," *IEEE Photonics Tech. Letters*, vol. 26, No. 21, pp. 2173-2176, Nov 2014.

MATEC Web of Conferences 59, 07008 (2016) DOI: 10.1051/matecconf/20165907008
ICFST 2016

Multiple target localization in WSNs using compressed sensing reconstruction based on ABC algorithm

Hao Feng[1,2], Junyan Chen[2] and Lei Luo[1]

[1]*School of Computer Science and Engineering, University of Electronic Science and Technology of China, Chengdu 611731;*
[2]*Guangxi Colleges and Universities Key Laboratory of cloud computing and complex systems, Guilin University of Electronic Technology, Guilin 541004, China*

Abstract. In allusion to the application of compressed sensing theory in multi-target localization in wireless sensor networks, the paper proposes a method of sparse signal reconstruction based on artificial bee colony algorithm and designs partial update strategy of location coding. A mode of disturbances mutation operation for food source exploitation is designed to improve the search efficiency in high-dimensional solution space; it also improves the global search efficiency of the algorithm. The comparing of simulation analysis with other algorithms verifies that the algorithm effectively solves problem of multi-target localization in wireless sensor networks and has better resistance to interference.

1 Introduction

Localization service as one of the important basic functions of the wireless sensor network [1], whether for the location information of the sensor node itself or for the targeting localization of the monitored area, is the basis of other applications such as environmental detection, target tracking, and structural health monitoring [2]. It can be said that the wireless sensor network will not have any effect without the location information [3]. Thus localization has been a challenging hot area on wireless sensor network. Common localization algorithms are those based on RSS (Received Signal Strength) and its improved algorithm [4], ToA [5], TDOA [6], AOA [7], and so on. Comparing with other algorithms the RSS-based algorithms has advantages as no need for special hardware, low cost, and easy access thus it has been widely studied and applied. In these localization algorithms, because of the limits of computing ability and energy consumption in sensor node, centralized algorithm has higher accuracy than distributed one. However, like other monitoring applications, in a centralized algorithm, each sensor node responsible for perception of location information needs to send the data of signal strength of each target to the sink, then the sink node with strong computing power runs centralized algorithms to recover the location of the target. In the application of multi-target localization, with a lot of wireless communication expenses, this results in rapid consumption of onboard energy in sensor node and decline in the performance of the communication network.

Compressed sensing is a new theory of data compression and reconstruction in recent years [8], the low requirements for sampling of compressed sensing system coincides with the characteristics of constrained resource in sensor nodes, resulting in its quick application in data collection, topology-aware and other aspects of wireless sensor networks. Meanwhile compressed sensing also offers new ideas and methods for the study of localization issues of wireless sensor networks. Cevher et al.[9] studied WSNs localization issues with CS, but this method requires a localization dictionary of sensor nodes, which increases storage expense of the overhead int the sensor node, thus it's unacceptable for the low-cost sensors with small memory. A grid-based WSNs localization problem is modeled as N-dimensional reconfiguration when K sparsity vectors are 1 [10]; compressed sensing is used to achieve multi targeting in a wireless sensor network; the shortcomings of this approach is that WSN data can't be further compressed, which is, to M sensor nodes under the situation of K targets, M×K perception data need to be transferred. In recent years, among some works [11] [12] [13] [14], target location has been studied via the model and solving method with compressed sensing respectively, and has made some progress.

This paper presents a reconstruction method of compression perception based on discrete artificial bees colony algorithm to localize multi-target in wireless sensor network. This method transfers the location information of target into a discrete encoding of food source in artificial honey bees system, obtaining suboptimal information of location reconstruction by foraging bees iterate operation, therefore achieves signal reconstruction with the unknown sparsity, consequently by constructing an effective measurement matrix and sparse matrix, the WSNs targeting problem is transformed into reconstruction problem of compressed sensing signals, which enables multi-target localization in WSNs. Finally, it shows better performance by verifying the effectiveness of the proposed method through simulation analysis and comparing with other recovery algorithms.

2 Problem Description and System Model

Assuming the square monitored region is divided into N grids, in this monitored region, we deploy randomly M position-known sensors. There are K (K is unknown) targets to be located in the entire monitored area. Sensor nodes are used to measure the signals of energy of the targets within the monitored region, finally to recover the position information of the targets based on grids in the fusion center.

The system works as follows: each target periodically transmits signal by broadcasting, the transmission cycle is T, and each target is independent without synchronization requirement. The sensor node periodically collects signals, the collection cycle is T as well, the node simply adds up the strength values of the received signals within the cycle. After the end of the time slice of the cycle, each sensor node sends the results collected to the fusion center, then the center runs signal reconstruction algorithm and gets the ultimate locations of targets in the grid.

Assuming the energy value emitted by K signals can be linear superposition, then the energy e_i received by i-th sensor is:

$$e_i = \sum_{j=1}^{K}(P_{i,j} + N_i),\ P_{i,j} = P_0 - 10n_p \lg(D_{i,j} / D_0) \quad (1)$$

$N_i \sim N(0,\delta_i^2)$ is the Gaussian white noise; $P_{i,j}$ is signal strength received of the k-th target by i-th sensor, the measure unit is dB; P_0 represents the received signal strength when the reference distance is D_0; n_p is the path attenuation exponent with typically values between 2 and 4; $D_{i,j}$ is the Euclidean distance between the target locations and sensor i.

For N grids, the target position is defined as the grid center, the available vector of target location can be represented as $s=[s_1, \ldots, s_i, \ldots, s_N]^T$. When there is targets in the i-th grid, $s_i=1$; otherwise $s_i=0$. Obviously s is the $N \times 1$ vector (K << N) when sparsity degree is K. Therefore WSNs multi-targeting problem can be converted to the sparse approximation problem with compressed sensing theory and sparse signal reconstruction method for the target's sparse vector in spatial location. Interference noise increases the uncertainty of signal reconstruction and challenges the compression-aware application in multi-target localization.

3 Compressed Sensing Localization and Measurement Matrix Construction

The general process of compressed sensing [15] [16] is: a known measurement matrix $\boldsymbol{\Phi}$ belongs to $R_{M\times N}(M<<N)$ and the linear measurement value $Y=R_{M\times 1}$ of some unknown signal $s=R_{N\times 1}$ when using the measurement matrix as formula (2):

$$Y_{M\times 1} = \boldsymbol{\Phi}_{M\times N} \times \boldsymbol{s}_{N\times 1} \quad (2)$$

Y can be seen as a linear projection of signal $\boldsymbol{s}$ in the measurement matrix $\boldsymbol{\Phi}$. The main objective of compressed sensing is to reconstruct unknown signal s from the measurement result Y. Obviously, as the dimension of s is far greater than the dimension of Y, (2) is an solution to the problem of underdetermined linear equations, there are infinitely solutions. This problem is an NP-Hard problem that has been proven already. However compressive sensing theory has proved that if the signal s is K sparse, and Y and $\boldsymbol{\Phi}$ satisfy RIP (Restricted Isometry Property) condition, signal s can be accurately reconstructed from the measured values Y by solving the minimization problem of l_1 norm as follow:

$$\boldsymbol{s}_{estimate} = \arg\min \|\boldsymbol{s}\|_1, \quad s.t. \quad \boldsymbol{\Phi s} = Y \quad (3)$$

(3) is a programming problem can be solved by a valid method of convex optimization, which greatly promotes the applications of compressed sensing theory. One of the key techniques of compressed sensing is the design of measurement matrix. In a grid-based localization model, assuming that the target is located in the center of the grid, the measurement matrix $\boldsymbol{\Phi}$ of each element $\varphi_{i,j}$ is defined as:

$$\varphi_{i,j} = P_{i,j}, \qquad i \in [1,M], j \in [1,N] \quad (4)$$

The definition of $P_{i,j}$ is the same as (1). Therefore, the sampling process of the system can be described with formula (5).

$$\begin{bmatrix} y_1 \\ y_2 \\ \vdots \\ y_M \end{bmatrix} = \begin{bmatrix} P_{1,1}s_1 + P_{1,2}s_2 + \cdots + P_{1,N}s_N \\ P_{2,1}s_1 + P_{2,1}s_2 + \cdots + P_{2,N}s_N \\ \vdots \\ P_{M,1}s_1 + P_{M,1}s_2 + \cdots + P_{M,N}s_N \end{bmatrix} = \begin{bmatrix} P_{1,1} & P_{1,2} & \cdots & P_{1,N} \\ P_{2,1} & P_{2,2} & \cdots & P_{2,N} \\ \vdots & \vdots & \ddots & \vdots \\ P_{M,1} & P_{1,2} & \cdots & P_{M,N} \end{bmatrix} \begin{bmatrix} s_1 \\ s_2 \\ \vdots \\ s_N \end{bmatrix} \quad (5)$$

Assuming the total number of the target is K, obviously, the dilution of N-dimensional vector s is K. the measurement result Y of the sensor is the product of measurement matrix and sparse vector s, the corresponding physical meaning of y_i is sum of signal intensities of multiple targets received by the sensor i in each period T. It can be proved that the measurement matrix $\boldsymbol{\Phi}$ satisfies the RIP condition. Because of the limited space of this paper, the proof is omitted; thus the multi-targeting problem in sensor network converts into a reconstruction problem of sparse signals of compressed sensing, due to the observed matrix meet the conditions of RIP, we can achieve multi-target localization function with l_l norm optimization model by using BP, OMP, GMP and other methods. This model works better in an ideal situation; but in a real environment, there is often a variety of interference affected by the deployment environment; so formula (5) in the real world is expressed as:

$$Y_{M\times 1} = P_{M\times N} s_{N\times 1} + N_{M\times 1} \quad (6)$$

$N_{M\times 1}$ is the Gaussian white noise superimposed on the signals received by the sensor M. Because of the existence of signal noise, the effect of solving l_l norm is not ideal with larger errors and false targets, so further research is needed to obtain location information with higher quality

4 Sparse Reconstruction Method Based on Discrete ABC Algorithm

The algorithm of sparse signal reconstruction is one of the core contents of the compressed sensing theory, the current research is moving towards stable, low computational complexity and a smaller numbers of observations [17]. ABC algorithm [18], as a new bionic swarm intelligence optimization algorithm proposed in recent years, has won the attention of academia and engineering with its good optimization ability in smaller populations, fewer iterations, less need of parameters setting in the face of high-dimensional and multi-mode optimization problem [19]. The basic ABC algorithm is designed for continuous optimization problem. In order to apply ABC optimization algorithm to the sparse reconstruction of signals, the paper proposes a discrete ABC algorithm, and proposes for food source development and exploitation strategies used in discrete problem while introducing a perturbation operator to enhance the ability to escape from local optimal algorithm. Here in DABCR (Discrete Artificial Bee Colony Reconstruction) algorithm, the objective function is defined as:

$$\min f(s) = \|Y - \Phi s\|_2,\ s \in \mathrm{R}^N \quad (7)$$

s is the demanded sparse signal, it is equivalent to food source coding in the algorithm; the coding rule is: when there is a target in i-th grid, the corresponding s_i is 1, otherwise is 0. The key to discrete ABC algorithm is to design the location updating strategy. Classic ABC algorithm uses formula (8) for localization updating.

$$v_{ij} = s_{ij} + r_{ij}(s_{ij} - s_{kj}) \quad (8)$$

s_{ij} is the j-th variable dimension of the i-th food source, k is the nectar obtained by random selection using formula (9). r_{ij} is a random number between [-1,1]. Formula (8) is for the optimization of continuous variables; in order to meet the issue model here, a food source localization updating strategy is designed as following. The definition of variables such as s_{ij} still uses that in formula (8), but the value of s_{ij} can only be 0 or 1, we redefine the operation rule of subtraction and addition as:

i. If s_{ij} and s_{kj} are different, then s_{ij}-s_{kj} equals to 1, otherwise the result to zero.

ii. If s_{kj} equals to 1, then s_{ij}+s_{kj} to 1; if s_{ij} and s_{kj} are the same, then s_{ij}+s_{kj} equals to 0, otherwise to 1.

iii. Considering the actual situations, s_{ij} value is restricted in three cases of -1, 0 and 1. When the number of 1 in s_i exceeds the threshold T_{max}, the position of 1 of T_{max}-$\|s_i\|_1$ in s_i will be changed randomly to 0.

$$P_i = fit_i / \sum_{j=1}^{SN} fit_j \quad (9)$$

To further enhance algorithm's ability of getting out of local optimization, an operator named *Perturb(s_i)* for escaping from the local minima is designed. If the searching operation in this iteration does not improve the quality of the location of food source, then disturbance operation would be executed with probability $P_{perturb}$. Disturbance process should be controlled at a moderate level, if the disturbance on the solution is too strong, then the algorithm degenerates into an arithmetic of multiple random starting points; weak disturbance on the solution is difficult to promote the search out of local optimum. The paper applies the following method: setting λ as a random integer in the interval (0, $\|s_i\|_1/2$), optionally chooses locations decoding λ as 1 in the solution s_i which would be replaced by the corresponding position value in current optimal solution, thus completes the disturbance.

In summary, this paper designs a discrete artificial bee colony reconstruction (DABCR) with disturbance mechanism to solve the problem of multi-target localization , its specific process is as follows:

Step 1 (algorithms initialization) Initialize each parameter of algorithm. To initialize SN food sources, and make the iteration number $t = 0$.

Step 2. Calculate fitness function, find the cumulative probability of each solution for observation of bee selection .

Step 3. (employed bee phase) Use formula (8) with redefined subtraction and addition for neighborhood search in each solution.

Step 4. If a better solution is found, the current solution will be replaced, simultaneously setting searching count to zero, otherwise using *Perturb(s_i)* function with probability $P_{perturb}$ for disturbing operation; if the resulting solution is better than the current one, then replace it, otherwise 1 will be incremented to the search count.

Step 5. (onlooker bee phase) Onlooker bees use cumulative probability to evaluate the selected solutions and chooses food source for neighborhood search in accordance with formula (9).

Step 6. To preserve the historic optimal solution.

Step 7. (scout bee phase) to analyze the searching accounts of all solutions; if the corresponding searching times of a solution are larger than the specified upper limit C_{max}, then re-initialize a solution to replace it.

Step 8. making t=t+1, to determine whether the maximum number of iterations reaches I_{max}, and if so, then output the optimal solution, the algorithm finishes; otherwise go to step 2 and begin the next iteration.

4 Equations and mathematics

The simulation runs on the Matlab2012b. The monitored zone is a 20m × 20m square area, the area is divided into 400 squares with side length of 1m; the parameter of measurement matrix is generated as: P_0=-40dB, N_0=2.5, D_0=1. Parameters used in DABCR algorithm are as follows: the sum of employment and scout bees SN is 100, and each takes up half. Maximum cycle count times C_{max} = 30, the disturbance probability $P_{perturb}$=0.1, the maximum number of iterations I_{max}=300. Figure 1 shows the use of 25 sensor nodes (M = 25), with six targets (K = 6), in the ideal environment with no interference, the target position can be accurately reconstructed. When SNR = 25dB and 10dB, the localization result gradually deteriorates along with the increase of noise.

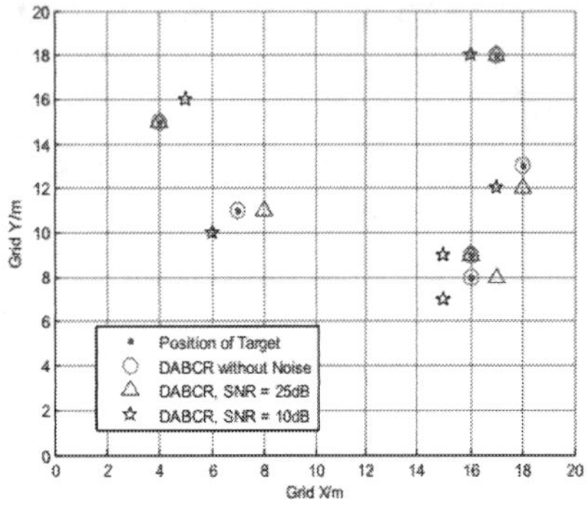

Figure 1. Location result of DABCR, with 6 targets

To test the error performance of node localization of different algorithms in different noise levels, the following parameter settings are used for the construction of test scenarios: M = 40, K = 10, SNR gradually reduces from 35dB to 10dB. When SNR is less than 10dB, false targets begin to appear in the reconstructing results, the algorithm fails. In each SNR level 500 independent Monte Carlo experiments were carried out. Figure 2 shows the average value of the various SNR. As can be seen from Figure 2, RMSE changes more when SNR <15dB, increasing with a larger margin. But in various noise levels the error of DABCR algorithm is still less than that of GMP and MLE algorithms.

Figure 3 shows a comparison of the performance of three algorithms with different objective numbers when M = 100.

As can be seen from the Figure 3, when the target number is less than 20, three algorithms get more satisfactory localization effect in the absence of noise and SNR =20dB; when the target number is greater than 20, the reducing magnitude of accuracy of GMP and MLE algorithms is greater than that of DABCR algorithm. Meanwhile DABCR exhibits better robustness than GMP and MLE.

5 Conclusion

This paper presents a signal recovery method of compressed sensing based on discrete artificial bee colony algorithm for multi targeting in wireless sensor networks, and proposes a new food source location updating approach on the basis of discrete coding mode, designs disturbing operators to be applied to improve local search efficiency and enhance the effect of the global search, and verifies the effectiveness of the algorithm with comparison of simulation experiments.

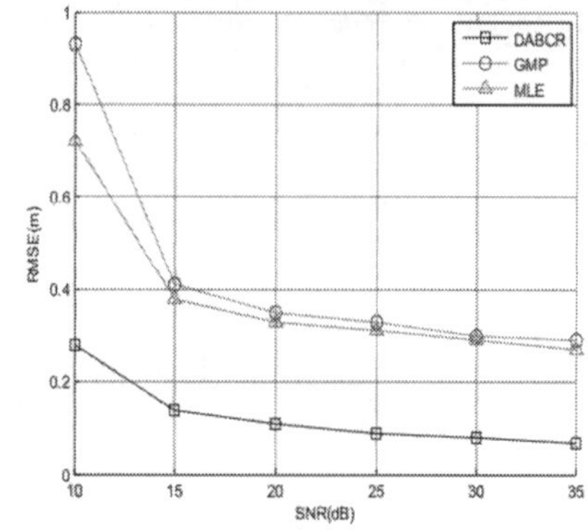

Figure 2. Comparison of RMSE under different SNRs

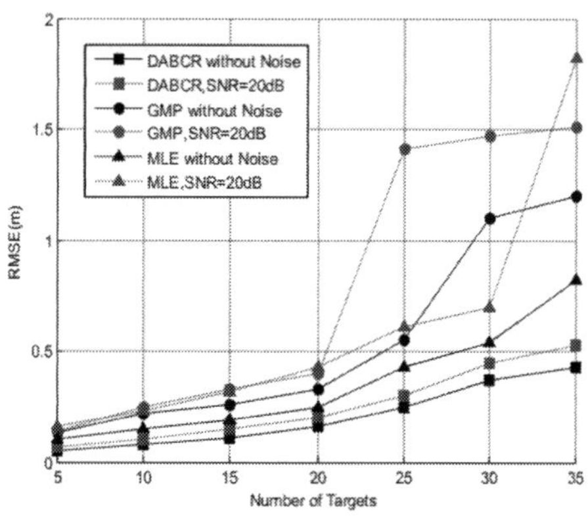

Figure 3. Comparison of RMSE for different target numbers and SNRs

Acknowledgment

This research work obtained the subsidisation of National Natural Science Foundation (Nos. 61563012, 61502290, 61363070); Guangxi Natural Science Foundation of China (No. 2014GXNSFAA118370); Guangxi Colleges and Universities Key Laboratory of cloud computing and complex systems (No. 15209). The authors would like to thank the (associate) editors and reviewers for their valuable and constructive comments and suggestions for improving this paper.

References

1. Akyildiz.I.F, Su.W, Sankarasubramaniam.Y, "Wireless sensor networks:A survey", Computer Networks, **38**(2002).
2. Patwari.N et al., "Locating the nodes: cooperative localization in wireless sensor networks",IEEE Signal Processing Magazine , 22(2005).

3. Chen.PC, "A nonline-of-sight error mitigation algorithm in location estimation" Proc. of the IEEE Wireless Communications Networking Conf., (1999).
4. Girod.L, Bychovskiy.v, Elson.J, Estrin.D, "Locating tiny sensors in time and space:A case study" Proc. of the 2002 IEEE International Conference on Computer Design: VLSI in Computers and Processors, (2002).
5. Girod.L, Estrin.D, "Robust range estimation using acoustic and multimodal sensing" Proc. of the IEEE/RSJ Int'l Conf.on Intelligent Robots and Systems(IROS 2001).
6. Aksu.H, Aksoy.D, Kurpeoglu.I,"A study of localization metrics: Evaluation of position errors in wireless sensor networks", Computer Networks, **55**, (2010).
7. Niculescu.D, Nath.B, "Ad hoc positioning system(APS) using AoA" Proc. IEEE INFOCOMM, (2003)
8. D. L. Donoho, "Compressive sensing ", IEEE Trans. Info. Theory , 52(2006).
9. Cevher.V, Duarte.M.F, Baraniuk.R.G, "Distributed target localization via spatial sparsity" Proc. of the European Signal Processing Conference, (2008).
10. F.Chen, S.Valaee, Z.Tan," Multiple target localization using compressive sensing" Proc. IEEE Global Communications Conference, (2009).
11. S.Nikitaki, P.Tsakalides, "Localization in wireless networks based on jointly compressed sensing" Proc. IEEE Signal Processing Conference,(2011).
12. Z.Bowu et al., "Sparse target counting and localization in sensor networks based on compressive sensing" Proc. INFOCOM, (2011).
13. W.Ju et al., "LCS: Compressive sensing based device-free localization for multiple targets in sensor networks" Proc. INFOCOM, (2013).
14. X.Qiang , Z.Wanting, "Adaptive Localization Algorithm Based on Distributed Compressed Sensing in Wireless Sensor Networks", Journal of Networks,**9** (2014).
15. E.Candès, J.Romberg,T.Tao, "Robust uncertainty principles:exact signal reconstruction from highly incomplete frequency information", IEEE Transaction on Information Theory,.**52**, (2006).
16. E.Candès, J.Romberg, "Quantitative robust uncertainty principles and optimally sparse decompositions", Foundations of Compute Mathematics, **6** (2006).
17. E.Candès, M.B.Wakin, "An Introduction To Compressive Sampling",**25** (2008).
18. D. Karaboga, B. Basturk, "A powerful and Efficient Algorithm for Numerical Function Optimization: Artificial Bee Colony (ABC) Algorithm", Journal of Global Optimization, **39**(2007).
19. D. Karaboga, B. Akay, "A Comparative Study of Artificial Bee Colony Algorithm", Applied Mathematics and Computation, **214** (2009).

MATEC Web of Conferences **59**, 07009 (2016)
ICFST 2016
DOI: 10.1051/matecconf/20165907009

A planning approach to monitor and control the satellite cluster

N. Gopalakrishna Kini and Ranjana Paleppady

Department of Computer Science and Engineering, Manipal Institute of Technology, Manipal University, Manipal, India

Abstract. There are many satellites in space that revolve round the Earth. Increase in the number of satellites leads to the importance of monitoring satellites for collision detection and avoidance. Collision of satellites can occur if their orbits intersect each other. In order to detect collision and to avoid the collision of satellites, the relative distance between any two satellites needs to be calculated as they revolve. As an objective to calculate the distance between any two satellites, parallel computation is required in order to detect collision. Our paper shows the prediction of collision possibilities between any two satellites and proposes at parallelizing the relative distance and position computation for a satellite of interest and its neighboring satellites.

1 Introduction

A satellite is an artificial object which has been intentionally placed into the earth's orbit. Increase in the number of satellites leads to the importance of monitoring satellites for collision detection and avoidance. Collision of satellites can occur if their orbits intersect each other. In some cases, a satellite may start approaching another satellite due to usual orbiting through its orbit which may destroy it or destruction may happen due to loss of control. The Radio Detection and Ranging (RADAR) system of a satellite provides information about the surrounding space debris and satellites [1].

As satellites orbit the earth they are pulled back in by the force of the gravitational field. If they did not have any motion of their own, they would fall back to Earth. For any given orbit there is a speed for which gravity and the centrifugal force balance each other and the satellite remains in a stable orbit. Satellites can have circular or elliptical orbits. The circular orbits of a satellite can be grouped based on their altitudes as Low Earth Orbits (LEOs) below 2000 km, Medium Earth Orbits (MEO) with altitudes around 20000 km and Geostationary Earth Orbits (GEOs) around 36000 km from the ground [2].

Monitoring the satellite for collision avoidance is a topic of increasing importance. The number of satellites in earth orbit is steadily growing and hence collision probabilities between two such objects can become critical as they may be either crossing through or resident in orbit. The surveillance of the space is a first step in monitoring the traffic of floating objects and has several applications under use for avoidance of collisions of satellites.

Satellite mission operators are usually well aware of the position of their space assets. Data from the *global positioning system* (GPS) can be used to perform orbit determinations and predictions. However, recent history shows that it is not sufficient to have knowledge of only one's own spacecraft. The term *space situational awareness* (SSA) describes the gathering and updating of knowledge of close approaches between space objects and their interaction with the space environment. SSA aims at gathering data from GPS, optical telescopes, and radar, to derive state information of space objects and be able to share warnings of close approaches between space objects [1].

There is a possibility of collision of any two satellites due to various factors such as if their orbits intersect each other or another satellite starts nearing a satellite due to usual orbiting through its orbit or another satellite starts nearing intentionally or due to loss of control on a satellite. If the distance between any two satellites at time t is obtained, then collision can be predicted and hence can be avoided.

2 Background

Satellites are usually semi-independent computer-controlled systems. A search of the National Space Science Data Center (NSSDC) listed 7,420 satellites launched into orbit since 1957 [3]. Among them some satellites put in earth orbit are no longer in orbit and some of them are inactive but still orbiting. According to the information provided [4], there are 1,071 active man-made satellites orbiting the earth and around 2,600 satellites still continue to orbit even though they no longer work. Hence it is required to monitor and control the orbiting satellites as there is a risk of collision between them.

A satellite may collide with another satellite due to internal error, external factors or intentionally to destroy a satellite. Sometimes, there are space systems composed

DOI: 10.1051/matecconf/20165907009

of multiple spacecrafts to achieve the same space mission. They are called spacecraft formation. In such situations these satellites are often prone to the risk of collision because the distance between them in a formation is closed, usually only a few meters to several kilometers. Due to too many retired and active spacecrafts in space, the formation is also prone to be collided by other spacecrafts. Collision will happen when two or more spacecrafts fly to a same position in the same time [5]. Studies have shown that a single collision within a cluster would not only drastically increase the risk to the other satellites within the cluster, but the debris would propagate and potentially render the orbit useless for many years to come [6]. Since there is a need to monitor the distance between more than one pair of satellites simultaneously, the concept of parallel computing needs to be used.

Parallel computing offer a great advantage in terms of performance for very large applications in different areas like engineering, physics, biology, chemistry, computer vision, econometrics [7]. To achieve parallel processing, it requires the development of more capable cost-effective computer systems. The ultimate goal is to achieve high performance at lower cost in performing large-scale scientific-engineering computing tasks in the various application areas.

3 Methodology

Concepts of Physics and Mathematics are used to simulate the distance computation between the satellite of interest, and its neighboring satellites [8]. An assumption is made that the parameters required for the initial calculations is available. As the satellites revolve, their position, velocity and relative distance keep changing. Hence, their relative distance needs to be calculated every time interval t so that the collision possibilities can be determined and avoided.

Here the software coded itself is capable to compute the next position and velocity of each satellite for every time interval. It is also capable of computing relative distance between any two satellites. When the relative distance is less than a threshold value, there is a possibility of collision. Once these parameters are known then it is possible to identify whether and when to monitor the position and velocity of satellites to avoid accidents in the space. Based on the alert message, satellites can be controlled.

3.1 Algorithm

The algorithm for monitoring and controlling the satellite cluster is as below:

• *Input*: Mass, time interval Δt, total time duration T, initial values for velocity and position along x-axis, y-axis and z-axis of each satellite.
• *Output*: Positions and velocities of collection of satellites over a period of time.
• Set Timestep = 0
• **for each** (Δt) **do**
• **if** *(Timestep <= T)* **do**
• Compute the distance between the satellite of interest *'q'* and rest of the satellites.
• **if** (*distance is less than threshold*) **do**
Take precautions to avoid collision.
else
Compute the total force exerted on satellite *'q'*.
Compute the next_velocity and next_position of satellite *'q'*.
Print the position and velocity.
end if
• Timestep = Timestep + Δt
• **end if**
• **end for each**

In our simulation, we have visually animated the path traversed by the satellites using Computer Graphics programming language like OpenGL. The revolution of satellites and possible collision is shown using OpenGL. To compute the next position, velocity and the relative distance between any two satellites a code has been developed using C programming language. Repetitive computation of the relative distance and collision prediction for two satellites is time consuming. Moreover, a satellite may be surrounded by many space objects. Hence, relative distance computation and collision prediction for the satellite of interest and its neighboring space objects needs to be done faster and in parallel. Therefore the concept of high performance computing and parallelism needs to be incorporated in the above problem. Thus the project intends to speed up the process of collision detection and avoidance. This will in turn ease the monitoring and control of satellites.

4 Result

In our work we have considered four such satellites in a spacecraft formation as in Figure1. These satellites do have their own revolving orbits and found to have intersecting orbits where there is a probability of collision of such satellites.

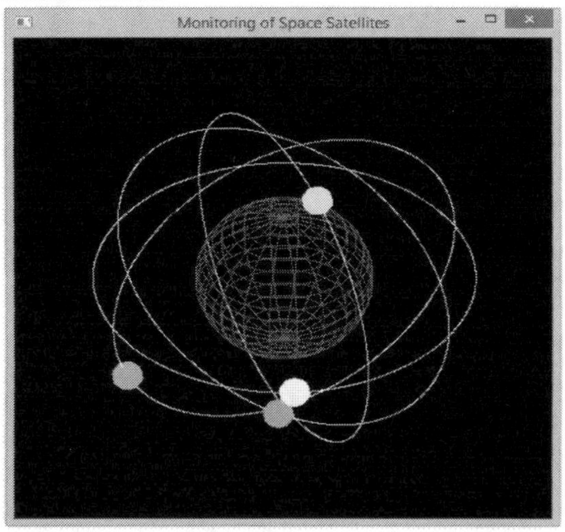

Figure 1. Satellites under consideration.

Figure 2 gives the initial values of satellite information. An assumption is made that these initial parameters such as mass, height above the earth, position and velocity of satellite under consideration is available for the initial calculations. These initial values are used to measure the position and the velocity of all the satellites at time *t* instances.

Figure 3 gives the position and velocity of satellite at time instance with 2 iterations. During every instance of time, if the distance is calculated between any two satellites and if the calculated distance is beyond the threshold value then the satellite system is safe. In other words to say, there is no probability of collision of any satellites.

Figure 4 shows the situation where in the future collision is predicted. If the distance calculated during any instance of time falls below the threshold value, it leads to a prediction of future collision. Here the threshold value is defined to be the sum of the radius of the two satellites under analysis in addition with a reasonable distance between them. In such case, the satellite under consideration needs to be controlled to avoid future collision. Such satellite's speed can be reduced / increased for the collision to be avoided.

Figure 5 shows the graph of distance between the two satellites and number of iterations. The distance between two satellites is plotted for a set of iterations in this graph. It is observed that the as the distance reaches to a low value near to the threshold, then there is a possibility of collision at these instances.

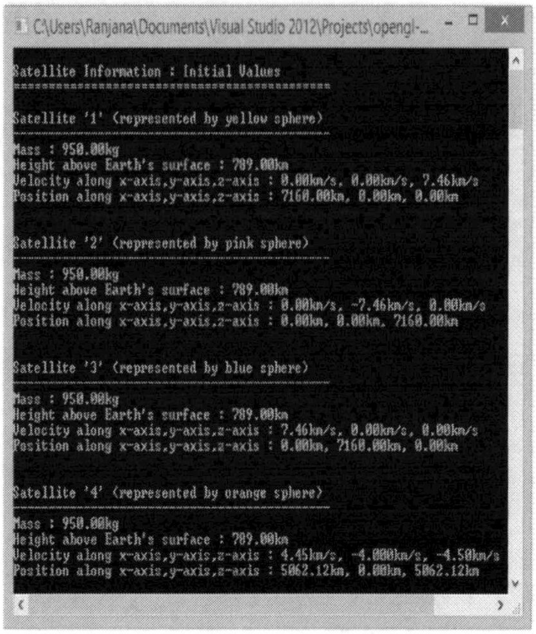

Figure 2. Initial known parameters of different satellites.

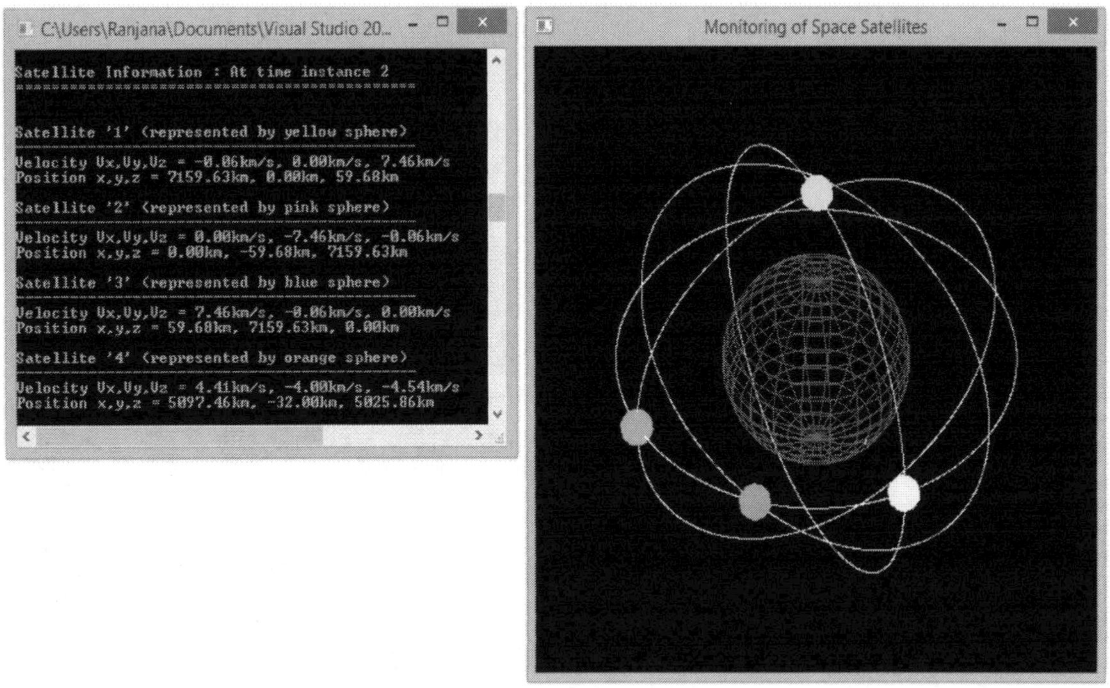

Figure 3. Monitoring the satellites by calculating their position and velocity at time instance with 2 iterations.

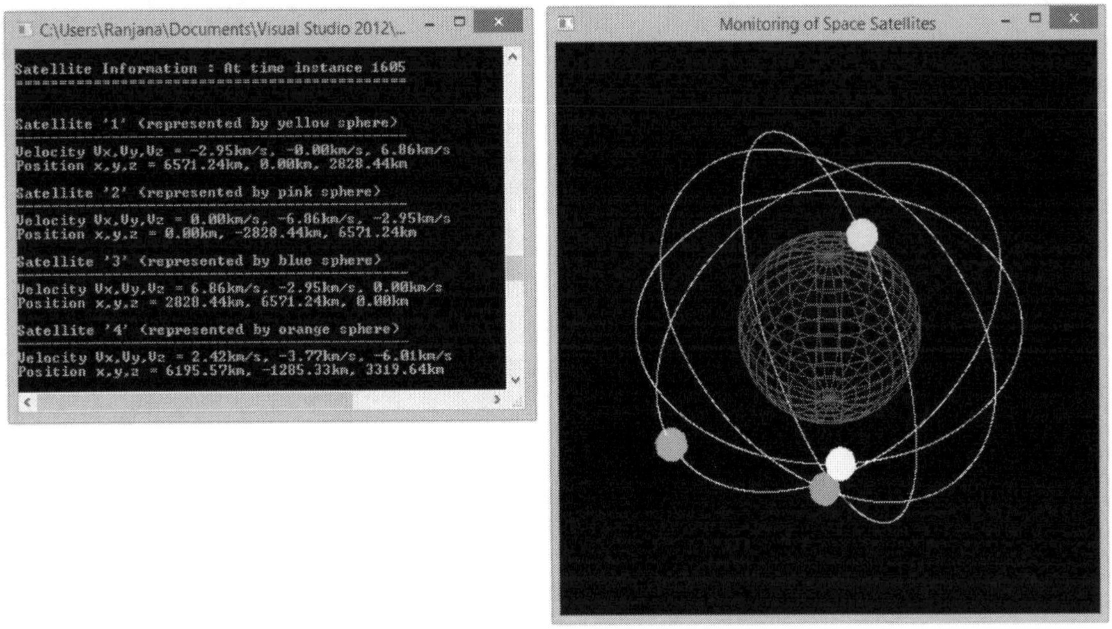

Figure 4. Two satellites are likely to collide and observed at time instance with 1605 iterations.

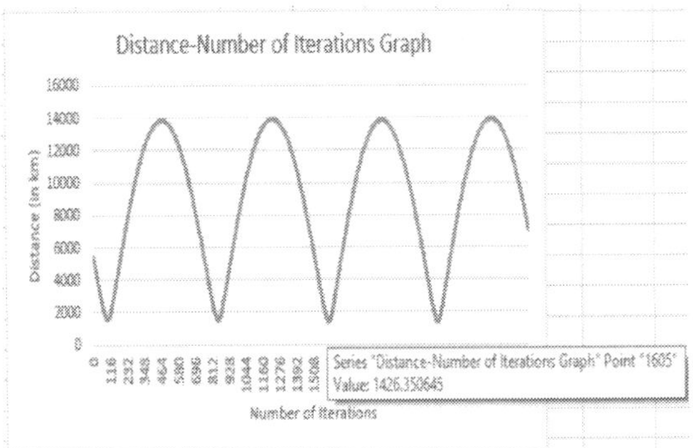

Figure 5. Distance between two satellites vs. number of iterations.

5 Conclusion

Collision of satellites can occur if their orbits intersect each other. In order to detect and to avoid the collision of satellites, the relative distance between any two satellites needs to be calculated as they revolve. In our work, we have shown the probability for the collision of satellites from which it can be monitored and controlled before this event happens.

As a future enhancement we would like to propose at parallelizing the relative distance and position computation for a satellite of interest and its neighboring satellites. As per NSSDC list it is found large number of satellites been launched and many more will get launched in near future by many countries. In this scenario it becomes an issue for a country to monitor and control their satellite if the orbits intersect each other or another satellite starts nearing a satellite due to usual orbiting through its orbit or another satellite starts nearing intentionally or due to loss of control on a satellite. In such situation there lies a large amount of frequent computations of relative distance, position and velocities computation which can be compensated by parallel computation.

References

1. E. Stoll, B. D'Souza, B. Virgili, K. Merz, and H. Krag, *Aerospace Conf. 2013 IEEE.*, pp.1-11, (2013).
2. A. Ciurte, A. Soucup, and R. Danescu, *Intelligent Computer Communication and Processing (ICCP), 2014 IEEE Intl Conf*, pp.179-185, (2014).
3. 3. NSSDC Master catalog, National Space Science Data Centre, NASA. http://nssdc.gsfc.nasa.gov/nmc/spacecraftSearch.do
4. http://www.universetoday.com/42198
5. D. Changping, R. Bo, Y. Hong, G. Pu, and T. Wei, *Guidance, Navigation and Control Conference (CGNCC), 2014 IEEE Chinese*, pp.1961-1966, (2014).
6. Mark E. Campbell, *IEEE Transactions On Control Systems Technology*, Vol. 13, No.1, (2005).
7. B. Oancea, T. Andrei, and M. Dragoescu, *Challenges of the Knowledge Society. IT*, pp.2026-2035, (2014).
8. P.S. Pacheco, An parallel Programming, Morgan Kaufmann Publishing, (2011).

Analysis of a UAV that can Hover and Fly Level

Ferit Çakıcı and M. Kemal Leblebicioğlu

Middle East Technical University, Department of Electrical and Electronics Department, 06800 Ankara, Turkey

Abstract. In this study, an unmanned aerial vehicle (UAV) with level flight, vertical take-off and landing (VTOL) and mode-changing capability is analysed. The platform design combines both multirotor and fixed-wing (FW) conventional airplane structures and control surfaces; therefore, named as VTOL-FW. The aircraft is modelled using aerodynamical principles and linear models are constructed utilizing small perturbation theory for trim conditions. The proposed method of control includes implementation of multirotor and airplane mode controllers and design of an algorithm to transition between modes in achieving smooth switching manoeuvres between VTOL and FW flight. Thus, VTOL-FW UAV's flight characteristics are expected to be improved by enlarging operational flight envelope through enabling mode-transitioning, agile manoeuvres and increasing survivability. Experiments conducted in simulation and real world environments show that, VTOL-FW UAV has both multirotor and airplane characteristics with extra benefits in an enlarged flight envelope.

1 Introduction

Innovative aerial vehicles have proved themselves in military and civil areas of different applications over a hundred years, by enhancing their capabilities over time, and adapting to new mission requirements. UAVs offer a unique set of advantages compared to piloted aircrafts with smaller, safer and lighter platforms. Future UAVs are expected to perform much more extended missions with higher manoeuvrability and higher degrees of autonomy.

Different capabilities like VTOL, hover, level flight, mode switching between hover and level modes, high endurance, long range and mechanical simplicity are expected from UAV platforms whenever mission requires. Comparison of capabilities of different types of UAV platforms (Table 1) provides insight about its mission profile. When VTOL and hovering are required, then rotary-wing aircraft such as helicopters, multirotors, ducted fans, tiltrotors and tailsitters are most optimal. However, if level flight, endurance or range is of priority, then a fixed-wing airplane type will most likely be preferred due to efficiency. When all of these features are desired, then VTOL-FW platforms becomes the best option, as an in-between platform among others. VTOL capability removes the need for runway or launch/recovery equipment and provides flexibility to operate in any theatre, whereas level flight capability allows efficient range and endurance flight. An aerial vehicle designed to possess the strengths of both a rotary and fixed-wing aircraft will have both of the advantages in one platform, with acceptable trade-offs in some capabilities.

Table 1. Comparison of UAV platform types.

Capability	Multi-rotor	Heli-copter	Airplane	VTOL-FW
VTOL	Good	Good	Bad	Good
Hover	Good	Good	Bad	Good
Level Flight	Bad	Bad	Good	Good
Mode-Switching	Bad	Bad	Bad	Good
Endurance	Bad	Good	Good	Neutral
Range	Bad	Bad	Good	Neutral
Simplicity	Good	Bad	Good	Neutral

Available studies [1-5] in VTOL-FW UAV platform category include tailsitters, tiltrotors and tiltwings. Although all of these platforms can perform hover and level flight, the main difference comes from the method of transition method (Table 2) between flight modes. These types of platforms suffer from difficult transition manoeuvres that operate the aircraft out of trim conditions and susceptibility to disturbances in transitions. However, a hybrid VTOL-FW platform asserted in this study, allows smooth transitions by being operated in an enlarged flight envelope.

Transition maneuvers between hover and level flight is of primary concern for VTOL aircrafts that are capable of level flight. T-wing tailsitter UAV with two counter rotating propellers was one of the pioneering studies; Stone [6] has developed a flight control system including low-level and mid-level guidance controllers, utilizing

 DOI: 10.1051/matecconf/20165907010

linear quadratic regulator and classical controllers, which were verified in flight tests [3]. Kubo [7] showed that a tailsitter UAV could achieve transitions between level flight and hover in shorter time using slats and flaps by using an optimal controller. Hogge [8] demonstrated transition maneuvers of a UAV with only one propulsion system using control surfaces. Tumble-stall maneuvers are implemented in achieving transitions by [5], [9] and [10] utilizing dynamic inversion methods. A state machine is designed by [11] and [12] for transitions between the flight modes, where the states were defined as hover, level, hover-to-level and level-to-hover. Backstepping control technique is studied by [2] for a coaxial-rotor tailsitter UAV and successfully simulated hover, level flight and transitions. Although available studies in this field are successfully implemented on different platform types, an aircraft that has physically separated multirotor and airplane control surfaces is not examined in demonstrating transition maneuvers.

Table 2. VTOL-FW UAV platforms' transition methods.

Photograph	Type	Transition Method
	Tailsitter	Tilts fuselage by control surfaces through stalling the aircraft.
	Tiltrotor	Tilts fuselage by tilting rotors that stalls wings.
	Tiltwing	Tilts wings that operates in stall region, while the fuselage remains parallel to Earth.
	VTOL-FW	Switches control elements between VTOL and FW, without stalling the aircraft.

In this study, design, analysis and implementation of control system of a VTOL aircraft with level flight capability is considered. The proposed method includes implementation of multirotor and airplane controllers and design of an algorithm to switch between them in achieving transitions between VTOL and FW flight modes. Thus, VTOL-FW UAV's flight characteristics are expected to be improved by enabling agile maneuvers, increasing survivability, providing redundancy and exploiting full flight envelope capabilities.

2 VTOL-FW UAV platform

VTOL-FW UAV platform is constructed by multirotor modification of a model airplane (Figure 1). Then, the platform is converted into an UAV by adding an autonomous flight controller (Pixhawk) and sensors like GPS, magnetometers, accelerometers, gyros, pitot-static system as in [13].

Main components of the aircraft (Figure 2) contribute to forces and moments acting on the vehicle in flight. Fuselage causes drag in negative direction of airflow, caused by linear motion. FW propulsion system (Prop.0) provides thrust to balance drag, while main wing provides lift to overcome gravitational force and ailerons, rudder and elevator provide roll, pitch and yaw motions as in a conventional airplane. VTOL propulsion systems (Prop.1-4) provide lift, roll, pitch and yaw motions by changing the rotational speeds of the propellers, as in a multirotor.

Figure 1. VTOL-FW UAV Platform.

Figure 2. Control Elements of VTOL-FW UAV.

Flight mode of VTOL-FW UAV is determined according to vertical and horizontal velocities of the aircraft in an enlarged flight envelope (Figure 3), that covers both VTOL and FW regions. When both of the vertical and horizontal velocities are small in magnitude, the aircraft operates in VTOL mode with VTOL control elements activated. As horizontal velocity is increased the aircraft enters the FW mode by enabling FW control elements. Intersectional region is used for switching between VTOL and FW modes, by changing active control elements.

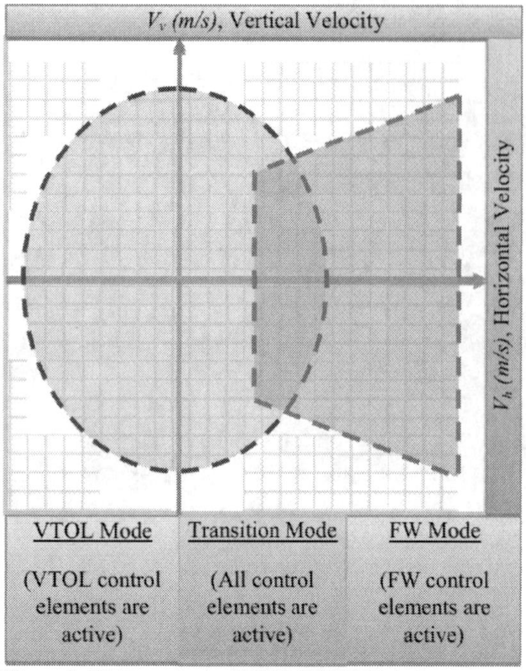

Figure 3. Flight envelope of VTOL-FW UAV platform.

3 Mathematical modelling

The complexity of dynamics of aerial vehicles, makes obtaining accurate mathematical models for a large portion of flight envelope a difficult problem. VTOL-FW UAV platform is modelled by using the real physical specifications of the aircraft in a MATLAB graphical user interface environment that is specifically developed for the preliminary design, analysis, control system design, mission planning and flight simulations of aircrafts. Initially, every main component like fuselage, wings, control surfaces and propellers are modelled using aerodynamical principles stated by Çakıcı [14], for the whole flight envelope including post-stall conditions. Then, these model's outputs are combined considering aircraft's geometry in calculating total forces (^{B}F) and moments (^{B}M). Equations of motion are formed as a set of nonlinear equations (1), using Newton's 2nd law of forces, Euler's formula for moments and kinematic relationships defined in body frame.

$$\dot{x} = f(x,u,t) = f(^{B}F(x,u), {}^{B}M(x,u), t) \quad (1)$$

where x: state variables, u: control variables, t: time.

4 Linear analysis

Linear analysis of VTOL-FW UAV is performed by examining stability and controllability of the linearized system dynamics for trim states.

4.1 Trimming

A trim state is defined as the equilibrium point, where the rates of the aerodynamic state variables are zero, when the resultant forces and moments are in balance. The trimming problem concerns the determination of control commands $[u_{rol}, u_{pit}, u_{yaw}, u_{thr}]$, which maps to control variables $[\theta_{ail}, \theta_{ele}, \theta_{rud}, \Omega_0, \Omega_1, \Omega_2, \Omega_3, \Omega_4]$, and aerodynamical variables $[u, v, w, p, q, r, \phi, \theta]$ that are required to hold the aircraft in equilibrium for a set of prescribed variables $[\psi, x_e, y_e, z_e]$. The trim conditions are obtained by solving a constrained optimization problem, defined as:

$$Minimize\ \|\dot{x}_a\|\ for\ x_p, u \quad (2)$$

$$subject\ to\ \dot{x}_a = f(x_a, x_p, u),\ given\ \dot{x}_p$$

where $x_a = [u, v, w, p, q, r, \phi, \theta]$, $x_p = [\psi, x_e, 0, z_e]$, $u = [u_{rol}, u_{pit}, u_{yaw}, u_{thr}]$ and $f(x_a, x_p, u)$ are the equations of motion.

Inspection of all prescribed operation points reveals that the trim condition could not be established for all of the points of interest, which reveals the limits of the flight envelope for each flight mode (Figure 4). An important observation is that the trim conditions of VTOL and FW flight modes intersect at level velocities between 12 and 16 m/s. This intersection region is used for transition maneuvers between modes. On the other hand, operation the in intersection region of the enlarged flight envelope allows inherent redundancy in control, which could save the UAV from crash due to failure of a control element, through switching to the other mode.

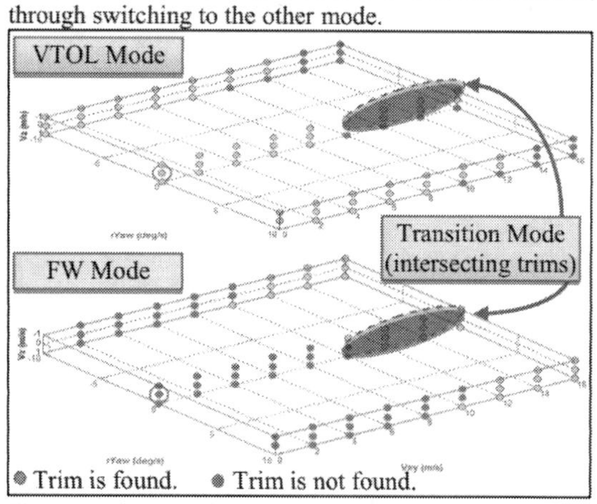

Figure 4. Trim conditions of VTOL-FW UAV.

Analysis of power requirements for an aircraft is important for achieving efficient flight. Assuming power dissipated on control surface servos is negligible and having lossless motors, electronic speed controllers, the major power consuming elements can be considered as propellers. Calculations of power required to fly (Figure 5) show that VTOL-FW UAV power consumption is similar to a comparable multirotor for hover. As level velocity is increased, multirotor power consumption increases due to the drag of the fuselage. As for VTOL-FW UAV in VTOL mode, power requirement decreases as wings start to provide lift. The required power starts to increase after some point since more lift means more

 DOI: 10.1051/matecconf/20165907010

moment provided by aerodynamically surfaces, where VTOL propellers consume more power in struggling with elevator moment. The steep increase in the power in VTOL mode, as the velocity is increased, is one of the major reasons of the need to transition to the FW mode.

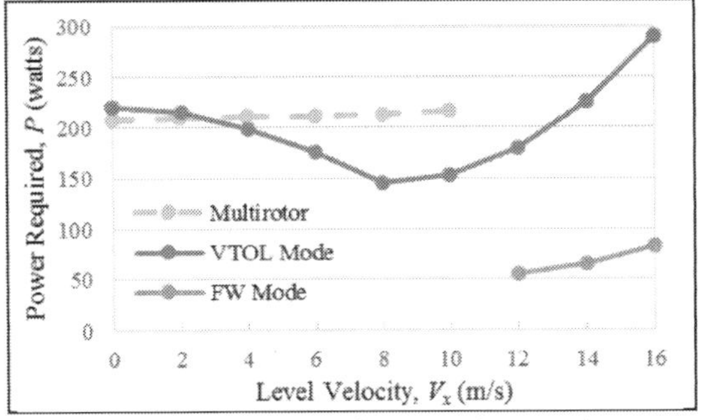

Figure 5. Power requirements of VTOL and FW modes.

4.2 Stability

Stability analysis (Figure 6) of the linearized dynamical system of VTOL-FW UAV shows that the aircraft is unstable in both modes. When a measure of stability is defined as the distance of the largest unstable pole in the right-hand s-plane to marginal stability line, the aircraft becomes more unstable in VTOL mode and less unstable in FW mode, as the horizontal velocity is increased. Analysis of the aircraft's dynamic modes proves that the aircraft demonstrates similar characteristics to an airplane in FW mode and to a multirotor in VTOL mode. Also, the controllability analysis shows that the aircraft is controllable in all of the trim points of both modes, which makes the controller design of the system possible. Eventually, stability analysis shown that VTOL-FW UAV demonstrates common aircraft characteristics in FW mode, and common multirotor characteristics with benefits in VTOL mode.

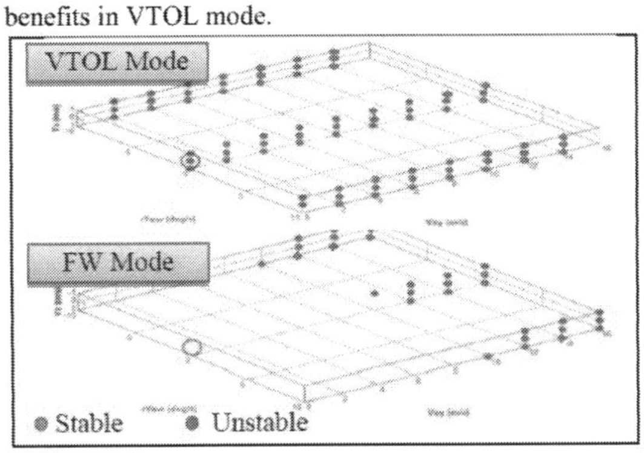

Figure 6. Stability of VTOL and FW modes.

5 Control system

A closed loop-control system is expected to stabilize the system, reject disturbances, reduce sensitivity to parameter variations, track reference, provide robustness to uncertainties, and be implementable for real world applications. Additionally, VTOL-UAV control system should make the aircraft follow guidance commands.

A control system architecture (Figure 7) is proposed that controls the aircraft in different flight modes. The inputs of the control system are obtained from guidance commands, and the outputs are defined as $[u_{rol}, u_{pit}, u_{yaw}, u_{thr}]$ which tell the aircraft to roll, pitch, yaw or change throttle, regardless of the active operation mode, which are then transformed into control element's commands through a control mixer.

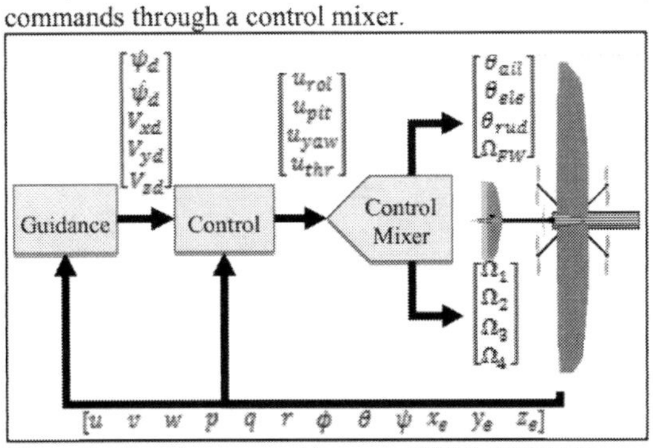

Figure 7. Control system architecture of VTOL-FW UAV.

6 Flight

A flight course of desired waypoints (Figure 8), forming a rectangle, is planned for flight testing. Initiating flight from waypoint 1, the aircraft took off vertically in VTOL mode and started ascending towards waypoint 2. After reaching waypoint 2, level velocity is increased in order to reach waypoint 3. As the velocity is increased further, the autopilot changed the mode to FW for level flight. When the aircraft approaches to waypoint 3, the autopilot changed the mode to VTOL again and the final waypoint is reached by vertical landing.

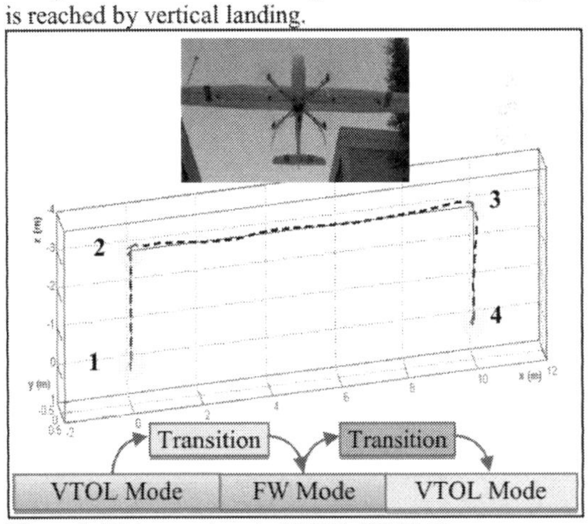

Figure 8. Flight experiment of VTOL-FW UAV.

The simulation results (Table 3) show that the aircraft completes waypoints (Figure 8) in shorter time with less power required when in "auto" mode, which automatically changes modes, compared to sole VTOL mode. The difference comes from the efficient flight operation through utilization of FW mode in auto so that the lift is mainly generated by the wings, instead of propellers of VTOL mode. These results prove the necessity and benefit of using different modes in an aircraft that can hover and fly level.

Table 3. Flight simulation results.

Mode	Flight Time (s)	Average Power (W)
VTOL	6.83	256.47
Auto (VTOL-FW)	5.21	163.81

7 Conclusion

In this study, design and analysis of a UAV with VTOL, hover, level flight and mode switching capabilities, VTOL-FW, is considered. The aircraft demonstrates distinct flight characteristics of a multirotor and a fixed-wing airplane on the same platform. The aircraft is modelled for analysis, simulation and controller design phases. Different PID controllers tuned for VTOL and FW modes performed well in flying the aircraft like a multirotor and airplane. Transitions between these modes are achieved through mode switching by applying smoothed control commands to control elements of different phases of flight.

One of the major problems in designing an air vehicle with VTOL and level flight capabilities is the location of center of gravity. An airplane requires center of gravity in front of the aerodynamical center, and a multirotor requires center of gravity close to aerodynamical center. Thus, this problem becomes a trade-off between better hover and better level flight. The aircraft fuselage remains mostly parallel to the Earth's surface in a large portion of its flight envelope. This allows both VTOL and FW mode flight envelopes intersect with each other, which makes the transition between these modes easier by having close trim state conditions. Also, the power consumption of VTOL-FW UAV in different modes reveals the necessity of switching modes for achieving efficient flight.

Simulation and real-world flight experiments proved the applicability of the proposed platform. VTOL-FW UAV is expected to find usage areas in missions requiring VTOL, hover and efficient level flight without the need for a runway and launch-recovery equipment.

Acknowledgment

This work is supported by The Scientific and Technological Research Council of Turkey (TÜBİTAK) under Grant 114E149 and 110E192.

References

1. T. Matsumoto, K. Kita, R. Suzuki, A. Oosedo, K. Go, Y. Hoshino, M. Uchiyama, 2010 IEEE ICRAA, 54-59, (2010)
2. X. Wang, H. Lin, "Design and Control for Rotor-Fixed Wing Hybrid Aircraft," *Proceedings of the Institution of Mechanical Engineers*, J. of A. **225-7**, 831-847 (2011)
3. R. H. Stone, P. Anderson, C. Hutchison, A. Tsai, P. Gibbens, K. C. Wong, J. of A, **45-2**, 673-685 (2008)
4. M. Hochstenbach, C. Notteboom, M. S. thesis, Ku Leuven Univ., (2014)
5. W. E. Green, P. Y. Oh, *Proceedings of the 2005 IEEE/ASME International Conference on Advanced Intelligent Mechatronics*, 693-698 (2005)
6. R. H. Stone, Institution of Mechanical Engineers, Part G: J. of A. E., **218-6** 417-433 (2004)
7. D. Kubo, *Proceedings of 25th International Congress of the Aeronautical Sciences*, (2006)
8. J. V. Hogge, M.S. thesis, Dept. of Mechanical Engineering, Brigham Young Univ. (2008)
9. N. Anathkrishnan, H. C. Shim, Idea Research and Development, (2008)
10. Y. Jung, D. H. Shim, J. of I. & R. S., **65-1-4**, 137-152 (2011)
11. S. R. Osborne, M.S. thesis, Dept. of Electrical and Computer Engineering, Brigham Young Univ., (2007)
12. F. Çakıcı, M. K. Leblebicioğlu, J. of D. M. and S., **9-4**, 335-345 (2011)
13. F. Çakıcı, H. Ergezer, U. Irmak, and M. K. Leblebicioğlu, "Coordinated guidance for multiple UAVs," Transactions of the Institute of Measurement and Control (TIMC), 2015.
14. F. Çakıcı, M.S. thesis, Dept. of Electrical and Electronics Engineering, Middle East Technical Univ., (2009)

Implementation of Low Power Multi-hop Wireless Sensor Network

Zhou Jianhong[1,2], Ma Maode[1], Fengyong[2]

[1] Chengdu Information Technology of Chinese Academy of Science Co. Ltd, Chengdu Sichuan 610041, China

[2] Nanyang Technological University, School of Electrical & Electronics Engineering, 639798, Singapore

Abstract. In this paper, the multi-hop wireless sensor network is created, which is realized by Micaz and Mib520 Board using TinyOS and Cygwin. The mote transmission mode is engaged and off-the-shelf protocols and algorithm, RSSI and PDR, are applied in this wireless network and the performance is improved. Several experiments are proposed to set standard RSSI threshold value to allow the user to send packet with efficient power level.

Keywords: Wireless Sensor Network; RSSI; PDR; threshold value; Transmission Power

1 Introduction

Wireless networks technology has been consistently improving with time and increasingly finding its way into all aspects of people's daily lives. Medical application is a key field where Wireless Networking has a promising future. In the healthcare field, access and cost saving are two of the hottest issues these days. Wireless Technologies can contribute towards helping with both of these issues which is in great demand [1].

Negligent mistakes by doctors and nurses costs hospitals, insurance companies and the government a large sum of money each year. These costs can be reduced by reducing the number of mistakes. Efficient and secure data handling and resource management is another area where wireless networks can help. Deploying huge machines around the hospital can be very expensive and time consuming. With Wireless Technology interfaces can be designed such that access to the machines can be provided from anywhere in the hospital. This allows for rapid and flexible deployment. By increasing the doctors' and nurses' efficiency, hospitals can provide care for more patients and increase their profits. On the other hand, Power efficiency is also a key factor for every wireless sensor network. Recently, Daniel Morrison et al. has a prototype design of a reconfigurable, low cost and low power miniaturized sensor node that interfaces with a variety of sensors, logs and transmits data wirelessly.[2] Meanwhile, a cluster based routing algorithm[3] has been proposed for WSN to minimize the power consumption.

In this paper, a reliable transmission of physiological data from ICU gateway to central server wirelessly in hospital environments was implemented by Crossbow products. Based on the analysis on the relation between message signal strength and transmission power, the off-the-shelf protocols and algorithms for TinyOs was applied and the performance was improved. At last, some possible methods were proposed to extend the usage of the battery energy.

2 Hardware And Software Review

2.1 Micaz

The Micaz mote platform is a third generation device used for enabling low-power, wireless sensor networks available in 2.4 GHz and 868/916 MHz. The Micaz Mote offers a 2.4 GHz, IEEE/ZigBee 802. The MICA Mote platforms are fully compatible with the Mote Works Software Platform and enable users to set up ad-hoc [4]. However, there're two improvements applying on the traditional Micaz mote platform in our system as shown in Fig. 1 and Fig. 2. The rechargeable battery is being used in order to decrease the total size of Micaz. Furthermore, the original antenna is replaced with a flat one which looks more professional and practical. From the experiment's result, the USB connector of MIB520 board will influence the packet delivery ratio with very high percentage, so a flat cable is designed to solve this problem. The Base Station and MIB520 board will be connected through the flat cable.

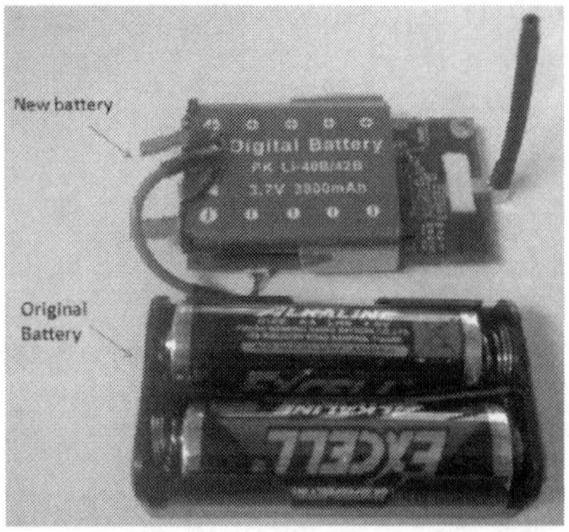

Fig. 1 Rechargeable Battery

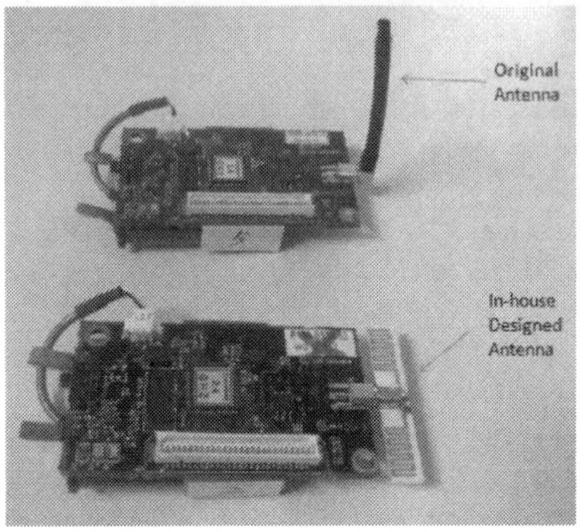

Fig. 2 In-house Designed Antenna

2.2 Mib520 Board

The MIB520CA provides USB connectivity to the MICA family of Motes for communication and in-system programming with attached 51 pins. Any MICAz node can be programmed as a base station when mated to the MIB520 USB interface board. In addition to data transfer, the MIB520 can also provide a USB programming interface.

2.3 TinyOS

TinyOS is a free and open source component-based operating system and platform targeting wireless sensor networks (WSN)[5]. TinyOS is an embedded operating system written in the nesC programming language as a set of cooperating tasks and processes. TinyOS programs are built out of software components, some of which present hardware abstractions. Components are connected to each other using interfaces. TinyOS provides interfaces and components for common abstractions such as packet communication, routing, sensing, actuation and storage[6].

2.4 Cygwin

It is free software that provides a Unix-like environment and software tool set to users of any modern version of MS-Windows for x86 CPUs (NT/2000/XP/Vista/7) and (using an older version of Cygwin) some obsolete versions (95/98/ME) as well. Cygwin consists of a UNIX system call emulation library, cygwin1.dll, together with a vast set of GNU and other free software applications organized into a large number of optional packages. Among these packages are high-quality compilers and other software development tools, a complete X11 development toolkit, GNU emacs, TeX and LaTeX, OpenSSH (client and server), and much more, including everything needed to compile and use PhysioToolkit software under MS-Windows[7].

Basically, the Micaz mote can be programmed and pre-configured by using the Cygwin interface under Microsoft environment. To figure out the best choice for a suitable power in order to save the battery energy, the maximum transmission range will be recorded. The Micaz mote is using the chip CC2420 which power index ranges from 1(Min.) to 31(Max.).

3 Protocol

3.1 Mote Transmission

TinyOS provides a number of interfaces to abstract the underlying communications services and a number of components that provide (implement) these interfaces. For demonstration purpose, two motes are being used to show how they can connect with each other. First of all, TinyOS library provides some very useful resource indicating some specific functions. For instance, the user can install two different programs into two Micaz mote named base-station and sending-mote separately. And then connect the base-station to PC through Mib520 programming board. The real code is designed and re-built based on the open source code provided by TinyOS Library, it allows user to define the program independently based on the specific requirement.

3.2 RSSI vs PDR Protocol

Each sending mote will send out 2000 packets to base-station with specific RSSI (Radio signal strengthen indication) value. Both of the RSSI value and Node_id (the node number of the mote which is used as a unique id for the mote tell the user where the received packet comes from) is assigned into the packet which can be sent to base-station, the system will help to create a file including these data in one default folder. Then the base-station can read the information by using the following *printf* command [8].

"printf("From node is %d \t RssiValue is %d \t seq num %d \t pkt rcv %d \n ",rms->nodeid,rms->RssiValue,rms->seqnum,counter); //rms->Tx);"

The RSSI vs PDR protocol builds and maintains the packet sending interval. The total number of packet can be pre-set from the code [9].

To compile the program of Micaz mote, basically it needs four files in the source folder: Makefile, Header file, Configuration file and Application file. The transmission power index, channel number and packet sending time interval can be set directly from Makefile.

One of the goals is to estimate the most efficient transmission power for each specific link, and it can be determined by setting the RSSI (radio signal strength indication) threshold value in order to keep the PDR (packet delivery ratio) higher than 90%.
For power control protocol, the basic procedures consists three steps. Firstly, the sending mote sends packet to base-station. The packet contains RSSI value, Node_id and other useful information. Then, the base-station will transmit these information to the sending mote again which is going to be compared with the standard RSSI threshold value. Finally, the sending mote will adjust its transmission power level based on these information.

4 Experiment and Discussions

4.1 Experiment Procedures

In order to avoid any unexpected influence, all the experiment including free space and on-body test is done in the chamber where can provide almost perfect environment and lowest impacts. The experiment procedure will be as following steps.

1. Nodes should be programmed as Base station and Mote, respectively.
2. The base station is connected to the work station through the programming board. It will receive packet and read the information attached.
3. Check the port number of Mib520 board from 'Device Manager' in 'My Computer' shown as Fig. 3. Need to check every time when connect it to the computer.

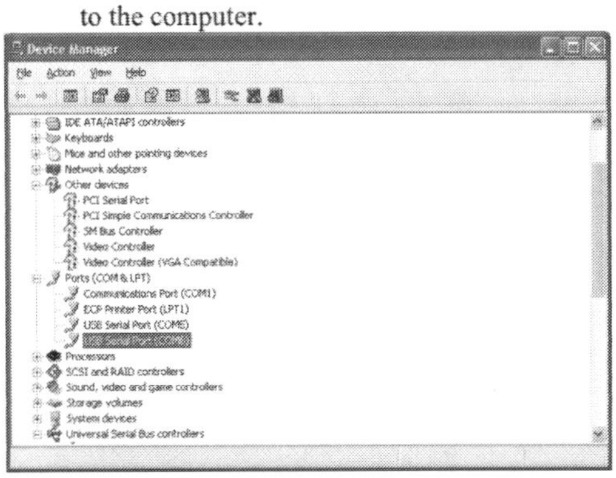

Fig. 3 Check USB Serial Port Number

4. Open the RSSI vs PDR folder and compile the Base station source code from Cygwin interface by keying in 'make micaz' (compile the program and check the errors). Then, the system will process in Fig. 4.

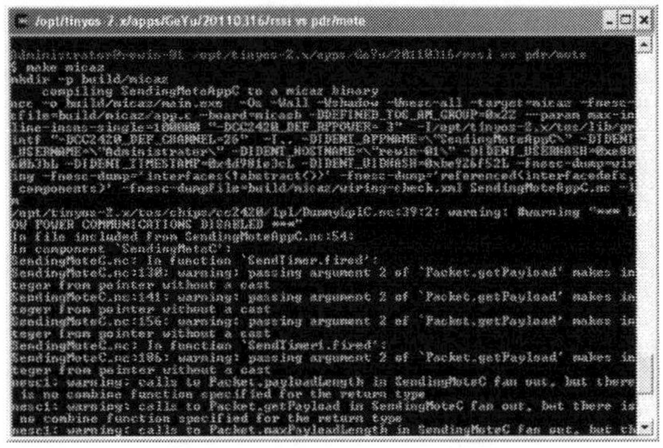

Fig. 4 Check USB Serial Port Number

5. Then the node_id should be assigned to 1 each time while installing the code by using the command as Fig. 5. (Note: ttyS'x', where x=COM-2)

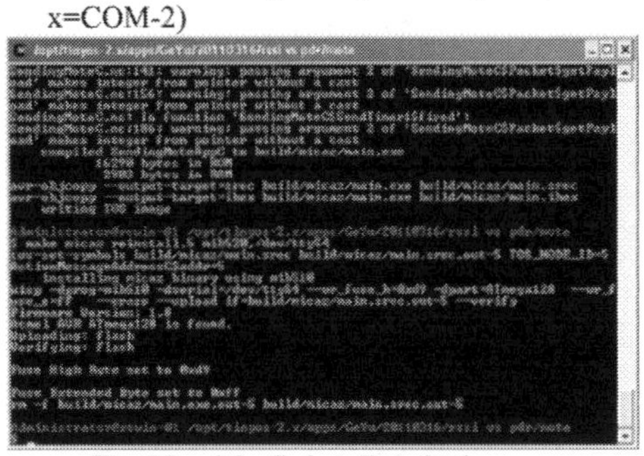

Fig. 5 Load the Code to Base Station

6. Change the transmission power in 'ApplicationDefinitions.h' in mote folder. Then repeat step 2 to step 5 with other motes, make sure the sending mote is assigned with specific node_id except 1 (1 is only for base station).
7. Then, place them on 2 tripods separately as Fig. 6.

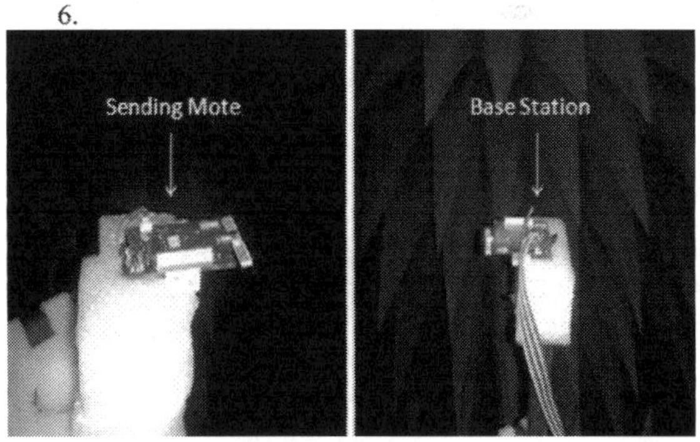

Fig. 6 Base Station & Sending Mote

8. Then, turn on the base-station and sending mote, use the command of '*java net.tinyos.tools.PrintfClient -comm serial@COM5:micaz > tx3_50cm_1.txt*' in Cygwin user interface, the RSSI value per packet and total number of packet received will be recorded and saved in .txt file. The base

station is fixed at one side of the chamber, for each transmission power it starts with the farthest distance. The sending mote is placed 250cm away from the base station. Once powered on, the mote will send 1000 RSSI packets to the base station.

9. The above experiment is conducted repeatedly with different distance.
10. The main purpose is to determine the RSSI threshold value for the power control strategy which will helps to save the power supply and extend the battery usage.

4.2 RSSI vs PDR Chamber Free-Space Result Analysis

In this experiment, the transmission power is pre-set to 3, and different distance was engaged, and then power 7 is set to different distance between two tripods.

The distance between two tripods is fixed at 100cm, 200cm and 180cm, and the result is attached in Fig. 7, Fig. 8, and Fig. 9 respectively.

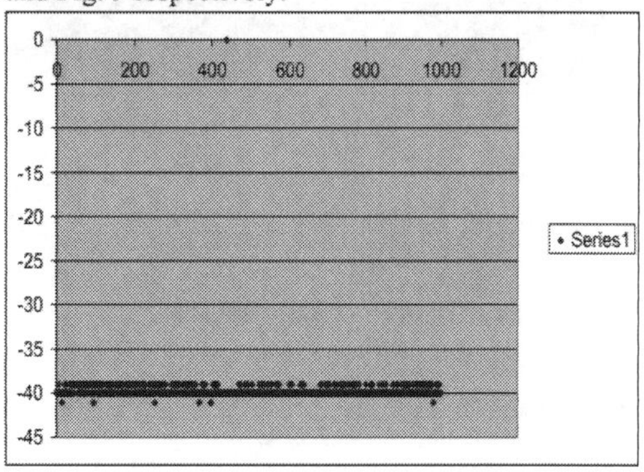

(a) RSSI Vs. Time

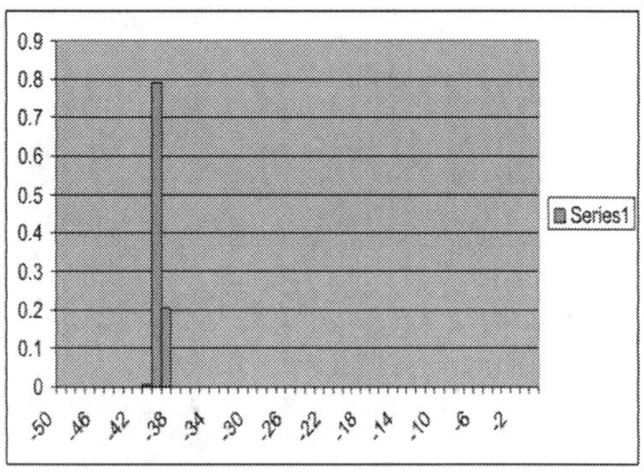

(b) RSSI Probability Distribution

Fig. 7 Result @Power 3, 100cm

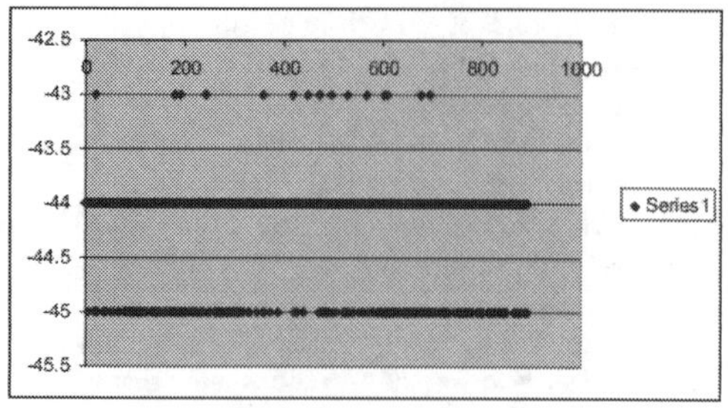

(a) RSSI Vs. Time

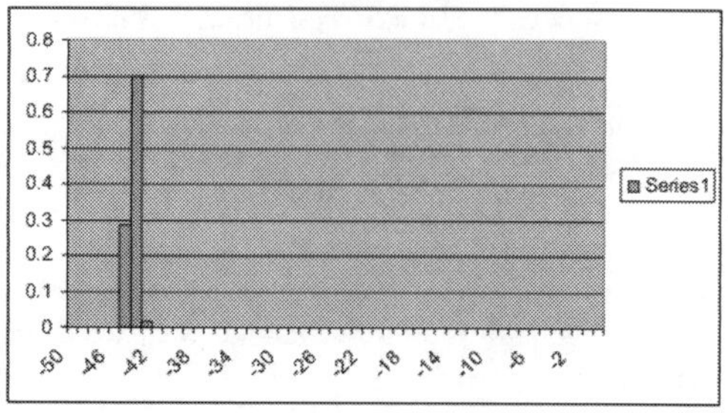

(b) RSSI Probability Distribution

Fig. 8 Result @Power 3, 200cm

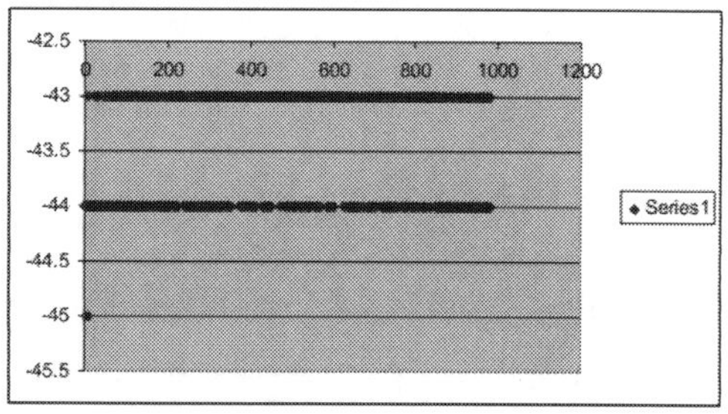

(a) RSSI Vs. Time

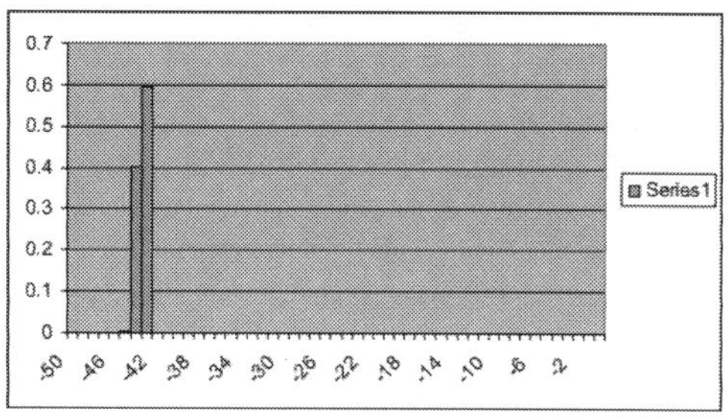

(b) RSSI Probability Distribution

Fig. 9 Result @Power 3, 180cm

The packet delivery ration is 100% in Fig. 7. The link performance is good, it can be seen that RSSI value is changing between '-40' and '-39' and majority of the RSSI values are distributed at '-40'. However, if the distance between two tripods is increased to 200cm, and the environment in chamber is clean, there will be no multipath. As shown in Fig. 8, the packet delivery ration will be 89.2%, which is unacceptable for a reliable transmission. Thus the distance between two tripods is decreased to 180cm as in Fig. 9. The PDR value achieves

to 98.9% which is higher than before, and majority of the RSSI values are distributed between -43 and -44.
Then the distance between two tripods is fixed at 300cm and the transmission power is changed and pre-set to 7. The result is as Fig. 10.

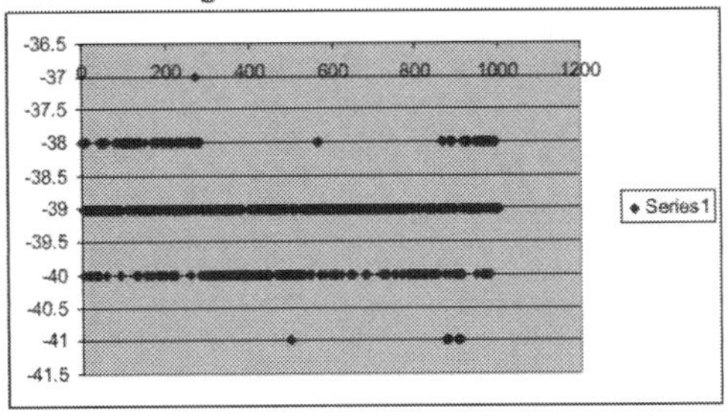

(a) RSSI Vs. Time

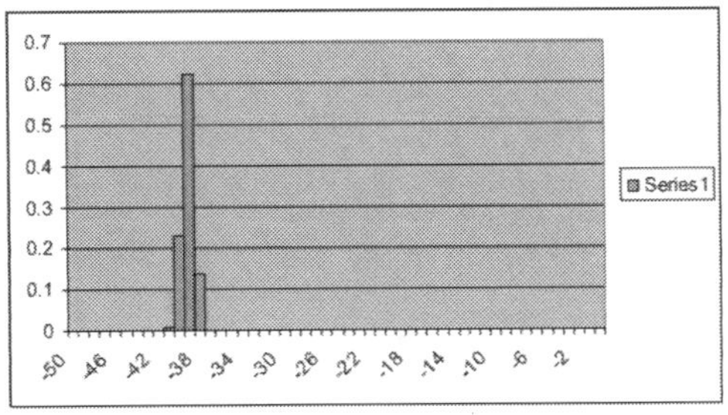

(b) RSSI Probability Distribution

Fig. 10 Result @Power 7, 300cm

From the results, it is clearly show that even the distance is 300cm, the PDR can still achieve to 100% with transmission power at 7.
In conclusion, for transmission power 3, the PDR value can be higher than 95%, restraining distance at 180cm, for RSSI value mainly distributed at '-43' and '-44'. As a result, the threshold RSSI value is determined at '-43' to keep the link transmission quality. It means that when the RSSI value of specific packet drop below '-43' or the distance become larger, the mote has to increase its transmission power to a higher level to meet the PDR requirement.

4.3 RSSI vs PDR On-Body Results Analysis

The PDR on-body experiment result conducted in chamber will be used as reference and the standard threshold value will be determined based on the link performance. All the Micaz motes should be wear properly on each specific position on human body, basically there are 9 Micaz motes functioning as sending mote and 1 as sink which is placed at the right-hand side of waist. Each sending mote will be placed on arm, head and ankle separately as Fig. 11.

Fig. 11 Node Position

The transmission power of base-station can be any value which will not affect the performance as receiver. It will receive packet from other sending mote. By using the '*printf*' function, the parent_id, RSSI (radio signal strength indication) and PDR (packet delivery ratio) can be displayed from the Cygwin interface. In terms of power saving purpose, the most suitable transmission power for each specific position can be chosen.
There are basically 10 links placed properly on human body with different location. All link performance for each node has been presented and evaluated as table 1.

Table 1. Performance Comparison of the 10 links

Node Number	Transmission Power	Majority RSSI Range	PDR
5	3	(-40,-39)	100%
6	3	(-18,-17)	100%
9	7	(-36,-35)	100%
10	7	(-40,-39)	100%
7	11	(-37,-36)	100%
4	11	-44	96.9%
2	15	(-38,-37)	100%
8	15	(-44,-42)	98.9%

In conclusions, the distance between two nodes and transmission power are two main factors for PDR, when the distance is farther, the transmission power should be increased to meet the PDR requirement. The majority RSSI range is referred to set the RSSI threshold.

4.4 Power Control Test

In this section, the link performance for each pre-designed RSSI value will be tested, and standard RSSI threshold value might be made based on the PDR vs RSSI results. There are several steps involved for this testing:

- Sending mote will stay at 150cm away from base-station for 1min,
- Move it to 180cm as fast as possible and stay at that point for 1min.

DOI: 10.1051/matecconf/20165907011

- Move to 200cm and stay for 1 min.
- Quickly move to 220cm and stay for 1 min.
- Finally, quickly move to 250cm and stay for 1min.

From the previous RSSI vs PDR testing for transmission power 3, the threshold value range (-44,-37), (-43,-37), (-43,-36), and (-43,-35) will be selected to run the procedure above. The comparison result is as Table 2.

Table 2. Performance Comparison of Different RSSI Threshold Value Range

RSSI Threshold Value Range	Power Probability Distribution	PDR
(-44,-37)	3(80%), 7(20%)	69%
(-43,-37)	3(58%), 7(42%)	81%
(-43,-36)	3(27%), 7(73%)	97%
(-43,-35)	3(19%), 7(81%)	100%

As shown in table 2 (-44,-37) and (-43,-37) consume less power, however, the PDR of the two modes are unacceptable, which may debase the link quality. Although the PDR value of (-43,-35) is 100% for, the RSSI threshold range is a little bit broad and seems over protected. As a result, the threshold range of -43 to -36 is concluded to be the most optimal range for future experiment.

5 Conclusions

In this paper, the multi-hop wireless sensor network is created, which is realized by Micaz and Mib520 Board using TinyOS and Cygwin. For the hardware, the rechargeable battery is being used in order to decrease the total size of Micaz. Furthermore, the original antenna is replaced with a flat one which looks more professional and compatible with Mib520 Board. The off-the-shelf protocols and algorithms for TinyOs, RSSI and PDR, are engaged in this wireless network. Several experiments are settled to set standard RSSI threshold to allow the user to send packet with efficient power level. In the medical market, this application can be customized and superior in the great saving of hardware cost with very high accuracy to fulfil divergent requirements.

Acknowledgement

Project supported by National Basic Program of China (973 program, project number: 2011CB302402) and National Nature Science Foundation of China (No. 11171053)

References

1. CHEN Tao, LIU Jing-tai, BIN Zhi-gang. Research on the Application of Wireless Sensor Network. Automation & Instrumentation, 2005(S1).
2. Daniel Morrison, Taran Ablitt, and Jean-Michel Redouté. Miniaturized Low-Power Wireless Sensor Interface, Sensors Journal, Vol. 15, Issue. 9, 2015, pp: 4371-4372
3. Kadir Tohma, Merve Nilay Aydın ve İpek Abasıkeleş Turgut. Improving the LEACH protocol on Wireless sensor network, 2015 23rd Signal Processing and Communications Applications Conference (SIU), 16-19 May 2015, pp: 240-243
4. Wang, Rui, Zhang, Lei, Sun, Rongli, Gong, Jibing, et al. A pervasive traffic information acquisition system based on wireless sensor networks. IEEE Transactions on Intelligent Transportation Systems; Special Issue on Artificial Transportation Systems and Simulation. 2011
5. Tan X, Li J. Cooperative positioning in underwater sensor networks. IEEE Transactions on Signal Processing. 2010
6. WU Xiaona, WANG Man. Introduction of Wireless Sensor Network Operation System TinyOS. Computer and Modernization. 2011(02)
7. Cygwin user guide. www.cygwin.com .
8. WU Gui-ping, YU Hui-qun, FAN Gui-sheng. A Sensor Network Allocation Algorithm based on RSSI. Journal of East China University of Science and Technology (Natural Science Edition). 2013(05)
9. Lee Yang Sun,Lee Ji Min,Yeo Sang Soo.et al.A study on the performance of wireless localization system based on AoA in WSN. 2011 Third International Conference on Intelli Kent Networking and Collaborative Systems. 2011

MATEC Web of Conferences **59**, 08001 (2016) DOI: 10.1051/matecconf/20165908001
ICFST 2016

Abnormal Appearance Detection of Substation Based on Image Comparison

Xu Zhang [1] , Li Li[2], Jianxiang Li[1], Juntao Lyu[3], Rui Huang[3], and Haiwen Xing[3]

[1] *Shandong Electric Power Research Institute, Jinan, China*
[2] *Shandong Luneng Intelligence Technology Co., Ltd Jinan, China*
[3] *State Grid Shandong Electric Power Company, Jinan, China*

Abstract. Based on image comparison, a novel algorithm for abnormal appearance detection of substation is proposed. Previous spatial states of an object are compared to its current representation in a digital image. Firstly, saliency maps are acquired using a fast implementation method of salient region detection. Based on saliency maps, image registration was completed by ORB (Oriented Fast and Rotated Brief). Then, sliding widow algorithm is applied to transform the whole image comparison problem into sub-image comparison problem. Textural feature and shape feature vectors (TSFVs) representing contents of images are generated by feature level fusion. Finally, decisions are automatically made as to whether or not change at the outline has occurred by the Euclidean distance of TEFVs. Experimental results show that the proposed method has good performance in abnormal appearance detection of substation.

1 Introduction

Inspection of electric power substation equipment plays an important role in daily maintenance of the electric power system. In recent years, robotics applications have attracted an increasing amount of attention in many research areas, especially in the construction, inspection, and maintenance of infrastructures and equipments in the power industry. Zhou et al. described the development and application of intelligent substation inspection robots, and pointed out that instrument recognition is important function of the substation inspection robots [1]. Currently, many works have been down in abnormal temperature detection based on infrared cameras, and less work for abnormal appearance detection using visible-light images. Most of equipment failure is mainly carried out by workers through observing pictures collected by cameras. Compared to other industries, research on image analysis and fault recognition of electrical equipments within substation was undeveloped. Li et al. proposed an image processing approach for abnormal appearance of electric power equipment using Mean Shift segment method [2]. A template matching algorithm based on the shape characteristics of image analysis is proposed to apply in the image recognition of power transformer [3]. But these methods obtain high precision only in the case that image matching achieves accuracy up to pixels. Meanwhile, different templates are needed to make artificially for different electrical equipments. To overcome above shortcomings of these algorithms, a new algorithm is proposed to detect different types of equipment failure according to the routinely inspection way of robots in the substation.

By automatic control and remote control means, substation equipment inspection robots developed by our laboratory, shown in Fig.1, can take infrared thermal imager and visible CCD to inspect the outdoor high-voltage equipments instead of human. While the intelligent robot is automatically moving inside the substation, it stops at the predefined stop point and turns the pan-tilt to acquire images at a fixed angle. Aiming at current image taken by camera, the other picture which has the same content as current one can be found in the database of last inspection. If image change is detected in multiple images of the same scene taken at different times, there is strong possibility that instruments abnormality occurred in this region.

Figure 1. Inspection robot developed by Electric Power Robotics laboratory of SGCC.

Hence, a new method suitable to detect abnormal appearance of substation is proposed in this paper based on images comparison. Fig.2 shows a flow chart of our approach to implement abnormal appearance detection of electrical equipment within substation.

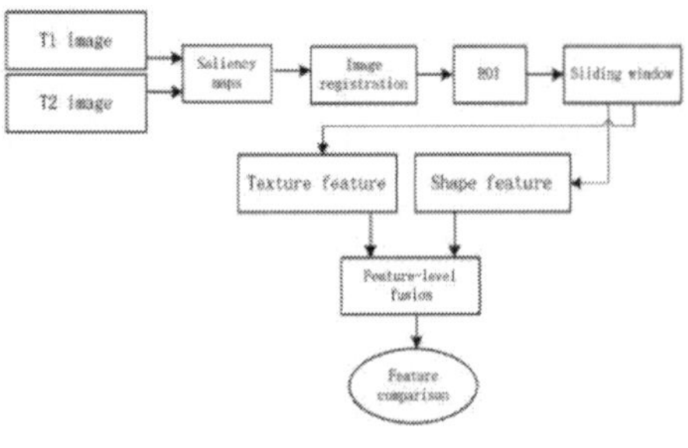

Figure 2. Flow chart of the proposed method.

First, saliency maps of two images with the same scene taken at different times T_1 and T_2 were obtained by local salient region detection algorithm. Then, image registration of salient maps was implemented based on ORB features. Sliding window algorithm divides the whole image into many blocks. Textual features and shape features are extracted respectively and fused at feature level to represent the content of blocks. Finally, we detect abnormal appearance through calculating Euclidean distance of fusion features of images participated in comparing.

2 Image Registrations

Scale invariant feature transform (SIFT) algorithm is widely used in image registration because of its invariance to image scale and rotation [4]. However, this method still needs to be improved, because threshold values are excess and uncertain, and computation is complex due to the high dimension of descriptor. A new descriptor named ORB with the same matching performance as SIFT and lower computation cost was proposed by Rublee et al. [5]. Here, image alignment is implemented based on ORB features extracted from saliency maps of two comparing images instead of SIFT.

2.1 Salient Region Detection

Salient region detection is useful in applications such as image segmentation [6], object detection and recognition [7], and region-based retrieval [8]. Salient regions are usually identified as those regions of an image that are visually more conspicuous by virtue of their contrast with respect to surrounding regions. Computationally detecting such salient image regions remains a significant goal, as it allows preferential allocation of computational resources in image analysis. In order to reduce time consuming and remove the uninformative regions of inspection images, salient region detection algorithm is applied here to separate foreground before image registration.

Achanta et al. proposed a method for finding salient regions using low-level features of luminance and colour [9]. Saliency is determined as the local contrast of an image region with respect to its neighborhood at various scales. The method is fast, easy to implement and generates high quality saliency maps of the same size and resolution as the input image. Hence, this algorithm is used here to get saliency maps of images shot by inspection robots. The specific step of salient region detection is detailed in paper [9]. Some results are shown in Fig.3.

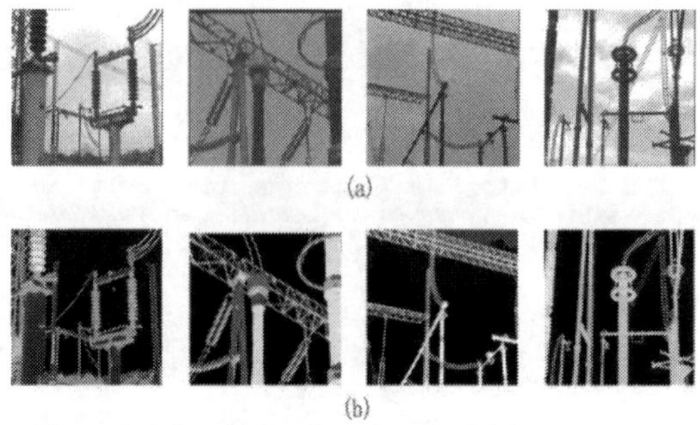

Figure 3. Salient region detection: (a) Original image; (b) Saliency maps.

2.2 Region of Interest

ORB is proposed as a very fast binary invariant descriptor based on BRIEF, it greatly speeds up processing procedure for feature extraction and matching compared to traditional key point descriptor, such as SIFT and speeded up robust features (SURF) [10]. Ref. [5] pointed out ORB is an order of magnitude faster than SURF, and over two orders fast than SIFT. Here, ORB algorithm is used to detect feature points and build descriptors of two comparison saliency maps. Then, RANSAC (random sample consensus) processing via transformation parameters estimation is employed to remove possible wrong matching points [11]. Transform matrix R is finally calculated by matching key points.

After getting transform matrix R, two corresponding images are finally converted into the same coordinate system. Four vertexes coordinate of current image mapped to the coordinate system of previous image. Four vertexes coordinate after space transformation can be expressed as top-left (x_1, y_1), top-right (x_2, y_2), bottom-left (x_3, y_3), bottom-right (x_4, y_4). According to the four vertexes coordinate, boundary coordinate of the regions of interest (ROI) are defined as:

$$\begin{aligned} ROI_{left}.x &= \begin{cases} Max(x_1, x_4) & Max(x_1, x_4) > 0 \\ 0 & otherwise \end{cases} \\ ROI_{right}.x &= \begin{cases} Min(x_2, x_3) & Min(x_1, x_4) < w \\ w & otherwise \end{cases} \\ ROI_{up}.y &= \begin{cases} Max(y_1, y_2) & Max(y_1, y_2) > 0 \\ 0 & otherwise \end{cases} \\ ROI_{bottom}.y &= \begin{cases} Min(y_3, y_4) & Min(y_3, y_4) < h \\ h & otherwise \end{cases} \end{aligned} \quad (1)$$

where w and h respectively represents width and height of previous image. Then ROIs of two images with the same size are extracted according to top-left corner coordinates ($ROI_{left}.x$, $ROI_{up}.y$) and bottom-right corner coordinates ($ROI_{right}.x$, $ROI_{bottom}.y$) of the rectangle.

3 Image Comparison

Detecting regions of change in images of the same scene taken at different times is of widespread interest due to a large number of applications in diverse disciplines, especially remote sensing methodologies [12] [13]. Here, abnormal appearance detection of equipment within substation, including appearance or disappearance of objects, or shape changes of objects, is implemented through change detection of two images taken at different inspection times. This is an image comparison problem, where an object is compared to itself, to identify whether or not it has changed.

For images comparison, the decision as to whether or not a change has occurred corresponds to feature difference. Aiming at images with rich contents, single feature has limitation in representing image content. In order to overcome the shortcoming of the low efficiency of previous single feature representing, a feature level fusion approach is proposed to implement a fusion strategy of textural features and shape features for change detection.

3.1. Salient Region Detection

The sub-image covered by the sliding window is divided into different overlapping blocks with the same size. Each block has 100×100 pixels. Step size of sliding window is 50 pixels so that adjacent blocks have 50% overlaps. The region with abnormal appearance may be divided into different blocks by sliding widow method with no overlap. If this is the case, change in image would be shared by different blocks. From the segmentation results in Fig.4, we can see that overlapped block can avoid the abnormal appearance region divided into different blocks. There are at least one block including most of the area of abnormal appearance.

Figure 3. Salient region detection: (a) Original image; (b)Saliency maps

3.2 Moment Invariants Features

Moment has been widely used in pattern recognition to describe the textural characteristics of different objects, and its calculation is related to the area, center, and breadth axel of the image or region. Invariant moment containing the global information of images reduce the sensitivity to the quality of pattern extraction. Hu proposed 2D moment invariants theory [14], and deduced seven moment invariants, which are of scale invariance, translation invariance and rotation invariance in continuous condition. In our paper, an improved computation technique [15] is adopted to make moment feature suitable to describe the texture feature of image. Then, we can obtain a 1×35 feature vector to represent the textual information of a sub-image.

3.3 Histogram of Gradient

Histogram of gradient (HOG) has been widely used for analyzing shape information, and has been demonstrated that it was powerful in object recognition and image classification [16]. In practice, each sub-image is divided to 25×25 pixel cells. For pixels falling into the same cell, the 9 bin histograms are summed up. Then each block contains a 36 dimension histogram vector of concatenated 9 orientation bins from 2×2 cells. In order to reduce time consuming and information redundant, blocks in HOG extracting process are taken with no overlap. Final a 1×144 dimensions shape feature descriptor is obtained.

3.4 Feature Level Fusion

Canonical correlation analysis (CCA) as a statistical tool has been proposed for identifying linear relationships between two sets of variables [17]. Given a pair of mean-normalized feature sets $U=[u_1,\cdots,u_n]$, $V=[v_1,\cdots,v_n]$, $u_i\in R_p$, $v_i\in R_q$, $i=1,\cdots,n$. Here U, V represents textural and shape feature codes respectively. In general, CCA can be seen as the problem of finding basis vectors w_u and w_v for two sets of variables U and V such that the correlations of projections of the variables onto corresponding basis vectors are maximized, while the projected variates within each data set are uncorrelated. More formally, CCA can be described as

$$(w_u, w_v) = \arg\max_{w_u, w_v} \frac{w_u^T U V^T w_v}{\sqrt{w_u^T U U^T w_u} \cdot \sqrt{w_v^T V V^T w_v}} \tag{2}$$

Solving (2) in an optimization manner, the vector pairs (w_{ui}, w_{vi}), $i=1,\cdots,d$ ($d \leqslant \min(p,q)$), can be obtained. More details about derivation and solution of CCA can be found in [18]. Let $W_u=[w_{u1},\cdots,w_{ud}]$, $W_v=[w_{v1},\cdots,w_{vd}]$, the feature fusion strategy is defined as

$$Z = \begin{bmatrix} W_u & 0 \\ 0 & W_v \end{bmatrix} \begin{bmatrix} U \\ V \end{bmatrix} = \begin{bmatrix} W_u U \\ W_v V \end{bmatrix} \tag{3}$$

While Z is called a textual-shape feature vector (TSFV). In general, shape features achieve higher efficiency comparing to moment features in change detection. To improve the discriminability of the fused feature vectors, feature fusion strategy here are modified as

$$Z_w = \begin{bmatrix} \alpha W_u U \\ \beta W_v V \end{bmatrix} \tag{4}$$

where α, β are the weights assigned to the two different features. Certainly, the values of parameters α and β

($\alpha < \beta$) can be determined by defining some learning strategies if the prior information about features is unknown. Convincingly, Z_w (called WTSFV) is better than Z in representing the properties of image. Final features Z_1 and Z_2 of two comparing sub-images are obtained. Finally, change detection is completed using Euclidean distance of Z_1 and Z_2 compared to a prior threshold value θ. Regions detected with change are labelled with colour boxes ultimately.

4 Experiments

We collected our own database to evaluate the performance of the proposed algorithm. The collected database contains two sections: locale images of substation photographed by inspection robots (Data-I) and man-made images with foreign bodies photographed in the laboratory environment (Data-II). Data-I has 320 pairs of images taken by different inspecting times and Data-II has 4 pairs of images taken by digital camera in different experimental environments. Every image in Data-II is resized as the same size 704×576 as images in Data-I.

From the results of image registration as shown in Fig.5, we can see obviously fewer matching points detected in saliency maps. 173 pairs of matching points were detected based on original images and the whole image registration processing is completed in 2.37 seconds. However, 57 pairs of matching points were detected based on saliency maps and the whole image registration processing is completed in 1.75 seconds including the time consuming of salient regions detection. The time cost of points matching and RANSAC processing based on saliency maps is fewer than that of original images. The same results were presented in other images registration.

(a)

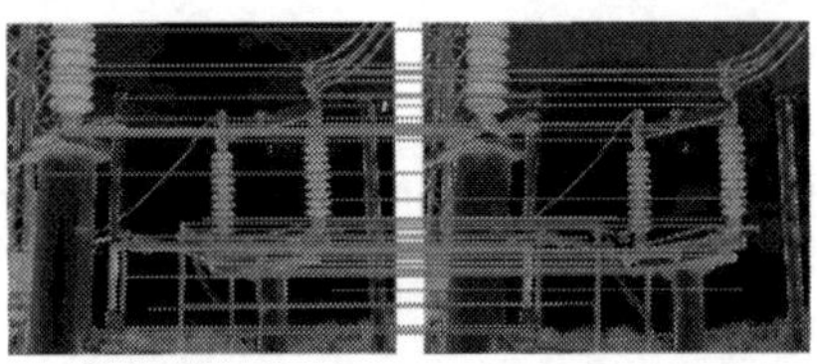

(b)

Figure 5. Feature points matching: (a) original images; (b) saliency maps

Data-I contains only one image with abnormal appearance. This abnormal region with foreign body was detected by our method as shown in the last column of Fig.6. Except this, the other fault regions detected by image comparison in Data-I are false alarm. This database is exploited to evaluate false positive rate (FPR) of our algorithm. The FPR is defined as,

$$FPR = \frac{\text{number of false alarm blocks}}{\text{total number of comparing blocks}} \times 100\% \quad (5)$$

Change detection based on WTEFV is implemented between blocks. There are 44320 pairs of blocks involved in feature comparison totally and 74 blocks in 19 images with change has detected by our algorithm. So FPR is approximately equal to 0.16%. Most of false alarms are caused by image shadow and image quality. How to reduce false alarms effectively will be addressed in our future work.

In order to evaluate the performance of abnormal appearance detection, a group of man-made abnormal images were acquired by camera in different scenes. The experimental results shown in Fig.6 show that every abnormal region was detected successfully.

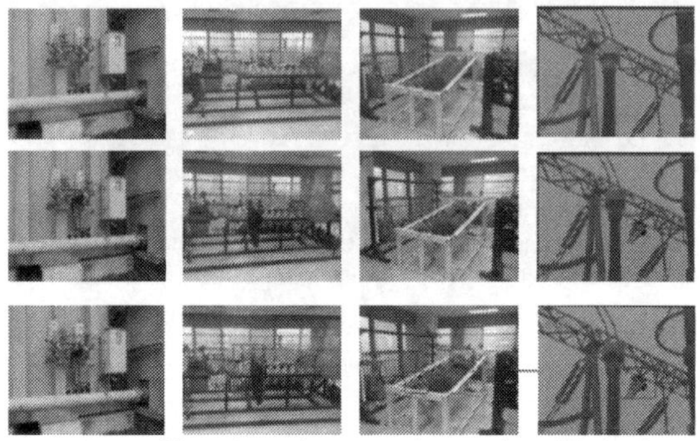

Figure 6. Given input images taken at time T1 (top); Given input images taken at time T2 (middle); Results of abnormal appearance detection by our algorithm (bottom).

Hence, using the proposed algorithm to implement images comparison has a high capability in abnormal appearance detection of substation. Besides, the proposed approach combined with autonomous robotic system improves the efficiency and reliability of substation inspection.

5 Conclusions

A new abnormal appearance detection of substation based on image comparison has been proposed in this paper. In order to reduce the influence of background and time consumption, image registration is implemented based on ORB features extracted from saliency maps. Then, sub-images were segmented from the whole image using sliding window algorithm. Moment invariants feature and HOG is extracted to describe the textural and shape information of sub-image respectively. Finally, Change detection based on fused features is implemented to detect abnormal appearance of substation. Experimental results have shown that the proposed method performed well in abnormal appearance detection of robots inspection in substation.

References

1. L. Zhou, Y. Zhang, Y. Sun, T. Liang, and S. Lu, "Development and application of equipment

inspection robot for smart substation," *Automation of Electric Power Systems*, **35**, 85-88, (2011).
2. H. Zhang, W. Wang, L. J. Xu, H. Qin, and M. Liu, "Application of image recognition technology in electrical equipment on-line monitoring," *Power System Protection and Control*, **38** (2010).
3. L. Li, P. Li, M. Yang, B. Zheng, and B. H. Wang, "Research on abnormal appearance detection approach of electric power equipment," *Optics and Optoelectronic Technology*, **8** (2010).
4. M. Brown, and D. G. Lowe, "Automatic panoramic image stitching using invariant features," *International Journal of Computer Vision*, **74**(2007).
5. E. Rublee, V. Rabaud, K. Konolige, and G. Bradski, "ORB: an efficient alternative to SIFT or SURF," *Proc.13th Computer Vision Conf,* 2564-2571, (2011).
6. B. Ko and J. Nam, "Object-of-interest image segmentation based on human attention and semantic region clustering," *Journal of the Optical Society of America*, **23** (2006).
7. T. Judd, K. Ehinger, F. Durand, and A. Torralba, "Learning to Predict Where Humans Look," *Proc.12th Computer Vision Conf,* Kyoto (2009).
8. T. Chen, M. M. Cheng, P.Tan, A. Shamir, and S.M. Hu, "Sketch2photo: Internet image montage," *ACM Trans*, **28** (2009).
9. R. Achanta, F. Estrada, P. Wils, and S. Sabine, "Salient region detection and segmentation," *Proc. 6th Computer Vision Systems Conf.*, 66-75, (2008).
10. B. Herbert, A. Ess, T. Tinne, and G. L. Van, "SURF: speeded up robust features," *Computer Vision and Image Understanding*, **110** (2008).
11. M. A. Fischer and R. C. Bolles, "Random sample consensus: A paradigm for model fitting with applications to image analysis and automated cartography," *Communications of the ACM*, **24** (1981).
12. L. Bruzzone, and D. F. Prieto, "An adaptive semiparametric and context-based approach to unsupervised change detection in multitemporal remote-sensing images," *IEEE Trans. On Image Processing*, **11** (2002).
13. A. Huertas and R. Nevatia, "Detecting changes in aerial views of man-made structures," *Image and Vision Computing*, **18** (2000).
14. M. K. Hu, "Visual pattern recognition by moment invariants," *IRE Trans. on Information Theory*, **8**(1962).
15. J. Yang, Y. Shi, J. Yang, and L. Jiang, "A novel finger-vein recognition method with feature combination," *Proc. IEEE International Conference on Image Processing*, 2709-2712, (2009).
16. N. Dalal, and B. Triggs, "Histograms of oriented gradients for human detection," *Proc. IEEE Computer Society Conference on Computer Vision and Pattern Recognition*, 886-893, (2005).
17. X. Zhang and F. Fang, "Multivariate statistical introduction," *Sciences Press*, **16** (1999).
18. D. R. Hardoon, S. Szedmak, O. Szedmak, and J. Shawe-Taylor, "Canonical correlation analysis: an overview with application to learning methods," *Neural Computation*, **16**(2004).

MATEC Web of Conferences **59**, 08002 (2016) DOI: 10.1051/matecconf/20165908002
ICFST 2016

A Novel Otsu Method Based on Prior Area Information for Concealed Target Detection in PMMW Images

Jintao Xiong, Simeng Li, Jianyu Yang, Xiaopeng Xue, Jingqi Mao and Zenan Mao

University of Electronic Science and Technology of China, Chengdu 611731, Sichuan, P. R. China

Abstract. A novel double-threshold Otsu algorithm based on prior area is proposed to improve the detection accuracy of small target in this paper. The proposed algorithm makes good use of the area information of each region in images to improve the accuracy of target detection, and can handle scenes containing small target which has low computational complexity, and achieves robust detection for concealed target in Passive Millimeter Wave (PMMW) images. Experiments are conducted on PMMW images collected by PMMW imaging system. The experimental results show that the proposed algorithm can effectively detect the concealed targets, and has better performance in target detection than other traditional threshold segmentation algorithm based on Otsu.

1 Introduction

PMMW imaging system obtain image through detecting the difference of millimeter energy that objects radiate, which has great performance in penetrating [1]. It not only can detect metal objects hidden in the fabric, but also can detect plastic pistols, explosives and other dangerous objects. These features make it ideal for battlefield reconnaissance and security of important places. Target detection is an important component of its application.

Detection has been studied for more than a decade, there are many detection methods such as target detection algorithm based on k-means[2], target detection algorithm based on mixed gauss model[3]-[5] and target detection algorithm based on active contour model[2], [6] and so on. These algorithms can meet the testing requirements of the PMMW images, but are computationally costly, cannot achieve the requirements of real-time detection. The target detection algorithm based on Otsu [7]-[8] method are widely applied in human image processing systems. It has simple principle, a small amount of calculation and stable performance in the detection process. But it will lose the small target. In order to solve this problem, we propose a novel Otsu algorithm based on prior area for target detection in PMMW images. The algorithm modifies the weighting coefficients in the objective function according to the prior area information of different regions of the PMMW images, obtain more accurate threshold and improve the accuracy in target detection. The experimental results show that the algorithm proposed in this paper has good effect, and meet the requirement of real-time detection.

2 Otsu method and its improvement

The Otsu method can meet the real-time of system and have good effect for target segmentation, but single-threshold Otsu method cannot effectively segment the concealed target in PMMW body images because the images which have concealed target can be divided into background region, human body region and target region. So we extend it to double-threshold Otsu method.

2.1 Single-threshold Otsu method

Let the pixels of a given picture be represented in L gray levels $[1,2,\cdots,L]$. The number of pixels at the gray level i is denoted by n_i, so the total number of pixels is $N=\sum_{i=1}^{L} n_i$, the probability of pixel at the gray level i is denoted by p_i.

$$p_i = n_i / N,\ \ p_i \geq 0, \sum_{i=1}^{L} p_i = 1 \tag{1}$$

In this method, the threshold operation is regarded as the partitioning of the pixels of an image into two classes by a threshold at level t. Then the probabilities of the first class occurrence, the class mean level and the class variance are respectively given by

$$\omega_0 = \sum_{i=1}^{t} p_i, \mu_0 = \sum_{i=1}^{t}(i*\frac{p_i}{\omega_0}), \sigma_0^2 = \sum_{i=1}^{t}[(i-\mu_0)^2 * \frac{p_i}{\omega_0}] \tag{2}$$

The probabilities of the second class occurrence, the class mean level and the class variance are respectively given by

$$\omega_1 = \sum_{i=t+1}^{L} p_i, \mu_1 = \sum_{i=t+1}^{L}(i*\frac{p_i}{\omega_1}), \sigma_1^2 = \sum_{i=t+1}^{L}[(i-\mu_1)^2 * \frac{p_i}{\omega_1}] \tag{3}$$

The total mean level of the original picture and total variance are respectively given by

$$\mu = \sum_{i=1}^{L}(i*p_i), \sigma^2 = \sum_{i=1}^{L}[(i-\mu)^2 * \mathrm{p}_i] \tag{4}$$

Let σ_w^2, σ_b^2 be the within-class variance and between-class variance respectively. An optimal threshold can be determined by maximizing one of the following criterion functions with respect to t:

$$\lambda = \frac{\sigma_b^2}{\sigma_w^2},\ \ \eta = \frac{\sigma_b^2}{\sigma^2},\ \ \kappa = \frac{\sigma^2}{\sigma_w^2} \tag{5}$$

Where

$$\sigma_w^2 = \omega_0\sigma_0^2 + \omega_1\sigma_1^2 \tag{6}$$

$$\sigma_b^2 = \omega_0 * (\mu_0 - \mu)^2 + \omega_1 * (\mu_1 - \mu)^2 \tag{7}$$

Of the above three criterion functions, η is the simplest and σ^2 is independent of t. Thus, the optimal threshold t^* is

$$t^* = \underset{t=1...L}{\arg\max}\, \sigma_b^2 \tag{8}$$

Because σ^2 is independent of t and the following basic relation always holds:

$$\sigma_b^2 + \sigma_w^2 = \sigma^2 \tag{9}$$

So the function (6) is equivalent to following function:

$$t^* = \underset{t=1...L}{\arg\min}\, \sigma_w^2 = \omega_0\sigma_0^2 + \omega_1\sigma_1^2 \tag{10}$$

2.2 Double-threshold otsu method

Double-threshold Otsu method is the improvement of single-threshold Otsu method. In this method, the threshold operation is regarded as the partitioning of the pixels of an image into three classes by two thresholds at level T_1 and T_2.

The probabilities, the class mean level and the class variance of each class are respectively given by

$$\omega_0 = \sum_{i=1}^{T_1} p_i \qquad \mu_0 = \sum_{i=1}^{T_1}(i*\frac{p_i}{\omega_0}) \qquad \sigma_0^2 = \sum_{i=1}^{T_1}[(i-\mu_0)^2 * \frac{p_i}{\omega_0}] \tag{11}$$

$$\omega_1 = \sum_{i=T_1+1}^{T_2} p_i \qquad \mu_1 = \sum_{i=T_1+1}^{T_2}(i*\frac{p_i}{\omega_1}) \qquad \sigma_1^2 = \sum_{i=T_1+1}^{T_2}[(i-\mu_1)^2 * \frac{p_i}{\omega_1}] \tag{12}$$

$$\omega_2 = \sum_{i=T_2+1}^{L} p_i \qquad \mu_2 = \sum_{i=T_2+1}^{L}(i*\frac{p_i}{\omega_2}) \qquad \sigma_2^2 = \sum_{i=T_2+1}^{L}[(i-\mu_2)^2 * \frac{p_i}{\omega_2}] \tag{13}$$

The optimal threshold T_1 and T_2 are:

$$T_1^*, T_2^* = \underset{1\leq T_1 < T_2 \leq L}{\arg\min}\, \sigma_f^2 = \omega_0\sigma_0^2 + \omega_1\sigma_1^2 + \omega_2\sigma_2^2 \tag{14}$$

3 Double-threshold Otsu method based on prior area

In a PMMW body image, concealed target account for small proportion compared with the human body and background, and the target will be undetected by single-threshold Otsu method. To solve this problem, we proposed double-threshold Otsu method based on prior area information to achieve the detection of small target.

The basic idea of target detection algorithm based on prior area information is describe as follows. Firstly, design objective function according to the area information of each region of PMMW image. Secondly, obtain the segmentation threshold which minimizing the objective function in accordance with histogram of PMMW image. Finally, use this threshold to segment the image and use the gray information of target to extract the target region.

Objective function of the algorithm is defined as:

$$\sigma_f^2 = \alpha_0\omega_0\sigma_0^2 + \alpha_1\omega_1\sigma_1^2 + \alpha_2\omega_2\sigma_2^2 \tag{15}$$

Where, $\alpha_0 + \alpha_1 + \alpha_2 = 1$, $\omega_0 + \omega_1 + \omega_2 = 1$, the meaning of $\omega_0, \sigma_0^2, \omega_1, \sigma_1^2, \omega_2, \sigma_2^2$ are same as the second portion, $\alpha_0, \alpha_1, \alpha_2$ are correction factor.

Let the area ratio of the region which at low gray levels is about P_0, the area ratio of the region which at high gray levels is about P_2, the others is about P_1. Because of adding to the prior information, it is assumed that the pixel values of image are independent of each other and to obey uniform distribution of [1 256] to simplify the calculation.

$$\omega_0 = \frac{T_1}{256} \quad \mu_0 = \frac{T_1}{2} \quad \sigma_0^2 = \frac{T_1^2}{12} \tag{16}$$

$$\omega_1 = \frac{T_2 - T_1}{256} \quad \mu_1 = \frac{T_2 + T_1}{2} \quad \sigma_1^2 = \frac{(T_2 - T_1)^2}{12} \tag{17}$$

$$\omega_2 = \frac{L - T_2}{256} \quad \mu_2 = \frac{L + T_2}{2} \quad \sigma_2^2 = \frac{(L - T_2)^2}{12} \tag{18}$$

Where L=256, then objective function can be expressed as:

$$\sigma_f^2 = \frac{\alpha_0 T_1^3 + \alpha_1 (T_2 - T_1)^3 + \alpha_2 (256 - T_2)^3}{256*12} \tag{19}$$

Derivative and make it equal to zero:

$$\frac{\partial \sigma_f^2}{\partial T_1} = \frac{\alpha_0 T_1^2 - \alpha_1 (T_2 - T_1)^2}{256*4} = 0 \tag{20}$$

$$\frac{\partial \sigma_f^2}{\partial T_2} = \frac{\alpha_1 (T_2 - T_1)^2 - (1 - \alpha_0 - \alpha_1)(256 - T_2)^2}{256*4} = 0 \tag{21}$$

By $p_0 = \frac{T_1}{256}$, $p_1 = \frac{T_2 - T_1}{256}$ and $p_2 = \frac{256 - T_2}{256}$, can be deduced

$$\alpha_0 = \frac{P_1^2 P_2^2}{P_0^2 P_1^2 + P_1^2 P_2^2 + P_0^2 P_2^2} \tag{22}$$

$$\alpha_1 = \frac{P_0^2 P_2^2}{P_0^2 P_1^2 + P_1^2 P_2^2 + P_0^2 P_2^2} \tag{23}$$

$$\alpha_2 = 1 - \alpha_0 - \alpha_1 = \frac{P_0^2 P_1^2}{P_0^2 P_1^2 + P_1^2 P_2^2 + P_0^2 P_2^2} \tag{24}$$

When α_0, α_1, α_2 taken (22), (23), (24), the double-threshold segmentation algorithm can obtain the optimal threshold.

4 Concealed target detection algorithm for PMMW image and experiment results

Otsu algorithm based on prior area information for concealed target detection in PMMW images is mainly divided into three steps: image de-noising, image segmentation and concealed target detection. Firstly, we choose the median filter to de-nose PMMW image to meet the requirement of real-time of system and maintain the image edge as possible. Then estimate the area of each region of the image according to the parameters of the imaging system such as the field of view, perspective, imaging distance, and use Otsu method based on prior area information to segment the PMMW image. Finally, extract the target by the gray scale range of the target occupies in PMMW image to achieve the purpose of detection. Flow chat of the algorithm is as Figure 1.

In order to verify the stability and effectiveness of the target detection algorithm based on prior area information, we use the PMMW images which imaging by PMMW imaging system to simulation. The simulation results are compared with single-threshold and double-threshold Otsu algorithm. The experimental results are shown in Figure 2-4.

According to the field of view and imaging distance of the PMMW imaging system, we can estimate the human region account for about 0.35 of the whole area of

 DOI: 10.1051/matecconf/20165908002

the PMMW image, target region account for about 0.02 of the whole area of the PMMW image.

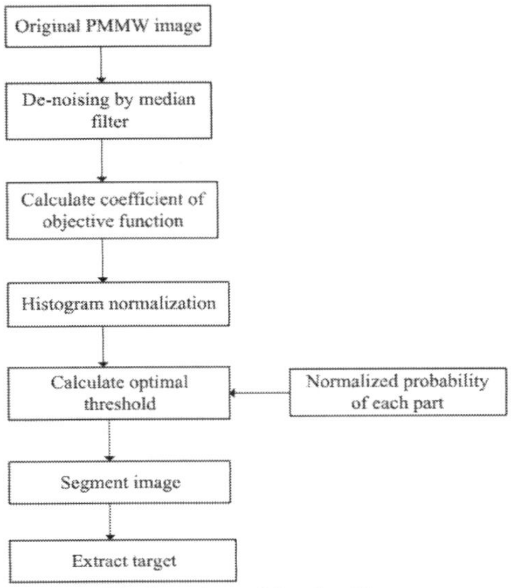

Figure 1. The flow chart of the algorithm

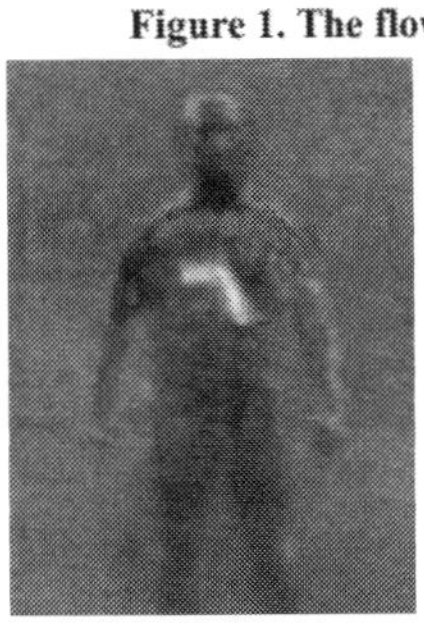

(a) PMMW Image

(b) The Segmentation Result of Single-Threshold Otsu

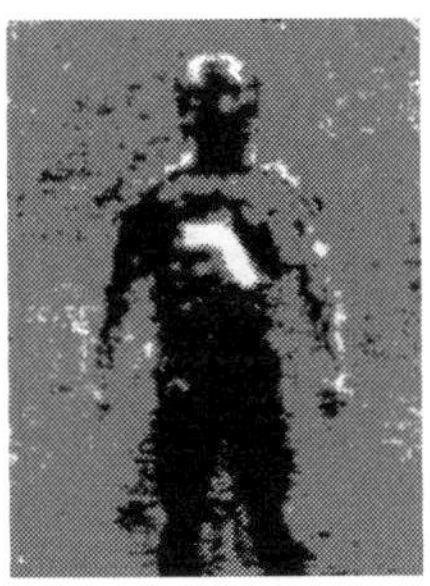

(c) The Segmentation Result of Double-Threshold Otsu

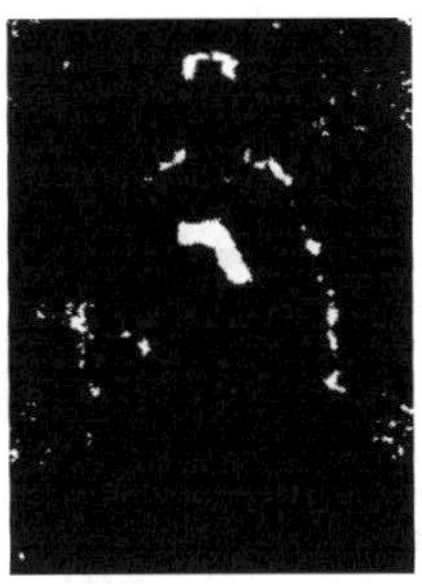

(d) The Detection Result of Double-Threshold Otsu

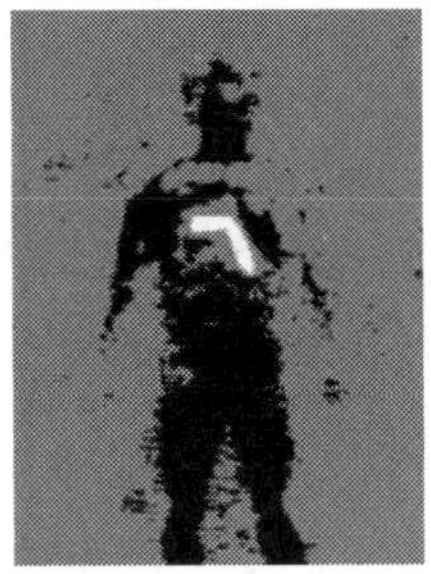

(e) The Segmentation Result of The Proposed Algorithm

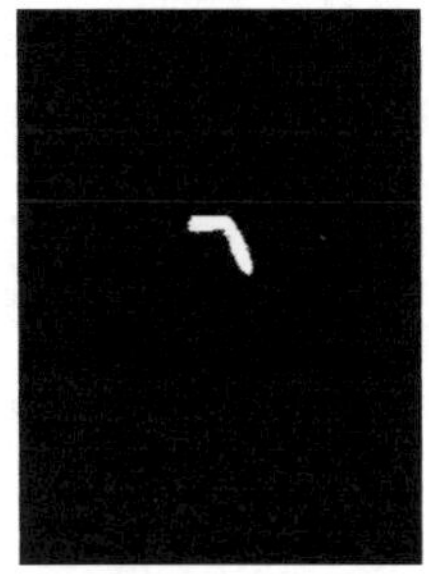

(f) The Detection Result of The Proposed Algorithm

Figure 2. Target Detection results of PMMW image

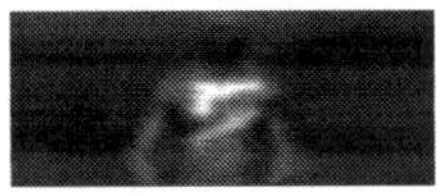

(a) PMMW Image

(b) The Segmentation Result of Single-Threshold Otsu

(c) The Segmentation Result of Double-Threshold Otsu

(d) The Detection Result of Double-Threshold Otsu

(e) The Segmentation Result of The Proposed Algorithm

(f) The Detection Result of The Proposed Algorithm

Figure 3. Target Detection results of PMMW image

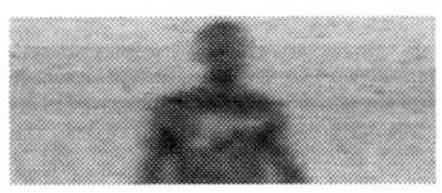

(a) PMMW Image

(b) The Segmentation Result of Double-Threshold Otsu

(c) The Detection Result of Double-Threshold Otsu

(d) The Segmentation Result of The Proposed Algorithm

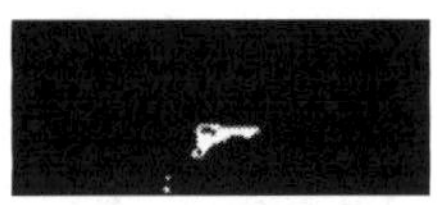

(e) The Detection Result of The Proposed Algorithm

Figure 4. Target Detection results of PMMW image

In Figure 1, we calculate coefficient by formula (22), (23), (24), normalized probability and calculate threshold by formula (11), (12), (13), (15).

It can be seen from the simulate experiments that the Otsu algorithm based on prior area information for target detection have a best effect. Although the double-threshold Otsu algorithm has certain effect, when the target is small as seen in Figure 4(b) the target is not detected. The Otsu algorithm based on prior area information for target detection we proposed makes good use of the area information of each region, which greatly improves the detection accuracy of the algorithm. The more precise of the prior area information, the higher detection accuracy we can obtain. For the complexity of the algorithm, the algorithm proposed in this paper can meet the requirement of real-time.

5 Conclusion

A novel Otsu method based on prior area information for concealed target detection in PMMW images is proposed in this paper, which can obtain a high performance on concealed target detection of PMMW images. The proposed algorithm estimates the area of each region of the image according to the parameters of the imaging system such as the field of view, perspective, imaging distance and makes good use of the area information to obtain a high quality of the target detection. The experimental results show that the proposed algorithm can achieve concealed target detection of the PMMW images, and has good adaptability and stability. The algorithm is practical and has higher real-time performance compared with the traditional Otsu methods.

References

1. L. Yujiri, M. Shoucri and P. Moffa. Passive millimeter wave imaging, Microwave Magazine, IEEE, **4**, 39-50(2003)
2. Haworth, D. Christopher, G. Gonzalez, ect.. Image analysis for object detection in millimetre-wave images, Passive Millimetre-wave and Terahertz Imaging and Technology, **5619**, 117-128(2004)
3. XL. Shen, CR. Dietlein, ect.. Detection and Segmentation of Concealed Objects in Terahertz Images, IEEE Transactions on Image Processing, **17**, 2465-2475(2008)
4. Lee, DS, Yeom, S, ect.. Real-time computational processing and implementation for concealed object detection, Optical Engineering, **51**(2012)
5. W. Dai, W. Mei and XD. Zhao. Detetion and Segmentation of Concealed Objects in X-ray Compton Backscatter Images, International Conference on Image and Graphics, 175-179(2013)
6. O. Martínez, L. Ferraz, ect.. Concealed Object Detection and Segmentation over Millimeter Waves Images, Computer Vision and Pattern Recognition Workshops (CVPRW), (2010)
7. N. Otsu. A Threshold Selection Method from Gray-Level Histograms, IEEE Transactions on Systems, Man, and Cybernetics, (1979)
8. P.K.SAHOO, S.SOLTANI, and A.K.C.WONG. A Survey of Thresholding Techniques, Computer Vision, Graphics, and Image Processing, (1988)

MATEC Web of Conferences **59**, 08003 (2016) DOI: 10.1051/matecconf/20165908003
ICFST 2016

Investigating camera calibration for eye tracking of the physically challenged

Zeenat Al-Kassim , Qurban Memon

EE Department, College of Engineering, UAE University, UAE

Abstract. This paper tries to address a problem faced by a proportion of community - the paralysed people, and investigates a solution in calibrating a camera used for the purpose of eye detection. The eye tracking is later used to read characters from a partner-assisted keyboard. As such, this keyboard solves the problem of immobility. In order to calibrate camera for the purpose of eye detection, two approaches are implemented-Artificial Neural Networks and Support Vector Machines in context of paralysed people, and are discussed in detail in this paper. The results of both approaches are compared in order to select the best approach to be used in keyboard that operates on eye tracking.

1 Introduction

According to National Health Service in the UK [1], Paralysis is loss of the ability to move muscles that may be associated with loss of feeling and other bodily functions. It can be seen as a form of nerve damage due to the inability of brain to control muscles, caused by damage in the nerves of spinal cord. Three main causes for paralysis in the U.S., according to a study done by the Christopher & Dana Reeve Foundation [2], are: stroke, spinal cord injuries and multiple sclerosis. Other causes include brain injuries, Parkinson's disease, Guillain-Barre syndrome, Spina bifida, Motorneurone disease [1]. Since sudden accidents are a major cause of paralysis, many people find themselves suddenly paralyzed. As such, the number of paralyzed people keeps on increasing every day. In the U.S., one person in every 50 suffers from paralysis [2]. This is around 6 million people (around 2% of the U.S. population) [2]. People with paralysis face many challenges in their lives. Paralysis can be in of the following four forms [1], as shown in Fig. 1:

- Monoplegia: where one limb is paralysed
- Hemiplegia: where the arm and leg on one side of the body are paralysed
- Paraplegia: where both legs are paralysed, or sometimes the pelvis and some of the lower body
- Tetraplegia/Quadriplegia: where both the arms and legs are paralysed

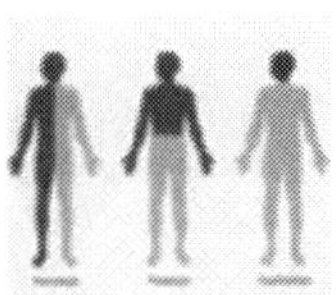

Figure 1: Types of Paralysis (grey: paralysed body part) [3]

People suffering from tetra/quadriplegia face difficulties even when performing simple daily life activities like communication. For instance, people during late stages of ALS [4] cannot move their entire body muscles except for the eyes muscles. As such, several techniques have been employed since ages to help those people communicate freely and easily. One such famous technique is Partner-Assisted Scanning. In this technique, the care giver presents a set of words or letters to the disabled patient, and the patient can select from among those letters or words in order to express his/her thoughts [5]. Selection, as in Fig. 2, might happen by pointing or blinking of the eye depending on the capabilities of the disabled person [5]. This method is used to help children as well as adults who suffer immobility due to late stages of ALS, multiple sclerosis, or severe injuries [5].

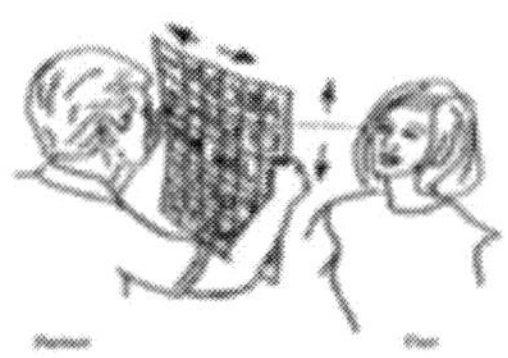

Figure 2. Partner Assisted Scanning [6]

In order to address this, technology has been exploited to build systems that try to simplify human life. These systems utilized concepts like human computer interaction, image processing, virtual/augmented reality, and so on, to build devices by which humans can communicate easily with the machines [7-8]. An example is the Kinect Sensor developed by Microsoft, and the Wii Remote developed by Nintendo. In this paper, a research work is reported that aims to utilise the advance technology of gesture recognition, or more precisely eye recognition to help such people. The work here is to build a keyboard, which exploits eye movement to implement the concept of Partner-Assisted Scanning on a digital screen. The first phase of this work involves camera calibration where the system needs to detect the user's eyes. Two commonly known methods (i.e., Artificial Neural Networks (ANN) and Support Vector Machines (SVM)) were used and compared to come up with a best method to implement in camera calibration of the user's eyes in such an environment..

2 Camera Calibration

The aim is to design a system that is able to detect the eyes of the user automatically. The attempt is to train the system to detect the face of the user prior to detecting the user's eyes. When the face is detected, the eyes' location can be extracted from the face. Since the system is intended for users with mobility difficulties, the user is assumed to have his/her head placed straight in front of the system screen. Head movements and tilting are not considered since users targeted are those with physical paralysis. The system captures images by a webcam with a capacity of 30 frames per second, at a resolution of 640 by 480 pixels.

Artificial Neural Networks (ANN): ANN [9] is a concept inspired from the central nervous system in a human body, as in Fig. 3, that consists of neurons in a similar structure to a network. ANN helps in machine learning and pattern recognition[10-12]. Thus, such networks have proved successful in hand gesture and face recognition. A neural network is trained to detect a face based on common dark spots in any human's face. The advantage of this method is that once the neural network is trained, it can automatically detect any face positioned in front of the camera. Thus, training is required only once.

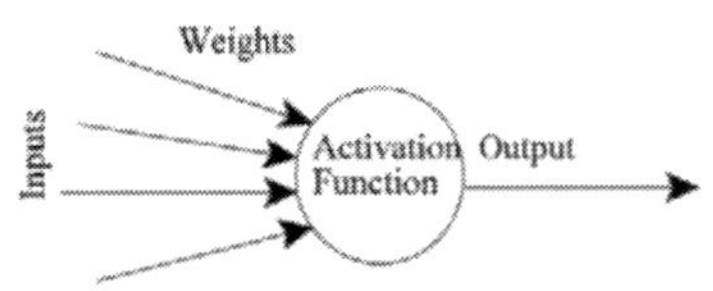

Figure 3. Artificial Neural Network

The neural network used is a two layer (one Hidden layer and one Output layer) feed-forward network trained with the Levenberg-Marquardt and scaled conjugate gradient back propagation algorithm. Thirty four (34) face samples were inserted for training: 27 (80%) samples used for training and 5 (15%) samples used for validation and the rest 2 (5%) used for testing. The hidden layer consists of 20 neurons. The hidden layer works with a sigmoid function and the output layer implements a linear function. The face samples were captured by the webcam in the original 640-by-480 resolution. The training samples consisted of faces of both male and female, adults (aged 20-50) and children (aged 5-8). In order to train the network to detect faces, some pre-processing on the face images was done, as follows:

- First, picture of the face of the training sample is captured in RGB at a resolution of 640-by-480 pixels.
- Captured RGB image is converted to grey scale.
- Adaptive Histogram Equalisation in all (R, G, and B) channels is performed.
- Cropping image to square and resizing image to 25-by-25 resolution (625 pixels).

The 625 grey scale pixel values (ranging between 0 and 255) of each image are normalised and inserted into the input file in a row. Each image occupies one row. As such, the input file for training the network is a 34×625 matrix consisting of 34 samples of 625 elements each. Participants were made to close and open eyes for capturing two images of each participant. With this, the network is trained to detect faces irrespective to whether eyes are opened or closed. The network is trained Supervised Training mode, which means desired output is provided to the network. The desired output values are four that indicate the position of the nose with respect to the eyes, as in Fig. 4.

Figure 4. Desired Output of Neuron (shown in yellow dots)

These four output values are desired since the area in the middle of the two eyes can then be segmented and used for skin colour segmentation in the tracking phase. Since the aim is to extract the position of the pixels and not the value, the output is converted from grey scale value to the pixel position by the following formula:

Output value =[(row no -1)x25]+column no. (1)

The output file is 34×4 matrix consisting of 34 samples of 4 elements each, which were normalised by division with 625. The network architecture consists of 625 input nodes in the input layer, 20 hidden neurons in the hidden layer, and four output neurons in the output layer, as shown in Fig. 5. The network was trained for 1,030 iterations.

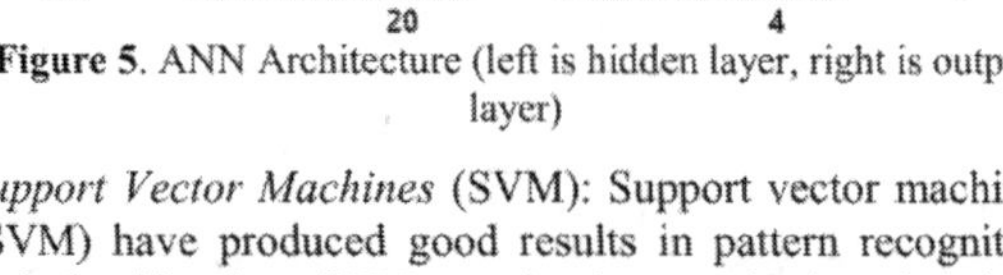

Figure 5. ANN Architecture (left is hidden layer, right is output layer)

Support Vector Machines (SVM): Support vector machines (SVM) have produced good results in pattern recognition and classification. SVMs are implemented when data is to be classified [13]. It searches for a best hyperplane, in Fig. 6, for the largest margin, to separate the two classes [13]. Support vectors are the points closest to the separating hyperplane [13]. The '+' points belong to one class, and the '-' points belong to the other. Mathematically, SVMs are implemented by a set of sequential equations. The separating hyperplane is given as:

$$\langle w, x\rangle + b = 0, \quad (2)$$

where <w, x> is a dot product of weights and data points, b is bias. To achieve best separating hyperplane, the following is satisfied:

$$yi(<w, xi> + b) \geq 1 \quad (3)$$

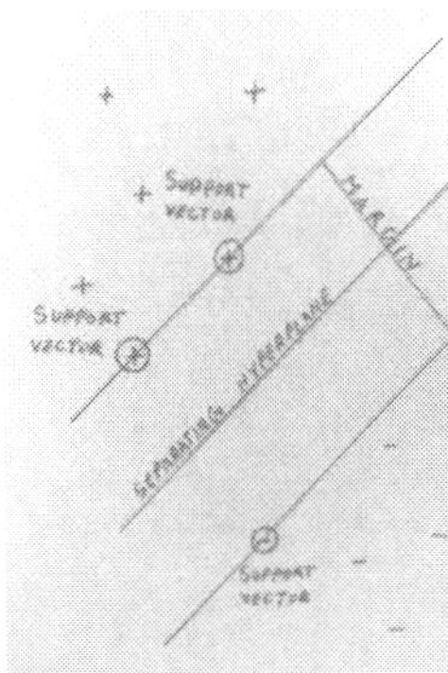

Fig. 6. Separating Hyperplane of SVM

where y is ±1 depending on the class. To simplify quadratic equations, Lagrange multipliers α_i are applied with the constraint and subtracted from the condition for maximal margin. Values of α_i lies in the range $0 \leq \alpha_i \leq C$, where C is a constraint which keeps values of α_i within a limited range. On the other hand, for nonlinear transformations, Kernels are often applied since a simple hyperplane cannot separate the two classes of data points apart. The concept of the Kernel method is that there exist a function φ that maps x *to a* linear space S, such that

$$K(x, y) = <\varphi(x), \varphi(y)> \quad (4)$$

This dot product takes place in the linear space S. This concept of kernels is implemented in training the support vector machines for face detection in the calibration. A polynomial kernel of order 3 was used to train the SVM. The input file is a combination of two files. One part is the positive samples consisting of 34 face samples of 625 elements (same as the training samples of the ANN discussed previously), with pixel values inserted column-wise. The other part is the negative samples consisting of 34 samples of 625 elements constituting of random non-face images taken from the internet. The output or grouping file consists of 1's for the positive samples and -1's for the negative samples. Out of the total 68 samples of the input file, 34 samples (half of which are 1's and the other half are -1's) were used for training, while the rest 34 samples were kept aside for validation of the training results.

3 Comparison: Training and Testing

Training: After 1,030 iterations in training the neural network, good results were achieved in terms of the mean squared error (mse), and the regression (i.e., correlation between outputs and targets. For the training samples, a mse of 0.00004633 and regression of 0.994613 were achieved. For the validation samples, a mse of 0.00872 and regression of 0.815132 were reached. For the testing samples, a *mse* of 0.00154 and regression of 0.999999 were achieved. Graphs showing the mse and regression numbers for all samples are in Figure 7 and 8 respectively. As for training SVM, 25 support vectors were used for training. An accuracy of 100% was reached with the 34 training samples; while an accuracy of 0.875 (87.5%) was reached with the 34 validation samples. Values of alpha were calculated during training by the kernel classifier, from which values of the weight matrix were calculated by:

$$w = \sum_i \alpha_i y_i x_i \quad (5)$$

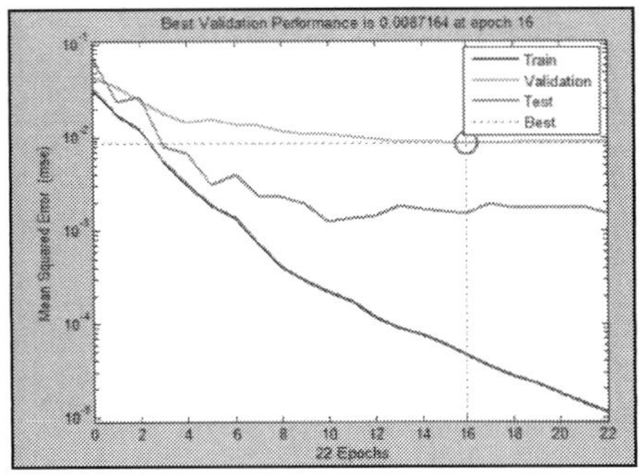

Figure 7. Error values of training, validation and testing samples

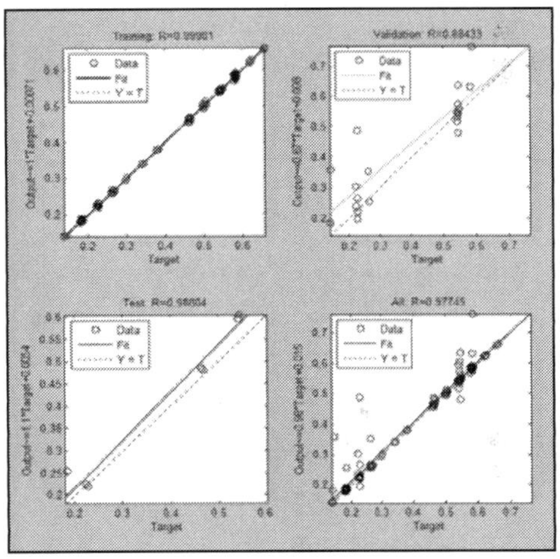

Figure 8. Regression values of training, validation and testing samples

The value of the constraint C affects the face detection results. A number of C values have been tried in training, ranging from 1000 to 0.00000001. When plotting the weight matrix after training with different values of C, it was noticed that lower the value of C, more the weight matrix looks similar to a face. Based on this, C=0.00000001 was selected as the final value for training a face detection SVM in the calibration phase.

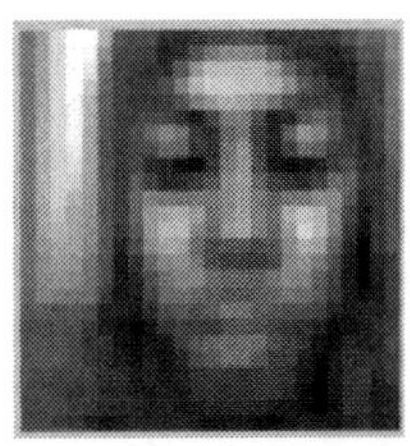

Figure 9. Plotting of average weight matrix of SVM

Testing: Testing for both ANN and SVM, was done by inserting a random face image into the system and comparing the results with the desired results. The test image needs to undergo a few pre-processing steps before insertion into both the ANN and SVM, as discussed before in section II. As for the ANN, the following are the results in Table 1.

Table 1. ANN Testing Results

Desired Output	0.2224	0.2256	0.5424	0.5456
Network Output	0.273832	0.276808	0.571432	0.574408
MSE	0.00224493			
Regression	0.99393			

As can be noticed from Table 1, a close precision is achieved but not good enough to detect the desired area in the face. This is because when retrieving back the pixel position, a significant difference is noticed. When applying the formula as given in (1), the network four output points are (6, 21), (6, 23), (14, 7), (14, 9), instead of the desired points (6, 14), (6, 16), (14, 14), (14, 16). The network outputs correct row number, but fails in recognising the correct column numbers which leads to a quite huge difference. In testing the SVM, a few more pre-processing steps are performed on the test image as shown in Fig. 10.

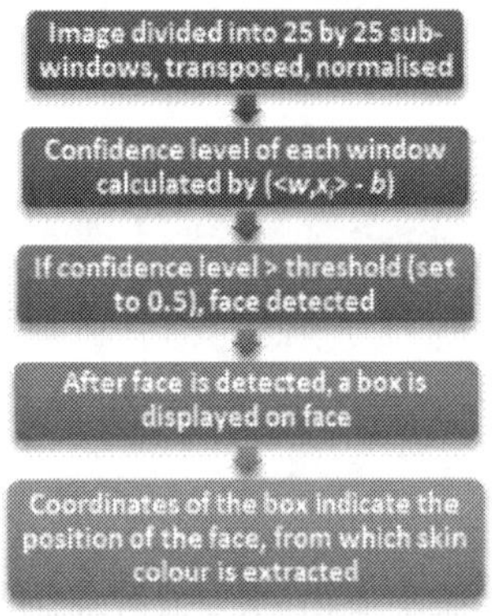

Figure 10. Steps undergone by test image before insertion into SVM

The face in the test image was detected with a confidence level of 0.73888, which is above threshold set (as 0.5). The box coordinates displayed on the detected face in the test image are [10.5000 5.5000 34.5000 29.5000]. Dividing the box width by half gives the exact position in between the two eyes (34.5+10.5 = 22.5). The 22nd column is in between the two eyes, which shows how accurately the box was placed around the detected face.

Comparing Both Approaches: While ANN and SVM are two different approaches, a comparison can be done between them based on some desired features like fast training speed and accurate results. Some features cannot be compared, like number of iterations, since SVM are trained in a different way. Alternatively, the time taken to train both ANN and SVM can be compared. This comparison is summarized in Table 2.

4 Conclusions

It was clear that SVM performed better than ANN in environment of partner-assisted eye scanning. Accurate results were reached in SVM when same training samples were used in both approaches. However, the performance of ANN can be improved because values of mean squared error reached were low, resulting in only producing accurate results in terms of row number. The performance of ANN can be improved by either increasing the number of training face samples, or by varying the training samples, as for instance, varying the lighting conditions for each training sample might produce precise column position. Overall, as a conclusion, the SVM was chosen due to its simplicity of training and good results in face detection prior to eye detection. These results are further used to build an eye tracking keyboard for people with physical challenges.

Table 2. Comparison of the Two Approaches

Feature	ANN	SVM
No. of training samples	34 face samples	68 face and non-face samples
Training accuracy	Samples: 99.4613% training , 81.5132% validation	Samples: 100% (training), 87.5% (validation)
Training time	1,030 iterations (5 sec)	No iterations, weight matrix calculated (0.6 sec)
MSE (testing)	0.00224493 (column position inaccurate)	(column position exact)

References

1. Paralysis, National Health System. Retrieved online from http://www.nhs.uk/Conditions/paralysis/
2. Reeve, C. (2002). One Degree of Separation: Paralysis and Spinal Cord Injuries in the United States. Christopher and Dana Reeve Foundation. Retrieved online from http://www.christopherreeve.org/atf/
3. Human Diseases and Conditions. Paralysis. Retrieved from http://www.humanillnesses.com/original/Pan-Pre/Paralysis.html
4. ALS Association. Online: http://www.alsa.org/about-als/what-is- als.html
5. Cincinnati Children's Hospital Medical Center. Communicating with Partner Assisted Scanning. Retrieved from http://www.youtube.com/watch?v=nGpSXQKrmR4
6. Retrieved online from http://www.surveymonkey.com/s/low-tech_AAC
7. Magrelli, S.; et al, "A Wearable Camera Detects Gaze Peculiarities during Social Interactions in Young Children with Pervasive Developmental Disorders," *IEEE Transactions on Autonomous Mental Development*, **6**(4), (2014), 274-285.
8. Xingjie W.; et al, "Dynamic Image-to-Class Warping for Occluded Face Recognition," *IEEE Transactions on*

Information Forensics and Security, , **9**(12), (2014), 2035-2050
9. Franceschini, N., "Small Brains, Smart Machines: From Fly Vision to Robot Vision and Back Again," *Proceedings of the IEEE* , **102**(.5), (2014), 751,781
10. Memon, Q., Shuja, M., "Crime investigation and analysis using neural nets," *IEEE Multi Topic Conference*, (2003), 346-350
11. S Khan, Q Memon, "Artificial Neural Network Approach to Camera Calibration and 3-D World Reconstruction for stereovision," *International Workshop on Recent Advances in Computer Vision*, (1998), 35-40
12. Memon, Q.; "Relation Based Clustering of Wear Particle Measurements for Industrial Automation", *International Journal of Automation and Control*, **1**(2-3), (2007), 207-219
13. Schölkopf, B., et al, "Estimating the Support of a High-dimensional Distribution," *Journal of Neural Computation*, **13**(7), (2001), 1443-1471.

AUTHOR INDEX

AUTHOR INDEX